PATTERNING OF TIME

PATTERNING OF TIME

BY LEONARD W. DOOB

NEW HAVEN AND LONDON, YALE UNIVERSITY PRESS, 1971

Copyright © 1971 by Yale University.
All rights reserved. This book may not be
reproduced, in whole or in part, in any form
(except by reviewers for the public press),
without written permission from the publishers.
Library of Congress catalog card number: 72-151572
International standard book number: 0-300-01454-6

Designed by Sally Sullivan
and set in Linotype Baskerville type.
Printed in the United States of America by
Vail-Ballou Press, Inc., Binghamton, N.Y.

Distributed in Great Britain, Europe, and Africa by
Yale University Press, Ltd., London; in Canada by
McGill-Queen's University Press, Montreal; in Mexico
by Centro Interamericano de Libros Académicos,
Mexico City; in Central and South America by Kaiman
& Polon, Inc., New York City; in Australasia by
Australia and New Zealand Book Co., Pty., Ltd.,
Artarmon, New South Wales; in India by UBS Publishers'
Distributors Pvt., Ltd., Delhi; in Japan by John
Weatherhill, Inc., Tokyo.

I remember deeply
Gratefully
My advisers
Guides who pointed
But did not command:
   Gordon W. Allport
   McQuilkin DeGrange
   William McDougall
   Franklin McDuffie
   James MacKaye
   Karl Mannheim
   Max Wertheimer
Were they alive they would smile
And I might too.

CONTENTS

Preface   xiii

PART I. PROCESSES

1. **FOCUS**   3
   - Patterning   5
   - Activity   10
   - Of principles and uncertainties   13
   - Methodological matters   17
   - ADDENDA
     1. *Quibbles*   23
     2. *Variations in the methods of measurement*   25
     3. *Evaluation of methods*   27

2. **POTENTIALS**   30
   - Primary and secondary levels   33
   - Over- and underestimation   37
   - The length and speed of time   39
   - ADDENDA
     1. *Explanation of "The Length and Speed of Time"*   43
     2. *Semantic problems*   44

3. **BEHAVIOR**   47
   - Universality and inevitability   48
   - Perspective   52
   - Modal values   56
   - Direction of energy   58
   - Temporal knowledge   60
   - Time as communications medium   63
   - Spirals   66
   - The biochemical clock   68
   - Personality   73
   - ADDENDA

	1. Variations in tenses and perspective	74
	2. Time reckoning in different societies	75
	3. Dispersion	76
	4. Intervention and perspective	77
	5. Pitfalls concerning modal values	78

4. TEMPORAL MOTIVES 80
Coordination and conformity 81
Power 86
Temporal factors in communication 87
Frustration 90
Deferred gratification 93
Achievement 97
ADDENDUM
1. Deferment and non-Western peoples 100

5. PERCEPTION AND SCOPE 101
Psychophysics 101
Salience 106
Event potential 109
ADDENDA
1. Filled vs. empty intervals 115
2. The contents of the interval 118
3. Space and time 121
4. Intervention 123
5. "Storage size" 127

6. DRIVES 130
Strength 131
Temporal hope and fear 132
Attainment 137
Unverbalizable cues 140
ADDENDA
1. The reduction of drive strength to events 142
2. Duration-dependent drives 143
3. Proverbs 145
4. Attainment vs. nonattainment of goal 146
5. Cues 149

7. SECONDARY JUDGMENT 152
The standard 152
Experience 157
Boredom and impatience 167

The translation 172
The passing of time 181
Feedback 183
Epilogue 187
ADDENDA
1. Experience without reinforcement 190
2. Experience with reinforcement 194
3. Compensating and counting 198
4. Watching the pot never boil 199
5. Introspective reports 200
6. Accuracy of sense modalities 201
7. Factors affecting the accuracy of measurement methods 203
8. The accuracy of the biochemical clock 204
9. Distortions in time 207
10. Sex differences 209

PART II. FUNCTIONING

8. DEVELOPMENT 213
   The progression 213
   The sense of time 219
   Personality traits 222
   Aging 234
   ### ADDENDA
   1. Learning to postpone 244
   2. Acquiring a temporal potential 247
   3. Consistency of temporal judgments 252
   4. Intelligence 255
   5. Traits and judgments 257
   6. Traits and orientation 261
   7. Attitude toward time 264
   8. Need for achievement 265
   9. Youth vs. age 267

9. DEVIATIONS 273
   Organic states 273
   Drugs 282
   Sleep and dreams 285
   Hypnosis 290
   Mental disorders 293
   Crime 301

School dropouts ... 303
Martyrs and mystics ... 304
   ADDENDA
      1. Brain localization ... 307
      2. Thyroid gland ... 309
      3. Temperature and related physiological changes ... 310
      4. Effects of drugs ... 312
      5. The duration of dreams ... 316
      6. Hypnotic time ... 317
      7. Abnormality ... 319
      8. Criminals ... 329

10. GROUPS ... 332
   Reference group ... 333
   Traditions and ideals ... 335
   Leaders and prophets ... 340
   Religion ... 343

11. WORK AND RISK-TAKING ... 347
   Productivity ... 348
   Monotony ... 349
   Decisions ... 352
   Diminishing risks ... 355
   Capital ... 360
   Conservation ... 362
   Occupation ... 364
   ADDENDUM
      1. Repetition and monotony ... 366

PART III. MANIPULATION

12. THE ARTS ... 371
   Orientation ... 372
   The arresting of time ... 374
   The suspension ... 377
   The portrayal of time ... 383
   Truth and wisdom ... 386
   On matters of style ... 388

13. CHANGING ... 392
   Planning ... 392
   Gradual change ... 397
   Rapid change ... 400

Concomitant change	403
14. ENVOI	407
References	413
Index of Principles and Hypotheses	447
Author Index	453
Subject Index	464

PREFACE

My aim is to understand time—and that is really all a preface needs to say about this book. No man can view himself or his universe without having exactly the same concern. The subject of time arises no matter where we turn: it bursts forth in conversation, it pervades the arts, it receives respectful and often agonizing treatment in all the sciences and humanities.

To understand time, or to try to do so, I have subjected myself to three torments. I have plagued myself with the problem for more than three decades, on and off and often also continuously. For I have sought to distill some guiding generalizations which, perhaps, suggest how we as persons confront time. Then, as the length of the bibliography suggests, I have exposed myself to the flood of writings on time. And the title of the book reveals the somewhat gigantic canvas I have tried to construct: don't be picayune, I have kept saying to myself, view time in very, very broad perspective.

Obviously I shall never know whether I have succeeded. Principles, hypotheses, and speculations are unveiled throughout the book: they help me, but will they help anyone else? I hope so. I have sought wisdom and data wherever I could—in fact, I say with guilty pride, I have failed to follow the academic custom of our time: all the references have been read not by a team of research slaves but by me. Doubtless I have failed to see many studies or viewpoints considered critical by someone else, and our knowledge about time will have further increased before this book leaves the printer. Of course, no synthesis is ever complete, but it can raise us a step higher. I realize, too, that my canvas will not receive universal acclaim: some must say that it is too broad, others that it is too narrow. The core problems of time and their implications, I maintain, are here portrayed if not resolved.

So be it.

Although ostensibly I have worked utterly alone and am therefore completely responsible for every line in this book, I am fully aware of my continual dependence upon scores of persons. I follow the hallowed custom of naming them here and indicating crisply the reasons for my gratitude.

Ideas: the splendid volume of Paul Fraisse (1963) which concentrates upon the perception of time; the trenchant analysis of Wilbert E. Moore (1963) which grasps time sociologically and humanely; the convenient collection of essays on time edited by J. T. Fraser (1966); the technical articles by Sanford Goldstone and his conscientious associates; and the imaginative, if not always feasible, suggestions of Henry Z. Abramovitch. *Encouragement:* Yale University which means in this context its liberal administrative officers, particularly William C. DeVane and Kingman Brewster, Jr., as well as chairmen of my department, especially Carl I. Hovland and Claude E. Buxton, who during these years have permitted me to work and think as I please in a stimulating, cordial environment; The Carnegie Corporation of New York which enables me to carry on research in Africa, sometimes related, more often not related, to time but always serving the function of compelling me to grope toward the unattainable; and Eveline Bates Doob who, unlike me, has always had faith in this enterprise. *Hospitality and friendship:* Anna Schwarz in Mölten, South Tyrol; Monique Bonte in Butare, Rwanda; Gordon M. Wilson in Nairobi, Kenya; Louis L. Klitzke in Dar es Salaam, Tanzania; and Marshall H. Segall in Kampala, Uganda. *Preparation:* Jane G. Olejarczyk, a chipper and efficient secretary, and Helene H. Fineman, a superb editor and stylist; the outcome of their efforts almost convinces me that an unwieldy manuscript has become a unified book.

April 17, 1970   L. W. D.

PART I: PROCESSES

1. FOCUS

Of course, of course, there are many, many ways to try to apprehend the nature of man and of the eternal, cosmic problems confronting him. But one view must provoke optimistic confidence unless we already adhere to a strongly fortified, highly rationalized position: both wisdom and goodness are elusive; no approach, no solution, is ever likely to be permanently or completely satisfactory or to transcend unchanged the time and culture of its origin. Each account, being never really new, twists what is already known in such a way that, hopefully, insight somehow is deepened and therefore truth and joy come just a bit closer, and ignorance and despair recede minutely.

Our concern, then, is with man and his problems. All of us have more or less unique devices to guide us toward some understanding of other persons and ourselves. That boy you know did this because—and the completion of the sentence contains some sort of implicit or explicit explanation applicable to him and perhaps to all human beings. It is possible to grasp great stretches of man by exploring some of his many activities. Clearly a knowledge of the society in which he lives, of the material forces which impinge upon him, of the communication forms he employs, of the government which controls him and receives his support, or of the metaphysics of his existence which impinges upon his awareness however dimly, is of capital importance. Having now alluded to other approaches, I turn to what I really wish to say about the sweep of this book.

Whoever seeks to indicate peculiarly human attributes can easily provide a list which almost certainly begins with language, probably includes a capacity to produce tools, and may possibly refer to an intimate connection with a deity. Many or all these attributes, we realize, also characterize other animals to some degree. Various species have rudimentary if complicated methods of communicating to one another and of manipulating the environment, and it may be, too, that they are not completely removed from divine concern. Here we concentrate upon another human propensity, shared to some degree with other animals: the capacity to feel, think, and behave in a temporal frame of reference, which means, we shall see, that the past, the present, and the future influence each other. Yes, the future affects the past.

On every level of existence we are preoccupied with time. Among the improvements men for centuries have striven to effect have been more precise or more convenient timepieces and calendars. Really inspired minds, in disciplines ranging from philosophy and physics to literature and religion, have been troubled by time both in the ancient and modern world (Nichols, 1891; Heath, 1936). Each science has various ways of measuring time (Gurvitch, 1963); thus a biologist determines whether he will employ the age of the organism, life expectancy, reaction time, chronaxie, or periodicity as a temporal frame of reference by noting which standard is most useful for the problem being investigated (Sollberger, 1967). Aside from expressing pride in our tendency to glance both backward and forward (Hearnshaw, 1956), various writers have insisted that time, like space, is "always present in sensory processes" and hence can be said to "invest all objectivity" (Jaspers, 1962, p. 79). Others proclaim that human feelings cannot be experienced, discussed, or analyzed without some reference to their temporal direction (Stern, 1938, pp. 547–74). It is not surprising, therefore, that temporal judgments have been investigated formally and artificially in psychological laboratories for more than a century. "Le temps, n'est-il pas le phénomène fondamental de l'existence?" (Minkowski, 1958b).

But erudite references are not needed to pay tribute to time. In every society men note its passing if only by observing the length of their shadows upon the ground or the phases of the moon. Simple proverbs portray and transmit the wisdom of the dead on matters temporal. "There is a time and a place for everything" suggests that one of the coordinates of orderliness and social living is the timing of an event, as a result of which we must judge time if we would be proper and respectable. While doing research in places scattered from North America to Africa I have always found that people, ordinary people, are able to reply to the question of whether time—which I do not define—"ever seems to pass more slowly or more quickly than usual." The question is sensible and intelligible, and it generally elicits affirmation. You yourself have rules suggesting your dependence on time which you also do not define explicitly but employ implicitly in many different ways. You know or think you know the best time to work, to seek advice from another person, to take a vacation, to learn a foreign language, to retire.

Finally, on a nonmetaphysical level it must be emphasized that we succeed, within limits, in changing most aspects of existence, whether it be our social status or our physical location in space, but we cannot affect the passing of time; at the very most we make only suitable adjustments to it. True, you likewise cannot stop snow or rain, though you may seek shelter, change your clothing, or move to another place in order to circumvent these natural elements. In contrast, there is absolutely nothing you can do to halt the coming of tomorrow—or, if I may cruelly remind you, to make yourself immortal.

PATTERNING

Quid est tempus? Si nemo a me quaerat, scio; si quaeranti explicari velim, nescio.

I follow the tradition of many writers on time by reproducing this bit of wisdom but, unlike most of them, I employ the language of the author, which I believe will be comprehended because the meaning of St. Augustine's dart is part of our common intellectual heritage. The same challenge, the same bewilderment, pursues us today; in the words of a thoughtful sociologist, "Time teases our minds when we think about it, but there is a reward in trying" (MacIver, 1962, p. xxiii). A novelist has a character exclaim, "I had now come face to face with the nature of time, that ailment of the human psyche" (Durrell, 1960, p. 12). An investigator has cried out from his laboratory where he has been conducting a series of experiments on time: "After ten years and almost 10,000 subjects the mysteries of human temporal functioning had multiplied more rapidly than solutions" (Goldstone, 1967). And of course philosophers pour oil upon the fire: "There is nothing in the external world corresponding to the distinctions we make between past, present, and future, which are therefore essentially subjective" (Whitrow, 1967a); "As we reach the end of this historical sketch of the attempts on the part of the philosopher to understand time, can we say that the problem has been solved?" and the answer given is "Certainly not" (Benjamin, 1966, p. 29).

Let us, nevertheless, try to banish some of the uncertainty about time by accepting past, present, and future as undefined terms and by then recognizing, as has almost everyone who gives a serious thought to time, that we are forced to live in a subjective or *psychological present*. Some philosophers have a low opinion of this concept, which they call a "specious" present because they believe either that it is not a basic constituent of our temporal experience (Mabbott, 1951) or that it often appears unmanageable conceptually and methodologically (Boas, 1950). But again our attention is only being called to man's cosmic plight—"all discussions of time lead sooner or later to fundamental problems of a metaphysical kind" (Gunn, 1929, p.

371)—which, we are forced to admit, is ultimately unavoidable and which sends us into other clouds perhaps more glimmering but also equally impalpable. No, we must disagree with these philosophers: we must pry ourselves loose from epistemological elusiveness and move on.

What we experience in the psychological present—what is on our consciousness; the ultimate stuff of our own private reality; the *"one* absolute instant" (Rubin, 1934)—happens fleetingly and, when appraised, has as its reference an experience that is in the past or the future. This aspect of time is perceived directly and, when conceived operationally as a continuous process, requires no additional response, such as counting or verbalization (Boring, 1933, pp. 133-4; 1936). It includes the ability to observe that two events occur simultaneously or almost simultaneously (Sturt, 1925, pp. 12-13). It refers, in short, to a peculiarly human capacity to halt our mental processes by an effort of attention—"La vie de la conscience devient punctiforme" (Minkowski, 1935)—and thus to make note of duration or succession while experiencing either the discreteness of events or their sweep both backwards and forwards. This attribute of which we speak has been expressed in the prose of a poet: "Time is our consciousness of the succession of ideas in our mind" (Shelley, 1813).

Even if we are willing to elude the philosophers and accept the reality of the present, we discover nonphilosophical and distinguished colleagues who also strive to delineate the past or the future. Historians, who are acknowledged, self-proclaimed experts on the past, for example, do not easily specify when the past is really past. They have a penchant to select for themselves some arbitrary line or point such as a war or a century, and then they resent having any other discipline, especially political science, capture the outlying domain containing participants and observers not yet dead. Most challenging of all is the contention that the historian's pure past cannot be grasped unless its images of the future are taken into account (Polak, 1961, pp. 20-33). And it is amusing to note that logicians until recently attempted to eliminate time in order to render a bit more manageable the complexities of human reasoning: the principal verb was placed in the present tense ("he came yesterday" could be transformed into "he is the person who came yesterday") and then their rules hopefully could operate in a timeless sphere (Prior, 1957, p. 105).

But let us go on to a deceptively simple example: *I touch my knee.* First I decided to make the move—the decision then was in the present because I was experiencing it, the intended action in the future because I had not yet experienced it. At the moment of touching, the action is in the present because I am experiencing it, the decision to do so moves into the past because that earlier experience has ended. Afterwards, both the decision and the touching are in the past because the experiences are over, the act of remembering then becomes a present experience. This mode of expression is not complicated when we agree, for practical and epistemological purposes, that our judgment about the direction of time depends not only upon the experiences we were having, have had, are having, and will, may, or should have but also upon the moment when the temporal judgment is passed. And that moment itself has duration and can be judged.

The vivid description of the dramatic incident of touching my knee makes patently obvious a fact which many students of time, especially psychologists, have neglected: in their understandably zealous effort to capture the elusive problems associated with time, they have often tried to conserve their scholarly and scientific energy by excluding the behavioral context in which human beings make temporal judgments. *This we dare not do.* The title of this book suggests that we are seeking to analyze time in a broad context: to comprehend both the parts of a pattern and their organization usually requires the utilization of many sources of wisdom. We must be somewhat megalomanical if we are to do justice to the complexity that a consideration of time involves, our research must be gargantuan if we are to include all the factors affecting temporal behavior and thus to feature the deviations as well as the general principles through which we would understand the phenomenon.

The canvas on which time is portrayed cannot be criticized for being oversize when we consider the unabashed tributes bestowed upon the subject. I could offer an anthology of additional superlatives to justify the size of the canvas, but I shall quote only two panegyrics, one from a venerable sage of modern psychiatry and the other from a psychologist specializing in geriatrics. The former: "The culminating feature of evolution is man's capacity of imagination and the use of time with foresight based on a corresponding appreciation of the past and of the present" (A. Meyer, 1952, p 88; italics omitted). The latter: Temporal orientation serves the function of liberating the individual "from dominance by his immediate concrete situation and . . . provides an alternative to impulsive action" and it offers "a framework within which self-identity develops, maintains, and transforms itself" (Kastenbaum, 1964). Words?

After this brief introduction, our central theses must be unveiled. We would accomplish two chores by simultaneously exposing and standardizing a segment of terminology which will be employed in the analysis of what hereafter will be called generically *temporal behavior:*

1. For people to live together, their interrelations, their activities, must be regulated. Restrictions are placed upon them by the conditions of their natural environment (and you may wish to be reminded that the word for

weather in Latin and the Romance languages is the same as that for time) and by their social milieu; some goals they must *renounce* in the present, others they may *anticipate* only in the future. These two activities depend upon learning in the past which in *recollection* is experienced in the psychological present and which most likely involves a system of time reckoning. The concept of activity requires analysis in its own right, and that will be done in the next section.

2. The awareness that an activity has ended, is being experienced, may have to be renounced, or can be anticipated is as good an operational definition of the *psychological present* as we can demand. "The entire life of human beings," one writer proclaims with enthusiastic exaggeration, "is permeated with the thought of things to come" (J. Cohen, 1964, p. 124); and he might just as well have added that each recollection may have some sort of date tag attached to it. On a broader level, it has been maintained by an awesome writer of the twenties that memory is "the real organ of history" because it "preserves as a constant present the image of one's personal past and of a national and a world-historical past as well, and is conscious of the course both of personal and superpersonal becoming" (Spengler, 1926, p. 132).

3. Such awareness involves or is equivalent to *temporal judgment* concerning either the *duration* of an *interval* or its temporal relation to other intervals *(succession)*. The interval refers to the event or series of events being judged or measured, the duration to the time (however conceptualized) elapsing between the onset and termination of the interval.

4. The direction of awareness at a given moment or characteristically over a long period of time is specified, respectively, as *temporal orientation* or *temporal perspective*. Orientation and perspective obviously demand a series of temporal judgments which are likely to be organized within the person; whereas by itself a temporal judgment may often be considered a discrete act momentarily detached from events which never halt. We are dealing here with phenomena which we shall call *subjective time* because any kind of awareness is a personal, not completely communicable, process.

5. Alongside subjective is *objective time* which refers to the more precise measures emerging from an hourglass, a watch, a clock, the movement of the earth or the moon or the stars, or from a reference to an interval or event on which there is common agreement in a society, such as the reign of a king long dead, the assembling of a crowd near a shrine, or the rising of the full moon after the melting of the snow. Originally or ultimately the conceptualization of objective time and the decision to employ a measuring instrument or an event as the reference point has been subjective too: finite men constructed clocks, agreed as it were to divide the week into seven

days, cued a ceremonial to the reappearance of a particular animal, etc. But thereafter the results of the subjective reactions have become objectified for those living in the society. If you are to cooperate with your fellow men, you must accept Greenwich Mean Time as your ultimate guide; you were not alive in 1884 when your country accepted this convention. The individual, however, may note that subjective and objective time do not always coincide at a given instant. "The time seems so short"—this is a temporal judgment involving only subjective time. "The time seems so short and I see that a full hour has gone by"—here subjective and objective times are being interconnected. The knowledge a person has concerning objective time, including the succession of events in the past and the future as perceived by competent observers, is *temporal information*. Common usage sometimes prefers the term "orientation" to refer to such information, but here—glance up at point 3—we have usurped that concept to indicate the direction of a temporal judgment. I see no great danger, however, in employing the verb *orient* or *disorient* in connection with information: if you say today is July 9 and it is in fact July 8 or June 9, then your temporal information is wrong, you are temporally disoriented, and your temporal orientation is in the present.

6. The duration of intervals may vary from milliseconds to billions of years. Nonprocrustean terms are needed to designate three lengths. That interval which can be immediately apprehended within the psychological present—or at least which gives that impression—we shall call *ephemeral;* the various speeds on a camera ranging, let us say, from 1/500 of a second to 1 second are all ephemeral intervals. Longer intervals whose duration can be remembered or anticipated clearly, distinctly, and easily are *transitory;* usually, but not always, the events of today or tomorrow fall within these boundaries. All others will be called *extended*.

7. Judgments of intervals are made partially on the basis of *standards* which may be either objective or subjective. An objective standard is obtained from consulting an instrument which measures objective time, whether it be the hands of a clock or the length of a shadow on the ground. A subjective standard is supplied automatically or deliberately by the person himself on the basis of experience in the past; an action, he intuits or calculates, takes a long or a short time to carry out.

These seven points embody my contention that the approach to time must be very broad. But now some system is needed to prevent us from going off onto too many relevant but complicated tangents. We may be told that an individual will suffer crippling anxiety if he has to face a severe crisis; yet we do not expect his psychiatrist to be able to predict on the basis of psychiatric principles the arrival of an economic depression which will cause him to loose his job and thus bring on the neurosis. In this book we could speak of celestial mechanics without being out-of-order because the sun and the stars have contributed norms for objective time, but we shall not go that far afield. Whenever it seems necessary or desirable to stop the analysis, I shall flag the decision thus: *Arbitrary Limitation*.

ACTIVITY

We turn to a description of the behavioral context in which temporal judgments are embedded and quickly note a need for concepts which dissect the trio of temporal orientations characterizing the psychological present. Otherwise challenging statements such as the assertion that "seemingly . . . the future determines the present, the present controls the past, but the past creates that future and so imposes its values on the present" are likely to be, as the oft cited author of that sentence suggests, "but juggling of terms and obscurantism" (Frank, 1939). Such a rich vocabulary is available that it is not necessary to try to achieve immortality by resorting to neologisms. Within the psychological present, we distinguish recollection from the past, experience in the present, and anticipation regarding the future. Of course these activities merge along a continuum, as the quotation above suggests. You recollect something from the past and then determine in the present to act in the future accordingly; you think of an old friend, for example, and decide to visit him—just like that. Or you participate in the present by recollecting something from the past or by anticipating something about the future. Unless you are self-conscious or for some reason are contemplating your temporal actions, however, you may or will not reduce your feelings to such neat, discrete categories; you are carried forward on a stream interrupted only by sleep.

The importance of recollecting and experiencing in human activity seems so self-evident that these concepts can fend for themselves. Anticipation, however, requires greater care since it is so critical in the learning process: it occurs in the present with reference to the future and is always affected by some experience that has been learned in the past. A dog is conditioned to salivate when a bell is sounded: the bell has been followed by food after the passing of an interval of time, hence later this stimulus indicates in the present the probability of food appearing in the near future (cf. Elkin, 1964). Whereas "the rat can sustain a delay of some 4 minutes, the cat 17 hours, and the chimpanzee 48 hours," it is suggested most broadly, "in man the horizon may reach far beyond his own brief existence" (Cohen, 1966). Unlike the salivating dog, moreover, a mature man can resist a communication in the present when he has the vague anticipation that the matter cannot be so easily settled or that there are grounds for another viewpoint. Long before the fashionable interest of social scientists in achievement motivation, solid studies had shown that a person's level of aspiration depended upon successes and failures experienced in the past and anticipated in the future. In general terms, hope among human beings must mean that "sometime in the future, the real situation will be changed so that it will equal my wishes"; the realization of many goals involves postponing action after imagining gratification in the future (Lewin, 1942; 1952, pp. 55, 82). Similarly an essential ingredient of anxiety is a future orientation, the "fear of something about to happen, rather than fear that something has happened" (Krauss, 1967). Wherever we look, therefore, even when efforts are merely made to change the attitudes of hapless undergraduates by rewarding them for expressing a viewpoint contrary to their own at different times during the painless ordeal (Rossomando & Weiss, 1970), temporal factors play a role in the learning or unlearning process.

It will be evident, I hope, in most of the chapters that follow that a great deal of the elusiveness of temporal behavior can be traced to the subtlety of the recollections and anticipations themselves. And they are subtle, I think, as a result of the intricacies of the learning process, one example of which can be quickly provided. By and large all organisms learn faster when they are constantly rather than intermittently rewarded; when the reward or punishment is immediate rather than delayed; and when the effort required to reach the goal is slight rather than great (e.g., Mowrer & Ullman, 1945). Reward under each of the first conditions is, respectively, more surely, more quickly, and more easily attained than under each of the second conditions. Paradoxically, however, these very conditions which supposedly favor learning have been shown (most thoroughly with white rats, though even here not completely convincingly) not to facilitate the retention of the habits to which they give rise *after* they are no longer reinforced by leading to the reward. The animal, for example, literally continues fruitlessly to look for food in the empty maze for a longer period if previously he has been intermittently rather than regularly rewarded. According to one view, the animal learning under *unfavorable* conditions has sought to reduce dissonance or discomfort resulting from being rewarded only part of the time, after a delay, or through exerting relatively greater effort; this he does by finding "extra attractions" in the maze ("aspects of the action itself, of the apparatus, and of the consequences of its behavior"); and that source of gratification persists even when the reward of the food is no longer present. Then, too, when there has been only intermittent reward during the original learning, the anticipation has been built up that sometimes there will be no reward; and so the nonappearance of reward cannot so quickly destroy the anticipation that maybe on later trials the reward will put in an appearance again (Lawrence & Festinger, 1962) It seems reasonable to assume that a great deal of human learning occurs under similarly unfavorable circumstances: people are not always praised, they often must wait to attain what they seek, and they usually exert themselves to be successful. Here perhaps is one of the reasons habitual ways of behavior persist even under unfavorable circumstances.

At a given instant, then, the psychological present moves backwards or forwards along a continuum ranging from the past to the future. Whichever point in time is involved requires that other points be at least momentarily excluded or *renounced*. You believe that a particular event occurred in the past, that it is occurring in the present, or that it will occur in the future; but you ordinarily do not hold all three beliefs simultaneously: you deliberately or unwittingly renounce two of them, you fuse your feelings into a temporally undifferentiated experience, or you pass no judgment whatsoever.

A special and popular form of renunciation is *deferred gratification:* abandoning a goal in the present or immediate future in the hope of increasing the probability of achieving that state of affairs or an even more satisfactory one in the future. Such gratification, therefore, always involves some temporal delay. In the experimental literature the animals or human beings who learn delayed responses do so because an immediate response produces no reward whatsoever: only waiting brings benefits. Similarly you do not spend the money you now have, instead you place it in a bank so that later you may enjoy both the principal and the interest it will have accumulated. But suppose you renounce your love because she will be happier with someone else. Do you do this only for her sake or are you worried about your own adequacy or the possibility that together you two would both be miserable? In any case your goal seems to be not deferred gratification but the avoidance of nongratification; unless you renounce her, you anticipate unhappiness for her (and that would make you unhappy too, or at least so you say), for yourself, or for both of you. In short, this kind of renunciation is always accompanied by anticipation; it is a present act oriented toward the future. But what about suicide, the renunciation of life? We shall deal with this too; we have only suggested how the processes interact.

Recollecting and experiencing have already been intimately connected with the problem of time by traditional writers on the subject (Boring, 1942, pp. 575–7). The other two parameters of temporal patterning, anticipation and renunciation, however, have usually been neglected. More should be said here about their content, but then it would be necessary to categorize most of human behavior and the environmental circumstances affecting that behavior. At this point I invoke the first Arbitrary Limitation: each process has some kind of relation to what we may easily call reality. More specifically, what we recollect, experience, and anticipate are *beliefs* or opinions concerning phases or aspects of reality as well as feelings or *attitudes* concerning their desirability or undesirability. The italicized words will be used extensively but unobtrusively throughout this book.

My principal thesis now emerges, it is hoped: we experience in the psychological present, even as we recollect the past and anticipate the future; and at any given moment, and sometimes over long periods, we renounce one of these activities as a result of one or both of the other two orientations. Temporal judgments which perforce occur in the psychological present are patterned by beliefs and attitudes which are recollected from the past and which affect future anticipations. We are always—or almost always—confronted with choices in time, and to choose we must renounce.

On a simple level, and only for purposes of illustration, let us note for the first but not for the last time that the watched pot never seems to come to a boil. Presumably it is being watched because the cook, having renounced all other activities (involving either the rest of the meal or her own ambitions and ego) believes and hopes that it will boil soon. Unless she can increase the heat, she is powerless to speed up the boiling. Under these circumstances, as she stands helplessly by, she is impatient and makes the momentous judgment: time is passing slowly or the water is taking a longer time than usual to come to a boil. Her temporal judgment has thus been patterned by her own impulses.

I do not at all anticipate that this set of distinctions has knitted together all the loose ends of time or that I have silenced objections to my approach. Some of the ends and objections (which I have disrespectfully titled "Quibbles") I shall deal with in an Addendum to this chapter. Indeed, the final pages of almost all chapters contain such Addenda. This device is employed as a way of elaborating the analysis by supplying evidence and citations whenever they seem necessary. I do not call them Appendices and I do not bury them at the end of the volume because that word and such a practice would suggest that they are dull and useless, whereas they are, all of them, fascinating and helpful. Read, therefore, *Addendum 1.1* and all its splendid successors.

OF PRINCIPLES AND UNCERTAINTIES

Analysis aims to achieve a set of principles through which past and present data can be understood and future data foreseen. With the help of his peers, as well as his enemies, each person evolves for himself a set of guides, hypotheses, or principles concerning his own behavior and that of other persons. Otherwise he would not know what to anticipate, he would find himself and them unintelligible, he would experience utter chaos. Science, being a cooperative enterprise of many experts, usually produces more objective, more general, principles than those formulated by a single individual, though they may not minutely include his nuances.

We think we know a great deal about behavior. Two courageous writers have boldly formulated no fewer than 1,045 generalizations which they suspect to be true of most, or at least of many, persons and peoples (Berelson & Steiner, 1964). It is not necessary to examine their propositions or any others too closely, nor for that matter is overwhelming perspicacity demanded to discover that undoubtedly no single statement is true under all circumstances. For principles about people come from people who have been examined or manipulated in particular situations; they are often based upon central tendencies from which inevitably there are major or minor deviations in individual uses. Rarely if ever can the behavior of a particular person be examined in toto; instead a segment of him is investigated and made to appear lawful; but at a given instant of time, what he does, what he feels or judges springs from a variety of tendencies and is therefore the reflection of multivariate phenomena.

This state of affairs is neither sad nor discouraging. Each solid piece of research reduces, however miscroscopically, the degree of uncertainty with which the same person or group under different circumstances, or a different person or group under the same circumstances, can be approached. What kinds of American women wear red earrings? If a doctoral dissertation based upon interviewing and observing 1,000 women establishes (a) that hair color, age, economic status, and political affiliation are not related to the wearing of these ornaments, red or otherwise, but (b) that a higher proportion of women who are only children in their families thus decorate themselves than those who have or have had one or more siblings, then surely no one would expect the exact finding to reappear in another sample of American women or one in a rural area of Africa. But the trivial knowledge gained in the investigation could serve at least as a first hypothesis in subsequent investigations and would certainly suggest that the presence or absence of siblings should be one of the factors to be noted and taken into account. No clinician ever expects a new case to be exactly like previous ones; yet the experience gained from the past gives him a preliminary and very valuable clue to the next patient.

This tedious, almost banal, methodological observation would not be necessary here were it more universally recognized. Poets and humanists are quite right when they stress the factor of uniqueness in almost all that they depict, love, and do; but they seem a bit silly when they gleefully point to the failure of a particular generalization or theory of science to be valid in every instance. Over against them are more systematic thinkers, especially contemporary social scientists, who are often (and the word *is* "often" and not "always") prone to commit their own professional sins. Some cling closely to data at hand, especially those they themselves have collected; and, lacking either the valor or acumen to indicate the more general implications of the research, they fail to suggest how their findings confirm or deny some principle from the past, or could be used to formulate one for the future. In contrast, others deliberately, illicitly, or unjustifiably leap from their own conclusions to other segments of the universe. An insidious form of leaping is to alter the tense of verbs from the past to the present—"being an only child *is* associated with wearing red earrings."

The literature on time is strewn with instances of non-generalizations and sly generalizations. It would be so easy to turn to the conversations on the magic mountain (Mann, 1928) and to whisper that the philosophically self-conscious author, while hiding behind the often ponderous reflections of his characters, in fact disclosed his own view of time. In exercising his artistic privilege, he was able to evade the empirical problem of deciding whether time is thus experienced by every man. Still, although he did not claim that those ill characters of his were representative of Europeans right before World War I, certainly as a philosophical novelist he wished to illuminate truths which, he assumed, transcended them and their idiosyncracies. Likewise we may feel somewhat patronizing toward a generalization about empty and filled time that emerges from an experiment based upon American college students performing a series of rather nonsensical tasks; how can anyone believe that people in real life would make similar judgments? I decline, however, to throw verbal darts, except in this rhetorical manner, and then only in passing. My purpose is not to crucify the guileless; rather I should like to comprehend, through all available means, the very phenomena the artist and the experimentalist are discussing. The conversation of patients in the Swiss sanatorium or the impressions of the lonely subjects in the artificial experiment are tidbits not to be disregarded but, if possible, to be added to the mosaic.

On the other hand, too much should not be expected from my analysis. We cannot precisely pinpoint complicated phenomena. We must anticipate, in fact welcome, exceptions to whatever generalizations appear. The anticipation comes from the realization that all generalizations must omit particulars, and the welcome from the good thought that truth is only temporary and, when revised through necessity, ascends perhaps a trifle closer to an unattainable ultimate. So I shall try to offer only important materials and observations leading to principles which will be explicitly stated, numbered conveniently for future reference, and italicized for emphasis. The alternative to principles or generalizations of this kind is either silence, which helps nobody, or a sly statement in ordinary type which may escape everybody; what I have to proclaim may be banal or involved, but at least it will be conspicuous. Some facts are well known and do not need to be replicated in the laboratory—perhaps—or they give rise to generalizations which startle nobody. We must make these generalizations explicit if our inventory of knowledge is to aim at completeness.

In the spirit of the last paragraph, the conjunction "but"—either at the beginning of or within a sentence—will be used again and again to indicate what I think to be the blatant exceptions, the ugly qualifications, the other things that are supposed to be equal and often turn out to be unequal. I shall try hard to squeeze principles out of data, however shaky or incomplete they are; and then I shall qualify them all with "but." The devil of non-generalization or of hasty generalization will have a sturdy, conspicuous advocate.

I shall be constructing principles and corollaries in Chapters 3 through 8 which represent the culmination of the analyses in Part I and Chapter 8 of Part II and which are reproduced seriatim in an index at the end of the book. They are at the basis of the remainder of Part II which describes the functioning of temporal behavior in various situations and under various circumstances. In Part III consideration is given to the manipulation of temporal behavior.

Better principles and corollaries can be formulated, uncertainties are reduced when as many errors as possible are removed from the data of observation or experiment. The pitfalls of research are of course numerous, too numerous to classify succinctly. Here one must be indicated because of the distinctive nature of the phenomena being examined: the danger involved in making inferences about how people "really" feel or think on the basis of what they say or do. Thus when asked whether they would prefer to be given a small sum of money now or a much larger one a year later, most African informants chose the smaller amount (Doob, 1960, p. 88). May the conclusion be drawn that they sought immediate gratification and hesitated to postpone their joys? That is a reasonable interpretation since many maintained, when questioned further, that the bird in the hand is more attractive than one in the bush. But others wanted the small sum right now, they said, not to expend it on carefree, riotous living in the present but to invest it in some small business enterprise which would yield, they hoped, more than the larger amount that was hypothetically proffered later.

Obviously, therefore, the original inference would have been partially incorrect or not completely true. But, if I may be trite, we do the best we can: if we worried about every possibility, we would be inhibited from risking any inference at all. Let us, however, at least recognize the peril.

The certainty or uncertainty of a principle or induction can best be indicated by specifying the data on which it is based and also the way in which the data themselves have been collected. Again the essential functions served by the Addenda should be evident. The reader has already noted in this very first chapter a tendency to make references to an anonymous "you" as an authority or a source of information; and the practice will have to continue throughout the book. This contrivance seeks to offer not evidence but clarification; and it is fully recognized that hypothetical illustrations or single examples prove little. I offer no apology for the use of "you": I am voluntarily driven to it when no other evidence is at hand, when the point being made appears to be self-evident (but when is anything self-evident and to whom?) or when a tentative guess is more provocative than a confession of ignorance or doubt.

METHODOLOGICAL MATTERS

We must now turn to the unpleasant problem of methodology before the analysis can move along. No special methodological difficulties intrude, however, when anticipation, recollection, and other psychological activities are appraised other than the usual ones associated with the behavior of men and beasts—and they are simply insuperable. But time is a somewhat unique villain requiring concentrated attention. Either in real life or in the laboratory temporal judgments are made in various ways. Terminology is not completely standardized, alas, even though or perhaps because for nearly a century time has been seized upon as a doctoral-dissertation subject. And so I conform to what appears to be the most popular consensus (Clausen, 1950; Bindra & Waksberg, 1956; Orme, 1962b; Wallace & Rabin, 1960) and emerge with this deliberately unoriginal, eclectic classification:

Chronometry: a measuring instrument, such as a clock, watch, chronometer, sundial, calendar, etc., is consulted in order to ascertain objective time, and what is perceived affects temporal judgment or, if the person has confidence in the instrument, determines it completely. The chief problem here is the accuracy of the instrument, although questions may be raised concerning the person's ability to perceive or report objective information without distortion. This method is so obviously useful and ubiquitous that it is usually not mentioned in psychology textbooks which prefer to describe the subtler methods favored in laboratories.

Verbal Estimation: the duration of an interval is estimated in terms of objective time or through some verbal reference to duration. For example, you are asked how many minutes have passed or will pass; or you indicate whether you consider an interval to be "long" or "short."

Comparison: two intervals in the external world are compared, and you then indicate which one appears longer (or shorter) or whether they seem equal in duration.

Production: in response to a verbal instruction or command, an attempt is made on some formal device (e.g., sounding a buzzer) or informally (e.g., by tapping or especially by counting aloud) to produce an interval equal to the one that is verbally indicated. "Push the button," you are told, "for exactly 7 seconds." Or: "count off 30 seconds."

Reproduction: after perceiving an interval in the external world, an attempt is made to reproduce one of equal duration. Ordinarily in the laboratory no word is spoken by the experimenter after the initial instruction has been given; a buzzer sounds, for example, and then you activate the same or another buzzer, or some other device, for what you believe to be the same duration.

Some of the variations in procedure and terminology are reported in *Addendum 1.2*. That skeletonized description immediately suggests that each method involves different psychological activities which, I think, can most easily be signaled by suggesting that an interval or a succession of intervals may be directly experienced or indirectly symbolized. You experience an interval when you in fact note its content or your own activities during its passing; you symbolize it by means of a conventionalized verbal utterance or thought. "Sit still, please, for two minutes"—the command contains the symbol of "two minutes," the actual sitting still is experienced. Now the methods can be briefly embraced by the dichotomy:

Method	*Stimulus*	*Response*
1. Chronometry	symbolized	symbolized
2. Verbal Estimation	experienced	symbolized
3. Comparison	doubly experienced	symbolized
4. Production	symbolized	experienced
5. Reproduction	experienced	experienced

All five methods make use of a standard, the objective or subjective unit with which ultimately, consciously or unconsciously, the perceived interval is compared. With Chronometry, the instrument itself is the standard, although I suppose we could say that it must be compared with some conventional standard (such as an official timepiece or the position of the sun) if its accuracy is to be determined. Verbal Estimation employs a standard supplied by the person himself. When the Method of Comparison is employed, either one of the intervals may be considered the standard; usually the investigator or the person passing judgment indicates which one is to be selected by the phrasing of the question. Thus if you are asked, "Is the second sound longer or shorter than the first?" then clearly the first is the standard. Reproduction and Production depend upon an objective standard: for the former it is the interval which is first experienced and is to be reproduced, for the latter it is the interval which is first indicated symbolically as the one to be produced.

In the laboratory it is of course the investigator who selects the method through which judgment is passed. In real life, as we shall emphasize later, the Method of Chronometry has priority whenever it is utilizable: your first impulse, when making a temporal judgment, is to consult a clock, calendar, or some kind of record. Ordinarily, too, Reproduction and Production are utilized to judge ephemeral or transitory intervals, not extended ones. After allocating an hour for a walk, you look at your watch to see how much time is left and thus you desert Production as such.

The methods have been described with reference to judging duration in the present, but some of them may also be employed for other purposes. Succession may be judged; in the laboratory, especially when ephemeral intervals are perceived (for example, in the study of rhythm), a variant of Verbal Estimation is most useful. You cannot, however, produce or reproduce in the future an interval you have not yet experienced, nor can you compare it with some other interval. But Chronometry is possible: you look at a calendar to discover when you have an appointment with the dentist. Through anticipation, moreover, another variant of Verbal Estimation may serve: you try to imagine what you must do in order to reach an objective, and then you make a verbal estimate of the interval needed to do just that —you perceive, you experience, imaginary activity.

Is one method superior to another? Superiority here means the accuracy with which judgments are made, with accuracy being defined in terms of objective time; the variability from person to person under similar conditions; the reliability of an individual's judgment, i.e., its consistency under similar conditions; and the objectivity with which the method can be employed. Probably, I suppose, all methods are quite objective since passing temporal judgment is seldom a delicate matter, such as providing information about income or sexual activity; hence the role of the investigator is likely to be minimal, or at least similar, no matter what method is employed. Otherwise, relevant research has produced regrettably inconsistent and contradictory results with respect to the remaining criteria. Under these circumstances, I can only squeeze and extract these two generalizations, the evidence and noteworthy exceptions to which can be found in *Addendum 1.3:*

1. Each method is likely to yield different results with respect to whatever aspect of time is investigated; correlations between the findings of two methods have often been found to be high, but they are never perfect. There is usually a high and negative correlation between the results obtained from employing Verbal Estimation and Production.

2. It is impossible to assert that one method is always more or less valid, reliable, or variable from person to person than another. The differences, it appears, depend not only on the methods themselves but also upon the duration of the intervals being judged and indeed upon the personalities of those passing judgment. This point is especially telling with reference to reliability: an individual's consistency may certainly be ascribed to the method of measurement or to his own ability. Responsibility is more likely to be a function of the method than of ability, however, when judgments are passed repeatedly under the same or similar conditions; ideally, this means that the same interval, or the same series of intervals, is repeatedly judged during the same experimental session.

When the investigation focuses not upon temporal judgment but upon other aspects of time—temporal orientation, perspective, knowledge, attitude, etc.—customary methods of research are employed and no really unique problems arise. But I would mention here one methodological distinction because it appears so frequently in connection with studies of temporal orientation: the questionnaire or the task to be performed is of the *projective* or *non-projective* type. Projective instruments do not reveal their purpose; the subject, for example, is told that he is to write a story in order to show how imaginative he is, when in reality the investigator is interested in the time span of what he writes as a measure of his orientation or perspective. Non-projective instruments, in contrast, come a little bit closer to revealing their true purpose; here the person may be asked directly whether he believes that time passes slowly or quickly when he is bored. In the latter case, even though he grasps the nature of the topic being investigated, he may not appreciate its guiding theory or the ways in which the information will be utilized.

Other methodological problems associated with measuring temporal judgment are, in my opinion, all variants of sampling: the kind of behavior and the number and type of subjects being examined. Is it possible to assume that the factors affecting temporal judgments measured under laboratory conditions are universal for mankind—such as the effects of filled and empty intervals—and that therefore the nature of the human beings used as subjects is within wide limits largely irrelevant? This certainly was the original assumption in the psychological laboratories in Germany and America: subjects were almost always university students, especially those concentrating upon the study of psychology; and nothing more was revealed about their culture or general background. My own feeling is that such an assumption is true only to a minimum degree and that therefore most of the laboratory work has led to results which, for all we know, may be of very limited generalizability. In less dull words: any principle we dare to formulate concerning temporal judgment needs to be tested—not only outside the laboratory and the clinic but also cross-culturally—and this, unfortunately, has rarely been done. Indeed it is quite possible that the failure of research on temporal judgment to produce powerful generalizations and to reconcile contradictory findings, as any review of the literature devastatingly demonstrates (Wallace & Rabin, 1960), results in part from the restricted sample of persons who have been tested. Other aspects of temporal behavior have received less parochial treatment either because they are less amenable to laboratory manipulation or because they so obviously have a potent cultural component.

Many of the early investigators deliberately used a biased sample of subjects because they wanted only persons trained in the method of introspection who therefore not only could pass temporal judgment but could also indicate how they arrived at those judgments. For this reason their subjects usually were either colleagues in the same university (and in many instances the experimenter himself served as one of the subjects) or university students. The sin of having too few subjects was almost always committed, understandably because often literally hundreds of judgments had to be made by each individual and also because, without benefit of modern statistics, it was assumed that almost all obtained differences were real ones and not due to chance. In some instances, moreover, adequate controls were lacking in the usual sense: a factor was designated as being critical even though the same subjects or a comparable group had not been tested in its absence. But these investigations, though imperfect, cannot be neglected completely: just as a starving man sloughs off some of his fastidious tastes if he would not perish, so we who require data on time must be grateful even for information which does not quite meet our approval. With love and respect, therefore, an investigation will be called "classical" when it stems from this tradition; and if its method or scope is considered almost but not quite satisfactory, it will be honored with the title of "semi-classical" —the love and respect will be shown by the absence of scolding italics, scornful capitalization, or condescending quotation marks.

A final methodological point involves the fact that temporal behavior itself, especially the passing of temporal judgments, has a temporal orientation. "One of the most frequently noted observations," a writer notes, "is that lack of overt activity on the part of the subject leads to overestimation" (Culbert, 1954). The validity of this observation cannot be tested unless we know when the judgment is passed: is the person recollecting, experiencing, or anticipating the "overt activity"? I have the impression that such careless statements usually assume that the judgment was made concerning an event in the past. To achieve methodological clarification, we must offer yet another concept and draw some easy temporal distinctions. Very simply the motivating process giving rise to temporal behavior is the *temporal motive*. Why do you want to know what time it is, when those two persons were married, whether you will arrive before or after he does? The temporal orientation of this temporal motive can be in any direction; therefore it may have any one of the following temporal relations to an interval whose duration is being judged:

1. *Prior arousal:* the motive is evoked before the start of the interval; you must note, you say to yourself today, how long it will take to climb the mountain tomorrow.

2. *Interim arousal:* the motive is evoked in medias res; you wonder, as you grow weary during the climb, how long the ordeal will last.

3. *Immediate subsequent arousal:* the motive is evoked right at the end of the interval; now that you have reached the summit, just how long was the climb?

4. *Delayed subsequent arousal:* the motive is evoked some time after the end of the interval; how long did it take you to climb that mountain last year?

These methodological strictures have been utilized throughout the analysis by seeking to ascertain for each study the subjects and method upon which it is based. How, then, can this information be conveyed with some stylistic grace? I have tested various devices in earlier versions of this book, but each has been found to have disadvantages. I have tried simple prose to set forth the information reported in a study, but the sentences grow long and repetitious. I have sought also to code the information which then could be placed concisely in parentheses; but, unless the code is memorized, the reader is faced with a combination of symbolic numerals and letters which look like jibberish until he has the patience to decode them. What I have finally done is to adopt a compromise: four bits of information, when relevant and when not already mentioned in the accompanying text, are supplied in an abbreviated form which, I hope, can be instantly understood:

1. The *country* in which the study was made; this provides a broad clue to the culture from which the sample or subjects have been drawn. The United States of America, where too many of the investigations have been carried out, becomes "U.S."

2. The *status* of the subjects or persons indicated crudely by a reference to age or education. So many have been U.S. college students or university students elsewhere that, again in the interest of conserving newsprint, the simple abbreviations of "college" and "university" will be employed.

3. The *method* of measuring temporal judgment; three of the five methods described in this section are each designated by a single, self-explanatory abbreviation (comp.; repro., and produc.), a fourth can be recalled by its distinctive adjective (verbal), and the fifth (chronometry) plays no role in research. The reader is warned, however, that these labels obviously cannot reveal the details of the method, which can be ascertained only by consulting the original reference. Occasionally the text will point out some of the details and in a few instances one of the labels attached to the variants will be employed.

4. The *time* at which the temporal motive has been evoked; here the four adjectives of prior, interim, immediate, and delayed suffice.

We have then two pairs, the first pertaining to the subjects and hence to the sample, the second to the method. The pairs are separated by a semicolon, the members of a pair by a comma. Finally, lest it be necessary to disrupt a second time the sentence to which this information applies, the conventional bibliographic data follow a second semicolon: name or names of investigators, date of publication, and page reference in the case of a book—the full citation will be found by consulting the reference section at the end of the book. Sometimes an appropriate adverb signifies politely that the methodological information is absent or ambiguous. A final detail: all italics in quotations are those of the author unless otherwise indicated.

ADDENDUM 1.1 QUIBBLES

The most glaring assumption of the present analysis is that for human beings any event has, or may have, a temporal dimension, just as it also has a spatial dimension: "if space be termed the breadth, time may be termed the length of the field of perception" (Pearson, 1892, p. 155). The dimension of time, therefore, is either potentially or actually part of consciousness and behavior; "it may perhaps be possible to conceive consciousness as existing without the space-mode of perception, but we cannot conceive it to exist without the time-mode" (ibid., p. 156). On occasion the temporal dimension may be unimportant: it is the glory of the ecstatic experience you note, not its duration or its relation to other events. But—at least, I suspect, in the West—more often than not that element is of great significance, especially in connection with remembering: you recall almost automatically not only the content but also the duration and date of an event.

In this analysis we are not at all concerned with conceptions of time that are necessary if the natural scientist or the philosopher is to function effectively in his chosen realm of discourse. I respect their problems so much that I exclude them lest I do them injustice. In declaring this Arbitrary Limitation I am opening myself to the charge I have leveled against those avoiding the social sciences when analyzing time. So be it. I am, however, only trying to set some boundaries for myself; thus I make only passing references to the problems of rhythm and reaction time, though these are topics par excellence for psychological treatment.

No serious problem, I think, is raised by the possibility that temporal judgments may be expressed in objective or subjective terms. Your journey from Aix to Ghent may be reported as having a duration of 3 hours and 20 minutes according to your watch or as being "very long" if those were your feelings while traveling the two places. In fact, so many perplexities associated with time arise because of the discrepancy between the two modes of measurement that the distinction should be accepted, provided we know in which universe of discourse the judgment is being passed.

I recognize the fact that temporal judgment itself may be considered to have a temporal attribute (Dunne, 1934, pp. 132–4). In recognizing this we feel the thud, for the first time, resulting from the infinite regress which compelled me, in the text, to offer time as a primitive concept. Yes, I am aware in the psychological present that I am thinking in the past; I am aware of my awareness of that thinking in the past; and so on. I find this kind of reflection fascinating, but on a nonmetaphysical, nonepistemological plane it seems to lead nowhere. I bid it farewell, or at least I try to do so.

The hazy distinction between past, present, and future should cause no distress when it is embedded in a behavioral context, since then it becomes most evident that any kind of behavior in one temporal mode can be affected by an orientation in a different one (Lewin, 1952, pp. 53–4). The past influences the present: that man's memories of the past prevent him from participating actively in his society. The past influences the future: his past accomplishments give him confidence that he can realize his anticipations. The present influences the past: he is so absorbed in his activities of the moment that he tends to forget what once happened. The present influences the future: he seeks gratification right now, he seldom worries about the future. The future influences the past: his ambition is so great that he falsifies what once happened to him. And the future influences the present: what he thinks and does now reflects markedly what he anticipates he will soon be able to accomplish. In these senses, then, time converges, or can converge.

Obviously the relation of an individual's temporal orientations which endure as his temporal perspective is of great interest from a philosophical or psychiatric standpoint. One writer, for example, avers that the "split man," the individual who is concerned with the future while functioning in the present, is "the *only* sound, healthy, mature, and integrated man" (Polak, 1961, p. 17). Maybe yes, maybe no, as we shall later in this book. But the very value judgment suggests that the issue involves more than a semantic distinction.

There is, I think, no need to be disturbed by the problem of the recurrence of time. Certainly, as Herodotus and most neo-savants before and after him have stated, intervals during which temporal judgments ars passed, like any other event, do not literally recur: the person, his milieu, and any river, are never the same from one moment to the next since they keep changing, however imperceptibly. In a psychological sense, therefore, time must be considered irretrievable. And yet is there not a rhythm in time as in music and in the recurrence of the seasons? Surely the tempo of those measures can be repeated, and obviously the winter always gives way to spring. But no, I must maintain, not exactly so, for the individual who experiences the intervals: the same tempo is different now that you have grown two minutes or two decades older, there may be from your vantage point lovely springs again but none so lovely as that spring a century ago when you were very young and so hopeful.

Throughout the analysis I shall not try to force other person's terms into the concepts that have been so far and will be later defined, unless clarity is thereby gained. For since time is so important, it has given rise to many different modes of expression and designation which are too numerous to catalogue. The most popular way, I note in passing, both in our society and elsewhere is to use the metaphor of space: we look backwards or forwards—and time is called short or long (Marshall, 1907). I shall do justice to this particular metaphor in the next chapter.

Finally and regretfully I must attend to those who charge that concepts such as those being used here are culturally loaded and hence are bound to produce a biased account. An anthropologist has written that *"Duration* is the most widely shared implicit assumption concerning the nature of time in the Western world" and hence he thinks it "seems inconceivable" to persons like me that "it would be possible to organize life in any other way." Time for the Hopi Indians "is not duration but many different things," such as "what happens when the corn matures or a sheep grows up"; and so some Hopi houses are "in the process of being built for years and years" since this activity is thought not to have "its own inherent time system" in the manner of corn and sheep (Hall, 1959, p. 171). I agree that life can be organized in different ways, which means only that there will be variations concerning the intervals whose duration will be judged, concerning the kinds of temporal judgments which are passed, and concerning the actions to which such judgments lead. But surely every interval can be described in terms of its duration in objective time, whether attention is focused on corn, sheep, or human pregnancy; and similarly through action the duration of an interval can be prolonged or shortened, whether that action involves constructing a house or a sonnet. Duration, in short, is a scientific construct utilizable to describe and analyze temporal activity. An anthropologist and I would be in perfect agreement concerning the fact that we in the West place more emphasis upon duration as an attribute than other peoples or that the temporal motive may be more frequently aroused in one society than in another. Of course, attitudes toward time vary greatly.

ADDENDUM 1.2 VARIATIONS IN THE METHODS OF MEASUREMENT

Unquestionably the Method of Comparison has received the most twists, in large part because it has been employed so frequently in the laboratory; and perhaps it has been popular there because it enables the investigator to control completely both the standard and the interval being judged. If the first interval is the standard and constant, if the second is successively increased or decreased until no difference between the two is reported (the person may consider them subjectively equal, though in fact they differ by a fraction of a second of objective time), and finally if the second interval is then gradually changed in the opposite direction until a difference once again is reported, the Method may be referred to as that of *Just* (or *Least*) *Noticeable Differences (Stimuli), Limits, Minimal Causes (Changes),* or *Serial Exploration.* The duration of the interval being judged and the standard with which it is compared may be varied not regularly but randomly or semirandomly; and, in order to avoid a so-called "time-order error," the order in which the two intervals being compared may be similarly varied; the labels applied to this Method are that of *Constant Stimuli, Constant Stimuli Differences, Right and Wrong Cases,* and *Frequency.*

Alternate names for the Method of Reproduction are the Method of *Average Error, Equation, Adjustment,* and *Production.* I have rejected the title of

Production in spite of its respectable proponent (Boring, 1942, p. 41), since "reproduction" seems to me to be a more accurate operational description, besides, it seems obvious, the word "production" is better suited to describe the Method of Production. A real variant of this method has been called *Successive Time Estimation,* the objective of which is to shorten the number of trials to reach a research objective, such as the interval that is judged most accurately. On the first trial the subject reproduces an interval which he believes approximates the standard; that reproduction of his (but unknown to him) becomes the standard which he tries to reproduce on the second trial; the second reproduction is then the standard for the third trial; and so on. This method is said to provide "automatic positive feed-back," but the information on each trial is obviously fed back not to the subject but to the investigator (Llewellyn-Thomas, 1959).

Other methods represent combinations of those already considered. When a subject in the laboratory in confronted with a series of intervals and is required only to say whether each one appears "long" or "short"—and this feat he can perform after perceiving one or two of them—the method is called that of *Absolute Judgment* or *Single Stimuli.* From one standpoint, this method is a variant of Verbal Estimation since the judgment is made in verbal terms; but from another, it is related to Comparison since the standard interval with which each successive interval is compared, though not provided by the experimenter, is derived from some "anchoring" effect evolved within the series itself. It should be immediately evident that the method is "absolute" only in the sense that judgment is passed without reference to an explicit standard; nevertheless, the person actually invokes his own internal standards and passes judgment relative to them. Another variant is for a standard or "anchoring stimulus" to be provided by the experimenter at first, as in the Method of Comparison; then after it has reappeared on each trial during a series of comparisons, it no longer is presented but still serves that function on subsequent trials by having been internalized.

"A modified method of limits technique" I consider to be a combination of Comparison and Verbal Estimation, with a touch of Production. The subject is asked to judge whether a given interval is "more" or "less" than 1 second; he is presented with a series of intervals that are progressively longer (or shorter) than 1 second until he consistently reports "more" (or "less"), and then the series shifts in the other direction. The measure thus obtained is the duration at which the subject reports "more" or "less" than 1 second 50 percent of the time. Clearly a comparison is then made between each interval and 1 second, but in a sense, too, the subject is using Production: he himself may not produce the intervals but he selects them, as it were, in terms of his own criterion of 1 second (Lhamon, Goldstone, & Goldfarb, 1965). So many fruitful experiments have been performed with this method that I shall refer to it with a distinctive name: *Limits-Verbal Method.*

The same investigators have introduced another modification which also relies upon Verbal Estimation. Instead of presenting the subject alternately with a series of intervals which descend and ascend from 1 second, they offer intervals whose duration, for example, range from 0.15 to 1.95 seconds; each interval is judged not in terms of "more" or "less" than 1 second but with reference to a 9-point scale which is either a qualitative one varying from "very, very short" to "very, very long" or a pseudo-quantitative one varying from "very much less than one second" to "very much more than one second" (Goldstone & Goldfarb, 1964a). Let this be called the *Rating-Verbal Method.*

An investigator in Sweden has performed a series of experiments by combining the Methods of Production and Verbal Estimation which she has labeled "the method of present- and past-time estimation" or, for short, the *PPT* method. According to her, such a method is needed in order to try to separate temporal judgments concerning an interval as it is perceived and as it is recalled; Verbal Estimation is "affected both by the amount perceived and the amount retained." A subject is asked first to read digits typed on a strip of paper at the rate of 1 per second; the number thus read during a given interval is the "present-time score." Then, with prior arousal of the motive to pass temporal judgment and after being told to stop counting, he is asked to make an estimate of the elapsed time based on "the immediate impression of how long the time felt"; this is the "past-time score." A "retained-time score" can be calculated by dividing the past-time score by the present-time score. In these experiments, it must be emphasized, the same subjects were used again and again under varying conditions, so that they obviously anticipated that they would be questioned concerning the elapsed time after being told to stop counting (Frankenhaeuser, 1959, pp. 37–40).

ADDENDUM 1.3 EVALUATION OF METHODS

Relatively few methodological studies have been made, in large part, I presume, because each investigator is usually interested only in justifying the method he has employed in his particular experiment. In no study, to my knowledge, has a perfect correlation been found between results obtained by different methods. When the same subjects judged the duration of a 5- and a 20-second intervals in three ways, for example, the highest correlations were between Verbal Estimation and Production (−.39 and −.59, respectfully, for the two intervals); those between Production and Reproduction were also significant (.56 and .25); but those between Verbal Estimation and Reproduction were negative and not significant (Israel, college; prior; Siegman, 1962c). Other studies have also revealed significant negative correlations between Verbal Estimation and Production (U.S., normal as well as schizophrenic and neurotic patients; prior; Warm, Morris, & Kew, 1963) (U.S., soldiers; prior; Cahoon, 1967).

But which method yields the most accurate results? In one of the two studies just cited (Warm, Morris, & Kew, 1963), the judgments with Verbal Estimation tended to be longer and more variable than with Production or Reproduction; but variability with all three methods increased appreciably as the duration of the intervals was lengthened. In another very careful study variability fluctuated insignificantly as a function of Method (Production, Reproduction, Verbal Estimation), modality (audition, vision, cutaneous), and duration (0.5 to 4 seconds) (U.S., presumably adults; prior, Hawkes, Bailey, & Warm, 1960).

On the other hand, many investigators have reported positive findings. The most conspicuous one suggests that Reproduction has more often than not led to the most accurate judgments; thus in American studies it has been found to be superior to: (1) Production and Verbal Estimation when schizophrenics judged intervals having durations of 5, 10, and 15 seconds (Clausen, 1950), or when normal children were instructed either to use or not to use counting as an aid in judging intervals ranging from 9 to 180 seconds (Gilliland & Humphreys, 1943) and (2) to Verbal Estimation when normal subjects and patients undergoing eye surgery judged an interval of 16 seconds (Ochberg, Pollack, & Meyer, 1965). Similar findings have been reported from three other countries (Canada, psychiatric patients and matched controls; prior; Cappon & Banks, 1964) (France, presumably normal adults; prior; Fraisse, 1963, p. 213) (Germany, normals as well as schizophrenics and patients suffering from cerebral tumors and senile dementia; prior; Kohlmann, 1950). In one somewhat unusual situation—the subject was given a signal by the investigator to produce an interval by pressing a dynamometer whose duration he was then asked to estimate—Verbal Estimation, however, tended to lead to "somewhat more accurate" judgments than Reproduction for the longer intervals of 24 and 48 seconds but not for the shorter ones of 6 and 12 seconds (U.S., university; Warm, Smith & Caldwell, 1967). It has been maintained that Production was "more sensitive to physiological change" than Verbal Estimation because results obtained from using that method at sea level and at a high altitude produced the highest correlation (.68), whereas the similar figure for Verbal Estimation (.25) was not significant; the two methods correlated significantly and, as ever, negatively with each other (−.37 and −.62, respectively) (U.S.; Cahoon, 1967).

Material on reliability is rather copious because most investigators routinely provide relevant information and thus suggest that their results are trustworthy in this respect:

Verbal Estimation: rank-order correlations between judgments concerning the total duration of a standard interview at approximately its midpoint and at its termination by schizophrenics and nonpsychotic persons were both in the middle .70s, indicating that "precision and direction of error in time estimation" tended to be "relatively stable" for the two groups (U.S.; immediate subsequent; Rabin, 1957); when thirteen intervals ranging from 0.1 to 30 seconds were judged twice during the same session, the correlations between the first and the second half ranged from .87 to .94 (U.S., college; prior; Zelkind & Spilka, 1965).

Production: correlations of .86 and .78 emerged when a 30-second interval was produced twice by patients, respectively, as they were seeking admission to a psychiatric clinic and a week after admission (U.S., adults; prior; Melges & Fougerousse, 1966).

Reproduction: correlations among eight traits were "consistently high, with a few exceptions"—most were over .80—even when two modalities, audition and vision, were systematically varied in presenting and reproducing intervals ranging from 1 to 17 seconds (U.S., college; prior; Brown & Hitchcock, 1965).

Few direct comparisons of the three methods with respect to reliability are available, and they have yielded inconclusive results:

1. All three were reliable: in a previously cited study, reliability was measured by correlating odd-even judgments during the same experimental session: the figures turned out to be high, most above .90, except for a low figure of .47 which emerged for the 1-second interval when it was reproduced cutaneously (op. cit.; Hawkes, Bailey, & Warm, 1960).

2. Reproduction was the least reliable: in a study, also cited above, in which subjects judged the duration of 5- and 20-second intervals by means of three methods, reliability was tested by comparing their first and second judgments during the same experimental session: Production was highest (.85 and .97 for the two intervals), then Verbal Estimation (.82, .84), and finally Reproduction (.59, .40) (Siegman, 1962c, op. cit.). Similarly Reproduction was less reliable than Production or Verbal Estimation among schizophrenic patients who judged intervals with a duration of 5, 10, and 15 seconds, with the criteria being the correlations among subtests and retesting over four months later (U.S.; prior; Clausen, 1950).

3. Reproduction was the most reliable: here Reproduction excelled both Verbal Estimation and Production in connection with intervals ranging from 1 second to 19 minutes even when its reliability was seriously impaired by anxiety (England, physicians; prior; Cohen & Mezey, 1961).

4. Verbal Estimation was the least reliable: that method produced the greatest inter-subject variability in comparison with Reproduction and Production among psychiatric patients and matched controls (Canada; prior; Cappon & Banks, 1964). Another group of psychiatric patients

(mainly melancholics) was retested after three weeks by means of the three different methods: that of verbally estimating 30 seconds was found to be unreliable (.36), that of producing 30 seconds and verbally estimating the 30 minutes during which they had been interviewed reliable (.75 and .76, respectively) (England, adults; immediate subsequent; Orme, 1964).

2. POTENTIALS

The aim of this chapter is lofty: to provide a very broad perspective for an analysis of time, to justify the breadth of that vision, and to outline and illustrate the mode of analysis. Whenever the exposition seems especially difficult or whenever the vision fades, some readers may wish to refer to the diagram titled "The Taxonomy of Time" which seeks to blueprint the overall schema. The taxonomy illustrated there will guide the analysis of the next five chapters, and thus becomes the foundation for the entire book.

I have already stated in the previous chapter that the analysis of time, if it is to be complete, must be offered in the context of human behavior. The moment human behavior becomes the object of attention, the serious writer or investigator finds it difficult or impossible to locate the boundaries within which he must choose to function. For so many different factors or variables are likely to play a significant role. Whatever you do at a given moment may result from a long history of events both within yourself and your society, and from whatever the circumstances are which provoke you to behave as you do.

The customary way to curb the faustian urge for completeness is to impose some Arbitrary Limitation which is signified by the declaration that other things must be equal even when it is perfectly clear that under most conditions they are certainly not equal. This is the method of science, and all the so-called principles in this book are based upon that procedure. But in dealing with time, which is so widely discussed and yet so imperfectly understood, we must allow ourselves the luxury of looking searchingly at the whole behavioral universe in order to locate significant factors.

These factors I am calling "potentials" because their function is thus clearly indicated: they may, but need not, exert tremendous influence upon temporal behavior. In the diagram the potentials are connected with single or double arrows in order to suggest the kinds of interactions that must plague and stimulate us if we would comprehend the intricate, multivariate phenomenon of time.

A few of these potentials and some of their components have been men-

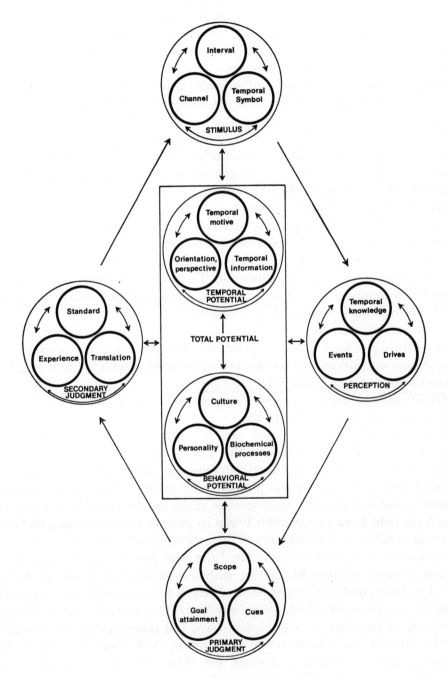

CHART 1. THE TAXONOMY OF TIME

tioned in Chapter 1. Now the entire schema must be more formally presented and discussed. First, there is a set of general tendencies previously or currently embedded in the individual, his *behavioral potential,* which is being described here in terms of recollecting, experiencing, anticipating, and or renouncing. These in turn reflect many dispositions, three of which must be specified for present purposes:

Culture: the modal beliefs and attitudes which prevail in a society and which each person acquires, however uniquely, during the process of socialization.

Personality: the actual, unique organization of traits (characteristic ways of behaving), cultural beliefs, and attitudes within the individual; included here is intelligence or skill as well as general or specific temporal perspective.

Biochemical processes: the individual's age, body temperature, etc., and what we shall call a biochemical clock.

The behavioral potential functions no matter whether the person is preparing to eat or to attend church, to have his patellar reflex or his mathematical ability tested, to judge the aesthetic qualities of an epic or the duration of an ephemeral interval. To analyze temporal behavior we postulate a more specific set of potentials, *a temporal potential,* which can be variously viewed in terms of the individual's.

Temporal motive: its time of arousal (i.e., its temporal relation to the interval being judged, whether prior, interim, immediate subsequent, or delayed subsequent) and its importance in connection with the goal or goals being sought during the interval.

Orientation or perspective: whether toward the past, present, or future and whether temporary or enduring.

Temporal information: defined very broadly to include data about the hour, date, age, etc. Knowledge of this kind in turn may evoke additional beliefs and attitudes related to time. You may know with precision the intervals separating you at the present moment from two widely separate events in the past (the last time you ate clams vs. the point in the past at which the light from a distant star began its journey to your retina), but you have quite different feelings toward these two elapsed intervals.

In outline, then, we have considered the *total potential* of the person as he goes forward to meet his destiny with respect to time. With this behavioral and temporal equipment he has been, is being, or will be confronted with an environment which stimulates him. The *stimulus potential* can obviously be described in a myriad of ways, but three seem most relevant:

The interval itself: its duration in terms of clock time, its content, its intensity, its organization or Gestalt, etc. The "etc." would cover every variation known to man, from the pause between movements in a symphony to the pace at which the seemingly perpetual glacier on a high mountain creeps forward or backward. Almost anything miscellaneous about a stimulus should be tossed here.

The channel: the way in which information about the interval reaches the individual, including the sense-modality being stimulated, the communicator, other sources of stimulation simultaneously affecting him.

The temporal symbol (if any): whatever stimulus or aspect of the stimulus provides a clue to objective time, whether the hands of a clock or the color of autumn leaves.

These three attributes of the stimulus potential are intended to be objective, it should be evident. They are the ways in which the investigator or the individual's qualified peers describe the environment. He himself then recollects, experiences, or anticipates them in ways derived from his temporal and behavioral potentials.

Both here and in connection with all the other principal variables (represented by the large circles of the diagram) exactly three factors are isolated. Why always a trio? I have no satisfactory reply: some must be specified and three seem convenient and sufficient.

PRIMARY AND SECONDARY LEVELS

Now that a hypothetical subject has faced, is facing, or will face a stimulus with all his psychological and cultural baggage we must describe in minute detail the process of passing temporal judgment concerning both duration and succession. The analysis is so detailed not because I like to count angels on the head of a pin but because it seems to me that time demands such an analysis. Unapologetically let me outline the rationale.

The need to distinguish various levels of temporal experience springs from the fact, in the words of a time-conscious novelist, that there is often an "extraordinary discrepancy between time on the clock and time in the mind" (Woolf, 1928, p. 91). This discrepany seems to be widely recognized and is noted variously: objective time is regular and uniform, subjective time often proceeds in jerks and spasms (Meyerhoff, 1955, p. 13); "time is long when we desire, it is short when we are afraid" (Paul Janet, 1877); "we are knowingly indulging in exaggerated talk and throwing metaphors about" when we assert that "delightful hours" spent with two old friends are followed by the declaration that we "do not know where the evening has gone to" (Priestley, 1964, p. 69); or contact with psychiatric patients suggests that "the present and the past are not uniform" (Lewis, 1932) or "four basic modes of experiencing the flow of time" must be distinguished (Lehmann, 1967).

William James also noted two phenomena which he considered curiosities and which illustrate the way in which time seems to fold back upon itself: (1) the "curious increase" in temporal judgment resulting from "hashish-intoxication," illustrated by the fact that "we utter a sentence, and when the end is reached the beginning seems already to date from infinitely long ago"; and (2) "a curious exaggeration of time-perspective at the moment of falling asleep," viz., losing consciousness between perceiving two events in a sequence (such as a person moving about in his room) which are separated only by "a few seconds" and then it seems as if "a long interval has passed away" (James, 1890, pp. 639–40, 641). Similar observations you have doubtless made about yourself: as you grow older, you continue to recognize that calendar years are of the same length but that birthdays recur faster than they used to, Christmas always seems to arrive sooner than you expect it. In short, we must become acquainted with the private, the intervening responses which you, poets, and I cherish and dread as we contemplate our own inner conception of time and its passing, or even if we would but appreciate the serious playfulness with time displayed by Tristram Shandy.

To work, then. First, the stimulus is perceived—and I use perception most broadly to include not only the stimulus which is present (you watch a flash of light or hear a bell ring) but the traces of one previously experienced or the cues concerning one that is anticipated. To comprehend the process of passing judgment—completely, utterly—two potential judgments are considered:

Primary judgment potential: the immediate, spontaneous, phenomenological judgment that is passed. "It must be three o'clock," you say when you wonder what time it is and before you have had an opportunity to consult a timepiece; or you simply feel that a long time has gone by or that time is passing quickly. The process may halt at this level: you judge the time or have a feeling about it and then go about your business without verifying what you have thought or elaborating what you have felt. A primary judgment may be so spontaneously reached that it seems to spring effortlessly out of the person or from the stimulus configuration being perceived; as a result he may not be able either to explain how and why he feels as he does or what standard he has employed; in fact he may be relying upon a Method of Absolute Judgment which enables judgments to be passed in the absence of a specifiable standard.

Secondary judgment potential: the judgment resulting either (a) from a reconsideration or re-evaluation of the primary judgment on the basis of experience, intuition, a principle, reflection, or a reference to some objective standard, or (b) from relying upon such a standard from the very outset. The second alternative circumvents almost completely the primary level:

you wish to know the time and you look at your watch; amen. But you may pass a secondary judgment without a timepiece: "It cannot be three o'clock," you think, "because surely I have not been working for two hours." Additional action may occur: having decided that two hours could not have passed, you may look at a clock and there discover that the time in fact is 2:40, from which fact you draw the dramatic conclusion that "time is passing more slowly than I thought." There are two other possibilities. You may be right: two hours may have indeed gone by, no more, no less, and your ego swells. Or you may adhere to your primary judgment by discrediting the measure of objective time: the clock must be wrong, you think.

Is it really necessary to postulate two separate judgment potentials? A moment ago I bombarded a paragraph with quotations indicating that literary persons, philosophers, and you and I distinguish different levels at which time is judged or evaluated. Now slightly more systematic defenses can be quickly considered. Time, it should be evident, is not perceived as such; rather, as one of the more ardent investigators has noted, it is "a concept . . . that is attached to perception only through a judgmental process" (Woodrow, 1951). If this is so, then the temporal process, like any judgment, may be rejudged: it folds back upon itself. Temporal estimates are so important, one writer points out, that "the individual gets a considerable amount of practice in adjusting his subjective impressions to physical standards" and he also "learns to make allowances for such factors as experience has taught him will distort the relation between subjective and objective time" (Frankenhaeuser, 1959, p. 18). Adjustment in any form is not likely to be spontaneous, hence there must be a difference between primary and subsequent judgments. A philosopher believes that a "radical change of situation" occurs when we shift from saying that the day "seems long" to suggesting that it "seemed long." For when making the judgment in the present "we are not really estimating duration at all," our judgment is not "an estimate of a time-interval but an aspect of certain physiological or psychological states"; whereas in recollecting the past we are concerned with "an apparent time-interval and its value depends on the number of events (in the ordinary physical sense) which occurred in it." Since "the same day can seem long in the first sense and short in the second" (Mayo, 1950), different processes must be involved: the primary judgment concerning the passing of time now leaps directly out of a current perception, the secondary judgment creeps out of the behavioral potential in which the recollection of that day (or its events) has been stored. A psychiatrist has expressed the same distinction by maintaining that temporal judgment has "two elements . . . (1) an immediate awareness of the magnitude of duration or of speed of succession; and (2) an ability to recall similar former experiences for comparison" (Lewis, 1932). From the standpoint of conditioning theory, moreover, two separate processes must be delineated: simple conditioning occurs as a primary judgment, higher order conditioning is reflected in secondary judgments.

Empirical facts, finally, demand that time be conceptualized on different levels. Ten subjects who ingested scopolamine, a drug which is a sedative and used as a so-called truth serum, were surprised to discover that four hours had elapsed: "the estimate of elapsed time based upon feelings was independent of experienced facts, e.g., the brightness of the room, and often seemed to contradict them" (Germany, presumably adults; Heimann, 1952, p. 34). In another study, all 12 college students who managed to endure 8 hours of sensory deprivation—they were strapped on a chair, with eyes covered and a masking sound in their ears—were very frustrated and reported that time had passed very slowly, but two of them gave quite accurate estimates of the hour at the end and 10 *under*estimated the interval by an average of 1 hour and 17 minutes (even though they knew the experiment would end 11 hours later and even though they had lunch): "the experience of *time in passage* is not to be identified with the usual operation of estimating time" (U.S.; Goldberger & Holt, 1958). Obviously the primary and the secondary judgments were different.

The secondary judgment potential, it can also be noted on the schema, may have two lasting effects. In a series of judgments, it may affect the appearance of the next interval or its duration. Or, the experience thus gained may be stored within the total potential, probably in the temporal potential, and then utilized in the future. For this reason arrows extend from the secondary potential, respectively, to the stimulus potential and the total potential.

The sequence from perception to primary to secondary potential as set forth in this conceptual schema, may include, it is hoped, any other mode of analyzing time, whether, for example, in terms of trace or delayed conditioning (Fress, 1962) or as "a statistical decision-making process" (Creelman, 1960). The events of course may be shortened as when a man for poetic or insane reasons clings to his primary judgment; then the secondary judgment may not be invoked, hence one arrow in the diagram goes directly from the primary to the total potential. Or those events may be prolonged, as when a geologist carefully verifies his first hunch concerning the age of newly discovered artifacts through carbon-dating. The possibility of shortening or lengthening is another reason why these judgments are called potentials. If there are qualitatively different manifestations of a time sense, as some writers maintain, then the different modes of passing judgment are to be found at some point in the indicated sequence.

In addition to these two potentials, we must also mention a third one which does not enter directly into the psychological picture. If you say it is 3 o'clock and another person (whether a friend, a psychiatrist, or an investigator) look at his watch and discovers that it is 2:40, he rather than you may pass judgment on your initial primary judgment before you yourself realize, if you ever do, your error. Here we have a judgment about the primary judgment, or it could well be also about a secondary judgment. Whether or not this external judgment is communicated to the person passing judgment—that is, whether there is feedback—is another problem which is either out-of-bounds for us or else highly relevant to other judgments in the future. In the first case, the outsider keeps quiet: he simply has ascertained how close to reality or how accurate the individual has been. In the second, he speaks out and reinforces the judgment either positively or negatively so that learning can occur and future judgments be improved or at least altered (see *Addendum 7.2*).

OVER- AND UNDERESTIMATION

The schema of judgment potentials must be utilized to extricate ourselves as quickly but as thoroughly as possible from a confusing problem of terminology: the question of defining over- and underestimation. Being confronted with this problem seems to be the destiny of anyone who would tidy the Augean tables of time (e.g., Bindra & Waksberg, 1956); for we are plagued with "considerable confusion" regarding the referents of those terms (Frankenhaeuser, 1959, p. 28). The reader may justifiably feel that I am belaboring this complexity in the present and the succeeding section; but I thus rid myself of the many frustrations I have endured in attempting to comprehend the literature which is so murky on this point and perhaps, more importantly, prevent him from experiencing the same misery.

Suppose, for example, you misjudge the duration of an interval you have experienced: 40 seconds of clock time you judge to be 30 seconds. Is your error one of over- or underestimation? Actually either description is applicable:

Overestimation of subjective time: 30 seconds of subjective time has been judged long enough to be applied to an interval of clock time that is actually greater by 10 seconds; it is as if you overestimate your own weight by believing that you are heavier than I am when a scale belies your belief.

Underestimation of objective time: 40 seconds of clock time has been judged short enough to be designated by an interval of subjective time that is actually 10 seconds too short; it is as if you underestimate my weight by believing that I am lighter than you when a scale belies your belief.

Clearly we must decide whether subjective or objective time—your weight or mine—shall be the standard for making the comparison. The decision in some instances could be left to the person passing judgment; ask him, for example, when he is reproducing an interval whether his subjective estimate or any of his temporal feelings are focused upon the interval he is trying to reproduce or the one he in fact reproduces. On the few occasions when this has been systematically tried (e.g., Triplett, 1931), the outcome has been fuzzy. No, as a few sturdy souls agree (e.g., Clausen, 1950), the more useful standard is the objective duration of the interval when one is available: it is more likely to be the conventional one in our, perhaps in any, society and certainly it enables people, including scientists, to communicate with one another. Is this decision anti-psychological? I think not: we evaluate behavior in terms of the person's relation to the external world and from that relation—which is all we usually can observe—we make inferences concerning his psychological state. In short, you have underestimated 40 seconds of clock time when you say the interval has lasted 30 seconds (Method of Verbal Estimation).

This procedure can be followed consistently. A subject in an experiment who presses down a key for 4 seconds when he has been asked to reproduce an interval which he has just heard, and which in fact lasted 3 seconds, is guilty of overestimation: the standard is the experimenter's interval of 3 seconds and the interval reproduced by the subject is 1 second longer than that standard. Another subject who is requested to count to 60 by seconds and actually takes 45 seconds to do so provides an illustration of underestimation: he has counted too rapidly, as a result of which he has underestimated the interval of 1 minute, the standard conveyed to him when he agreed to count. In this instance one could say that he has overestimated the time it took him to count to 60 which he considered long enough to be called 1 minute, or that he was similarly overestimating his rate of counting. We must resist, however, the temptation to use either of these locutions, if we would avoid confusion, by remaining consistent: the standard is clock time of 1 minute for which 45 seconds is an underestimation.

It is to be noted that these illustrations make use of three methods of judging time, Verbal Estimation, Reproduction, and Production. For the Method of Comparison ambiguity can be removed simply by indicating which of the two intervals being compared is to serve as the standard. In experiments the subjects may simply declare which interval seems longer, the first or the second; then the experimenter or investigator decides whether they are over- or underestimating intervals of particular duration by arbitrarily calling one of them the standard.

Unfortunately we are not yet in the clear even after the long trek already endured in this section. In common speech and in the absence of a measure of objective time, the terms over- and underestimation are seldom employed, instead there are metaphorical substitutes too numerous to list. But two pairs, though ambiguous at first glance, can induce us to make distinctions which will be useful throughout this book. Time, it is asserted, seems short or long; and I suppose most of us would agree that a great deal of temporal imagery is spatial in nature (U.S., college; Guilford, 1926). Then time is said to pass quickly or slowly, again expressions related to space. We may be dealing here only with metaphors, but it is noteworthy that the Kachin in Burma have distinctive terms for the word *time* when used with reference to clock time, short time, long time, present time, etc.; in fact, they have "no single word" corresponding to time as such (Leach, 1961, p. 124). Otherwise I spare the reader and myself additional equivalent expressions for length or speed of time in English (e.g., it may seem constricted, contracted, condensed, dilated, expanded), but I respectfully recognize their glib utility in dealing with specific phenomena such as hypnosis (Weitzenhoffer, 1964).

THE LENGTH AND SPEED OF TIME

More than a moment's reflection reveals that there is no easy way to establish equivalence among the terms representing length, speed, and estimation of time. As an illustration let us consider an interval being judged by two methods:

Verbal Estimation	*Production*
\<center\>*Primary Judgment*\</center\>	
1. You judge that an elapsed interval is *long* (e.g, 60 seconds) and that it appeared *longer* (e.g., 80 seconds) or *shorter* (e.g., 40 seconds) than you might have anticipated or hoped; hence you feel that the interval passed, respectively, *slowly* or *fast*.	1. You produce an interval that seems *adequately* to represent one previously designated by the symbolic words of "60 seconds"; in this situation you do not decide whether time is passing *fast* or *slowly*.
\<center\>*Objective Time*\</center\>	
2. You discover that *less* objective objective time (e.g., 50 seconds) did in fact elapse.	2. You discover it took you *less* objective time (e.g., 50 seconds) to produce that interval.

Secondary or External Judgment

3. During the elapsed interval subjective time was *longer* than objective time (or objective time was *shorter* than subjective time), objective time passed more *slowly* than you had judged; therefore, you *overestimated* the interval.

3. During the production of the interval, subjective time was *longer* than objective time (or objective time was *shorter* than subjective time), objective time passed more *slowly* than you had judged; therefore, you *underestimated* the interval.

Two important lessons may be learned from the above exercise. In the first place, the concepts of length and speed, when applied to intervals, do not have a one-to-one equivalence. The first statement to the left indicates that a primary judgment of "long" can mean that time is passing or slowly or quickly, and the decision rests on the anticipations or hopes concerning the interval. Second, and more importantly, the paradigm suggests that these two pairs of concepts are ambiguous unless the level of judgment and the method of measurement are clearly specified. Thus with Verbal Estimation you may have thought that time went by quickly as you passed primary judgment (the interval seemed long but not as long as you thought it would); but, lo and behold, your watch then indicated that not as much time had elapsed as you imagined, so you revised that judgment and on a secondary level concluded that time was creeping.

To avoid the ambiguity just mentioned, it is essential to know whether the judgment is primary or secondary. "Time seems to be going by slowly" cannot be interpreted until we are told whether the judgment stems from feeling on the primary level (and even then we remain ignorant concerning whether an apparently long or short interval appeared, respectively, equal or just longer than had been anticipated or hoped for) or one on the secondary level derived from consulting a timepiece. "As I grow older, the years seem to become shorter." Here it is useful to assume judgments on both levels, certainly as a primary judgment because of the mode of expression and equally certainly as a secondary one because people in our society are not likely to fail to perceive calendar time. Or: "it seemed to take a long time to walk from here to there, but later I discovered that time had indeed passed quickly." The difference in the two judgments must arise from different temporal motives: as an interim motive, apparent time on the primary level seemed longer than what might have been anticipated from the distance being covered; but afterwards knowledge of actual elapsed time and a recollection of the primary judgment indicated, perhaps, that less objective time had elapsed than had been previously imagined.

For quick reference I have summarized and extended the foregoing analysis in the large chart which carries the same title as this section. A more detailed explanation of that chart appears in *Addendum 2.1*. I trust the reader will find the schema more than a whimsical bit of staggering verbiage. By organizing the concepts around acceleration and deceleration I think I have achieved a way of unifying the data we obtain in the laboratory or real life. For example, the person whose judgments have been placed in two columns at the start of this section believed that 60 seconds had passed during 50 seconds of clock time (overestimation) and produced a designated interval of 60 seconds in 50 seconds (underestimation); consequently, he has a consistently accelerated sense of time. One assumption is herewith made explicit: there is no point in worrying about these details unless there is a discrepancy between primary and secondary judgments.

The chart may be entered at whatever point best describes any statement concerning time; and then the consequences of that statement under the specified circumstances can be unearthed. Let us assume, for example, that we think an individual's timing is accelerating. We begin with row 1, where we must choose from among four columns, nos. 1, 2, 5, and 6. The choice is made on the basis of Method, row 2. If the Method is that of Verbal Estimation concerning an elapsed interval, we then follow column 2 down to the bottom of the chart. En route we find that the primary judgment is likely to be "long" (row 3a) which, depending on his anticipation or hope (row 3b), may lead him to feel that time is passing slowly or quickly (row 3c). Under these circumstances "less" objective time probably will have elapsed (row 4), in which case we may say that his subjective time is longer than the objective time (row 5a), etc., until we emerge with the conclusion that he tends to "overestimate" elapsed time (row 5c). The chart may also be employed in reverse if it is found that a person over- or underestimates time: here we start from the bottom row and work upwards in whatever column is specified by the method of measurement.

Other loose ends concerning this chart and its implications I have tied together in *Addendum 2.2*. And so now the discussion of methodology in this and the previous chapter yields, I think, an operational principle which will serve us in the subsequent analysis:

2.1 *Different methods of assessing temporal judgment produce different judgments which may be, however, psychologically consistent; judgments from Verbal Estimation and Production are likely to be negatively correlated.*

CHART 2. THE LENGTH AND SPEED OF TIME

	(1)	(2)	(3)	(4)	(5)	(6)	(7)	(8)
1. Timing	ACCELERATE	ACCELERATE	DECELERATE	DECELERATE	ACCELERATE	ACCELERATE	DECELERATE	DECELERATE
2. Method	comparison, reproduction	verbal (elapsed)	verbal (pending)	production	production	verbal (pending)	verbal (elapsed)	comparison, reproduction
3. Primary judgment	II > I; II = I			adequate	adequate			II < I; II = I
a. length		long	long			short	short	
b. anticipation, hope		longer shorter	longer shorter			shorter longer	shorter longer	
c. speed		slow fast	slow fast			fast slow	fast slow	
4. Objective time	II ≦ I; II < I	less	less	more	less	more	more	II ≧ I; II > I
5. Secondary or external judgment								
a. Subjective time	longer	longer	longer	shorter	longer	shorter	shorter	shorter
b. Objective time								
i. length or subjective units	shorter	shorter	shorter	longer	shorter	longer	longer	longer
ii. speed	slower	slower	faster	faster	slower	slower	faster	faster
c. Conclusion	OVER-ESTIMATE	OVER-ESTIMATE	OVER-ESTIMATE	OVER-ESTIMATE	UNDER-ESTIMATE	UNDER-ESTIMATE	UNDER-ESTIMATE	UNDER-ESTIMATE

Note: For a detailed explanation, see Addendum 2.1.

ADDENDUM 2.1 EXPLANATION OF "THE LENGTH AND SPEED OF TIME"

The paragraphs below follow the numbering of the rows on the chart. A glance at the last row (5c) reveals that the first four columns culminate in overestimation and the last four in underestimation.

1. We begin with the assumption that the individual's timing is accelerating or decelerating; unless there is this kind of change which produces *inaccurate* judgments from the standpoint of objective time, none of the problems in the chart are relevant.

2. The methods of passing judgment are listed, except for Chronometry. Verbal Estimation appears in two different forms because judgments of elapsed and pending intervals result in different outcomes. It is assumed that normally or usually the other three methods are used only for elapsed intervals. Comparison and Reproduction are combined: the interval coming first, designated as I, which for Reproduction must be the standard, is assumed here also to be the standard when two intervals are being compared.

3. In those columns involving the Method of Verbal Estimations (nos. 2, 3, 6, and 7), the Primary Judgment is first expressed in terms of "long" or "short" (row 3a) and then as a function of the anticipation or hope (row 3b) the corresponding equivalent in terms of speed is suggested (row 3c). This phenomenology, it is assumed, is not applicable to the other three methods. Instead for Comparison there are three possibilities: the second interval II may appear greater or less than, or equal to the first one, I. For Reproduction, the reproduced interval II appears equal to the preferred one, I. The concept of "adequate" is used to designate the primary judgment for Production, because presumably the individual really thinks he has produced an interval equivalent to the one symbolized in the instructions.

4. Objective time refers, as ever, to some conventionalized system of reckoning, and again the assumption is made that there is a discrepancy between it and the primary judgment. For Comparison and Reproduction, the primary judgment may be wrong for one of two reasons: if II has been called less or more than I, it is, respectively, more or less than I with Comparison; if II has been judged equal to I, it is then more than I with both methods. For Production and Verbal Estimation the individual or the investigator discovers that the primary judgment has been in error, because the objective time of the produced interval is actually "more" or "less" than what was intended. A special problem arises in connection with Verbal Estimation which is signified by the dotted line connecting columns 2 and 3 with 6 and 7. For the judgment of length on the primary level may turn out, when objective time is consulted, to have been in error not because it was "more" or "less" than that objective time but because it was, respectively, even less or more than it should have been. Thus you think that the time has been short (e.g., 30 minutes) and then you learn that less time has gone by than you imagined (e.g., 20 minutes); under these circumstances, the description of your behavior begins in column 7 but then, after row 3, it must move over into column 2.

5. The secondary or external judgment takes the objective time into account, and then in restrospect the subjective time revealed by the primary judgment is viewed as "longer" or "shorter" than that objective (row 5a); or the length and speed of objective time with reference to that subjective time may be appraised (rows 5b, i and ii). The "subjective units," also in Row 5b, i, refer to whatever measures have been employed to estimate subjective time. Thus if you verbally overestimate an elapsed interval (column 2), your subjective time on a primary level was "longer" than clock time (row 5a) and, since it was longer, you may have been thinking in

terms of "shorter" units; calling a 60-second interval 80 seconds means that you extended the 60 seconds (i.e., you made them longer) and hence, as it were, you squeezed the subjective units, 80 of them, into a short interval, 60 (i.e., you made them "shorter"). As a conclusion it is possible to indicate whether the judgment represents an over- or underestimation (row 4c). The vocabulary of the rows under 5, it must be emphasized, may be used either by the person himself as he passes secondary judgment or by an outside observer or investigator to appraise his accuracy; in each instance, a knowledge of objective time is presumed.

The columns headed "Comparison, Reproduction" (1 and 8) require two related comments. The categories of "subjective time" and "objective time" refer to the comparison of interval II with interval I or the reproduced interval II in comparison with the model I. Thus with the Method of Comparison interval II may be judged on the primary level to be longer than I; if it turns out to be shorter or equal to I, then we could say that during II subjective time seemed longer (row 5b, i) or that objective time during II was in fact less. Then, secondly, the heading of "Comparison" serves a double duty in our analysis. It refers to the Method as such: two intervals are presented to a subject and then compared. But the comparison may also be made as an external judgment by an investigator when in fact the subject employs a method other than comparison to pass judgment. The interval a person produces by sounding a bell may be objectively longer or shorter than the one he produces by lighting a light. Here there is no ambiguity concerning the meaning of long and short because they are defined objectively: not the experience of the subject but the objective duration of the interval is involved. When a person maintains that one interval seems to have elapsed more quickly or slowly than another, however, the ambiguity reappears.

ADDENDUM 2.2 SEMANTIC PROBLEMS

The alternative to conceptualizing over- and underestimation by using objective time as the reference point for the Methods of Verbal Estimation and Production, the interval being reproduced for Reproduction, and the other interval being compared for Comparison is to appraise consistently the interval that is experienced, whether the interval is produced by the investigator or the person himself (Bindra & Waksberg, 1956). Thus, if you judge that 40 seconds of clock time which you experience is 60 seconds, you have overestimated the interval—both approaches agree. But if you produce what you think to be a 60-second interval in 40 seconds, I would say you have underestimated the interval which it was your task to produce; but the other approach would suggest that you have overestimated the interval which you produced because you made it equivalent to a larger one, the 60-second interval which had been presented to you symbolically. Obviously the latter statements are equivalent, they simply have different reference points.

The problem of reference points recurs when intervals are compared not directly by the individual passing judgment but by an investigator. In one experiment, for example, the task of the subjects was to judge whether or not a series of intervals was "more" or "less" than 1 second; this is the Limits-Verbal Method described in Addendum 1.2 of the previous chapter. The medium judgment was 0.81 seconds when the intervals were filled with mild visual stimulation and 1.0 seconds when they were unfilled. The investigators draw the following conclusion: "More unfilled clock time was judged equivalent to the same temporal concept than filled clock time, reflecting greater relative overestimation of filled time" (Goldfarb & Goldstone, 1963). There can be no quibble with the first part of the statement since the figure for unfilled time was indeed greater than that for filled time. When they state that there was "greater relative overestimation of filled time," however, they are saying in effect that the subjects tended to stretch that subjective interval of filled time to make it cover a longer period of objective time than indeed it covered (they made, on the average, an interval lasting 0.80 seconds equivalent to 1 second of symbolized time). On the chart of this book, I come out with the same statement, provided II symbolizes filled time and I unfilled time; then column 1 is used because the primary judgment is that II equals I (since each is said to represent a second) and objective time indicates that II is less than I; hence II represents an overestimation with relation to I (row 5c). But reversing the symbols for the two intervals would land us in column 8 which eventually would suggest that unfilled was underestimated with relation to filled time, a formulation which the investigators obviously could also have made.

This petty semantic problem does not arise in connection with the Method of Verbal Estimation. If you think 40 minutes have passed and discover that in fact the elapsed time is a full hour, nobody is likely to say that you have overestimated the time by making 40 minutes equivalent to an hour; no, indeed, you have underestimated clock time. The villain is generally Production, and here I shall always demonstrate my terminological preference by indicating that the method is being employed and that over- or underestimation refers to the objective time that is supposed to be produced. The guide, I believe, can be the principle suggesting the inverse relation between Production and Verbal Estimation, splendid support for which I derive from the following statement:

> Faster subjective time relative to a control condition can be inferred from both greater underestimation of the standard using the P [Production] method, and greater overestimation of the standard using the VE [Verbal Estimation] method relative to the control condition. . . . Therefore, it is likely that the results will be in opposite directions for the two methods and that the methods may differ in sensitivity to treatment effects [Lockhart, 1967].

It is certainly true that each method measures not only temporal behavior but also different psychological processes. But these processes may depend in part upon some central process within the individual (such as a sense of timing or a biochemical clock) which functions similarly no matter which additional processes occur. And this is another reason to heap lavish praise upon our mode of organizing temporal behavior around acceleration and deceleration.

I do not think that the Methods of Comparison and Reproduction can be profitably employed to make inferences concerning timing. With Comparison it must be assumed that timing or the clock functions differently during the two intervals being directly compared, perhaps because one precedes the other. While it is true that so-called time-order effects often appear, it seems impossible to select one interval rather than the other as the basis for making the inference. For Reproduction an admittedly dubious assumption has to be made when that method is utilized, viz., the "internal clock [is] slower [or faster] than external clock *during reproduction*" (Bindra & Waksberg, 1956; italics theirs); neither the authors nor I can think of any good reason why this should be so.

I am simply insisting in this Addendum, I suppose, that we must standardize our terminology and that my chart offers at least a consistent solution. If for a split second the reader thinks I have been engaging in a fruitless digression, I point to a first-rate study involving the effect of drugs on timing. At the outset of her report, the investigator assumed that, if time seemed to be passing "more quickly than usual" as a result of a drug, the person would count more *slowly:* "the interval of clock time which he normally thinks to last one second will now appear to him shorter than previously and he will therefor wait longer before judging one second to have elapsed." Later on, however, she states that "an equally plausible case may be made out for the opposite relation": with time passing more quickly, the person might count more *quickly* because "it would appear to him that the moment for counting the next second was already there when according to clock time it has not yet arrived" (Steinberg, 1955). First we have here an empirical question involving this Method of Production: does he count more slowly or quickly after ingesting the drug? The chart suggests (column 5) that a quicker count would mean that "less" objective time had passed and hence, if he discovers that this is so, then objective time is passing "slowly," from which might be inferred an accelerated timing mechanism; and a slower count (column 4) would mean the opposite. But what if he said that objective time in general seemed to be passing "more quickly than usual" and yet, contrary to the chart, he counted faster? This would have to mean a lack of coordination between his general impression and the specific act of counting.

3. BEHAVIOR

This chapter and the next are devoted to an examination of the center of the taxonomy guiding our analysis: the behavioral and temporal potentials composing the total potential. The two potentials interact, as the double arrows on the taxonomy diagram would signify, but for the time being the behavioral potential is considered the independent variable, the temporal potential the dependent one. How, then, is the temporal potential affected by the behavioral potential; under what circumstances do individuals pass temporal judgments, orient themselves in one direction rather than another, and acquire adequate or inadequate temporal information? Occasionally, moreover, the reciprocal relation between the two potentials will also be indicated: the temporal potential often influences the behavioral potential.

We begin with a deceptively simple proposition: a person judges or evaluates an interval in the past, present, or future only when he is motivated consciously or unconsciously to do so. What is gained by thus placing temporal judgment in a deterministic frame of reference which is in fact what we do when we assume that all action is motivated? Viewed thus, for example, subjective time becomes not an epiphenomenon, a psychological luxury, even though human goals are ordinarily achieved through the clues offered by objective time: it serves some function which must be discovered if analysis is to be complete.

You are walking along a street, you are stopped by a stranger who asks you the time of day, you consult a watch, and then you tell him. If you do not have a watch, you apologize or offer your best estimate. Your motive has been the conventional one of being polite. On another occasion you are happily engaged in some kind of activity, you forget what time it is, you are not aware of the passing of time; if asked, your judgment may not be very accurate. Here your motive involves the activity, not time-telling. But if you know that you have only a limited amount of time to devote to that activity —you must keep an appointment or you have promised not to be late to dinner—then you may be quite conscious of time, you may frequently look at your watch and silently note that time seems to be or is passing too quickly. In this situation the goal is the keeping of the appointment or the promise, and the temporal judgment is definitely related to its attainment. Or while walking down another street you happen to notice a clock because it is large or unusual; now your motive is not very important since the clock is the critical stimulus. Still, if fully preoccupied, you may not see the clock or, having seen it, you may not remember or respond to the information it forced upon you. Even the subject in a psychological experiment has a motive: he passes temporal judgment because he would please the experimenter or not make a fool of himself; if an American undergraduate, he may be lured into the laboratory by the fee being paid or by a requirement of his psychology instructor that he submit himself to a certain number of experiments per term. Ordinarily, therefore, we may not be aware of time even on a primary level unless some purpose is thereby served. Uncovering the motive for passing temporal judgment elicits new, relevant information.

Motives, however much they are biologically embedded, function in a social setting—"culture" in our schema—to which one must turn, consequently, for at least a partial explanation. The systematic glances of historians at the past and of anthropologists at societies outside the Western orbit have revealed again and again the diversity of human behavior. Wherever we turn, we discover peoples performing more or less uniquely with respect to the language they speak, the values they hold, the goals they seek. Infinite ingenuity, it seems, has been unwittingly and wittingly expended to devise different ways to achieve similar ends, whether it be the technique of drying hay upon the fields where it has grown or of killing men during battles in which, voluntarily or not, they participate.

There are, however, limits to cultural variability, and hence amid the diversity common human denominators can be discovered. Sexual beliefs, attitudes, and behavior indeed fluctuate markedly from society to society, we now know; but, as an academic wag once stated to a female anthropologist whose pride in the varying statuses of women in different societies is boundless, show me a society in which men and not women bear the children. The basic drives of the organism, no matter how they are conceptualized, must be satisfied: men must eat, drink, protect themselves, etc. The "etc." gracefully and unsubtly suggests that an analysis of human nature is beyond our present scope, a not too Arbitrary Limitation, I think. Let us, therefore, first note uniformities which affect the temporal potential.

UNIVERSALITY AND INEVITABILITY

Among the human universals is surely a sense of time. This capacity inevitably is learned and activated by external conditions in any milieu. Men live in a physical world in which change forever confronts them. The sun rises and sets every day in most areas; even in the far north or south when for brief periods its rising and setting are not visible, changes in the amount of daylight and darkness can be detected. Virtually everywhere there are seasons, not always as dramatically different as in temperate zones but at least fluctuating with respect to sunshine, rainfall, humidity, and temperature. The surrounding environment is altered: rivers shift their beds, soils erode, crops are planted and harvested, the skin of people grows wrinkled and splotched, and living matter eventually matures and perishes. Women normally menstruate on regular monthly cycles; and their bodies, when pregnant, undergo violent and perceptible upheavals. The perception of such changes may point to events that have occurred and never will occur again as when a person long known eventually dies; it may indicate the possibility of the recurrence of events once or now experienced, as when night comes and the next day is anticipated; or it may also suggest that events never heretofore experienced will eventually take place, as when a child realizes that some day he will become an adult. In short, an anthropologist suggests, everywhere there is repetition as well as aging or entropy which give rise to experiences related to time (Leach, 1961, p. 132); and an historian guesses that an awareness of the passage of time is "the point of departure for human evolution," at least as a "preliminary step" thereto (Kahler, 1956, p. 40).

3.1 Periodic changes in the external milieu inevitably and everywhere provide the potential for acquiring knowledge concerning the duration and succession of intervals and for the arousal of temporal motives.

Every society, it appears, has some system of reckoning time which usually stems from changes associated with one of the natural phenomena mentioned above (Nilsson, 1920). The presence of clocks, almanacs, or calendars—or their equivalents—obviously means that temporal judgments are passed by those consulting them and that simultaneously temporal information is thus obtained. Why, however, do persons everywhere make note of the changes? The fruitful and the naive assumption is that thereby certain purposes are achieved, certain goals are attained. If the purposes and goals in turn are to be identified, attention must be directed toward other forms of behavior which appear both universal and hence inevitable. And here we find that, just as the animal in the simplest conditioning experiment anticipates the arrival of the unconditioned stimulus (the food) while perceiving the conditioned one (the bell), so human beings in all societies must live in both the past and the future. To anticipate correctly is to profit from past experience, and, born as we are with so many potentialities and so few perfected mechanisms, we survive or live successfully by salvaging what is useful from the past. To anticipate incorrectly is to court grief and frustration, especially when the anticipation, as psychiatrists suggest (e.g., Arieti, 1947), springs from a neurotic tendency. Undoubtedly anyone capable of speech is able to give some kind of answer to the question, "What is going to happen . . . ?" if only by making a prophecy concerning the impending weather. The moon, according to an archaeologist, was "the first continuous time-measurer for periods of time within the year" not for reasons of caprice or aesthetics but because "almost everyhere the primitive woman knows that ten moons (that is, about nine months) after her periods have stopped her child will be born" (Breasted, 1935). The components of the temporal potential, no matter whether viewed through society or the individual, spring from remembering experiences in the past and from anticipating their utility in satisfying future needs; and the eternal achievement of the human being is his awareness of this temporal relation as well as his ability to verbalize and express it.

All, or almost all, the regulations of a society can be effective only when people possess relevant anticipations which, more often than not, turn out to be correct. If you are driving on a main artery, you expect another driver coming from a minor road to halt, and he knows (or is supposed to know) you have that expectation. If one of you is wrong, there may be a collision. Policemen in a modern society are said to be approached "for assistance or with a complaint" only when ordinary citizens are able to anticipate the treatment they will receive (Banton, 1964, p. 167) and if, of course, from their standpoint, favorable or at least not arbitrary treatment is anticipated. Much of the formal and informal education of any society is devoted to teaching the inexperienced generation what they may and must anticipate. In many societies initiation ceremonies for youth are prolonged so that the initiates may become acquainted with adult responsibilities and privileges. Traditionally midwives not only help deliver babies but may also provide prenatal instruction to mothers concerning what they may expect during childbirth and later (Read, 1966, pp. 66–8).

Anticipation must be acclaimed as a key variable in temporal patterning because it also gives rise to another universal, inevitable activity, viz., intervention. Animals hunt, or at least seek out food, and in the higher species the instrumental actions to attain that goal often require the passing of relatively long periods of time and the carrying out of complicated activities. Among human beings intervention is obviously of supreme importance. Without food we shall starve, and therefore men everywhere intervene to avert that catastrophe: their actions range from cultivating plots of land next to their homes to formulating master plans on a national scale to prevent the erosion of the country's soil. People ordinarily do not sit idly by as they postpone gratification; they intervene in order to attain goals more rapidly or efficiently.

Intervention is usually based upon the belief that gratification is attainable not at present, but in the future. Whether realistic or not, such a belief dimly or vividly guides the individual's behavior. He thinks his skill is adequate, he believes he can accomplish what he seeks, and then he sets to work. Or he realizes his own inadequacy, but in sheer desperation tries anyway to achieve the goals he would attain. In either case he patterns future time for himself: with varying degrees of rational or non-rational precision, he anticipates the intervals in which he will perform certain acts, he plans his action. Intervention may arise through necessity: the present state of affairs is impossible. Or it may involve renunciation: for the sake of the future gain, some satisfaction at the moment is sacrificed. The act of intervening may also bring satisfaction either in its own right or because it seems to be leading to the goal.

Clearly cultural factors play the decisive role in determining which aspects of existence are considered to be beyond human intervention and which are subject to control; but in any society some intervention is assumed to be efficacious, however dimly or unconsciously the assumption is perceived. The ceremonies of nonliterate tribes serve to transmit and to emphasize the importance of various customs from the past for the present and future. Ancestors are venerated, otherwise their wrath is anticipated. Or, the upholders of tradition may fear that, unless there is respect for the past, young men will intervene and change the traditions of the fathers in the future. Communications, whether from paramount chiefs, politicians, or public relations counsels, are transmitted in the belief that people can be altered, that the "way of all flesh" can be affected by intervening with promises or exhortations. The man of God speaks up anywhere, even when he ostensibly subscribes to some kind of doctrine involving predestination: he hopes his words will influence the audience at hand in the direction of his own goals. And so, no matter how sophisticated the formal dogma happens to be, more than a touch of a belief in free will pervades all people; otherwise they would passively respond without ever seeking to kick or nudge aside the forces ostensibly engulfing them.

An especially intriguing variant of intervention is the self-fulfilling prophecy. The person anticipates a state of affairs which he hopes will come to pass or which, quite contrariwise, he wishes to avoid; then the fact of having so anticipated induces him to behave in such a way that he realizes his hopes or banishes his fears. If confident, he struggles and achieves what he has believed would be realized; if timid, he defeats himself by acting in the very way he has anticipated and would avoid. In the first instance the original anticipation inspires, in the second it depresses, and in both it is fulfilled because, having been evoked, and having come into consciousness, it affects behavior. Or of course the prophecy may be unfilled: the desirable is not attained, the dreadful is avoided.

Since present behavior results from past experiences or from future intentions—the bent twig is bent again in advance—recollecting, anticipating, and intervening are universal and inevitable; therefore:

3.1.a *All persons everywhere are oriented periodically toward the past, the present, and the future.*

The most facile support for the proposition I believe to be the fact that, so far as we know, three tenses, or their equivalent, exist in all languages. Certainly languages have nuances in this respect, but the idea of the past, present, and future can somehow be expressed. Since speech both reflects people's needs and affects their modes of perceiving and behaving, the tenses must be the fruits as well as the seeds of human experience with three modes of temporal orientation. Other evidence of a non-linguistic sort is not difficult to find (see *Addendum 3.1*).

But the statement about temporal orientation is flabby: surely we ought to be more precise and not simply say that three orientations exist. Which one will be emphasized? We shall see, for example, that differences in orientation characterize the mentally ill. For the moment, however, let us observe differences in culture.

PERSPECTIVE

To comprehend temporal orientation and perspective, the expository device of drawing contrasts between traditional peoples and those living in the West will be employed (Doob, 1960). The method is admittedly risky and unfashionable because obviously each society is unique and hence any generalization about many societies perforce is an abstraction which disregards uniqueness. But we discuss the societies only to gain insight into time, not necessarily into them.

Perhaps the outstanding characteristic of a traditional society is the restrictive security its people enjoy and endure. Their community provides frequent face-to-face contacts within a limited group. Loyalty tends to be centered primarily in the nuclear or extended family. Larger units, such as the clan, the moiety, or the entire tribe are recognized, but their importance is likely to be somewhat less compelling, except under special circumstances such as an initiation or a war. Primary groups provide for the normal exigencies of life, and indeed many, perhaps most traditional groups achieve within their family or neighborhood a very high degree of self-sufficiency. Under these circumstances, in the deservedly popular terminology of our era (Riesman, 1950, p. 9), there is a very strong tendency for behavior to be tradition- and other-directed rather than inner-directed; ancestors and peers, as it were, discharge the role of the internal conscience.

Embedded as they are in a network of mutual obligations and with fewer choices available, traditional peoples are likely to pay relatively little attention to the future. The role the individual is destined to play in his society is more or less determined at his birth, not simply by his sex or family background (which is true to some degree everywhere), but also by the fairly rigid nature of the social structure and the few opportunities offered in the society or its environment to pick and choose among careers. When queried, therefore, he may be able to anticipate the future with accuracy, but his anticipation tends to be passive and usually to involve little intervention. Thus the elders of the tribe receive respect and homage; but these honors are more or less automatically bestowed, they may be anticipated without the need to do very much except to grow old. Anticipation in these societies, it seems, resembles weather forecasting in Western society: the clouds, the humidity, the wind direction, and the temperature may be noted, but intervention of a direct sort cannot alter the anticipated change. In most other respects, however, it has been suggeststed that we in the West seek and achieve variety; "few of us," for example, "can say what we are going to have for lunch or dinner three days from now, let alone next year," whereas "there are millions of [traditional] people in the world who know exactly what they are going to have, if they are to have anything at all" because "they will eat the same thing they had today, yesterday, and the day before" (Hall, 1959, p. 179). Under these circumstances we probably think more about the future, and maybe, too, we are convinced that we intervene more frequently than our remote ancestors. A delusion?

Some degree of renunciation, however, occurs everywhere. Don Juan cannot love all the women in the world, or be faithful to more than one; and the excited traveler cannot live for any length of time in all the places he considers beautiful or stimulating. Choices must be made, if only not to intervene; the unchosen possibilities are renounced. And renunciation, especially when standardized within the society, is likely to be accompanied by some kind of substitute gratification that is anticipated in the future and hence by an appropriate temporal orientation. The substitute may eventually become real, as when married persons are permitted freely to make the kind of love that is prohibited premaritally. Or the rewards appear symbolic or metaphysical—and yet no less real—as when in certain Islamic countries the parents of an "innocent" child who dies believes that they will be supported by him in paradise (Read, 1966, p. 73). The ultimate renunciation of all is suicide, for then the miserable person abandons his own life not because he has satisfying anticipations concerning the future—unless "he suffers or enjoys exquisitely" the suffering of his survivors in advance (Meerloo, 1954, p. 97)—but because he has only gloomy anticipations for himself or none at all, so that he rejects the future forever.

Although differences in anticipation between traditional and modern societies are not absolute, the fluctuation from society to society must be clearly stated:

3.2 Modally within the person, within significant groups, and within the society as a whole, one temporal perspective rather than another is likely to be facilitated.

Illustrations of the principle crowd in upon us (Frank, 1939), especially when a community or a society is sensitively investigated. In a Canadian suburb, for example, the inhabitants as a whole "live almost entirely *in* the present but *for* the near future (with the past largely obliterated)." Life there reveals "a pendulum-like swing . . . between tension and relaxation, preparation and realization, work and leisure, application and dalliance." Most of the formal and informal groups adhere to a rigid schedule; at a service club there are "so many minutes for lunch—the late-comers must hurry, or sing with their pudding; so many minutes for singing and announcements; and a strict twenty minutes for the speaker." Even on weekends there is a schedule for disregarding time. In addition, "women predominately think in the long range, almost *sub specie aeternitatis,* in terms of ultimate effects," whereas men, "much more earth-bound and datum-driven, take into consideration an evanescent present or, at most, a very short-run future, in which things will be much as they are now and have always been" (Seeley, Sim, & Loosley, 1963, pp. 5, 63, 65, 389). Like these Canadians, most of us living in a "modern" society are likely to be punished when we do not conform to the modal temporal schedule of our milieu.

The Tiv of Nigeria have temporal concepts derived from natural phenomena (the sun, the moon, the seasons) and important social events (especially markets); but the key to understanding their way of thinking about time is reported to be their reliance upon "a direct association of two events." Thus the sentence, "he came the day I left" is a way of pinpointing time by means of such an association. For them, however, an association does not imply a causal relation and hence is reversible. They might say, as we would, that "we cut the guinea corn when the first harmattan comes," but the reverse is just as likely to appear, viz., "the first hamattan comes when we cut the guinea corn." In addition, the various time systems attached to phenomena or events are not "interdigitated," so that the Tiv "make no attempt to correlate moons with markets or either with agricultural activities, or seasons; this is a way of saying that these people do not measure time, they simply indicate it in various useful social contexts" (Bohannan, 1953). The traditional system of another African society is examined in *Addendum 3.2*.

With rare exceptions (cf. Murdock, 1934, p. 218), a sense of history seems to be another universal. For the presence of the aged is a continual reminder of the past; and so the eternal cosmological questions concerning the origin and development of men and their land also arise. The use to which history is put, however, varies considerably. For non-Western peoples traditions from the past may be not only recalled but also venerated: we do this, in effect, because our fathers did. Their history, consequently, is likely to be considered complete and not subject to revision resulting from empirical investigation: a king once reigned and the events allegedly occurring during his life are unalterable in people's minds. Such history includes a new chapter only when there is a new king or a new event. In contrast, all school children in a New England town may be acquainted with aspects of their country's glorious history, but most of them now have, and later as adults will continue to have, no more than a very superficial interest in the past which is clearly less potent than the belief in the intervention of ancestors. In fact the Historical Society of their community, which investigates and preserves local details, mostly trivial, from the past, may draw most of its members from the upper classes whose affluence enables them not to have to concentrate on the future (cf. Warner, 1959).

The reference to modal tendencies in the principle must be violently emphasized, lest a large variety of facts be invoked as exceptions. Thus the tendency for traditional peoples to have a past perspective does not mean that from time to time, as when they plant crops or look at their children, they do not also look toward the future. A traditional person who allegedly worships his ancestors may conform to the rules of his group because most of the time he believes those ancestors are about and will punish him if he misbehaves; and he knows that they are about because they stir in the wind or because they are represented by some image or animal which he can touch and feel. The ensuing fear of his, however, is certainly internalized, and so functions as remote from perceivable reality as the modern man's conception of a tempting devil or an avenging or loving God. The principle, therefore, is to be treasured because it is useful to detect modal emphases which differ from society to society; but simultaneously deviations can be cheerfully recognized, as further emphasized in *Addendum 3.3*.

MODAL VALUES

We do not need to debate further the point that each society has modal ways of behaving which stem from, and give rise to modal preferences or values, whether pertaining to food or beauty. More specifically, for present purposes we have another interaction between the behavioral and temporal potentials:

3.2.a *The modal temporal perspective of a society reflects and affects a modal philosophy of values pertaining to other behavior.*

One respect in which modal personalities vary certainly involves the general belief concerning the efficacy and desirability of intervening in the affairs of nature (Kluckhohn & Strodtbeck, 1961, pp. 15–17). On the one hand, people may feel passive: they accept their fate, good or bad, and do not seek actively to alter what they know must come. A touch of this approach exists everywhere, no doubt; nobody sensibly struggles against gravity on earth which cannot be overcome or against gods to whom absolute, unyielding power is attributed. Everyone wonders about the weather the next day, but most of us are content merely to anticipate what will occur on the basis of various signs such as clouds or air pressure. On the other hand, men may seek to combat natural forces over which they believe they can triumph; perhaps profound or complete fatalism is felt only with reference to very central metaphysical problems. Our efforts to affect directly tomorrow's weather must be mediated by prayer or hope; or we accept the forecast and in anticipation intervene by changing not the weather but our plans for the immediate future. The pioneer or planner who anticipates that he will be injured by a landslide quickly moves away or employs a bulldozer to set up an embankment.

The attitude toward activity-passivity glides into a conceptualization of the relation between the past or future and the psychological present. Even as subjective or objective time is conceptualized in terms of change or repetition, so men believe in progress or cycles. Those with faith in progress conceive of the past as an important stepping stone to the future which differs from it while embodying some of its components. At the opposite end of some sort of continuum is the cyclical view: the future is just a recurrence of the past, perhaps only in a very slightly different form. I rather think that the division is not so sharp when the person anywhere, by some miracle or compulsion, reflects upon his own existence. For the staunchest advocate of progress must recognize that the future is not altogether different from the past: there are alternating periods of depression and prosperity, even though—he may contend—the latter may occur more frequently and the former less so. And the advocate of cycles perceives that at least on the surface what recurs has its distinctive qualities: that lean period, though sharing some basic characteristics with previous eras of misery, is also different from what is taking place at the moment. Undoubtedly the ensuing mixture of progress and cycles has some effect upon the extent to which the person or society feels that intervention is possible and desirable, and each is therefore valued accordingly. The relation between perspective and attitude regarding temporal perspective (it can be observed in *Addendum 3.4*) appears on many different levels.

Persons in the West tend to be activists in almost every area—a few of us directly, and most of us vicariously, move mountains and conquer space. For this reason it is not easy to understand or sympathize with other approaches to intervention which emphasize passivity and resignation. If you are a Christian or a Jew, you are baffled by people who often accept poverty, misery, and suffering by saying that all this is the will of Allah. The referents over which control can or cannot be exercised and hence concerning which intervention is or is not considered feasible include literally all aspects of existence. Can better crops be grown? Can floods be prevented? Must we become ill, or can preventive measures be taken to ward off illness; when sick, must the malady run its course or can physicians and medicines facilitate recovery; can the life span be prolonged? Can a person's character be changed? Can I be less moody, can I make myself happier? The very raising of these questions indicates a bias toward activism: the person resigned to his destiny or fate probably would not think along such lines.

The value placed upon time also has a strong cultural component. Some goals are worth seeking only if the interval that transpires in reaching them is not considered excessive, and the norm for excess is likely to be a reflection of a conventional standard. You know how much extra time you are willing to spend to attain an extra reward or bonus. Conversely, the value of a goal may be affected by the length of the interval required to achieve it. There is the common belief that the view from atop the mountain is enjoyed more by those who ascended the height by foot than by those using a motor car. Then time as a concept or metaphor, including the very word itself, has in a society as diverse as ours a host of emotion-arousing connotations. A psychiatrist has listed "some symbols of time frequently used by my patients"; time is or can be one of the following: a figure that eats itself away, an arrow that moves on, a two-faced Janus that looks both to the past and the future, submission to or revolt against paternal command, an obsessive repetition compulsion, money, boredom, the desired parent, the beloved, creation, waiting, punctuality, fate, a loan, or even an opportunity (Meerloo,

1970, pp. 31–49). Clearly, just as patients are only exaggerations of normal persons, so these symbols can be found, in greater or less degree or more or less frequently, among all of us who inherit the social and literary traditions of the West.

But, just as deviations from modes dare not be overlooked, so great care must be exercised in ascribing modal values—or perspective or behavior—to a society: sweeping statements which appear cosmic and sagacious are not necessarily valid. We quoted, in the previous section, an anthropologist who asserted that "few of us can say what we are going to have for lunch or dinner three days from now. . ." Most of us, however, drink coffee or tea and perhaps have exactly and habitually the same food for breakfast, day after day, year after year—and all of us know what we are not going to eat (e.g., elephant or zebra meat which are great delicacies in parts of Africa) the rest of our lives. In another section (and in *Addendum 3.1*) reference was made to differences in the tenses within the languages peoples speak; yet from these linguistic facts one should not recklessly suggest, even when the thoughts are self-admittedly derived from "scientific vision" rather than from "a long period of toil," that "the sense of time in a primitive society is reduced to a minimum" (Haas, 1956, pp. 40–1). For non-Western peoples may possibly pay less attention to the dimension of time when they perceive and arrange the outside world, their systems of reckoning the duration and succession of events may not be so elaborate or based upon quantification, and indeed their temporal orientations may be somewhat limited; nevertheless such a temporal framework is very far from the kind of "minimum" sometimes displayed by children and the mentally ill. The way in which a people or a person divides subjective or objective time may actually be quite precise and adequate in terms of the values that are important in everyday life. In addition, socialization everywhere demands anticipation of the future to some extent. "Remember The Day After Tomorrow, my son and daughter," a Yoruba novelist writes. "That was how our father used to warn us whenever Aina and I offended one who was older than us" (Tutuola, 1967, p. 12). Finally, slippery and tricky methodological problems arise whenever the attempt is made to characterize the modal values or perspective of a society (see the solemn warning in *Addendum 3.5*).

DIRECTION OF ENERGY

While discussing the relation of culture to the temporal potential, I must raise a somewhat impertinent question: why does a society favor one modal perspective and an appropriate system of values rather than another? I am quite aware that the respectable reply of modern social science is to suggest that any culture trait represents the historical outgrowth of a long process, partially trial-and-error, partially deliberate, through which a way of behaving best suited to people's needs under the environmental and social circumstances evolves. The modal temporal perspective, therefore, serves a function within the society. I would not gainsay this reply, but I would add to it a statement which follows from the definition of perspective:

3.2.b *The stronger the temporal perspective, the weaker the orientations in other directions.*

By strength is meant the frequency with which the orientation affects behavior and the significance of the perspective within the society or person. Admittedly adequate evidence is lacking to test, no less prove this principle. But when the "conception of the nature, duration, and process of time" which characterizes the world's great religions is conceptualized in broad, sweeping strokes—"in the philosophical and religious speculation of India . . . an annulment of time and change is sought" (Cassirer, 1955, pp. 118–40)—then I wonder whether concentrating upon one kind of symbolism really does leave little energy for, or interest in, another kind. Similarly, I am slightly impressed by the negative correlation that was once found between the ability of Ph.D. candidates to recall picayune details from the recent and remote past on the one hand and a measure of future orientation on the other hand. Recall was appraised through direct questions (e.g., "How tall were you when you were 13 years old?") and orientation by ratings of instructors concerning their effectiveness as students (presumed to be a symptom of future orientation) and by a projective test (U.S.; Goldrich, 1967). It looks as though these students had a tendency either to dwell on the past or future, but not on both.

Perhaps the principle suggests one of the reasons why persons in traditional societies experience great difficulty in moving into the "modern" world: they must divert some of their energies away from the past and into the future. Or, it could be that in general, as volumes of psychiatric cases suggest, too great an emphasis upon renunciation for the sake of the future leaves some energy unexpended in the present, with the result that repression leads to numerous psychological difficulties. Of course "energy" as here used is really a metaphor, and it is difficult even to try to imagine how one would try in the interests of measurement to strike a balance between the immediate frustration of a renunciation and the deferred gratification being anticipated.

Again I would only argue in behalf of the contention that temporal perspective is a matter of emphasis, and that emphasis on one subject implies the neglect of another. It is all very well for philosophers and others to make refined distinctions concerning time—one psychiatrist distinguishes no less than seven, extending from the "remote past," which he calls an "obsolete" zone, to a "remote future," which is the zone pertaining to prayer and "ethical action" (Minkowski, 1933, pp. 95–120, 138–58)—but it is these sages and not ordinary persons in society who have the time to make such fine secondary judgments. The rest of us pick and choose among temporal orientations until we achieve a perspective, or else we scatter our interests superficially in all directions.

TEMPORAL KNOWLEDGE

3.3 Each society provides appropriate information for passing temporal judgment.

This principle also stems from the general nature of culture: uniformities of behavior must be learned and elicited. The information may have a direct or indirect relation to the temporal potential. Directly, we are provided with dates and timepieces which become standard equipment. Indirectly, a series of secondary judgments may be involved before the information can be utilized. Many temporal judgments, it will be shown, depend upon the age of the person who makes them; in part, age is important because it indicates the individual's distance from death, which in turn influences the way in which he has learned from experience and his society to view the future. What is the outlook of a man of 45? His age is likely to be evaluated quite differently in Western society, where the life expectancy may be around 70, from the way it is in most nonliterate societies where before the advent of modern medicine the expectancy might well have been below his present age. The assumption here is that the facts of life-expectancy are vaguely or clearly intuited by almost every one in society.

Other information is sufficiently uniform for members of a society to be able to communicate and cooperate with a minimum of error. The sensational variability in methods of time reckoning is usually to be found between societies, and not within a particular one. Some of this standardization results from the imposition of cultural rules upon climatic and astronomical factors. Spring in the northern hemisphere must always begin on March 21 or 22 and end exactly three months later, regardless of the weather. In Western society, no doubt because of the figures on clocks and watches, and perhaps because of the number of fingers and toes, there is a strong tendency to use rounded numbers or to think in multiples of five or ten. You say we shall meet in five minutes, not in six or four. This tendency is marked even under laboratory conditions with American and European subjects whenever the Method of Verbal Estimation is employed (Yerkes & Urban, 1906; Urban, 1907; Axel, 1924; Berman, 1939).

In our society, too, the hour is the standardized unit of temporal information in so many different situations that we accept it as if it were as natural and inevitable as the lunar cycle or the daily tides. Who, even for non-scientific purposes, would dare measure speed in units other than miles or kilometers per hour? Obviously a conventional measure is convenient. But, a moment of reflection suggests, why follow the custom in the West of having most segments within the school day divided into hours or into the slightly shorter academic hour? Why devote the same amount of time to subjects diverse as history and mathematics? It looks as though convenience rather than reason or a set of facts is the guide. Can the same explanation be given for the fact that we in the West almost always know the year, the month, and the day of the month of our birth but not the day of the week, whereas in many traditional societies in West Africa only the day of the week is noted and may in fact be used as the individual's name?

With one exception, our knowledge of time-reckoning in various societies is most inadequate: available are a broad survey of the variations and detailed descriptions of the devices employed in particular societies (see *Addendum 3.2*). The exception is a careful study of 58 non-Western societies, scattered throughout the world, for which adequate data could be obtained (Zern, 1967, 1970). First, the investigator has shown (by means of a so-called Guttman scaling technique) that the methods of time-reckoning in these societies can be arranged in an orderly progression so that, by and large, a society possessing one of the following is likely to have all of the succeeding but none of the preceding ones: a calendar for the year, division of the months, names for the days of the equivalent of the week or month, elaborate differentiation of parts of the day, a series of annual dates of ritual significance, names for different months, and crude differentiation of parts of the day. Here is evidence that the cultural imposition of time-reckoning upon the environment follows a principle which seems to begin, in the author's words, with "gross natural phenomena" and to end, in my words, with temporally distant, more abstract events. Then, secondly, the kind of time-reckoning system of these societies was found to be significantly related, not to a measure of their complexity, but to a particular modal socialization practice: the larger the number of temporal divisions in the above list, the greater the tendency *not* to overindulge infants during the first year of life. Although the relation was imperfect—the overall correlation was .42—it is impressive for two additional reasons. It appeared when the societies were examined separately in four regions of Eurasia, Oceania, North America, and South America (but not in Africa). In a longitudinal study of 60 essentially middle-class American children, the same investigator established a similarly significant, if not impressive, connection between *non*-indulgence (as observed and also as ascertained from mothers during the first year of life) and the "salience of time structuring" inferred from the stories they constructed in response to standardized pictures (the TAT) at the age of 14½. Thus the temporal knowledge prevalent in a society could be viewed more or less as an expression of a single principle: both generally within the sample of societies and specifically within these particular Americans, it was related to the same aspect of the socialization process.

Some of the units employed in perceiving the order or succession of intervals or events do not need to be standardized since perforce they can almost always be perceived in no other way. Lightning reaches our ears before thunder, and that's that. A sequence once learned then functions as an anticipation evoking in most instances a future orientation; after you see lightning, you expect thunder. Culture, however, plays a most significant role in the production of order or succession of many events other than those inherent in nature. You begin with soup and end with nuts, you read from left to right (unless your language is one like Arabic or Chinese), you know the succession of notes if you are to produce the desired melody—everyday instances such as these are sufficient to establish the point.

The units of historical time may be similarly viewed. Non-Western societies with no written records generally can look back at their history in very restricted and vague terms. In the West, the conception of the past has varied with the state of knowledge: "since 1800, the discovery of unlimited time has fractured the foundations of earlier world views even more irreparably than the earlier discovery of unbounded space," with the result that "the static framework of the natural world has been dismantled" and our view of that world has "dissolved into a temporal flux billions of years in length" (Toulmin & Goodfield, 1965, p. 266). The "sheer extent of the past," epitomized by the technique of carbon-dating, may not be known in accurate detail by most persons, but they acquire some knowledge thereof from the mass media, as a result of which their temporal judgments concerning the past are bound to be judged on a scale different from that used by their ancestors or people outside the West. Probably there are similar differences with respect to units for judging future intervals—and I imagine that they are affected by ones directed toward the past—but here society can provide a less precise guide: anticipations cannot be plotted so easily as memories.

TIME AS COMMUNICATIONS MEDIUM

All the aspects of time that are standardized within a society—the modal information, standards, perspective—may be utilized as clues to understand the significance of many forms of behavior either by any individual therein or by an outside observer in the role of traveler, journalist, or social scientist. Clear-cut criteria for establishing that significance exist:

3.4 The value of an activity is positively correlated with its temporal priority, the objective time devoted to it, and the frequency or precision with which its duration is judged.

By value is meant whatever feelings and actions are preferred, demanded, or sought by people; the needs thus satisfied range from the biological and mundane to the aesthetic and sublime. Again and again in everyday life an aspect of time becomes "the silent language" (Hall, 1959, pp. 165–85) for communicating significant information: the actions which speak louder than words include not only what you do, but also when and how long you do it. Priority can be indicated by the order in which different activities are performed: first things first. When awake, where are most persons in the West likely to be found at a given moment or, at least, where would they prefer to be? The answer must be: at work or with their families. The executive or laborer values his occupation, for this is the role which he wittingly or unwittingly discharges to earn a living, and hence both of them will and must spend most of their lifespan in that role; and each hurries home—or, if a laborer, tries to—when an emergency arises there. Almost all married women in the West, for better or worse, have no choice except to renounce whatever activities conflict with their responsibilities to their husband and children. A telephone call in the middle of the night suggests to the poor man who is awakened that his caller has a vitally pressing problem, or is insane. The time elapsing between asking for and being granted an appointment may suggest the value a lawyer places upon the client who would seek his counsel. An individual arrives late for a conference or makes his visitor wait in the outside office in order deliberately to insult him. A last-minute invitation to a somewhat formal event is usually considered to be unflattering. In many countries you are supposed to be in your seat before the important official arrives to address the meeting, and you know he is important because of this regulation. The encoding or decoding involved in these situations may be quite deliberate or it may be so habitual that it becomes almost unconscious, but the communicator in some manner passes temporal judgment before using an aspect of time as a medium. The ensuing communication can be correctly apprehended when its recipients also pass temporal judgment *and* appreciate the significance of the communicator's previous decision.

The amount of time devoted to any activity is another clue to its importance; in this context time is a "resource" that is carefully distributed (Heirich, 1964). Involuntarily we pay tribute to sleep by devoting a large portion of our lives to it. The length of the interview granted by a leader suggests how much he values a lesser man. Most schools and textbooks in a formal system of education concentrate upon the history of their own country and tend to neglect the development of the rest of the world. As a result, the citizens of the land not only learn more about the accomplishments of their ancestors but may also have the feeling that their history commands more time (another reason for considering it important) since they are better acquainted with its events: as will be noted subsequently, crowded events may have a tendency to seem longer than empty ones. A clear-cut illustration is provided by the calendar of the Catholic Church: "The four weeks of Advent condense the thousands of years of expectation of a Messiah since the fall of Adam and Eve, the six weeks from Christmas to February 2d cover the childhood of Christ, and the Holy Week has the same duration as the events it recalls" (Fraisse, 1963, p. 169). Here the correlation is perfect: the more important the historical event, the more calendar time devoted to it by the Church. On quite a different level it has been shown that young women in an American college gave different and in a sense more primitive or basic responses—more "vulnerable areas of the normal personality" were revealed—when they were deliberately made aware of the passing of time and were under "pressure" to respond quickly than when they were not subjected to such temporal coercion (Siipola & Taylor, 1952; Siipola, Walker, & Kolb, 1955). Students who generally worried a great deal (according to their responses to a questionnaire) performed more poorly on a standard intelligence test when they knew they were being timed than when they believed they had unlimited time at their disposal; the reverse tended to be true of those low in worry (Morris & Liebert, 1969). Thus time allocation reveals personality in diverse ways.

The frequency with which temporal judgment is passed also provides a clue to the value of an activity. The parts of the day, the days of the week, and the months have distinctive names in our society with the result that we can make easy reference to them: our lives tend to be regulated within these frameworks. Every holiday, both religious and secular, has its unique date, which reminds people of its annual occurrence. When the activity considered significant is the passing of time itself—the "time-is-money" approach under special circumstances—then temporal judgments must be frequently made: otherwise the precious commodity might be wasted or improperly allocated.

The three temporal criteria mentioned in the principle are certainly interrelated, though not perfectly so: if you grant an activity priority, you probably will devote considerable time to it, and during a long interval you are also more likely to pass judgment frequently—whether you are stimulated or bored—than during a brief one. The other reciprocal relation which, though also imperfect, appears again and again is that between the communicator and his audience. You know that a letter you must write is important; hence you arrange your time so that you can devote a long interval to writing it; and afterwards you impress yourself and those who knew what you were doing (including the recipient if you tell him how much time you spent or if its length conveys that fact) with the value you attach to what you have written. Or if you and your correspondent are very sophisticated, you spend considerable time condensing your cosmic message into a brief, neo-poetical form and he appreciates the temporal significance of that brevity.

The attitude toward time, especially the conception of the time needed to achieve a goal, is likely to have a marked effect upon behavior. How much time, for example, should you devote to preparing a two-minute speech on a simple subject when the arguments you are to employ are provided by someone else? One group of subjects was given five minutes to do just that and another—because the investigator, by design, was called out of the room—an extra 10 minutes. It was then found, as Parkinson's so-called Law suggests, that the second group spent appreciably more time than the first one preparing another similar speech, even when some of them had the added incentive of being able to leave and still receive academic credit for the work as soon as they were finished (U.S., college; Aronson & Gerard, 1966; cf. also Bryan & Locke, 1967). Speech-writing had thus become associated with an interval having a specific duration either because of different work habits acquired during the first task or because of differential importance attached to it as a result of the interval's length. But, at least among American college students, this quickly acquired feeling about time was confined to a rather particular task: in a similar experiment, temporal experience in ranking photographs in terms of various traits, with 5 or 15 minutes being "accidentally" allocated to the problem, tended to generalize to an identical task but *not* to the somewhat similar one of rating arguments against the prohibition of cigarette advertising or the quite different one of preparing a talk on intercollegiate athletics (Aronson & Landy, 1967). In real life, of course, we are accustomed to being told repeatedly how long it takes to accomplish a given objective.

Why, it is not impertinent to ask explicitly, why is there a correlation between the value of an activity and the various temporal factors? In some instances the relation seems straightforward: basically organic processes, such as sleeping and eating, have survival value and hence consume huge slices of time. In others the reverse may be true: an action consumes so much time—working for a living—that thereafter the activity is considered important.

The principle being discussed does not specify in detail how time functions as a communications medium. Are the norms and meanings attached to the language of time, for example, as clear and as well known as the more conventional language of speech and writing? I suspect that the language of time is used less deliberately than the other language: when you do not admit to yourself that you dislike a person, you somehow never have enough time to see him or, when you meet him by chance, you believe you are in a hurry to do something else. Of course the same process can occur on a verbal level: your tongue slips and you say something nasty when you are thinking you wish to be pleasant.

A final point: a close relation sometimes exists in the mass media between space and time as ways of conveying information indirectly. In the West, for example, the most important story in a daily newspaper is usually accorded not only the most prominent position but also the greatest amount of space; from the reader's standpoint, the position and the space communicate its importance, and the latter attribute is transmuted into time when he begins to read. On radio and television, significance during programs when time is limited (such as a newscast) is communicated, among other ways, by the quantity of time devoted to an event or a description of an event, but time means space on the speaker's or actor's script.

SPIRALS

The last principle introduces us to a problem perpetually plaguing the analysis of human behavior: it suggests a correlation between two variables and hence does not specify definitively whether the value of an activity is a cause or an effect of temporal factors. A spiral relation between two sets of factors becomes evident. On a very broad, virtually philosophical level, we note that the modal value which people in a society place upon time and the precision with which they pass temporal judgments serve as vital ways to communicate values which then have profound effects upon their behavior. If they are told very often that "progress" is important, they must learn to have a temporal orientation with heavy emphasis upon anticipations concerning the future, and they are likely in consequence to pass temporal judgment continually in order to determine how far they or their country have advanced to the specified goals. In contrast, if they become convinced by the formal and informal teachings of their society that the future will be like the past and hence that change is either illusory or impossible, their temporal orientation may be deficient with respect to future components; and then temporal judgments are passed less frequently, or, when passed, have less significance for a life plan.

Similarly temporal judgments in modern society are so essential for economic and social reasons that clocks and calendars have been perfected and can easily be consulted; and the fact that they are more or less perfect and can be so easily consulted makes it easier and more tempting to pass temporal judgment. If it is true (Hall, 1959, pp. 174, 176) that "the basic vocabulary of informal time" among Americans is based upon eight or nine distinctions (instantaneous, very short, short, neither noticeably short nor long, long, impossibly long, and forever) and that among Arabs of the Eastern Mediterranean upon only three (no time at all, now or present, and forever or too long), then here and now Americans will make finer distinctions than Arabs because their vocabulary to do so is more salient. Historically one factor, but not the only one, which makes the former vocabulary more precise must be the sweep of American civilization and especially its industry which has required finer temporal discriminations.

Temporal landmarks within a society obviously affect behavior. When individuals know that the weekend or a festival is close at hand, they may or must behave differently. A very dramatic discovery has been the location of a "dip" in the mortality curve among Americans before a person's birthday, and before a Presidential election; and among American Jews the same tendency has been noted before the sacred Day of Atonement (Phillips, 1970). It would appear that temporal information and the desire to participate in the anticipated events enables some persons, as it were, psychosomatically to postpone their own deaths.

3.4.a *Temporal information is likely to have an appropriate effect upon behavior.*

This corollary would suggest only that, because it is so highly valued in whatever form it exists within a society, temporal information cannot be easily disregarded. The word "appropriate" refers to actions related to the goals being pursued when the information becomes available. Illustrative examples in everyday life and in the laboratory are numerous. You see a clock, and the resulting knowledge causes you to drive more slowly, more quickly, or at the same speed, provided reaching your destination is linked to some temporal consideration. College students, with a "shock electrode" strapped to their right forearms, were warned in advance that allegedly they would be shocked within a specified period of time. That waiting period, they were told, would be 5 seconds, 30 seconds, 1 minute, 3 minutes, 5 minutes, or 20 minutes. They all experienced some stress as indicated by physiological measures, ratings, and interviews, but the greatest stress was associated with the 1-minute interval (Folkins, 1970).

As indicated in connection with variations in verb tenses (*Addendum 3.1*), there is generally a spiral relation between linguistic and cultural factors. At first glance it appears as though a language must have terms referring to events that arise for natural reasons (sunrise, sunset, death) or are considered socially important (harvests, initiations). Vocabulary and structure, however, may have only a rough correlation with actual conditions in the milieu or society. Temporal concepts can diffuse from one group where they are appropriate to another where they are not needed; for example, the vocabulary of clock time has reached many traditional societies before temporal precision has become necessary. Then, and for an opposite reason, linguistic habits may linger longer than the conditions they once represented; thus words such as *spell* and *flash* persist in English as informal units of clock time but are employed only in very special but appropriate linguistic contexts. The decision to change or not change aspects of language is seldom made consciously or deliberately, the way a law is changed or repealed or a clothing style accepted or rejected: over generations men, who are only dimly satisfied or dissatisfied with features of their language, interact and thus rather blindly through trial and error promote or hinder linguistic change. Finally, linguistic structure may have no ascertainable effects upon behavior beyond the use of language itself. Clearly the use of a three-syllable word in German (*vorgestern*) and of four separate words in English (*the day before yesterday*) to designate the same day scarcely differentiates Germans from Americans and English with respect to the significance attached to that unit of calendar time.

Although the spiral compels any discussion of psycholinguistics to be somewhat inconclusive, one principle seems clear:

> 3.4.b *Appropriate linguistic concepts and structures mediate a positive if imperfect correlation between the temporal and behavioral potentials.*

Or, less elegantly: know a man's mode of expressing time and you have some preliminary insight into aspects of his personality and behavior.

THE BIOCHEMICAL CLOCK

Up to this point we have been considering external factors in the society which affect the behavioral and hence also the temporal potential. We must now shift the focus of our attention somewhat and examine processes operating within the organism whose present functioning (if not whose origin) depends only in part or not at all on external factors. For many temporal judgments seem at first glance to be or to become relatively independent of the external milieu (that is, to be endogenous) and to spring from deep-seated internal cues or rhythms which have been variously designated: internal clock, chemical pacemaker, Kopfuhr, and—the term arbitrarily selected in this book—biochemical clock. You probably awaken each day at a more or less fixed time that is to some extent not a function of surrounding noises, the coming of daylight, etc.: you open your eyes *before* hearing the alarm clock. This dependence upon some sort of internal cue has been explicitly recognized since ancient times and numerous references appear in historical records (Clauser, 1954, pp. 5–9). In the modern world travelers transported in jet planes across many time zones usually discover that their bodies continue to some extent to operate on the clock time of the region from which they have come. A person, especially an older one, who is accustomed to arise at 8 A.M. each day in Rome is likely to find himself wide awake at 2 A.M. in New York until days later when, as it were, his biochemical clock has been reset. In one study subjects flown from Oklahoma to Tokyo demonstrated not only a lag in body temperature but also, at least during the first day after the flight, a deterioration in reaction and decision time (Hauty & Adams, 1966). This "physico-physiological day-night cycle asynchrony," as it has been called, is estimated to affect about 70 percent of travelers, to last from three days to one week, and to produce such obvious difficulties as feeling hungry, sleepy, or awake at the wrong time (Strughold, 1962). In the Oklahoma study, the older men were more affected than the younger ones; and the end of the boring 10-day testing period in Tokyo apparently led the men to report little or no fatigue after the return flight home (Hauty & Adams, 1965).

There seems to be no doubt that all living organisms exhibit some temporal reactions which, whether or not their origin is internal or external, persist in the absence of "normal" environmental cues (Bünning, 1967). It has been noted for more than two centuries that plants which unfold their leaves in daylight and fold them at night continue this alternation for several days even when they have been placed in total darkness (Hamner, 1966). For a period of two weeks oysters opened their shells when it was high tide in New Haven, Connecticut, after they had been transported to a dark room in Evanston, Illinois, where, of course, the moon's zenith and nadir are reached at different times (Cloudsley-Thompson, 1966).

Clearly such deep-seated reactions serve a vital function under normal conditions: "The adaptive significance of circadian rhythmicity is that it enables the organism to master the changing conditions in a temporally programmed world—that is, to do the right thing at the right time" (Aschoff, 1965). Among human beings the most dramatic and commonplace bit of evidence on this point is offered by sleep. One summary of available research on that subject suggests that "the development and maintenance of 24-hour sleep-wakefulness and body-temperature rhythms stem from being born into, and living in, a family and community run according to alternations of light and darkness, resulting from the period of rotation of the earth around its axis" (Kleitman, 1963, p. 147).

Even among plants and lower animals whose periodic behavior is inherited, the biochemical clock is modifiable within limits by changes in the environment (Bünning, 1967, pp. 16–17, 62–4, 103–5): organisms respond to external cues which have temporal significance. Usually the research problem is to ascertain just what those cues are and how the organism is able to respond to them. When exposed to an appropriate length of sunlight during the day, for example, the leaves of plants transmit "messages" to buds causing them to form flowers instead of leafy shoots (Hamner, 1966). Although caged finches continued to display cyclic activity under the conditions of constant illumination in the laboratory, just as if they were responding to alternations in light and darkness in their normal environments, they were affected by this artificial condition: when the constant light was intense, they functioned on a 22-hour cycle (Aschoff, 1965). Many birds, at almost exactly the same calendar-date each season, migrate away from a region about to endure winter and then return after the passage of a relatively constant period of time as the season of that area turns toward summer; possibly some factor, like the exhaustion of food in the north, may initiate the flight southward with the approach of winter (Wing, 1962). In the laboratory it has been shown that pigeons can detect intervals of time with sufficient accuracy to account for their ability to navigate by means of temporal differences in latitude (Meyer, 1966).

Frequently the functioning of the biochemical clock can be most parsimoniously explained by finding external cues on which it is, to some degree, dependent. Any kind of delayed response may serve as a paradigm. In the laboratory the animal is taught to wait a specific period of time until it is fed: it is alerted to the presence of a meaningful reward (the food) by a conditioned stimulus (the sound of a bell), but it is punished (by an electric clock) if it seeks that reward before an interval of time (20 seconds) has elapsed; after the training period, the environment is changed by eliminating the punishment, but the animal continues to wait the allotted time. How does it know that it must wait before it can safely reach out for the food it craves? It has no stop watch to consult, and presumably it is not counting silently to itself from 1 to 20. Somehow as a result of the shock it has learned to inhibit the impulse for the proper length of time. We then may say, if we wish, that during that period of inhibition the biochemical clock is ticking away, even though of course we do not know what ticking means in physiological terms. Like the oysters in faraway Evanston, the animal may learn to adapt to its new environment: the delayed response eventually is extinguished, if it is no longer shocked for seeking the food too soon: the clock ceases to function.

Soviet physiologists and psychologists have extended this paradigm and suggest that "time, as a unique exciter of reflex activity, is capable (in conjunction with other stimuli) of creating and strengthening a definite course in physiological processes, of giving them a definite rhythm, a definite interrelationship in time." Their evidence comes not only from dogs, the favored species for investigation in Russia, but also from a large variety of other animals, including man. One study, for example, has shown that "in many people the number of leucocytes [in the blood] increases at the usual hours of eating"; when the eating time was changed, this reflex for food took 6 to 7 days to establish, 2 to 4 days to extinguish, and the longer it took to establish, the slower was the extinction (Dmitriev & Kochigina, 1959). The Soviet scientists write in physiological terms without being able to specify either the precise locus of the storage process in the nervous system or the details of how the process is stimulated to occur with such regularity. In fact, it must be added, it is difficult for us to verbalize how non-verbal information is stored—verbal communications, we say glibly, are stored in sentences, as if that were an explanation—although we certainly know that we do retain the non-verbal. Just how, for example, do you store your ability to swim during months or even years when you do not swim and then utilize that ability when the occasion arises?

Some human phenomena depend upon timing mechanisms which seem to function "involuntarily." The largest number of young women in a large Czechoslovak sample stated that their menstrual periods began between 4 A.M. and noon; very few indicated the hours from 8 P.M. to 4 A.M.; and those beginning between 8 A.M. and noon had a shorter period than those beginning in the afternoon (Málek, Gleich, & Malý, 1962). Most normal births have been shown to occur during the eight hours after midnight, with the peak between 3 and 4 A.M., fewest between 2 P.M. and 8 P.M., and the trough between 5 and 6 P.M. (Kaiser & Halberg, 1962). The internal and external cues for menstruating and giving birth at a particular time are clearly not conscious, just as—on quite a different level—we cannot always say how we know the approximate time of day without consulting a clock. Cues of this kind I shall hereafter call *unverbalizable*.

It must be assumed that physiological processes determine, or at least affect, temporal judgments stemming from these unverbalizable cues. The judgments of six graduate students concerning the duration of intervals ranging from 30 to 120 seconds fluctuated with the time of day; the investigator also notes that this diurnal variability "tended to resemble" changes in blood pressure (U.S.; prod., prior; Thor, 1962a); and other investigators have also reported diurnal fluctuations, though there is disagreement as to the precise time at which the clock is going fast or slow (cf. U.S., college; prod. & verbal with prior; Pfaff, 1968). Presumably subjects in these experiments were unaware of changes in the clock, and obviously they did not make note of their blood pressures, even though that pressure may have affected their conscious judgments.

For human beings, however, it is sometimes necessary to look for other than physiological processes to account for the rhythms of the biochemical clock and the unverbalizable cues it provides. There may be, for example, an important ingredient of anticipation and hence a prior arousal of the temporal motive, at least on some occasions. Originally and continually the person has decided that it is best for him to awaken at a particular time: oversleeping is inconvenient, it means the disruption of daytime responsibilities. Even in the absence of a conscious decision, a regular waking time serves the same social function. In addition, some alternation between sleep and waking activity is biochemically essential for survival, and fatigue can be as effective a condition for promoting sleep as the alarm bell is for waking up. Thus we possess, innately, the capability of acquiring a circadian rhythm and, when one is established, we achieve some degree of independence from the outside environment by being able to respond to a self-regulating, internal mechanism which, nevertheless, continues to be related to, and frequently dependent upon, changes in the physical and social environment. It is also interesting at least to raise the question as to how the biochemical clock in turn mediates its time-telling proclivity. One writer, relying precariously on self-observation and other anecdotes, argues that during sleep the clock sets off a "shocking" dream which awakens the person (Bond, 1929).

The discussion, being lengthy, inspires a lengthy principle:

3.5 *Changes in the external or internal milieu (especially periodic ones) of significance to the organism give rise to internal processes which function more or less in phase with those changes, which are not completely independent of them, which may persist or not be immediately extinguished when external conditions are altered, and which provide unverbalizable cues for arousing the temporal motive.*

Anyone finding it difficult to accept the postulate of a biochemical clock for human beings need only be reminded of the immediately perceptible rhythms which engulf him: he breathes and his heart beats more or less regularly; he can observe the swing of a pendulum and the flow of the tides; he hears the songs of birds and the noises of engines.

PERSONALITY

Although a broad insight into the behavioral and temporal potentials is obtained by examining the underlying cultural and biochemical factors, specific information is also needed regarding the person passing judgment: if he is the focus of attention, the uniqueness of his personality must be taken into account. Is he, for example, compulsively or obsessively punctual when the regulations of his group permit much wider latitude regarding the keeping of appointments or the adherence to some kind of schedule? On a preliminary basis, to be sure, it is fruitful to treat people's attitudes toward time, their motive to pass temporal judgment, their temporal perspective, like all other behavior, and to look for its genesis in the culture. If you insist on knowing exactly how long it takes to go from point A to point B, my first assumption will be that you are a German or an Englishman and not a Bushman or a Sioux. But if I know that you are a German, I would then wish to determine whether your demand is "typically" German, that is, whether most of your compatriots would make the same demand. Then if I discover that they do not do so, I would begin delving into your personality to try to account for your insistence. In addition, it is useful and necessary to consider any kind of behavioral factor on an individual level, for both culture and the biochemical clock are internalized more or less uniquely within each person. All or almost all the citizens of a society feel within themselves the conventional attitude toward time and they consequently behave temporally according to the pattern as they have internalized it. It seems strange, you say, to be going to work on a Sunday when an emergency requires you to do so: this is the day of the week when other activity is normally scheduled, and you experience the strangeness because you know the norm.

The role of personality in affecting the temporal potential may be of two kinds:

3.6 The temporal potential of a person is affected by personality traits either when those traits pertain specifically to some aspect of temporal behavior or when they influence aspects of his behavior potential which in turn modifies that temporal behavior.

Traits are considered to be predispositions which, when evoked, produce some degree of consistency in behavior. Proust may serve as an example of someone who, at least on the surface, was markedly affected by his temporal perspective: he organized himself and the writing which brought him fame around a past he felt impelled to recapture. In contrast, Whitman pushed his poetry, and many of his own impulses, toward the present and the future. A person in psychic travail, on the other hand, may be motivated by his desire to be healed and this he would accomplish, with or without assistance from a psychoanalyst, by recovering the memories of his wounds in the past; or he tries to learn how to adjust to, and communicate with his fellows in the present (through so-called sensitivity training, for example). It is an empirical question whether ordinary persons possess such enduring perspective or other traits with a temporal content. The evidence on this latter point will be examined in a later chapter more gracefully when the theme is the development of the temporal potential. Here only a claim in behalf of personality as a critical determinant has been staked out.

ADDENDUM 3.1 VARIATIONS IN TENSES AND PERSPECTIVE

There is probably an important but not a one-to-one correspondence between linguistic and cultural practices. "In the classical Indian languages," it has been suggested, "there are no words which corresponded to the concept 'to become'"—and then this linguistic fact is used as evidence that the speakers of the language possessed a static notion of time (Nakamura, 1966, p. 77). Undoubtedly there must have been some relation between the absence of this category and the prevailing conception of time, but surely a cultivator could somehow have expressed the notion that eventually his crop would ripen. Maybe all that is meant by the statement is that the idea of becoming could not be easily, and hence was not frequently expressed.

In the Bantu languages, a distinction is often made between two kinds of present and two kinds of future by means of different infixes which are inserted into a verb: one refers to the immediate and the other to a less recent past or future. European languages lack this grammatical nicety, but it would be difficult to contend that Europeans are less concerned with drawing fine temporal distinctions than traditional Africans. One African scholar maintains, however, that the idea of a far future cannot be expressed in many African languages. In two of them (those of the Kamba and Kikuyu peoples), for example, there are three future tenses: (1) action in two to six months, (2) action that will occur immediately, and (3) action "in the foreseeable future, after this or that event." Then he comments:

> You have these tenses before you: just try to imagine the tense into which you would translate passages of the New Testament concerning the Parousia of Our Lord Jesus Christ, or how you would teach eschatology. . . . If you use tense no. 1, you are speaking about something that will take place in the next two to six months, or in any case within two years at most. If you use no. 2, you are referring to something that will take place in the immediate future, and if it does not take place you are exposed as a liar in people's eyes. Should you use no. 3 you are telling people that the event concerned will definitely take

place, but when something else has happened first. In all these tenses, the event must be very near to the present moment: if, however, it lies in the far distant future—beyond the two-year limit—you are neither understood nor taken seriously." [Mbiti, 1968]

But again the analysis is restricted to tenses; both African languages can express the idea of a far future through more involved verbal explanations. Perhaps, more generally, the ineffable simply requires us to be long-winded.

I suspect that slight but distinctive differences in the use of tenses appear whenever two languages are compared but that they do not necessarily have psychological significance. The ease with which the present tense in German can indicate action in the future probably has no great effect upon Germans and hence does not distinguish them from the English whose language does not possess this feature so markedly. The differences may well begin to affect perception and expression in important ways when the languages being compared belong to diverse linguistic groups, such as Hungarian and French (cf. Sauvageot, 1936).

Obviously we judge others and ourselves both backwards or forwards (Kastenbaum, 1965); "the adjustment to time is always an extremely difficult, even insolvable task which mankind has to carry on" (Dodge & Kahn, 1931, p. 64)—these are the kinds of assertions which students of behavior are prone to make. From archaeological remains, such as tombs or tools, it seems reasonable to assume that men showed some concern for the future in prehistoric times (Whitrow, 1961, p. 54). Similarly since "stories of the creation of mankind appear to be universal" (Kluckhohn, 1960, p. 48), the inference may be drawn that mankind must always look backwards; or, for that matter, the frequency with which the "myth" of paradise appears, either one that has been lost in the past or one to which man may look forward under specified conditions (Eliade, 1960), suggests one or both of these orientations. I suppose that every anthropologist who has cast a systematic or unsystematic glance at a variety of societies would agree that, in spite of considerable variations, "as far as we know, all societies have concepts of time as well as views as to its use and importance" (R. J. Smith, 1961).

ADDENDUM 3.2 TIME RECKONING IN DIFFERENT SOCIETIES

A very extensive survey of time reckoning throughout the ages and the world by and large is disappointing: it only documents the great variability which appears in connection with every temporal unit (Nilsson, 1920). But there is a limit to the variability: "the phenomena of the heavens . . . and the phases of Nature . . . are the same for all peoples all over the globe, and can be combined only in a certain quite small number of ways" (p. 2). Some of these phenomena can be observed more easily than others; thus "an observation of the annual course of the sun . . . unlike that of the stars—which everywhere, no matter where, can be performed immediately—demands a fixed place and special aids to determination," for which reason "the observation of the solstices and equinoxes belongs to a much higher stage of civilization than does that of the stars" (pp. 311–12). The one generalization which the compiler himself emphasizes again and again is what he calls the *pars pro toto* principle: the "deep-rooted tendency" for many societies not to count units of time (such as days, months, or years) but rather to keep track of the concrete events (dawn, moon, snow), the parts of the totality (p. 358). Nonnatural or "artificial" events (such as, for example, holding a market every fifth day) may also be used for measuring time but these, our encyclopedist thinks, belong "to a highly developed stage of time-reckoning."

From the standpoint of the present analysis, the examination of time-reckoning systems is valuable in two respects: it suggests the occasions on which the temporal motive is evoked and it indicates, very broadly, features of the environment as well as natural and cultural events that are of special importance within a given society. Thus the traditional system of time-reckoning among the Kaguru, a Bantu group in east-central Tanzania, concentrated upon cultivating crops:

1. There were two terms for abstract time, one referred to a period less than a day and the other to one longer. The former could indicate even a shorter period when a diminutive prefix was added.

2. The historical past was designated not through the use of units such as a year but through references to important events. The referent seemed to have been a period of time, such as the interval of the last great famine. Genealogical descent was rarely traced beyond three generations: "within this narrow scope, passage of time is meaningful only in terms of helping to explain the alterations in various social relations and allegiances, such as the distribution of bridewealth, loyalty to kin, responsibility and obligations in paying and receiving property."

3. The most frequently used terms were those relating to days which, however, had no traditional names; instead there was a distinctive name for "today," and then time was calculated therefrom either backwards or forwards ("three days ago," "four days from today"). The day itself was divided into daylight and dark; but the latter, being undifferentiated, was not subdivided. In contrast, the daylight section was divided into periods with distinctive terms.

4. Months were noted through the phases of the moon.

5. The year was divided on the basis of the rains, with distinctive terms indicating whether they were light or heavy. For example, the heaviest one in April was called "downpour."

6. Except for Venus in its role as evening and morning star, the stars were not employed as time-reckoning devices.

Clearly, as the investigator points out, we have here "a vague sliding scale focused on the near present in which the past and future are of relatively little concern"; emphasis is placed upon the agricultural cycle on which life depended (Beidelman, 1963).

ADDENDUM 3.3 DISPERSION

Any survey, any investigation involving behavior undoubtedly reveals deviations from central tendencies. This truism is elevated to the status of an *Addendum* in order to give further emphasis to a multivariate approach to the problems of time. I could choose any public opinion survey remotely related to time and make the same point again and again, which I shall now do with a few haphazardly selected illustrations.

The description of the orientation of men and women in the Canadian suburb, cited in the text, conveys the fact of variability by means of a simple adverb, "predominately": "Women predominately think . . ." and "men predominately are . . ." (Seeley, Sim, & Loosley, 1956, p. 389). A sample of American adolescents was confronted with the statement, "Planning only makes a person unhappy since your plans hardly ever work out anyhow." The percentage disagreeing with this view was 90 for those of Jewish origin, 62 for those of Italian origin. The difference is statistically significant, hence shows a trend for more Jews than Italians to believe they could assert "mastery" over the external world (Strodtbeck, 1958). Obviously this overall conclusion expresses only a mild central tendency to which exceptions must be noted: the differences between the two groups were not absolutely, but only statistically, significant; virtually identical percentages, 80 and 79 respectively, disagreed with another statement appearing to be within the same frame of reference ("Nowadays with world conditions the way they are, the wise man lives for today and lets tomorrow take care of itself"). In a study of American college students—a relatively homogeneous group if ever there be one—it was found on the basis of replies concerning 11 public issues that 27 percent seemed to favor a future orientation, 43 percent one in the present, and 30 percent one in the past (Anast, 1965). The perspective of American high-school seniors was judged by noting whether or not they spontaneously referred to the past or the future in stories which they constructed in order to complete simple sentences such as "Wally and Carol were at the dance together. . . ." The result: 92 percent of the stories involved the future, 40 percent the past; the future was mentioned more frequently in connection with neutral than affectively toned themes, and the reverse was true of the past which, moreover, tended to be mentioned more frequently in pleasant than in unpleasant contexts (Kastenbaum, 1965).

ADDENDUM 3.4 INTERVENTION AND PERSPECTIVE

Texans and Mormons who considered activity rather than passivity as characterizing man's mode of self-expression tended to be oriented toward the future rather than the present or the past; Navaho and Zuni also favored activity rather than passivity but tended to be oriented away from the future toward the past and the present; and Spanish-Americans who favored passivity rather than activity tended to be oriented toward the past but placed more emphasis on the future than the present (Kluckhohn & Strodtbeck, 1961, p. 351). Then, in more general terms, fathers in the West who have achieved or wish to achieve middle-class status assume that their sons will progress at least as far if not farther than they themselves. In consequence they intervene by trying to provide a better education than they themselves have had; when appropriate they offer their sons the opportunity to carry on the business or professional organization they have established or helped establish for themselves. In contrast, according to a broad generalization—probably much too broad—about African societies in Tanzania, "the social pattern of each generation starts afresh in its acquisition of resources and position." African fathers, its is contended, assume that "a youth must make his own way in life" and therefore "even though parents may have special advantages—businesses, a large farm, etc.—the trend is towards using these advantages to give youth an equal *start* rather than handing over the enterprise to be enlarged and managed by the next generation" (Armbrester, Baughman, & Moris, 1967, p. 65).

On the level of the simple survey the same kind of relation appears. During the early fifties, for example, an association was found between the willingness to risk an all-out war with Russia and personal outlook concerning the future: the less favorable that outlook, the greater the willingness to court disaster for oneself and the world (U.S., college; Farber, 1953). Two investigators suggest, moreover, that some of the variability in opinion associated with census variables such as education and age may be mediated by what they call "the apocalyptic-serial dimension" which is a way of dichotomizing the content of future temporal orientation and has implications for the attitude toward intervention. An individual, they say, may believe that events are never likely to be repeated in the future (the apocalypse) or that they will be repeated (the series). They assume that the apocalyptic view stems from a very short or a very long temporal orientation which stresses the passing value or the futility of most activity: little is to be gained by intervening. Then a simple "Futility Index" based upon four paper-and-pencil items (e.g., "Since life is so short, we might as well eat, drink, and be merry and not worry much about what happens in the future") yielded scores which were related to the desire to find immediate and extreme solutions to problems, such as agreeing in the year 1948 that the United States must "fight Russia before they get any stronger" (Back & Gergen, 1963).

Clinical and semiexperimental studies also indicate the important effect upon behavior of the feeling toward future intervention. A simple paper-and-pencil test was once utilized to divide subjects into two groups: those believing they could control their environment and those holding the contrary viewpoint. More of the former than the latter were "alert" to environmental cues which could help them solve future problems, more were trying to "improve" their position in society, more showed concern for their own ability to cope with failure, and more were "resistive" to the influence of others (U.S., college; Rotter, 1966). On an even more basic level, a survey of the literature suggests that in general persons who believed they could influence events by intervention tended—under conditions of total reinforcement (i.e., being rewarded always for a correct response)—to acquire habits that resisted extinction to a greater degree, to have lower thresholds for perceiving stimuli that would bring pain, to run fewer risks in gambling, and to be more conformist than persons who believed they were relatively powerless to affect their destinies. May we have confidence in these findings based in large part upon college populations because analogous results have been extracted from rats? I just do not know. In real life, too, the same writer points out, differences in action and values appear as a result of differences in anticipation: in the early sixties in America, whites were found to be more confident that they could affect the external world than Negroes, and middle-class persons similarly more so than those from the lower class—and certainly these groups behaved differently (Lefcourt, 1966). Another investigation portrayed the reverse sequence: intervention strengthened anticipation concerning the future and hence a future orientation. The experimenters asked some of their subjects to memorize definitions in order to prepare themselves for a so-called intelligence test which they might or might not be called upon to take; others were given hocus-pocus instructions of such a nature that they expended very little preliminary effort while waiting for the great decision as to whether they would be similarly tested or not. The former who had been given the opportunity to intervene by being so active tended to believe more firmly that they would be given that test than did the latter (U.S.; Yaryan & Festinger, 1961).

ADDENDUM 3.5 PITFALLS CONCERNING MODAL VALUES

It is really precarious to make inferences concerning the modal values or perspective of a society without a knowledge of the interpretation the inhabitants themselves place upon their own beliefs. This caution seems especially necessary in connection with one of the most widely held beliefs throughout mankind, the view that the society once had a Golden Age which may or will never return: paradise is probably lost forever. At first glance the prevalence of the belief suggests an orientation toward the past: men look backward, glorify what once existed, and hence are less concerned about the present and may deprecate the future. But praise of the past may also spur people on to restore the glory that is gone, and so they look backwards in order to push forward. Then, as is said to be the case in many African societies (Mbiti, 1968), individuals may feel that they can or will return to the Golden Age after death, so that even the future is oriented toward the past. They will be absorbed into the kind of life their ancestors once led and even now as spirits continue to lead. Another mode is to accept the Golden Age and to believe it will or can be restored not after death but during life: the millennium will come, perhaps must come, it is hoped, and thus emphasis upon the past elicits a future orientation.

It seems so easy to say that traditional peoples, with limited knowledge and few records at their disposal, are able to peer backwards at a shorter distance than modern men. But the skeptical whisper in the text must be raised to the level of a shout: how profound is the knowledge which people in general, traditional or modern, have of their past (or for that matter of their present and future) if they are not professional historians? Likewise it is so easy to assert that traditional peoples have no other guide, no system of science that can extrapolate a future differing, somewhat or markedly, from the past. Yes, life looks stable—people do in fact grow into the roles they anticipate, so that one can have faith in the anticipation of repetition; but has that confidence been measured? Inferences about our own society demand similar reserve. A direct survey of American college students once revealed, not unexpectedly, that about all of this mobile, ambitious group, while considering the present 1.2 times more important than the future and 12.7 times more important than the past, claimed to be hopeful about the future and to "shape" their plans in advance (Israeli, 1932); but can we take their replies to questions at their face value? It is interesting, though not as "fascinating" as another investigator claims, that a group of students ranked Sunday below Saturday in order of preference for the days of the week because, although both days afford leisure, Sunday carried with it thoughts of the next day's responsibilities (Farber, 1953); just what does this isolated datum indicate about the perspective or values of these students or about Americans in general? In another study American children and adults listed the items about which they had allegedly spoken or thought during the preceding two weeks: references to the future far outweighed those to the past (Eson, 1951). Have we here simply a reflection of the American culture which either compels us to emphasize the future, or distorts our recollection of conversations and thoughts in that direction; or could it be that references to the past, though less frequent, exert more of an influence on actual behavior than those pointing forward?

We must also hesitate when we contemplate the possible causal connection between perspective and values. For historical reasons, we would guess, both a philosophy of values and a temporal perspective develop side by side and continually interact. At any given moment the observer may try to say that for his generation one variable rather than the other is more important or determines the other, but such a contention is difficult to establish or validate. A "concern with the future . . . is most characteristic of present-day western European and American concepts" and hence "much of our life is concerned with getting rewards for achieving goals and with setting up new goals around which to organize the behavior and expectation"; whereas Hindu philosophy seeks to develop an ego which "extends infinitely into both past and future." The same writer maintains that Western perspective affects behavior markedly but that of the Hindu has "little direct influence" since man has been "granted eternity to work out his faith" according to this view (M. W. Smith, 1952).

4. TEMPORAL MOTIVES

Under what circumstances is the temporal motive likely to be evoked? The question is essential when we assume, as indeed we do, that temporal judgments involving subjective or objective time on any level occur not spontaneously but emerge from goal-directed behavior whether conscious or unconscious, whether reflecting the past, the present, or the future. After having thus recommitted our souls to the devil of determinism, we reap the benefit of the alliance and are in a position to provide an instant reply:

4.1 A prior interim temporal motive is likely to be aroused or temporal information to be sought whenever the duration of an interval has significance for the achievement of a goal.

The clearest commonsense illustration of this principle is the situation in which the individual is working "against time": here the motive for passing judgment is strong because only in that fashion can progress be assessed. Noting how many minutes, hours, days, or months have passed provides the necessary information; in the absence of a knowledge of objective time or for some other reason, such as a proclivity to reflect or introspect, subjective time may be appraised. Similarly, attention may be paid to time on a primary or secondary level whenever there is motivation to have an interval pass quickly. The person or persons who are ahead in a game bounded by objective temporal intervals (such as soccer or American football, but not tennis or bowling) seek to retain their advantage, when they have the initiative, by delaying their own actions and thus they "eat up" the clock to prevent their opponents from scoring. At the other extreme is the person who wants time to stand still because he knows he has a disease which will cause him to die in the near future; here the goal of not reaching the end of the interval is unattainable, hence repression may cause him to avoid or to try to avoid passing temporal judgment.

Patients, suffering from tuberculosis and confined to a hospital for an indefinite and varying interval, are continually motivated to pass temporal judgment: they use every possible clue to try to estimate when they will be released, inasmuch as they usually conceive of their treatment "in terms of putting in time rather than in terms of the changes that occur in [their] lungs." They query physicians and nurses, they observe the course of the disease in other patients, they note the treatment they are receiving, they try to believe that they have been assigned by the authorities to a classification that means they are close to recovery, etc. (Roth, 1963). Similarly, the same writer suggests, individuals in some occupations and many professions in the West are very time-conscious; they seek to learn how long an interval of time must pass before they can expect to be promoted and they measure their progress, and hence evaluate themselves appropriately, by the rate at which they do or do not pass through the various stages. Patients and men pursuing careers usually also make an effort to intervene in their own behalf to affect the duration of the pending interval. Those suffering from tuberculosis, for example, may remind physicians that it is time for them to receive surgery, a stepping stone to eventual discharge; and ambitious workers may curry favor with superiors from whom they might obtain more rapid advancement. Both groups believe in effect, therefore, that the "one way to structure uncertainty is to structure the time period through which uncertain events occur," particularly "by drawing, when possible, on the experience of others who have gone or are going through the same series of events" (pp. 93, 115).

We shall now consider in more general terms some of the circumstances under which the durations of interval have significance for goal achievement.

COORDINATION AND CONFORMITY

In some traditional societies men hunt alone, in others they are organized into bands. The decision to begin the chase in the first instance may depend upon perceiving a need for food or the presence of game, but it may also require a temporal judgment cued, for example, to the rising or the setting of the sun. For a band of hunters to cooperate, however, some synchronization is essential; thus a drum is sounded to summon them to a central place. Among the components of a "syndrome of modernity" established by interviewing 6,000 men in six different countries ranging from Argentina to Nigeria was the desire to have other persons "be on time and show an interest in carefully planning their affairs in advance" (Inkeles, 1969). In the West, "the proportion of people who wear a watch in any town increases as its population grows" (Fraisse, 1963, p. 389). The assertion is not documented, but it certainly sounds as if it ought to be true, either because the greater coordination of more people required by a larger community demands more precise temporal judgment or, quite simply, because the central timepiece, such as a village or church clock, is no longer visible or audible to most of the inhabitants. Do you wonder whether the latter change heralds and helps produce the alienation associated with urban living? In the United States, however, when adults were unexpectedly asked in natural settings to estimate the time of day, no difference was found between those in six urban and two small-town areas; their accuracy, moreover, was "remarkable," with the median error being "only around two minutes" (Lowin et al., 1971).

4.1.a *The duration of an interval is likely to have significance when goals are achievable only, or more easily, through coordinating the activities of more than one person.*

The word *coordination* is used broadly to include any form of activity, whether synchronized at a given moment or over time. Science requires the coordinated activity of many scientists; hence when Galileo and others introduced time as a variable to comprehend the regularity and lawfulness of physical phenomena, this factor became significant to note, and thereafter measuring precise instruments were devised which scientist and others now consult in order to achieve uniformity and coordination (Millikan, 1932, pp. 13–17). It would be gratifying if in turn we were able to indicate when coordination is required within a particular society. But without an Arbitrary Limitation we would thus be encompassing a universal of human behavior and be leading ourselves back to Adam or Aristotle. This would be as if we asked when people notice color, when they go swimming, when they are impolite, when they pray for forgiveness. These are sensible problems, though a trifle broad, and better left to sociologists (Moore, 1963) to analyze, since they have assumed the tasks of indicating the conditions under which coordinated activity comes into existence and of indicating the kinds of timing those activities require.

One point, then, seems clear: almost any activity in which people engage, other than a few bodily processes such as blood circulation or respiration under normal conditions, potentially demands some interpersonal cooperation. In theory each individual could eat separately whenever hunger pangs become intense, but instead people assemble at specific times and eat with one another (though the sexes and age groups may not necessarily mingle as in the West). Actual timing of meals in most societies is so regular that hunger or appetite may come to be cues not only for a biochemical clock activated by stomach movements or rumbles but also for a real clock or the time of day as culturally determined. Through simple or complicated conditioning men have also taught such cueing to those higher animals, the pets and beasts of burden and most of those providing food, with which they have close contact. In our very civilized world it is often necessary to arrange to detach oneself from other people in order to think or relax in private reverie. Even before the appearance of Friday, Robinson Crusoe followed the proper European's temporal scheduling which lingered on within his nervous system and philosophy. It was a criminal offense in *Erewhon* to have a watch, not because that society's inhabitants did not need an objective way of passing temporal judgment to achieve coordination of their many activities, but because they placed timepieces in the category of machines and they hated machines of all kinds which, they felt, could overthrow them (Butler, 1917, pp. 63–6, 163, 239).

All or most of the activities associated with an institution must be analyzed and identified when the coordination achieved or sought by timing is described or appraised. Timing refers to two different but related kinds of temporal judgment. First, there is the act of measuring an interval so that greater efficiency of effort can be achieved. The expert in industry determines how long it takes for a group of men to complete a given task. Or a psychoanalyst suggests that in running an institution it is necessary to discover empirically how long disturbed children can be permitted to engage in an activity, such as a competitive sport, before it must be terminated for their own welfare. Thereafter, equipped with this empirical experience, the leaders can develop a second kind of timing which involves controlling duration; they learn to anticipate trouble and pass appropriate temporal judgments: "all practitioners in recreation, groups leadership, and teaching develop some kind of hunch of just when a game has lasted so long that it has 'worn out' or when it has become so overstimulating that from now on a loss of control would be expected" (Redl and Wineman, 1957, pp. 358–63). Obviously cues must be at hand which may be only partially verbalizable before passing judgment: the players may seem to be getting too rough, they may be permitted to continue for a moment or two, and finally they are brought to a halt.

Timing, then, is an integral part of social behavior. American parents renounce in part the joys of spending here and now in order to save money to pay the college fees of their children years later; and of course this timing may bring them joy in the present as they feel proud of their sacrifice or anticipate the outcome in the future for themselves and their children. The purchasing departments of big manufacturing plants must see that raw materials arrive at the time they can be optimally utilized in production, not too early lest storage facilities be crowded, not too late lest the work of the plant be delayed. Intervening at the right psychological moment is an intuitive art not easily acquired or exercised by the most eager psychiatrists or the most ardent lovers. On an international level, leaders try to time their maneuvers to make the best or the most lethal possible impressions upon their friends and foes. A good sense of timing in a comedian or a politician means, I suppose, that the quip, the communication, the gesture is delivered and received at the very moment that produces the greatest impact or laughter. Many interdependent actions are facilitated by tacit agreement concerning their temporal succession: we usually applaud at the end of a performance, we reluctantly acknowledge that work takes precedence over play, we bravely assert that women and children come first in an emergency.

Coordinated activity, in addition, demands conformity: people seek the same or similar goals in the same or similar manner. The customs and rules of a society specify goals considered traditionally to be taboo, optional, or mandatory. You are not supposed to murder your neighbor except under rare circumstances such as self-defense; you may try to grow wealthy but you will be neither punished nor disgraced if you do not make the effort, or if you fail; and you must discharge certain obligations toward your family. The ways of achieving optional and mandatory goals in turn are also indicated; likewise some are taboo, others optional, and others mandatory. If you decide that you would become wealthy, you know, too, that you should not achieve your wealth by stealing, you have the choice of joining a profession or going into business, and you must give an accounting of your earnings to the tax collector. Children, therefore, are taught that there is both a time and a place for all things, and thus they come to appreciate both the temporal and spatial requirements of social life: they may or must engage in activities only under specified circumstances. You may work or whistle during certain hours and in certain places, but you must work some of the time.

Another reason for the close link between coordination and temporal behavior stems from the fact that objective time, as the last chapter emphasized, is a scarce commodity which cannot be accumulated and stored (Moore, 1963, pp. 4–8). The day always ends, and darkness comes (except, of course, for brief periods in the far north and south). Within a given interval of clock or calendar time, only a limited number of actions can be performed. The time "saved" through time-saving devices also disappears. The free time owed to you because you postponed your vacation is not retained but is expended by another person who was once a younger you and is now the you grown older in the interim.

To achieve both coordination and conformity there are formal and informal regulations specifying the purposes temporal judgments can or must serve, their frequency, their orientation, and the circumstances under which temporal motives are aroused. Temporal information, consequently, is transmitted and reinforced by each society or group. The temporal aspects of some events must be remembered (the exact dates, in Western terms, of historically noteworthy anniversaries), others anticipated (the approximate period for behavior associated with adulthood or marriage), and still others experienced (the duration of an important assembly); and the time to be allocated to an activity must be renounced (recreation during an emergency, such as a war). As ever, anthropologists can arrange societies along a continuum: at one end are those showing relatively little concern for the passing of time, at the other those more or less obsessed with the problem. Examples of cultural norms requiring such conformity abound wherever one looks, or wherever a cultural anthropologist looks who is inspired by the doctrine of cultural relativity (Hall, 1959, pp. 23–41). When will an interval be judged with reference to the criterion of punctuality? It all depends— on the culture. The polite latitude of delay permitted in keeping an appointment, when the status of the two people is roughly the same, is no more than 5 or 10 minutes under normal circumstances in the United States, but it may be up to an hour or even more in some Latin American countries and still be within the bounds of etiquette. Indeed, much of etiquette in general is concerned with defining not only what behavior is proper but also when such behavior shall and shall not be exhibited. The dependence of social arrangements upon temporal schedules and agreements is illustrated most vividly when a worker suddenly shifts from a day- to a nightshift: both his biochemical clock and his social life are disrupted, and he must wonder whether the extra pay he receives is worth the renunciations he must make.

4.1.b *The duration of an interval is likely to have significance when people are seeking the same or similar goals and must cooperate or coordinate their activities to achieve them.*

In honor of this corollary I give the last word—and a pungent one it is too—to an anthropologist who says, in connection with festivals:

> We talk of measuring time, as if time were a concrete thing waiting to be measured; but in fact we *create time* by creating intervals in social life. Until we have done this, there is no time to be measured [Leach, 1961, p. 135].

This dictum will not be tossed aside as a dramatic exaggeration if a central thesis of this book and its mode of analyzing social and behavioral science are convincing.

POWER

In every society people segregate themselves and others into distinctive groups which, more often than not, result in a designation of superior and inferior. Differences in status in turn have many consequences. Those in a superior group, for example, have higher prestige, which may mean that they have greater access to the goods and services prized within the society or, at a minimum, that they have the satisfaction which comes from the deference of the inferior. Here, denotatively, a partial definition of power emerges: one group dominates another in some respect and hence, in effect, regulates its behavior in material or symbolic ways. Many aspects of power so conceived stem formally or informally from custom; the person born in one caste, for example, learns to anticipate or renounce certain privileges. The one institution that is charged with the overall responsibility for regulating the power structure of a society is obviously government.

The power regulated by government is of many sorts. In a modern society there are rules and laws concerning the right to vote and to hold public office; the need to pay taxes and to drive below a specified speed; the definition of what is considered a criminal offense; the conditions under which people may marry and divorce; the rights and responsibilities associated with property; etc. Many of these governmental actions serve the functions of indicating how, where, and when people should allocate their time. *How* suggests the kinds of activities that are encouraged or discouraged; *where* the situations at which some activities are appropriate or inappropriate; and *when* the intervals which may be devoted to them. The person convicted of running afoul of the law is either sent to prison or fined. A prison represents the ultimate in the regulation of time, for inmates have very little choice as to how they may allocate their activities; thus, to give the simplest possible illustration, in many prisons the lights go out at night at a fixed time so that all activities requiring vision must cease and be allocated to a different period of the day. Except for the very wealthy, a fine means that the person must earn or replenish the sum of money and hence that he must reallocate or postpone in some respects the ways in which he spends time. The ultimate, the most drastic, decision of government with relation to time and all existence is the imposition of the death penalty. Compulsory military training, which to our sorrow persists in most modern countries, compels young men and their families, first of all, to make temporal judgments concerning plans which come to depend upon when that service will begin and when it will end. Obviously the obligation itself has temporal priority over all other activity, except when deferments as a result of certain preoccupations (such as being a student) are included in the regulations. In fact, the extent to which the prerogatives of the state enable it to regulate the temporal activities of its citizens is an index of its power.

The regulative aspect of government can be expressed in various sociological and psychological ways. When it is suggested, for example, that the individual is subject to both residual and preemptory claims upon his time (Moore, 1963, p. 86), the point is being made that persons in our society feel obliged to spend their spare time in some groups (the family) but are compelled to spend certain large slices of it in others (the work group). Ultimately government makes such a distinction, for example, in determining the length of the working day or week, and the attitudes and values of those concerned are affected by the degree of compulsion that is involved.

This discussion of government may sound a trifle myopic. It is not only government which regulates time but also a host of other institutions, especially those called economic. When men are imprisoned or drafted into an army, their spatial as well as their temporal behavior is severely curtailed; thus, time plays a role in collaboration with other factors and processes.

Power relations exist between individuals as well as between groups. Parents have power over their children: they tell them when they may get up in the morning, what they should do throughout the day, when they should go to bed. The hypnotist has power over his subject: the person in the trance remains passive or in a sleeplike state until the hypnotist gives him another command. Love is a powerful drive, by which is meant, I am told, not only that the divine emotion is strong and hence tends to dominate behavior, but also that persons in love, sweetly or unsweetly, are able to control the ways in which they allocate their time, with the supreme instance of conceivable reciprocity appearing in the intimacy of love-making. Non-love, one index to which is criminal behavior, may also have a temporal component arising from the fact that it is more likely to be generated or expressed during some intervals than during others. Homicides, for example, have been reported to occur most frequently on weekends and holidays and, in one statistical analysis of sane and insane homicides in New Jersey, "murder at breakfast-time is nearly always the act of the insane or by a man so weighed down by circumstances that he is seriously ill" (Gibbens, 1958). So we conclude that:

4.1.c *The duration of an interval is likely to have significance when it involves the power relation between groups or individuals.*

TEMPORAL FACTORS IN COMMUNICATION

Any kind of communication involves temporal factors which are important in determining its effectiveness. A broad analysis has shown that these factors play a significant role when the communicator designs the content of the message (especially when that design is deliberate rather than unintentional), when the audience perceives the message, and when information concerning that audience's reaction is fed back to the communicator; in short, they may crop up in connection with the entire communication process (Doob, 1961, pp. 214–22, 229–38, 340–7). The length of a communication, its duration, may be related to the extent to which the communicator's original or unwitting objective can be transmitted; it may indicate to the audience the importance of the communication while simultaneously inducing interest or boredom. A variant of length is repetition: the number of times a theme is repeated within a single communication, or in a series of communications, may convey to the audience its importance in the judgment of the communicator and may also facilitate or reinforce learning. Likewise the position of a particular message within an oral communication—whether it is the first or last of a series or in the middle—may influence both the impression it creates and the ease with which it is learned and remembered.

When a communication reaches people, their ongoing mood is likely to have a marked effect upon receptivity. And that mood is certainly dependent, or may be dependent upon the precise time at which the communication arrives. Early in the morning vs. late at night, Monday morning vs. Sunday morning, the week before vs. the week after Christmas, the middle of winter vs. the middle of summer, the year when a person celebrates his 16th birthday vs. the year of his 60th—simple reference to these contrasting intervals suggests to anyone geared to the temporal schedule of the West how moods or prevailing predispositions fluctuate with intervals that can be designated by means of the conventional labels of objective time. The mood may be influenced by temporal factors within the communication itself. In a series of items, those coming first establish a frame of mind which can affect those coming later. Here arises the temporal problem of recency vs. primacy, for which there is no overall solution: either position may have an advantage, depending upon the kind of response evoked by the parts of the communication occupying the initial position and the interest of the audience in the entire communication. The mood of an audience may also harken back to their prevailing conception of the optimal duration of an interval devoted to a certain kind of communication; thus a speech of the same duration and content may bore an audience in one society and thrill one in another because of varying convictions concerning how much time such a communication may properly consume.

In modern society the mass media and other groups compete with one another for the attention of large audiences; hence the amount of time people devote to a particular medium is an index of the importance (whether for information or entertainment) they attach to it. There has been a slight tendency for the less well-educated Americans to be more dependent on aural and not on visual media as a source of news than the better educated (Hovland and Janis, 1959, p. 12); they, therefore, renounce less time or effort to learn the news or the simplified essence of a particular account, or at least seem satisfied with less. Comic strips are the medium per excellence which enable an audience to grasp the communicator's message with little delay and usually with less enlightenment.

The effectiveness of a communication, finally, may be markedly affected by the information which the communicator has at his disposal concerning audience reaction. Such feedback may have various temporal relations to the communication. It may be simultaneous or almost simultaneous: the speaker observes what effect his words have upon an audience and alters his theme appropriately or places a different emphasis upon it. It may occur subsequently: the speaker on a radio or television program discovers that he is gaining or losing in popularity by reading monthly reports he receives from a market-research organization. Or there may be advanced feedback: a speech is pretested on a small sample before it is delivered, or the feedback from one speech is used to improve a later one. Finally, the receipt of feedback information itself involves a temporal factor: does the communicator receive information rarely, sporadically, or continuously?

One other aspect of communication reveals an interrelation between effectiveness and feedback: the so-called sleeper effect. Experimental and empirical studies with Americans suggest that a communication which at first seems to have been ineffective may be found later to have influenced them. The explanation may be that the source of a communication tends to be forgotten more rapidly than its content, especially when the source lacks prestige with the audience (Hovland, Janis, and Kelley, 1953, pp. 253–9).

Temporal factors, while affecting the success of a communication, also influence the temporal potential: they are likely to require temporal judgments and at least the arousal of a temporal motive. Undoubtedly communicators who are deliberately communicating in order to achieve certain quite conscious objectives—such as specialists in advertising or public relations in modern society—consider such factors in planning their communications, and hence pass appropriate temporal judgments concerning the length of the communication and the optimum moment to reach the relevant audience. Members of the audience in their turn decide whether the speaker is talking too long, when a particular program is available on television, etc. Both communicator and audience are thus enmeshed in time.

4.1.d *The duration of intervals involved in planning and receiving communication has significance, respectively, to the communicator and the audience.*

FRUSTRATION

So far consideration has been given to positive circumstances which lend significance to intervals. We now examine the negative side. Ordinarily, since a frustrating state of affairs is intolerable, the frustrated person is likely to engage in activity which will reduce his tension by seeking a substitute goal, by resorting to repression, or by being aggressive, or by all three or in any combination. From experience he gradually learns that this discomfort, too, will pass: he anticipates or hopes that in the near or far future he will be able to achieve the goal now eluding him. Under these circumstances the passing of time becomes significant: the individual seeks to rid himself of past and present misery and to achieve a happier future:

4.1.e *The duration of an interval is likely to have (or to have had) significance when it is anticipated (or recalled) that its continuation or termination will (or did) bring the relevant goal closer or will (or did) reduce the frustration.*

Does frustration, then, inevitably evoke a temporal motive? And is the converse of the proposition true, is the temporal motive a function only of frustration? I think the answer to both questions is negative: there can be frustration without a temporal motive, and a temporal motive without frustration. It is well to be reminded that "the word *duration* comes from the Latin durus, meaning hard," but this bit of etymology does not fully justify the contention that "duration only becomes a psychological reality when the present action does not bring immediate satisfaction" (Fraisse, 1963, pp. 199–200). While it may be generally true, as the same writer contends, that "a child at play, a lover carried away by passion, a writer at his work, all are unconscious of time for long moments," it is also possible for the individual during these moments of non-frustration to pass temporal judgment in order to remind himself of his bliss or to be certain that his joy does not overflow into another interval. Yes, we keep looking at our watches when the plane is late because we would make another connection, or not be late for an appointment; but we notice the time in the first place to see whether the flight is punctual, a desire resulting perhaps not from so-called idle curiosity but from those same needs and anticipations.

Persons undergoing frustration, then, need not be conscious of time. On the commonsense level, think of the child at play who is quarreling with his companions, the lover who finds his passion unrequited, the writer who runs out of ideas. These pitiful creatures do not inevitably pass temporal judgment; they try to solve their problems without necessarily noting the passing of time. Thorough absorption in an activity produces no judgment unless or until—as the principle states—the duration of the interval becomes relevant to the attainment of the goal producing such intense behavior.

A poignant illustration of the principle, according to a psychologist, is the therapeutic situation in which the psychiatrist and the patient have different attitudes toward the duration of that interval. The psychiatrist is time-conscious, he has other appointments to keep, he wants the patient to hurry along and emit significant information about himself. But the patient may be unconcerned with the passing of time, he is so preoccupied with his own problems, so great is his misery, that for the moment he has a timeless value system unconcerned with the duration of the session. "The result, on both sides, is a feeling of frustration, of not being understood" (Burton, 1960). Presumably the specific drive of the therapist regarding the duration of the interval is stronger than that of the patient.

Anyone steeped in the tradition of the West is likely to be so ethnocentric that it never occurs to him, I think, to challenge the statement that the temporal motive is likely to be evoked under virtually all circumstances. For any event obviously occurs in time, and therefore both its component stimuli and the symbols which come to represent it can induce a temporal reaction. Just as it is useless to try to conceive of an object which is not located somewhere in space or which cannot be classified with respect to color (including the limiting category of colorless), so some attribute of time must always be present when there is contact with the external world. No matter what you perceive, you have the potentiality of noting its age, you may remember the previous experiences you have or have not had in its presence, or you may anticipate the changes in the future it will undergo, or those you will undergo, in connection with it. Even abstract statements which would transcend the momentary may provide temporal cues: "A thing of beauty is a joy forever . . ." or "We hold these truths to be self-evident. . . ." But surely, the devil's advocate must say, perceiving time is not like perceiving color: duration is a dimension more like size which is perceived only under conditions of adequate motivation. I agree with the advocate, and therefore place so much emphasis upon the relation between behavior and the arousal of the temporal motive.

Unquestionably there are conditions in any society which facilitate or inhibit the arousal of the temporal motive. In developing countries, both during the struggle for independence from colonial domination and thereafter, stress has been placed upon "education as a panacea that would lead to higher standards of living and to the material benefits deriving from technological progress" (Cowan et al., 1965, p. 17). With such a future orientation, the inhabitants are stimulated to pass judgments about the passing of time. Perhaps fewer judgments are made when there is a belief in predestination: anticipation, planning, and hence temporal estimates seem less compelling. Or, to avoid a dismal possibility in the future, such as nuclear war, there may be a tendency to avoid thinking of it (Withey, 1964), unless one can intervene to try to avert the calamity.

There is one other link between the temporal motive and frustration which is embarrassingly intimate: the arousal of that motive often leads to a temporal judgment which itself is frustrating. "Time" is irreversible, as is so often and sadistically indicated. You will never be young again, yes, and you know that your body will age and that other persons and circumstances will change. If you wish to summarize all this with the concept of "time," well and good; but to some extent you are oversimplifying, and are personifying the concept. Many of the occasions, therefore, on which time must be noted are the ones which cannot be reversed, and to that extent frustration and temporal judgments of despair are involved. In addition, the limiting case of death faces us all; and hence, noting our age—after, let us say, the bloom begins to fade—is bound to be frustrating to some extent, from which circumstance generalization to many other aspects of time may well occur. Evidence that "time" as such is hated and feared appears among the broadest and often the gentlest of thinkers, philosophers. They are said to have a real horror—"une horreur particulière pour lui"—when confronted with the problem, with the result that "they have all sought to suppress time" (Pierre Janet, 1928, p. 496). When one philosopher, with the cool detachment and quiet equanimity of his craft, observes that "time is the arch-enemy of all living" and "the broader the consciousness, the more it perceives with despondent clarity the negative aspect of time," we can immediately recognize that he is using a hyperbole and that his comment about consciousness either has no meaning or else is an oblique way of referring to the rational tradition of the West (Haas, 1956, p. 40). Another philosopher is reminded of the story told about Nicholas Berdyaev who, after pleading passionately for the insignificance of time, would suddenly stop and look at his watch with genuine anxiety at the thought that he was two minutes late for taking his medicine (Whitrow, 1967b). Perhaps our philosophical friends would punish Chronos for the ancient crime of eating his children. Immediately, however, we must recognize that philosophers like Kant have diligently sought to come to terms with time. Bergson made duration the key concept in his philosophical system and has had an important effect not only upon philosophy but also upon psychiatry; in addition his analysis is said to have "brought about a new conception of character in much of modern fiction" (Mendilow, 1952, p. 149), to which—in my opinion—film could also be appended.

An alternative to irreversibility is a belief in cycles or some other form of recurrence. But the senses and experience too often belie the faith: the dead simply do not return. Under these circumstances "time" again becomes the archenemy, for the interval is too short to achieve a goal and, as even a laboratory experiment demonstrates (U.S.; Lichtenberg, 1956), action must be appropriately modified. You know that the journey will take a specified period of time because the timetable tells you so, but you may not undertake the journey either because you cannot spare that time or because you are not confident that the carrier will adhere to its schedule. I would end this unpleasant section by observing that sometimes we are ahead of schedule, and then pending time can seem ample to achieve our objective.

DEFERRED GRATIFICATION

One other circumstance or condition inevitably evokes the temporal motive:

4.1.f *The duration of an interval is likely to have significance whenever gratification is deferred.*

A considerable body of research exists which suggests that persons following non-Western traditions have a strong tendency to prefer immediate rather than delayed rewards (see *Addendum 4.1*). Such research must be interpreted most cautiously, although it is in line with a previous assertion, viz., that traditional peoples have a strong penchant for the present and the past and hence are inclined to have a temporal orientation or perspective inclining them away from the future. If we can tentatively accept such a tendency as substantiated, then we must also assume the following relation between temporal motives and the behavioral potential:

4.1.f.i *Gratification is likely to be deferred when the temporal orientation is toward the future, and vice versa.*

The troublesome "vice versa" in the principle merely suggests that probably the two phenomena are positively correlated. The correlation, moreover, is not perfect, inasmuch as no society can foster one temporal orientation to the complete exclusion of all others (Principle 3.1.a). Although one's ancestors may be worshiped with awesome and profound reverence, the cultivator who is faced with the unalterable fact that he must work for years before some trees begin to produce their crop—cocoa from 3 to 4 years, rubber and coconut 5 years or more—and even longer to reach full production, dare not dwell only in the past: in the present he has no alternative except to defer gratification and in the meantime patiently cultivate and wait until the crop can be consumed or sold. And the very same persons who exhibit such patience may also be unable or unwilling to anticipate gratification from present reununciations: they may seek aid from a dispensary, for example, to relieve a pain, but—because they cannot foresee the efficacy of the measure in the future or because momentary demands are too pressing—they may be less willing to be inoculated or to dig a latrine, either of which may well prevent the recurrence of disease in the future. Or, as was once the case in Ghana, they may be unwilling to destroy diseased cocoa trees to prevent the spread of the infection.

The empirical evidence seems to be overwhelming that people do not automatically defer gratification—renunciation is a painful process—and hence they must somehow be indoctrinated with the belief that deferment and renunciation are either necessary or desirable. Many, perhaps most of the ceremonies of a society, especially a traditional one, seem deliberately to glorify the past. People assemble either in the spirit of religion or recreation, and then are reminded of historical or religious events in song, dance, pageant, etc. In some instances the rites may in fact revivify, or at least pacify, ancestors by ostensibly giving them assurances that in effect traditions are not being violated. Under such circumstances there may be a fear of deviating from the past and, with it, a belief concerning the futility of anticipating the future or taking any risks to intervene.

Many of the reinforcing agencies within a group or society transmit values concerning deferment and renunciation most informally. An American inevitably learns "a penny saved is a penny gained" as a proverb or as a harmless banality. The crude plea to set money aside so that it can earn interest or be spent more wisely later, as a philosophical sociologist in Germany recognized long ago (Weber, 1930, esp. pp. 158, 172), thus gradually becomes part of an intellectual orientation toward time which emphasizes the far future rather than the present or the near future and which urges renunciation and postponement instead of immediate gratification. Parents in some societies or in some groups or classes within a society suggest to their children from the outset the importance of education; schooling, they say in effect, may be painful and boring, but it leads to long-run compensations in terms of world position or spiritual satisfaction. They also point out that it is all very well to give vent to impulses, especially sexual ones, while you are young, but this sort of thing may spoil your chances for a successful or satisfactory life or marriage.

Some of the attitudes involving present and future goals are so deeply embedded that they function more or less automatically or unconsciously.

How, for example, do individuals react to pain? The responses of patients in a New York City hospital were shown to be related to their ethnic backgrounds. Both Italians and Jews tended to be violent and emotional, whereas the Irish were more quiet and stolid. Although on the surface the first two groups seemed similar, in fact their goals were quite different. The Italians were oriented toward the present and sought immediate sympathy and consolation; in their agony, they cried out for pain-killing drugs. In contrast, the Jews were worried about the future course of the illness and its permanent consequences for them; they tended to eschew the immediate gratification from drugs for fear they would become addicted to them (Zborowski, 1952).

The renunciator may view the renunciation as a form of gambling based upon a hedonistic calculation, and investigators may analyze it in terms of game theory: a small but certain gain in the present as against a large but uncertain one in the future. For the latter alternative there must also be immediate pay-offs which bring minimal though sufficient gratification in order to encourage risks and to alleviate the discomforts of sacrifice. While gaining the penny that is saved, the provident person obtains interest which he may spend; or else, with another pinch of renunciation, he may add it to his capital and thus increase his future gain. School children in the South Tyrol are given a holiday in the spring to tramp into the mountains and there plant seedlings which decades later replenish the forests as a source of lumber. They thus observe and are reminded by their teachers that they are contributing to future gains which in actual fact they themselves may not enjoy—they may die before the trees reach maturity—and they derive immediate gratification from the holiday-picnic and from their elders' praise.

I think this simple illustration suggests that in general renunciation may often turn out to be a special instance of deferred gratification which demands in addition some immediate gratification. At the very least, renunciation removes whatever pangs would result from non-renunciation. A mother sacrifices her own ambitions or comfort for her children, without anticipating any immediate or future rewards (is that really so, is she that selfless, does she not think that they will be grateful to her now or later and that such gratitude will be rewarding to her?); but not to sacrifice might bring her both pain and guilt. An external reward for sacrifice is more likely. The children thank the mother in words or demeanor; but she may continue to make sacrifices, even when they are ungrateful. Or the bank provides a passbook in which the principal and interest are recorded; thus there is a concrete token of the amount of postponement, a recognition which can bring immediate fiscal joy and serve to suggest the security and independence to be anticipated in the future. Banks in modern America would encourage the postponement through momentary seduction; they offer an immediate inducement which itself also promises future gratification, such as a clock or painting, to those opening new accounts.

The principal point of this discussion may now be summarized:

> 4.1.f.ii *Gratification is likely to be deferred when the frustration from renunciation seems either necessary or less than what may be attained in the future and when there is appropriate reinforcement in the present.*

A tendency toward deferred gratification is probably inevitable in all persons since socialization always demands some renunciation. Conceivably living organisms could give expression to every impulse and not favor one at the expense of another, but this condition is not likely to exist for self-evident reasons. Many impulses are usually aroused more or less simultaneously, not all of which can lead to action without interference or conflict. Some sort of compromise results, or one or more impulses must be repressed or abandoned for the moment, for some period of time, or even forever. You cannot gather all the rosebuds in the garden if you would avoid fatigue and get about your own affairs; some you simply must renounce. There are, moreover, uncontrollable events in any society, ranging from human tempers to staggering lightning—and so the universality of some resignation, some sacrifice: "God's will be done." An extreme illustration: the wife in one traditional society must have observed "strict sexual abstinence" if the battle wounds of her husband were to heal (Read, 1966, pp. 71–2). Gradually, therefore, people come to appreciate the need for some kind of renunciation, otherwise there will be chaos. You slow down your car, and thus renounce reaching your destination as quickly as you might, because you know the car to your right has precedence—and you anticipate not only that he will not slow down but also that he will anticipate conventional conformity by you because, obviously, you both are acquainted with the same regulations and customs.

There is one important alternative which damages the analysis so far advanced: the act of renunciation may itself be gratifying. On the basis of his rough-and-tumble experience, a psychoanalyst suggests that "the compulsive man, desperately desiring to create a happy future but also in deadly fear of it, always postpones, and the postponements add to the conflict"; in addition "man has a special delight in anticipation," inasmuch as "it is a typical masochistic reaction in which one pays the price for desire in advance" (Meerloo, 1954, pp. 88–9). So-called deferred gratification, whether neurotic or normal, may thus become immediately gratifying as the temporal motive is evoked and the temporal judgment reveals that the future is far off or close-by.

ACHIEVEMENT

The arousal of the temporal motive has now been linked to the duration of the interval. One other general circumstance can be specified when that motive is likely to be evoked, in this case whether it be prior to, during, or after the interval or event. Without the arousal of the motive, I continue to assume, no judgment will be passed.

4.2 *A temporal motive is likely to be aroused, or temporal information to be sought, when a temporal judgment or such information has instrumental value in attaining a goal.*

The concept requiring explanation in the principle is "instrumental value": any activity, covert or overt, which has aided, is aiding, will aid in the attainment of a goal. You must have some reason for wondering how old I am. You agree, for some good or bad motive, to serve as a subject in an experiment; then you are asked questions concerning the duration of an interval; your reply serves the function of indicating your cooperativeness or of demonstrating your genius at passing judgment. You do not know how long it took you to cook that meal and you may not find out unless you believe the information will be useful in helping you to plan the time to be allocated in the future to a similar meal.

We can immediately note, as it were, a negative relation between the passing of temporal judgment and relevant temporal information:

4.2.a *After the arousal of a temporal motive, judgment is not likely to be passed when relevant temporal information is available.*

If the individual wishes to know the time of day and a watch readily supplies the information, he will not make an estimate of subjective time—unless, of course, he wishes to compare judgments on the primary and secondary level or unless he lacks confidence in the accuracy of the timepiece. Similarly, as we shall emphasize later, previous temporal judgments concerning the duration of an interval are likely to be stored and then utilized again in the future. You remember how long the holiday lasted not by reliving the experience but by recalling what you knew then to be its duration.

A challenging question to raise, I believe, concerns the effect of the achievement or non-achievement of a goal in the past upon the subsequent arousal of the temporal motive. The answer to this provocative phrasing can be in either direction. On the one hand, non-achievement can lead to mild or severe repression, with the result that the interval may be relatively quickly forgotten and hence not judged temporally; in contrast, achievement is satisfying, and so becomes an experience to be remembered. On the other hand, with equal psychological plausibility, it can be said that achievement brings closure, hence no impulse to judge or evaluate; whereas non-achievement may be challenging and hence may induce the individual to try to profit from the experience. Are you more likely to determine how long it took you to go from your home to the station if you miss the train than if you catch it? My first impulse is to avoid the madness that might result from trying to determine what we can or cannot remember by invoking another Arbitrary Limitation. On reflection, however, I think one important point can be rescued by applying the principle (no. 4.1), with which this chapter has begun, to the past and the future: a useful function is likely to be served by passing subsequent temporal judgment when the duration of the interval appears to have been relevant for the achievement or non-achievement of the goal that was being sought and likewise for a similar one that will be sought.

The answer to the question about catching or missing the train thus miraculously emerges: neither event per se will lead the person to judge the interval unless its duration seems to have determined the outcome and unless he wants to catch that train again. The realization of an anticipation is likely to be gratifying even when the event itself is not as satisfying as had been believed or hoped, indeed even when it turns out to be about as disagreeable as had been forecast: at a minimum the person can be pleased by his powers of prognostication, he has been able to demonstrate his partial mastery of the universe by foreseeing an event before its occurrence. Under these circumstances the tendency to anticipate must be strengthened, a reward is present. If anticipation is reinforced, the inclination toward future temporal orientation may generalize to many forms of activities: you plan your existence in more detail, you make subsequent judgments about duration in order to behave even more adequately in the future—having missed the train, you allow more time to reach the station on another occasion. It is conceivable, however, that correctly anticipating the unpleasant may weaken the tendency to anticipate in the future; why think about trouble if it is going to come one's way anyhow? One answer can be: anticipating gloom is useful, the individual may intervene and thus prevent the occurrence of what he foresees.

Incorrect anticipations in the past, moreover, may eventually lead to adjustments ranging, once again, from not anticipating at all to anticipating, and behaving, differently. Immediately, when an anticipation is not realized, the person feels uncomfortable or frustrated, and—as the various sects of psychologists who investigate, and quarrel about, human dissonance have astutely indicated (Cardozo & Bramel, 1968)—the uncomfortable state of affairs must be removed. Either to himself or to others he respects, he may deny that he ever believed the outcome would be as it turned out to be, or he can try to interpret it as a vindication of his expectations. You never really thought it would happen that way, you claim you just said so without believing it. Or you say it only looks as though you are wrong; wait and see, you maintain, after a while I shall turn out to be right. It may also appear that nonanticipation in the form of a surprise may be gratifying. More careful observation, though, reveals perhaps that the source of gratification is not the unanticipated as such but the content of the surprise itself. Nobody, presumably, likes an unpleasant surprise. Whether or not the anticipation and then the realization of a gratification is more or less than the non-anticipation and the surprised realization thereof is another problem in hedonistic calculation better left to philosophers. I imagine that, as ever, both cultural and personality factors are involved, for there must be cultures or at least individuals for whom even the most gratifying surprise is simultaneously disturbing because of its very unexpectedness.

The anticipation involved in renunciation and deferred gratification, moreover, often cannot be easily validated. You renounce one career in favor of another because you anticipate that you will not be successful or happy if you were to pursue it; but then you will never have any certain way of comparing what you do achieve with what might have occurred if you had selected the alternative. Ordinarily, therefore, we are doomed to survive in the absence of adequate experience. Other anticipations are unprovable in a metaphysical sense, such as those rejecting the joys of this world in favor of an afterlife.

Another factor determining whether an interval in the past is relevant for one in the future is the intervention that has or has not taken place. For intervention demands some sort of anticipation and temporal judgment. In contrast, passivity makes the past largely irrelevant; the experience gained therefrom is likely to be vague. This is sheer speculation; I know of no evidence for the following which I therefore offer most brashly and without italics: the likelihood that an interval is considered instrumental in achieving or not achieving a goal is positively related to the degree of intervention during that interval.

A fitting ending to this chapter and the preceding one, concerned as they have been largely with anthropological and sociological matters, is to say again that, while temporal motives may not be evoked as quickly or as automatically as those pertaining to size or color during simple acts of perception, Western society places so much emphasis upon temporal judgment that for us, I think, it is difficult, perhaps often impossible, to remember or anticipate an event without some reference to a temporal dimension. Ordinarily, therefore, you are not the least bit embarrassed when you ask yourself, or someone else asks you, when an event occurred or will occur. Our memories and anticipations have actual or potential date and time tags attached to them, not because of our nature or the nature of time itself, but because of our particular culture.

ADDENDUM 4.1 DEFERMENT AND NON-WESTERN PEOPLES

Research on this subject has been conducted with a number of methods. There has been straightforward anthropological observation. Navajo Indians, for example, are said to prefer a broken-down horse immediately rather than a prize animal months later: "only the immediate gift has reality; a promise of future benefits is not even worth thinking about" (Hall, 1959, p. 33). As indicated previously, samples of Africans tended to choose the lesser of two hypothetical sums of money when the greater was obtainable only in the future; but many of the more acculturated preferred, they said, to wait (Doob, 1960, pp. 88–9). Similarly children have been offered a choice of a real or meaningful reward (such as candy or money) immediately or after a delay, with the quantity being greater in the latter case. It has been found that the following percentages of samples of children in the indicated countries selected the immediate rather than the delayed reward: 44 percent, Arab refugees in Palestine (Melikian, 1959); 53 percent, Negroes in Trinidad (Mischel, 1961b); 63 percent, another sample of Negroes in Trinidad (Mischel, 1958); 25 percent, Negroes in Granada (Mischel, 1961b); 33 percent, Indians in Trinidad (Mischel, 1958); and 39 percent, Australian aborigines (Bochner and David, 1968). For yet another sample of Negro children in Trinidad, 27 percent consistently chose the immediate reward and also responded to two questions indicating they were intolerant of delay; that tendency was weakly associated with a milder need for achievement as inferred from TAT drawings and with an impulse to comply as measured by a primitive, semiprojective question (Mischel, 1961a). In all instances save the Australian sample, the tendency to prefer the delayed reward increased with age or schooling. Great care, obviously, must be exercised in interpreting these results. As previously indicated in connection with my own African research, a risky inference may slip in: some of the informants choosing the lesser of two preferred or hypothetical rewards intended, not to obtain gratification immediately, but to use the sum to invest in a small business and hence to obtain greater rewards in the future. The results are expressed in modal tendencies from which there were individual, and sometimes group, deviations. In the Australian study those choosing the delayed reward may have done so because they knew the delay would be slight or because they trusted the white person who was the investigator; in fact, the choice of the reward by young American boys in a similar situation depended upon whether the promise of the greater reward was made by one adult with whom they had had previous experience in receiving or not receiving delayed rewards or by another adult who was a complete stranger (Mahrer, 1956). Finally, the usual question must be raised concerning the sample of behavior chosen for investigation: we really have no way of knowing whether the way Africans reply to questions, or children respond to bits of candy or money, are typical of the ways they would behave in real-life situations.

5. PERCEPTION AND SCOPE

The subtle, sensitive, complicated process of passing temporal judgment demands the detailed analysis on which we now embark. The basic factors determining whether or not a judgment motive will be aroused have been examined, but the behavioral and temporal potentials, as they have been defined and appraised, serve only as the essential introduction to the judgmental process. Step by step, therefore, we shall swing ourselves around the outside circle of our guiding taxonomy. We need only one assumption to begin: for some good reason the temporal motive has been aroused, the individual would pass judgment concerning the duration of an interval or its temporal relation to other intervals. This chapter and the next explore the process of perceiving the interval and the primary judgment potential, Chapter 7 the secondary judgment potential.

The perception of an interval whose duration has been, is being, or will be judged, whether subjectively or objectively, whether on a primary or secondary level, depends, as does all perception, upon an interaction between the predisposition of the person passing judgment and the nature of the stimuli in the external environment. The stimulus potential, as indicated in Chapter 2, may be conceptualized in terms of the actual duration and the content or composition of the interval itself, the channel through which information about that interval is transmitted, and whatever temporal information is symbolically present. The individual is likely to be affected by all three: even if a timepiece indicates the exact duration of an interval, he may be convinced that it "seems" longer when his ears rather than his eyes are the receptors or when considerable information is conveyed.

PSYCHOPHYSICS

One branch of psychology, psychophysics, has attempted to free itself from the nasty array of predispositional and stimulus factors by excluding the former and by rigorously controlling the latter. The individual's own idiosyncratic predispositions are largely eliminated, it is hoped, by conducting the investigation in the laboratory. There the experimenter's instruction makes only one predispositional tendency dominant: the subject is told to pay attention to the duration of an interval or a set of intervals and nothing else—often he has been previously trained to do just that—and hence to suppress or eliminate as far as possible any other impulses or distractions. Watches are removed and no temporal information is provided; the duration and mode of stimulation are regulated and measured in units designated by the physicist; and usually there is interest only in primary judgments which are to be expressed quickly and without reflection. The effort is thus made to control what is perceived, hence almost exclusively to relate changes in the stimulus to changes in the primary judgment, and ultimately to state that relationship in mathematical form.

Clearly psychophysics adheres to the tradition of the controlled experiment: all other things are declared equal and the attempt is made to have that perfect state come to pass. The tremendous body of research has been competently summarized again and again (e.g., Woodrow, 1951; Fraisse, 1963, pp. 99–132; Schaltenbrand, 1967). Here I would quickly raise the principal questions of primary interest to psychophysicists and suggest some of the tentative answers that have been supplied principally in the areas of vision and audition:

1. How long must an interval be in order to be perceived however briefly and then have its presence reported verbally? This is the problem of threshold and is often considered under the rubric of sensation. *Answer:* perhaps slightly over 0.01 seconds for sound and between 0.07 and 0.10 seconds for light.

2. What is the maximum duration an interval can attain and yet remain within the psychological present by being perceived and reported as a unified or single event? This is the problem of duration or unity. *Answer:* from 2 to 12 seconds for simple stimuli.

3. What must be the minimum duration of the interval separating two stimuli before they can be perceived as separate entities or events? This is the problem of succession or rhythm. *Answer:* possibly, if only the perception of interruption between the two stimuli, and not their order, is considered, 0.001 seconds for noise, 0.002 for tones, and 0.16 for visual flicker. But when it is necessary to distinguish which stimulus preceded the other and not simply whether there is a single stimulus or two stimuli, the two must be separated by 0.02 seconds for the judgments to be correct 75 percent of the time; it matters not whether the two activate different modalities or, if only one modality is involved, whether the same or a different sense organ (e.g., the right eye only or the right and then the left one) is

stimulated—at least that was the outcome in one series of clean-cut experiments (U.S., young adults; prior; Hirsh & Sherrick, 1961).

4. What is the optimal duration of an ephemeral interval if its duration is to be judged accurately? This is the problem of the "indifference point" or "zone," defined as the interval which is neither overestimated or underestimated most of the time. *Answer:* traditionally around 0.75 seconds, or within a range of from 0.50 to 1 second, though reference has also been made to an upper limit of 5 seconds (Orme, 1962b).

5. How great must the difference be between two intervals before they can be accurately perceived as different? This may be viewed as the problem of Weber's so-called Law that "two sensations are just noticeably different as long as a given constant ratio obtains between the intensities of their stimuli" (Boring, 1942, p. 35); in the present context, substitute durations for intensities. *Answer:* perhaps around 10 percent (which would mean, for example, that the second interval must be 10 percent greater or less than the one with which it is being compared if it is to be distinguished, regardless of the absolute magnitude of the two intervals), but some variation from that modal ratio always appears which depends upon the duration of the intervals, the mode of measurement, the number of trials, etc. (Henry, 1948; Treisman, 1963). In passing, I would like to raise a question of my own to which, I think, the literature of psychophysics to date has no reply: does the tendency to overestimate or underestimate intervals also adhere to some kind of mathematical formula?

6. What is the relation between objective and subjective time? *Answer:* the relation is clearly not one-to-one which means, as just suggested in connection with Weber's Law, that a change in the duration of an interval is not necessarily perceived or that an increase of a certain magnitude in that duration is not judged with complete accuracy. The goal here is to state the relation in mathematical form; but so far no single, simple formula has been evolved. The investigation of the problem is made difficult by the fact that results are affected so markedly by the method of measurement. One investigator, after surveying the possibilities, concluded that "the method of reproduction most directly attacks the problem of time perception." Then over a period of 4 months he had two adults and two children reproduce intervals whose total range was from 2 seconds to 80 minutes for the former and from 2 seconds to 5 minutes for the latter. Even with such a small sample—which yielded, however, thousands of judgments—he is forced to the following summary: "the time estimates can not be described by a single monotonic function of the actual time," but "for each individual, a family of power functions having the same constant exponent can adequately fit the data" (U.S.; prior; Richards, 1964). It appears, there-

fore, that only on an intra-individual basis can the relation between the two kinds of time be expressed mathematically through the use of a power function (Stevens, 1962).

The figures so glibly given in some of the answers to the above questions are only very, very general guides to the truth, inasmuch as all of them can be drastically altered by changes in the stimulus potential or in the predisposition of the person passing judgment. There may be one indifference point or zone, but schizophrenic patients when confronted with three intervals (5, 10, and 15 seconds) tended to overestimate (but only slightly) the shortest interval and to underestimate the longest one with the Methods of Reproduction and Production, and they very significantly overestimated all three with Verbal Estimation (U.S.; Clausen, 1950); in the very careful psychophysical investigation already mentioned above, it was found that "there is more than one time interval which can be reproduced most exactly by any individual" (Richards, 1964). The proponent of the power function used to describe the psychophysical relation has himself indicated that the precise formula varies with the modality; and we know, too, from him and others that it fluctuates as well—probably not randomly, but lawfully— with the method of passing judgment (U.S., presumably students; verbal, prior; Stevens & Galanter, 1957) (U.S., college; repro. & verbal with prior; Chatterjea, 1964) and even with the duration of the clusters of intervals being tested (Netherlands, presumably adults; verbal, prior; Michon, 1967). There is classical or solid experimental evidence demonstrating that in at least one study one or more of the following stimulus-potential factors has produced marked fluctuation in the half-dozen replies to the psychophysical questions: the mode of presenting the interval (e.g., the intensity of the stimulus being judged; the intensity of the stimuli bounding the interval; the relative intensity of the stimuli composing the two intervals being compared; the pitch of the auditory stimulation; the position of the stimulus or stimuli in space; the duration of the interval, if any, between two intervals being compared); the order of presenting the two intervals being compared (often referred to as a time-order error because, unless that order is randomly varied, a systematic error is likely to be introduced into the results); and the sense modality or modalities through which the interval or intervals are perceived. Variations in predispositional factors, ones that cannot always be excluded even when psychophysicists do their best to control and purify their subjects, likewise may have marked effects upon the outcome: the individual's attention span; his attitude toward the sources of stimulation; his fatigue during a given experimental session (some sessions require hundreds of judgments, one right on top of the other); and idiosyncratic differences among subjects or, as these differences were called during the early part of this century, the personal equation.

As a result of such multivariant factors, contradictions almost leap at you wherever you turn in the psychophysical or at least in the neophychophysical literature. In one experiment employing an unusually large sample, for example, the indifference point was found to be not the 0.75 seconds mentioned above but 2.48 seconds with "empty" intervals and 1.30 seconds with intervals which the subjects "filled" by counting (U.S., college; prior, absolute judgment; Ferrall, 1935). In another, from a different country and with different intervals and a different method, that point was somewhere between 30 seconds and 1 minute (U.S.S.R., children; repro., Elkin, 1928). Even the outstanding historian of psychophysics, a man who always sought to find generalizations as seemingly impeccable as his own prose tried to be, has written these discouraging words concerning the indifference interval:

> The discovery of a physiologically absolute duration was a reasonable aspiration in the days of the new psychology when the mental chronometry of the reaction times seemed also to be providing natural durations for appreciation, discrimination, cognition, and choice. Nevertheless, these men were wrong. The indifference point is variable and has no regular periodicity [Boring, 1942, pp. 580–1].

In virtually every psychophysical investigation based upon more than a handful of subjects, the fact of interpersonal variability looms large. With restraint I confine myself to a single illustration, the conclusion drawn in a thoroughly classical experiment which employed eight subjects judging 13 intervals with a range of from 0.2 to 30 seconds:

> Some subjects underestimated short intervals and overestimated long ones. Some did the reverse. Some varied from overestimation to underestimation irregularly with change in the length of the interval. Some subjects overestimated all the intervals. . . . The subject sometimes overestimated an interval one day and underestimated it on a different day. . . The hypothesis is suggested that the constant errors and individual differences therein are mainly the result of differences which the individuals take in listening to, and in reproducing, the intervals [U.S., college & graduate students; repro., prior; Woodrow, 1930].

It seems reasonable to assume that predispositional factors springing from the behavioral potential of these students produced the variations.

This brief excursion into psychophysics—and, incredible as it may seem, it has been brief for anyone interested in psychophysics per se, though the temporal judgment of the less-intrigued reader may be different—seems to suggest an enormous distance separating the laboratory from real-life situations. Consider an example of a reasoned rather than a psychophysical approach to temporal judgment. What time do you think it is? No timepiece is available, you have no external cues to serve the same purpose, and your intuition fails you. But you remember that when you left your home it was 2 A.M. You then performed an errand which, you think, must have taken about half an hour. Next you drove a distance, and—since traffic was heavy —that journey must have accounted for about three-quarters of an hour. You have been where you now are, you estimate, for at least an hour. And so you have a sum to calculate: $½ + ¾ + 1$ which is over 2 hours, and then you even out the sum and say it must be 4:30 P.M. This sensitive, most hypothetical reasoning of yours makes no reference to an indifference point, to thresholds, or to any of the niceties of psychophysics. Should we, therefore, claim—as many writers have—that psychophysical and reasoned judgments are qualitatively different? I do not think so, if only because, as we shall see later, the psychophysicists and classical experimenters who collected introspections from subjects passing judgment concerning even ephemeral intervals quickly discovered that these subjects were mentally active. Your reasoning about the time of day, moreover, included references to what you did; and what laboratory subjects do also affects their judgments. I feel, therefore, that psychophysics provides a basis for principles concerning temporal judgments but that even the purest experiment is likely to be plagued by the same annoying psychological factors involved in more complicated, more realistic situations. In slightly different words: predispositions and secondary judgments play some role in psychophysical experiments, but the weights attached thereto are likely to be much less than in most other situations.

SALIENCE

When contact has been established between the individual and the external world, exactly what in fact does he perceive? The content of his perception will affect either one or more of his judgments, if there is prior or interim arousal of the temporal motive; or his memory of the interval, if there is immediate subsequent or delayed subsequent arousal. I think we must agree that "it is not time itself we perceive, but what goes on in time" (Whitrow, 1965). Fine—but then, since we time intervals as diverse as the crying of a baby or the separation of a lightning flash and its clap of thunder, we find ourselves confronted with a staggering challenge to classify human experience.

We return to the safety of abstraction and there make use of three variables which we have had to unearth for this purpose because, except for a suggestion here and there (e.g., Belyaeva-Eksemplyarskaya, 1962), we cannot discover a previous attempt to designate ones relevant to the problem of temporal judgment:

> 5.1 *The perceptual potential is a function of the temporal knowledge, the events, and the drives perceived or associated with the interval.*

Temporal knowledge comes from whatever temporal symbols are provided by the stimulus and may or may not be affected by the store of general information in the temporal potential; events refer to the segments of the experiences identified by the individual as distinctive, organized wholes; and drives include all goal-directed tensions motivating behavior. These three variables may be variously weighted as a result of the stimulus potential or the total potential. Temporal knowledge? The normal person may be guided by the clock which he sees during the interval, but the paranoid may believe that it cannot be trusted because the enemy has tinkered with its mechanism. The events? The musicians produce the music, but your mood (part of your behavioral potential which, in accordance with my policy of Arbitrary Limitation, you and not I must account for) makes you perceive what you hear either as a melody which is an integrated, single event or, in terms of its components, as a series of related events. The drives? You are anxious before the interval begins, or an event during it elicits the anxiety. The same situation exists with respect to determining the onset and termination of the interval itself: the person may mark off his own boundaries or accept ones from a conventional source within the stimulus potential. You yourself may decide whether to consider the composition an event or a series of events, but you have virtually no alternative except to perceive that objectively a sonata has ended when the orchestra stops playing.

The next problem is to consider the circumstances under which each of these variables is likely to be salient. It seems reasonable to assume, in the first place, that always, or almost always, a temporal symbol when present as part of the stimulus potential and when considered trustworthy will have an overpowering effect upon perception. The person looks at a clock, and that's that:

> 5.1.a *Temporal knowledge, when communicated as part of the stimulus potential by a source considered reliable by the individual himself, is likely to be the most salient component of the perceptual potential.*

A person with valid and adequate temporal knowledge, we have previously emphasized, may also pass some sort of primary judgment. He may know the exact time of day or the current year, but still feel that time is passing quickly or slowly. If he is deranged, moreover, he may reject temporal information from his milieu. For reasons such as these our analysis must continue.

What factors determine whether the events or drives associated with the interval will be salient? Two assumptions burst out: strong drives, and drives in general, are likely to be more readily perceived than weak drives and events. For strong drives have greater significance for the individual than weak ones, and are indeed more insistent; and drives, being processes within the organism, are more absorbing, ordinarily, than external events.

> 5.1.b *When temporal knowledge is not salient, the stronger the drive, the more likely it is to be perceptually more salient than the events associated with the interval.*

The principle does not specify when a drive is weak or strong: this is another boundary which an analysis of time cannot transcend. I would only insist again that temporal patterning is affected by all that transpires within the individual. Other factors determining whether the drive or the event attribute of an interval is perceived, I think, can be isolated: the point in time of the temporal motive and the duration of the interval. The feelings and emotions associated with drives are likely to be frequently forgotten because, being largely unverbalizable, they are difficult to store. Thus hunger may be compelling at the moment that food is sought, but not when the activity is recalled: the drive has been satisfied and then, as happens with some drives, repression may set in. In contrast, the events leading to satisfaction have instrumental value if the action is to be repeated, and those leading to dissatisfaction have similar value if the action is to be avoided. Such an experience, as well as a deep conviction that events are more highly correlated with objective duration than drives, is probably reinstated less impulsively and a bit more deliberately after the interval than at any other time. Then the duration of the interval plays a role: the drive attribute is not likely to be adequately perceived during ephemeral intervals. My best hunch for such an interaction is embodied in the following principle:

> 5.1.c *When temporal knowledge is not salient, and when the drives associated with the interval are not relatively strong, events are likely to be more salient than drives under either one of the following conditions: (1) when the temporal motive is delayed subsequent (rather than prior, interim, or immediate subsequent) provided that the interval*

being judged is transitory or extended (rather than ephemeral); or (2) when the interval is ephemeral regardless of when the temporal motive is aroused.

EVENT POTENTIAL

Now we carry on by assuming the conditions just indicated: events alone are salient or more salient than drives affecting the perceptual potential, and temporal information is lacking. Virtually every interval is capable of being perceived as a series of events. The separate auditory vibrations which can be detected during the most ephemeral of intervals constitute events when they are perceived as such, an interval of that composition is different from one containing only silence and bounded by signaling stimuli. It seems reasonable to postulate that every sentient being inevitably has certain experiences that affect the way in which he perceives and evaluates the events within intervals and therefore influences the standards he employs for passing judgment, whether on the primary or the secondary level. These inductive generalizations, hypotheses, or principles I shall hereafter call *Guides* in order to emphasize their informality, and I assume they are likely to be conscious influences upon secondary rather than upon primary judgments.

One of these Guides stems from the experience of every person that he is able to pay attention at a given moment only to a limited number of objects, persons, or events. Occurrences outside his immediate span of attention take time to apprehend and hence are longer in duration. Eventually the guiding standard emerges, that the greater the number of events, the longer an interval is likely to be. There are exceptions, to be sure; a great deal can really happen in the twinkle of an eye. Usually, though, it takes longer to walk six steps than it does to go only one. The shout from the devil's advocate that a runner can traverse a longer distance faster than a man on crutches moves a shorter one, I shall impolitely ignore as a deviation from the mean. An illustrative experiment: American undergraduates were told they could view or manipulate various objects as long as they wished and that they "would not be asked to remember anything"; the objects differed in complexity, ranging from the simple (e.g., a plastic sponge) to the intricate (e.g., a partially disassembled clock); the more complex the object, the longer the time spent inspecting it (Gaschk, Kintz, & Thompson, 1968).

The outcome of the discussion ought to be that, *with the event potential salient, timing acceleration tends to vary positively with the number of discrete events associated with the interval;* hence *the greater the number of events, the stronger the tendency toward overestimation of an elapsed interval with Verbal Estimation or Reproduction, and the stronger the tendency toward underestimation of a pending interval with Verbal Estimation.* Before elevating the generalization to a principle, however, we must pause and reflect at some length.

For on any level we find evidence for and against the postulated tendency. If asked "to estimate the relative length of two historical periods," it is asserted with the challenge of commonsense, the person "is likely to judge that period to be longer about which he knows the most" (Sturt, 1923). Presumably the more we know about an historical period, the more crowded with events it seems to be, and hence the longer. On the other hand, as scores of writers in English have pointed out, the German word for boredom is *Langweile*—a long while, a long period of time. You are bored when nothing happens, time passes slowly or time seems long since so few events are taking place. We are confronted, therefore, with a contradiction that we must try to resolve.

"The issue of filled vs. unfilled time still remains unsettled," two writers stated after surveying the experimental literature up until 1960, "though the bulk of the evidence shows little significant difference in time estimation under the two conditions" (Wallace & Rabin, 1960). This and other evidence is examined in some detail in the *Addenda* attached to the present chapter; there the reader may examine, if he wishes, the large variety of situations which have been investigated. Here we attempt to supply a compact Baedeker for the jungle.

First, we find studies contrasting empty intervals and those containing one or more events that are so perceived—the variable here has been called *filling* in the old literature and *input* in the new. As *Addendum 5.1* demonstrates, there are many studies which support the generalization concerning a positive relation between length of primary judgment and the number of events. In some instances, however, no difference at all appears. Whether or not a filled or crowded interval that has elapsed is judged longer than an empty or less crowded one has been shown to depend upon factors other than the number of events taking place therein: the duration of the interval, the order of presenting the two intervals being compared, the method of judging the interval, the sense modality being stimulated—all these have played a critical role in one or more experiments. In addition, the behavioral potential must be invoked: some subjects have veered fairly consistently in one direction, others in the opposite direction.

The components of the interval as well as the context in which it is perceived also have been shown to affect temporal judgments (see *Addendum 5.2*). Studies have concentrated upon the number and organization of events, the intensity and regularity of stimulation, the organization of the components, the sense modality, and the bounding stimuli, in effect the quality as well as the quantity of the events. Again the results are inconsistent, and often contradictory, and cannot easily be subsumed under an abstract concept such as complexity. A special problem involves the relation between space or distance and temporal judgment: the latter tends to be longer when the former is experienced as greater, and thus more time seems to pass—it is said—on a journey which covers a longer distance than on one covering a shorter one in the same period of time, perhaps because the lengthier trip stimulates a greater number of sense impressions than the shorter one or because we invoke a Guide which suggests that ordinarily long distances take more time than shorter ones (see *Addendum 5.3*).

Finally, events may be defined in terms of output which means the actual activity or intervention of the person passing judgment. Perception is also an activity, but it involves only the sense organs and the responses to which they immediately give rise. The classical experiments began the tradition of attempting to ascertain whether a difference in judgment would arise when a piece of prose was read by the experimenter or by the subject during the interval. There is substantial evidence indicating that intervals of relative activity are judged longer than those of relative inactivity, but as ever exceptions also appear. Likewise there are trends—with exceptions—suggesting that temporal judgments are longer when the activities during the interval are less complex, and when the amount of expended energy increases. On the whole, inadequate evidence suggests, periods with input are likely to be judged longer than periods of output, and attention to the former rather than the latter may have a similar effect (see *Addendum 5.4*). The issue here is confounded by the fact that intervention usually involves a change not only in the number and quality of events but also in the drives that are evoked.

This survey of the experimental evidence presented in the various Addenda to this chapter indicates beyond doubt that our original generalization concerning a positive relation between length of temporal judgments and number of events, even under the specified conditions, is not completely valid. The first difficulty which confronts us when we seek to improve the generalization is the definition of the "discrete events" composing an interval. Most modern investigators, it has been pointed out (Fraisse, 1963, p. 128), have decided whether an interval is filled, crowded, or empty on the basis of a physical description which they themselves provide. Any principle, however, must assume that the person passing judgment reacts to his own conceptualization of the events. Both a French and a Dutch investigator have shown that increasing the number of events during an interval had no appreciable effect unless the additional stimulation was perceived by the subjects (university; repro. with prior, verbal & immediate subsequent; Fraisse, 1961) (young adults; prod., prior, Michon, 1965). From a psychological standpoint, therefore, an interval considered empty by the investigator may be filled with the subject's own thoughts, and hence "the difference between filled and unfilled intervals . . . is really a difference between different kinds of filling" (MacLeod & Roff, 1965). Waiting for an unpunctual friend makes an interval appear longer when one has nothing to do except to wait. For then such an interval is not empty but crowded with unpleasant internal events resulting from "thwarted conation": the person longs to start toward his destination, he wants to begin to talk with his friend, he notes that his own pride is hurt, he considers the words of reproach he may eventually use, he thinks up alternate plans, etc. This "very frenzy of whirling thoughts" makes the waiting seem long: "The apparent length of time . . . is due rather to the number of impressions we force upon ourselves than to the unpleasantness of the experience" (Sturt, 1923). But different people "force" different numbers of impressions upon themselves. I would think that *for me,* for example, an interval of intense sound or light would be more eventful than one filled with mild sound, for I can imagine forcing myself to think of many events, both pleasant and unpleasant, when intensely rather than when mildly stimulated. Am I typical or not? The evidence in *Addendum 5.2* provides no certain reply, leaves me filled with suspense.

There is no easy way out of this psychological situation, but it is helpful, I think, to lay stress upon the organization of discrete events or stimuli by the person passing judgment (cf. Fraisse, 1963, p. 224). Let us agree that either because of the nature of the music or of his own predilection—or for both reasons—the individual perceives the melody not in terms of separate notes but as a whole, as a single event. When he himself plays an active role in the organizing of discrete components into a whole, he is not only diminishing the number of physical events but is also intervening more vigorously, both of which may favor a decrease in temporal judgment. But tendencies may operate at cross purposes; thus listening to a slow metronome involves less activity and less organizing than the arithmetic task of performing long division, though the stimulation from the instrument might be said to transmit more objective events, at least when and if the beats are perceived discretely. In this fashion it is perhaps possible to resolve some of the contradictory findings of research: increasing the number of objective events may increase temporal judgments when the events are perceived as discrete entities but will decrease the judgments when the result is tighter organization or, as has been suggested (Friel & Lhamon, 1965), stronger cohesiveness. Another concept, that of "storage size," seeks a similar solution (see *Addendum 5.5*).

Activity during an interval, intervention, means almost always that some kind of goal is being sought: a drive is aroused. In contrast, drives play a less important role under most circumstances when only perception is occurring. Here is yet another reason why research on filled and empty intervals does not lead easily to a flawless generalization: the event potential may be salient, but the drive potential also plays a role. Classical experimenters certainly tried to be methodologically clear-cut when they sought to eliminate drive components by varying, for example, only the number of clicks within intervals, inasmuch as clicks are apparently such innocent stimuli. We now know, however, that they are not that innocent. For example: in an experiment mentioned in *Addendum 5.1* (Siegman, 1966) it was noted that higher verbal estimates for intervals between 2 and 120 seconds were obtained when they were bounded by clicks than when they were filled with the tone of a buzzer. Here we would anticipate that the event potential would not be completely salient since the temporal motive was not delayed and since some of the intervals were not ephemeral. In this instance we have some insight into the functioning of the drive component because the difference did not appear when the subjects were delinquents who, perhaps, responded more sluggishly to the sound of the buzzer than did the nondelinquent controls.

Whether the mode is to over- or underestimate intervals of a particular kind, there are always some subjects who do the opposite. Differences in drive or drive strength, rather than the event potential, I suspect, may account for this inter-subject variability in various studies, whether the subjects have been pretrained adults (France; comp., prior; Gavini, 1959) or schizophrenic patients (U.S.; repro., prod., & verbal with prior; Clausen, 1950). The fact that the monotonous task of copying the same proverb again and again produced greater under- *and* overestimation than less monotonous tasks involving manual work may mean, as the investigator himself virtually suggests, that the underestimators responded in terms of the event potential and the overestimators were influenced by the drive potential (France, students & lab. assistants; verbal, prior; Korngold, 1937). When it is suggested that "a period spent in success induces a greater degree of organization than a period spent in anticipating failure" (Harton, 1939b), the implication is clear: the drive associated with the anticipation may affect the temporal judgment via the organizational factor.

Not simply intervention but also feelings may be involved when a person compares his subjective impressions of duration concerning a crowded historical event (which he has experienced only second-hand or symbolically, with or without arousal of passion) and a relatively uncrowded event in his own life (obviously experienced, and again with or without passion). Any kind of intervention, we have been suggesting, must mean that a drive has been aroused, since—we remind ourselves—we operate under the deterministic assumption that all behavior is motivated. As many others have concluded in effect, especially in connection with experimental studies (e.g., Gilliland, Hofeld, & Eckstrand, 1946), this drive factor cannot be adequately controlled and hence some interaction between it and the event potential is always, or almost always to be anticipated.

In many situations it is extremely difficult to argue that the event potential has a greater effect upon temporal judgment than the drive potential. One experimenter in particular subscribes to this viewpoint, and she has designed ingenious experiments which seem to substantiate her view. Two examples must suffice. In one, the control series consisted of intervals between 13 and 72 seconds bounded by two clicks, the experimental series of the same intervals whose ready signal was also a click but whose stop signal was a mild electric shock. There was no difference between the rate at which the subjects counted during the two sets of intervals, but in retrospect the duration of the intervals followed by a shock was judged significantly longer than the control series, a result "consistent with the well-known experience that a time period spent in anticipation seems long"; hence it is claimed that there must have been "an increase in amount of events retained under conditions of sustained attention" (Sweden, medical students & lab. technicians; verbal, prior; Frankenhaeuser, 1959, pp. 82–6). In the other one, it was found that intervals filled with the performance of different tasks tended to be judged longer than those filled with similar ones; "since a high degree of similarity between successive items usually has a detrimental effect on learning processes," it must follow, the investigator argues, that more of the events in the interval with the diverse tasks was retained and so in retrospect they seemed longer than those which had been filled with similar activities (ibid., verbal, prior; pp. 87–95).

From this long discussion, it should be clear that external events, no matter how they are conceptualized, provide only a preliminary clue to the individual's reaction: important is what he actually perceives, his reactions to those perceptions, the way in which he organizes the perceptions, and his action or intervention. Having said this, I hope we can push the analysis ahead by using a term which refers simultaneously to the three components of the events potential, viz., the number of events, their organization, and the intervention. Nothing fancy is needed, I shall embrace all three with the word *scope,* and immediately put it to good use in a principle which seeks to be in line with most of the evidence presented in the various *Addenda* to this chapter:

5.2 When temporal knowledge is absent and the event potential is salient and hence affects perception, timing acceleration tends to vary at first slightly negatively, and then positively, with increases in the scope of the interval.

The relationship thus takes the form of a U-shaped curve. With an elapsed interval and Verbal Estimation, the following would be the ideal form of that hypothetical curve: when empty the interval seems longer than when it contains few events, but as the number of events increases the interval once again appears longer. Under these circumstances and on the level of primary judgments, an interval seems long and time passes slowly when that interval is empty or when it is crowded; the interval seems shorter and time passes more quickly when it contains some events—but not too many nor too few. Or, an empty interval is judged longer or shorter than a filled one, depending on whether the number of events has placed the filled one, respectively, to the right or the left of the nadir of the curve, viz., the point at which the number of events produces the lowest temporal judgment.

I am well aware that this principle is not a panacea: it does not account for all the variations in results that have been obtained (e.g., the dependence of some outcomes on the method of measurement). It certainly does not relieve us of the responsibility of determining the meaningfulness of the scope of the interval from the standpoint of the persons passing judgment. Left open, moreover, is the question as to whether or not the content has different effects upon intervals of varying length; for example, it has been stated that "in short periods of time the filled-up interval is as a rule estimated to be longer, and in long periods of time the vacant interval is generally estimated to be longer" (Roelofs & Zeeman, 1949)—presumably during a short interval there is less time for organizing the events than during a long one, and hence its scope appears greater. The principle, however, at least suggests the psychological foundation for appraising the event potential and inhibits, or ought to inhibit, an easy, a priori comparison between two externally different events such as performing long division or listening to a metronome.

ADDENDUM 5.1 FILLED VS. EMPTY INTERVALS

Some of the early research comparing two kinds of intervals must be quickly brushed aside because of imperfections in research design; e.g., one investigator compared different kinds of filling or input but he used a different method of measurement for each filling (verbal and production) and evoked the temporal motive for one interval prior to stimulation and for the other immediately afterwards (U.S., college; Myers, 1916). Support for the tentative generalization that filled intervals tend to be judged longer than unfilled ones comes from many of the classical experiments (Germany, university; comp., prior; Brown, 1931) (Germany, university; comp., verbal and prior; Jaensch & Kretz, 1932) (Germany, university & staff; comp., prior; Langen, 1935, p. 23). Filling in this kind of investigation ordinarily consisted of the sensory stimulation provided by a light or buzzer, but in one instance temporal judgments concerning intervals ranging from 2 to 10 seconds increased when the interval's boundaries were indicated not by taps on a table but by having the investigator simply place his hand on the subject's hand during stimulation and by having the latter reverse the process as the form of reproduction (Wallon, Evart-Chmielniski, & Denjean-Raban, 1957). An interval bounded by light flashes was judged longer when a single additional was inserted at the start of that interval than one without the addition or with a flash appearing later in the interval (U.S., college; repro., prior; Israeli, 1930); thus an additional event in one place but not in another was effective in the expected direction. And in real life: men working for 10 minutes judged the interval to be longer when they heard a noise than when there was quiet (U.S.; verbal, immediate subsequent; Jerison & Smith, 1955).

No differences between filled and unfilled intervals have also been reported under a variety of circumstances: intervals of 5-, 10-, and 15-seconds duration, with "unfilled" signified by illumination from a lamp and "filled" by the sounds of a buzzer (U.S., adult schizophrenics; verbal, prod. & repro. with prior; Clausen, 1950); intervals of 30 seconds, with idleness constituting no filling, and reading or hearing prose or poetry, as well as taking down dictation, designated as the filling (U.S., college; repro., prior; Spencer, 1921); intervals ranging from 0.5 to 8 seconds, with simple sounds constituting the boundaries when empty and sound the filling itself (France, adults; comp., prior; Gavini, 1959). One experiment has revealed inconsistent results for the subjects as a group but a fairly consistent tendency for some of them to judge filled intervals to be longer than unfilled ones and for others to show the reverse trend (U.S., graduate & college; repro., prior; Triplett, 1931).

There is a considerable body of literature seeking not simply to compare results obtained when comparable intervals are filled and unfilled but also to isolate one or more of the conditions associated with the differential effects. It seems to me to be significant that one of the first and certainly the most widely-cited series of classical experiments—and one of the most difficult to summarize—did emerge with the conclusion that the shorter, but *not* the longer, intervals ranging in duration between 0.1 and 10 seconds tended to be judged longer when they were filled than when they were not filled; for those longer intervals the reverse tended to be true. In addition, the experimenter offered data from his nine subjects demonstrating that these tendencies could be affected, under circumstances he tried to specify, by the order of presentation—he almost always used the Method of Comparison—as well as by the modality, the kind of filling, the "attention energy," the "pleasurable character of the activity," etc. (Germany, adults; prior, Meumann, 1896). Since that time, similarly complicated interaction of variables has been demonstrated, only two of which will be cited here. In one study, the duration of the interval and the method of measurement played crucial roles. With the Method of Verbal Estimation there was a consistent but not a significant tendency to judge intervals with a duration of 15, 30, and 45 seconds to be longest when they were empty, slightly shorter when filled with a nonrhythmical buzzer, and still shorter when filled with music—accuracy followed the same downward progression. The Method of Comparison produced similar results when the comparison was between the empty or the buzzer-filled and the music-filled intervals having a duration of 10 and 15 seconds, but not between the unfilled and the buzzer-filled intervals or between any of the other pairs having a duration of 5 seconds (U.S., college; prior; Whitely & Anderson, 1930). In another investigation there was a tendency for filled intervals to be judged longer than unfilled if the modality was audition but not vision; this tendency was affected by the order in which the filled and unfilled intervals was presented as well as by the mode of rating (U.S., college; rating-verbal, prior; Goldstone & Goldfarb, 1963).

In recent years a new variable, that of sensory deprivation or isolation, has been appraised under laboratory and realistic (or semi-realistic) conditions; intervals are thus rendered empty or relatively empty. On the whole there does seem to be a tendency for such intervals to be underestimated as our tentative generalization would lead us to expect. All eight subjects, for example, underestimated the duration of 1.5 hours of sensory deprivation to a greater degree than roughly the same amount of time spent reading or filling out questionnaire forms; and in all except one instance their judgments were more accurate in the latter situation (Canada, not specified; verbal, presumably immediate subsequent; Banks & Cappon, 1962). These findings in effect have been replicated with the Method of Production (England, presumably adults; prior; Reed & Kenna, 1964). In addition, American graduate students were confined from 8 to 96 hours "in a dark, lightproof, soundproof cubicle (9' x 4' x 8')" which contained only "a bed, a food box, and relief bottles for urination" (a chemical toilet was located in a dark antechamber right outside the cubicle); they of course had access to food. Under these circumstances—long, relatively uneventful intervals—the tendency was to underestimate the number of hours as each one seemed to pass by as well as the total time spent in the cubicle (U.S.; prior or immediate subsequent; Vernon & McGill, 1963). A French speleologist spent 59 days alone in a cave with one-way communication to the surface; he underestimated that period of time by 25 days because he miscalculated his "working" days during which he had relatively little to do other than to attend to the chores of existence, to read and listen to music, to explore his surroundings, and to report to his colleagues. In addition, it appears as though a more general sense of time was affected: his last conversation with those colleagues lasted 20 minutes but he estimated that "five or ten minutes" had gone by (Siffre, 1964, pp. 188, 189, 222, 225). On the other hand, nurses and hospital workers, confined individually to a soundproof room for from 5 to 92 hours but visited four times a day at "slightly varied" hours, overestimated clock time by from one-half to 3 hours; one-third of them underestimated the time at first, but then later overestimated it (England; verbal, prior; Smith & Lewty, 1959). The judgments of adult men concerning the duration of intervals ranging from 2 to 32 seconds were not affected when they were isolated by being suspended in the middle of a tank of water, but their isolation lasted only long enough for them to become adapted to this condition, viz., from 10 to 20 minutes (U.S.; three forms of repro.; Marum, 1968).

When U.S. Air Force subjects were isolated in an almost soundproof room up to 48 hours, one kind of filling made no difference in their production of intervals ranging from 1 second to 2 hours, viz., illumination of the room. This result is somewhat surprising since the isolation as such terrified some of the men and hence darkness might have been expected to increase that reaction and affect temporal judgment. Under these conditions, the shortest interval of 1 second was overestimated, 5 seconds tended to be judged accurately, and the remainder were underestimated (Mitchell, 1962). We are not given, however, comparable data concerning these subjects under normal conditions, so that the effect of this kind of emptiness cannot be appraised; but the presence of marked individual differences suggests that more than emptiness was influential.

Finally, in one experiment, previously

cited, a personality trait became significant: 10 extreme extroverts and 10 extreme introverts did not differ under normal conditions when an interval of 15 minutes was produced; but with sensory deprivation the mean of the former was 29.8 minutes and that of the latter 10.9 minutes (Reed & Kenna, op. cit.). This greater tendency for the extroverts rather than the introverts to overestimate the intervals—equivalent to underestimation for the Method of Verbal Estimation—resulted, the investigators speculate, from the fact that extroverts were more dependent upon external events than introverts, the very events that were absent or diminished during the period of deprivation. If the likely assumption of personality differences between delinquents and non-delinquents is accepted, then this factor must account in some way or other for the fact that empty intervals bounded by clicks were judged longer than those filled with a constant buzzer-tone by Army recruits but not by delinquents (Israel; verbal, prior; Siegman, 1966).

ADDENDUM 5.2 THE CONTENTS OF THE INTERVAL

Our tentative generalization suggests a positive relation between temporal judgments and the number of events. The operational question involves both the definition of number of events as well as the context in which the events are presented. Thus in one well-designed series of experiments, temporal judgments of auditory intervals increased as the number of clicks within the interval increased (or at least until the clicks followed one another very rapidly), but the effect was greatest at the shortest interval of 8 seconds and thereafter decreased as the intervals lengthened to 250 seconds (U.S., college; prior, verbal; Jones & MacLean, 1966). At first glance the next experiment seems to offer a clear-cut result: when subjects read off digits at a rate set by a metronome, and when afterwards they estimated the number of digits they had read and the elapsed time, the two sets of data tended to be positively related—as the estimate of elapsed time increased, the estimate of the number of digits tended also to increase. The relation here between judgments and events, however, was not so straightforward: some subjects consistently gave lower estimates of digits than of the elapsed time, others did the reverse (Sweden, medical students & lab. technicians; verbal, prior; Frankenhaeuser, 1959, pp. 71–4).

One of the few consistent effects to appear in a well designed study which varied the modality of presentation and of reproduction and the duration of the intervals was a tendency for intervals with alternating stimulation to be reproduced as longer than those with constant stimulation (U.S., college; Brown & Hitchcock, 1965). In other studies, subjects estimated the time to be less when they learned a single maze than when they learned a number of mazes during the equivalent interval of clock time. Possibly the difference may have reflected the varying difficulty of the task, for there seemed to be a tendency for that difference to become less when the shorter mazes were too easy or too difficult; but such an explanation could not be validated when the subjects' own evaluations of the tasks were analyzed (U.S., college; verbal, immediate subsequent which probably became prior; Harton, 1939a & 1942). Then the attempt by the investigator himself to substantiate his hypothesis that "a greater unity of organization of the subjects' activity" must have characterized the goal of solving one maze rather than several mazes ran into difficulties when he sought to replicate his findings in "life situations," by which he meant arousing the temporal motive not prior to, but after, intervals filled with various kinds of activities; and he pays homage to the drive element by noting "complicating factors . . . especially of an emotional nature."

It might be presumed that intense stimulation is more eventful than less intense and that therefore intensity ought to be positively related to temporal judgment. But here, too, the results have been conflicting. An increase in the sound filling the interval from 40 to 60 decibels led to a significant increase in the reproduction of intervals having durations of 0.35 and 0.70 seconds, but a further increase to 90 decibels produced no such lengthening, nor was intensity at all effective when the interval being reproduced lasted 1.4 seconds (France, presumably adults; Oléron, 1952). Similarly the estimates of auditory intervals ranging from 0.5 to 8 seconds was a positive function of stimulus intensity, but accuracy and variability were affected by method of measurement (U.S., presumably college; verbal, comp., repro., and prod. with prior; Zelkind, 1969); and an increase in intensity with which the index finger of the left hand was stimulated by a vibrator, which thus signified the two intervals being compared, produced a logarithmic increase in temporal judgments concerning three different ephemeral intervals (Sweden, students; comp., prior; Ekman, Frankenhaeuser, Berglund, & Waszak, 1969). On the other hand, an extremely careful psychophysical investigation of ephemeral auditory intervals has demonstrated that "the effect of signal level [i.e., stimulus intensity] on duration discrimination became negligible as soon as the signals were made highly detectable" (U.S., students; Creelman, 1960). Finally an increase in the intensity either of irregularity presented sounds preceding the interval or of the sounds forming the interval produced no differences (Sweden, medical students & lab. technicians; verbal & comp., with prior; Frankenhaeuser, 1959, pp. 69–81).

The experimental literature also suggests that manipulating the composition of the interval, only on the input side, can make a significant difference. *Longer temporal judgments* have been found to be associated with the *first* of the two variations in each instance rather than with the second:

1. Regular vs. irregular sounds as filling (Germany, university students; comp., prior; Grimm, 1934); here one might have speculated incorrectly that irregular sounds would be more outstanding as events.

2. Tones of constant vs. decreasing or increasing intensity during the interval; but the differences were statistically significant only for intervals of 2, 6, and 8 seconds when intensity was increased and for intervals of 2, 4, and 6 when intensity was decreased; changing intensity had no effect upon judgments concerning a 12-second interval (South Africa, European university students; prod., prior; Danziger, 1965).

3. Tones considered "passive" by comparable subjects vs. ones considered "active" or "neutral"; the tones were activated by the subjects to signal the start and end of the 10-second interval they were trying to produce (U.S., high school; prior; Butters, Jones, Hoyle, & Zsambok, 1968).

4. Noninterruption vs. interruption of the interval during its exposure; for example, two intervals tended to be judged equal in duration when both were produced by "luminous squares" and when the first was exposed for a total of 0.42 seconds, but with an interruption of 0.14 or 0.35 seconds, and the second was exposed continuously for 0.394 seconds (Netherlands, unspecified; comp., prior; Roelofs & Zeeman, 1949). This appears to contradict the previously cited studies on the regularity and intensity of stimulation.

5. The presence of a "small red neon tube of comfortable brightness" vs. flashes; dim vs. bright lights during the interval (U.S., college; limits- and rating-verbal, with prior; Goldfarb & Goldstone, 1963; Goldstone & Goldfarb, 1963). Flashes or bright lights would appear to contain more "events" than constant or dim stimulation, but again

the consequences are not the ones anticipated.

6. Auditory vs. visual, "white sound" vs. "white light," auditory vs. whole-visual-field stimulation (U.S., college; limits- or verbal-rating with prior; Goldstone, Jernigan, Lhamon, & Boardman, 1959; Goldstone, Boardman, & Lhamon, 1959; Goldfarb & Goldstone, 1964). This tendency for auditory stimulation to give rise to longer judgments than visual ones in a variety of contexts has been replicated when the standard given the subjects was not one of "more" or "less" than a second, but a 9-point scale involving either the symbol of second (range: "very much less than one second" to "very much more than one second") or qualifying adjectives (range: "very, very short" to "very, very long") and when the intervals were of different duration (U.S., college; prior; Goldstone & Goldfarb, 1964). Auditory intervals were judged longer than visual ones regardless of the filling which was sounds, lights, or nothing (U.S., college; rating-verbal, prior; Goldstone & Goldfarb, 1963). A different group of investigators, however, has reported that visual rather than auditory stimulation increased temporal judgment, but only very significantly when an interval of 0.5 seconds (and not with ones of 1 and 1.5 seconds) was employed and really after the first 100 trials (U.S., college; comp., prior; Tanner, Patton, & Atkinson, 1965).

Yet another kind of manipulation has been investigated, viz., the stimulus context in which the interval is produced; an *increase in temporal judgment* has been found associated with the *first* of the two conditions in each instance rather than with the second:

1. Sounds signaling the start and end of the interval with an intensity roughly equivalent to those filling the interval vs. those with an intensity markedly higher or lower (Germany, university; comp., prior; Grimm, 1934).

2. An increase vs. no increase in the duration of the bounding sounds, more especially when the sounds were signals at the start vs. the end of that interval. This conclusion emerged in a variety of experimental situations involving an interval of 0.5 seconds and both the Methods of Comparison and Reproduction. An example of the former: comparing an interval bounded by a long and short stimulus with one bounded by a short and stimulus. The investigator explains the greater influence of the initial sound by suggesting that it cannot be ignored "because one has to 'catch' its ending in order to apprehend the beginning of the interval," whereas after the terminal sound begins, "no matter what its length, the observer can turn his attention away, as the interval is then complete" (U.S., college; prior; Woodrow, 1928). In passing, it must be noted that the bounding sounds themselves tended to be overestimated when they were reproduced.

3. The first vs. the second of two ephemeral intervals; the second vs. the first of two transitory intervals. This time-order error is labeled positive in the first instance and negative in the second. When the Method of Comparison is employed and the intervals are ephemeral and equal in duration, the first one is likely to be called longer; under the same conditions, but with Reproduction, the second is reproduced as longer than the first to compensate for its apparently shorter duration. In psychophysical investigations, as indicated in the text of this chapter, attempts have been made to discover the precise duration of intervals producing negative and positive time-order errors. One experimenter used a method of fractionation in connection with tones: the subject was presented with a standard interval (which may come first or second) and was required to keep adjusting the duration of the variable interval until, in his opinion, its duration was one-half of the standard. The resulting data revealed whether an interval was being over- or underestimated, for it could be noted whether the mean judgments were, respectively, more or less than half the standard interval. In this particular study, there was a positive error for intervals of 0.4, 0.8, and 1.6 seconds; for

2.4 seconds there was in effect no error resulting from order; and a negative error occurred for 4.8 seconds (U.S., college; Gregg, 1951). These results are roughly in line with ones previously obtained with a more usual method, that of Comparison (U.S., college & staff; prior; Stott, 1935).

4. Reproducing a tone or light in the presence of noise after the stimulus had been perceived in quiet conditions vs. that some reproduction under quiet conditions after the stimulus had been perceived in the midst of noise. No difference in reproduction appeared when the same tone or light was perceived with the visual background being changed from light to dark or vice versa. The authors conclude that these results suggest "a strong relation between apparent or psychological time and level of auditory stimulation" and the absence of such a relation with visual stimulation (U.S., presumably adults; prior; Hirsch, Bilger, & Deatherage, 1956). Eight subjects were involved in this experiment; we have no way of knowing, therefore, whether the presence of this auditory effect, and the absence of the visual one, reflect some kind of cultural conditioning to which they have been unwittingly subjected.

5. Words considered "passive" (e.g., snail) vs. those considered "active" (tornado) or "neutral" (scene) which served as the signal for producing an interval of 10 seconds; words with connotations that were "savory" (color), "tasteless" (dirty), or "neutral" (deep) did not influence the productions (U.S., high school; prior; Butters, Jones, Hoyle, & Zsambok, 1968).

ADDENDUM 5.3 SPACE AND TIME

From the standpoint of physics, there is an intimate relation between objective time and space since, according to the standard formula, velocity or speed is a function of space or distance divided by time. All three of these terms are of course objective, but the formula has also guided psychological research in a direction which indicates that objective space is another of the contextual factors affecting temporal judgment. An early experiment demonstrated that distance perceived for a long period of time tended to be judged greater than the same distance perceived more briefly; likewise an interval whose stimulation involved a greater distance tended to be judged longer than one of identical duration involving less distance (Germany, university students; comp., prior; J. F. Brown, 1931). The first phenomenon has been labeled the *tau-movement* effect, the second the *kappa-movement* effect. An example of tau: if two parts of a journey cover the same distance but one part takes perceptibly longer than the other, the one taking longer seems to have covered more ground. An example of kappa: if two parts of a journey take the same length of time but more distance is covered in one part than in the other, the one covering the longer distance may appear to take longer.

The tau effect was originally demonstrated by marking off three *equidistant* points on the subject's forearm: if the interval of time between the stimulation of points 1 and 2 was less than that between the stimulation of points 2 and 3 (i.e., the first interval was less than the second), the distance between points 1 and 2 tended to be considered less than that between points 2 and 3 (U.S., graduate and college with one staff member; Helson & King, 1931). In a later experiment subjects were stimulated by flashes of light located at different distances from each other. They were asked to adjust the appearance of the middle flash so that it would mark off equal temporal intervals between that flash and the other two flashes: the duration between the first flash and the middle one was to be made equal to the duration between the middle and the third one. It was found that the interval between the two flashes more widely separated was made *shorter* than the two which were closer together; thus an interval created by flashes appeared longer (since it was called equal to a longer one) when the flashes were further apart rather than closer together. "The space-time events about which judgments are made," consequently, "are psychologically interdependent" (England, presumably adults; repro., prior; Cohen, Hansel, & Sylvester, 1953). Kappa has been demonstrated when the distances were exposed successively but not simultaneously (U.S., college; repro., prior; Parks, 1968) and has been shown to have a proportional effect upon the indifference point (U.S., college; prod., prior; Earl, 1969).

Kappa has also been investigated in an extra-laboratory situation. Subjects journeyed either blindfolded or enclosed in a car from which they could obtain no clues from the outside for a distance of about 12 miles. At the temporal midpoint of the journey a bell rang. On some of the trials, the car accelerated when the bell sounded; on others, it decelerated; and on the remaining ones its speed remained the same. At the end an estimate was made of the duration of the intervals before and after the bell as well as of the distances traversed or the speed. Again the tau and kappa phenomena appeared, and there was also a combination of the two: if two distances were covered at the same speed but if the first was shorter and covered more quickly than the second, that first speed tended to seem greater than the second. But, as ever, temporal judgments usually spring out of multivariant situations and hence the conclusion is drawn that, although it has been shown that time, space, and speed are obviously interrelated, an estimate of a third factor cannot be made with complete precision from estimates of the other two; the subjects "believe they have travelled for a longer time and at a faster speed as judged by the derived estimates of duration and speed respectively" (England, unspecified; presumably immediately subsequent; Cohen & Cooper, 1962). From a practical standpoint it would seem that in these experiments passengers could be said to have escaped from both kappa and tau when the vehicle was traveling at the rate of about 30 miles per hour, for then their judgments of speed were most accurate; otherwise slower speeds were overestimated, higher ones underestimated. Many drivers, preliminary results indicated, were victims of kappa: they underestimated the speed at which they were traveling (Cohen, 1967, p. 49).

In addition, the problem of individual differences intrudes into this trio of stimulus factors, which again suggests the need to consider temporal judgment in relation to the behavioral potential. In one part of an experiment, subjects walked, and then ran, the same distance (150 feet) without knowing that the two distances were equal; afterwards they estimated those distances and the time taken to cover them. In the other part, the subjects first walked the distance of 150 feet and then were asked to run the same distance; afterward they estimated the time consumed for each stretch. With both methods, half the subjects first walked, and then ran; the order was reversed for the other half. In the first part of the study, there was a decided tendency for the distance walked to be considered longer than the distance run. In the second part, the order of running and walking was important: subjects traversed a distance longer than 150 feet if they walked first and hence spent less time running, and shorter than 150 feet if they ran first and hence spent less time walking. In both instances, however, a good proportion showed the opposite effect, and so the investigators—with brilliant ad hoc originality—divided their subjects into Tortoises and Hares. The Tortoises, they think, were "primarily conscious of the passage of time" and were "scarcely aware of speed as such"; hence long time meant long distance. The Hares must have been concentrating upon speed; hence fast speed meant long distance (England, high school students; verbal, immediate subsequent; Cohen, Cooper, & Ono, 1963).

Another way to investigate the relation between space and time is to have subjects view an object moving at a constant speed through space for a given period of time; then the object disappears beneath a screen and the subjects are asked to indicate when it has reached a given point which is visible to them on the screen. In one experiment it was found that subjects consistently and significantly overestimated this interval while the object was occluded but that overestimation was not a function of the way in which they viewed the object, background factors in the stimuli, or the size of the moving object (U.S., college; Reynolds, 1968).

ADDENDUM 5.4 INTERVENTION

This Addendum is concerned primarily with output: the individual passing judgment is engaged in some kind of activity during the interval other than merely perceiving it; when the Method of Reproduction or Production is employed, he may also be similarly active while reproducing or producing the requested interval. In fact, it may well be argued that the subject is more active when he is employing the Method of Production or Reproduction than when he uses Verbal Estimation: instead of passively paying attention, he must at a very minimum indicate the end of the interval he is producing or reproducing. In one experiment in which the same subjects estimated intervals varying between 2 and 29 seconds, Verbal Estimation led to longer judgments than Production (U.S., college; Horstein & Rotter, 1969). The investigators themselves suggest that the greater activity of producing made waiting "less boring" and hence the intervals thus measured appeared shorter.

Real-life situations have been examined. Passive spectators overestimated the interval in which they had merely observed a basketball game before time out was declared as a result of an injury to a player (U.S., college; subsequent; verbal; Myers, 1916) or in which they watched a motion picture (France, adults; verbal, subsequent; Fraisse & de Montmollin, 1952). The assumption here is that such passivity was associated with long judgments, but no control situation involving activity was examined.

In the laboratory, it is possible to compare judgments under conditions of passivity (i.e., with the interval empty) and activity (i.e., with the interval eventful). A fairly clear-cut tendency becomes evident: under these conditions objectively empty intervals seemed longer than those filled with activity (U.S., children, college & university; verbal, prior Axel, 1924, p. 43) (U.S., college; repro; prior; DeWolfe & Duncan, 1959) (U.S., college; verbal, prior; Gulliksen, 1927) (U.S., college; verbal, prior; Hawkins & Meyer, 1965) (U.S., college; verbal, immediate subsequent; Miller, Frauchiger & Kiker, 1967) (U.S., college; verbal, prior; Swift & McGeoch, 1925) (France, children & adolescents, repro., prior; Wallon, Evart-Chmielniski, & Denjean-Raban, 1957). Obviously these results contradict the tentative generalization advanced in the text since presumably activity increased the number of events during the interval without a corresponding effect upon temporal judgments. In addition, slight changes in procedure may upset the central tendency. Thus a decrease in the number of nonsense syllables which subjects heard and wrote down during intervals of 1-minute duration was found to be associated with an increase in temporal judgment, but this relationship appeared only in a very attenuated form when words "with high associational value" rather than syllables were employed (U.S., adults; prior, repro.; Friel & Lhamon, 1965). Or patients in a hospital significantly overestimated to a far greater degree intervals of from 30 to 120 seconds while simply judging their duration than when trying to solve stylus mazes; but in the first instance the temporal motive had been aroused prior to the interval and in the latter immediately subsequent to it, so that the order in which the two kinds of intervals was presented was also statistically significant (U.S., adults; verbal; Cohen & Parsons, 1964).

Many investigations have sought to determine the kinds of activities which affect temporal judgment; representative studies are summarized below by listing the actions in order of *increasing* temporal judgment (i.e., the one judged shortest or shorter comes first):

1. Performing long division, copying from dictation, reading directions in a mirror, holding the palm on a thumb tack, holding arms extended from the side, and complete rest (U.S., college; verbal, prior; Gulliksen, 1927). So far as one can tell on an a priori basis, the progression seems to have been from the complex to the simple in terms of the required skill or patience, with less rather than more activity or events producing longer judgments.

2. Solving anagrams, printing letters of the alphabet upside down and right to left, and resting (U.S., college; repro., prior; DeWolfe & Duncan, 1959). This order from complex to simple is in line with the first experiment just cited, but it was obtained only when the subjects performed those actions during the initial interval which they were then to reproduce: the experimenter told them when to begin and when to stop, after which they tried to reproduce an interval of the same duration. In contrast, exactly the opposite results emerged for those actions when they were performed not during the interval controlled by the experimenter but during the reproduction itself: here the subject signaled when he estimated that the right amount of time had elapsed. These trends appeared whether or not the same or a different activity was performed during the initial period and the reproduction.

3. Analogies or completion of number series, cancellation of numbers, tapping or empty (U.S., college & university; prior, verbal; Axel, 1924, p. 43). The same investigator used a similar technique with children between the ages of 9 and 14:
arithmetic or cancellation of numbers; writing a letter; empty. Both investigations are said to show that "the higher the level of behavior during a given time interval, the greater is the tendency for underestimation, and vice versa. . . ." (p. 68).

4. Intervals filled with similar activities; intervals filled with different activities. Similar tasks consisted of adjusting visual or auditory signals to an agreeable rate; tapping with one hand; or multiplying or writing. Different tasks consisted of combinations of the similar ones: adjusting both kinds of signals; tapping with alternate hands; or multiplying and writing (Sweden, medical students & lab. technicians; verbal, prior; Frankenhaeuser, 1959, pp. 87–95).

5. Counting irregularly presented sounds, counting regularly presented ones (Germany, university; comp., prior; Grimm, 1934).

6. Being able to finish a test of mechanical ability; not being able to finish it (U.S., college; verbal, prior; Hawkins & Meyer, 1965).

7. Locating a relatively small number vs. a relatively large number of specified 2-number digits during an hour (U.S., college; verbal with prior & immediate subsequent; Bakan, 1955).

There are, however, complications. The stimulus context may play a role. One study involved three intervals with a duration of 3, 5, and 7 minutes, during which the undergraduate subjects engaged in three tasks: adding numbers, crossing out letters, and supplying missing letters in words. The intervals and the tasks were counterbalanced. The only significant, or almost significant, variation turned out to be the tendency for the interval in the second position to be overestimated to a greater degree than those in the first or third positions, regardless of duration or activity (U.S.; verbal, prior; Postman, 1944). Then, as ever, negative findings intrude: tapping a telegraph key once or twice in order to reproduce intervals with durations varying from 0.5 to 8 seconds had no effect upon the accuracy or consistency of temporal judgments (U.S., college; Doehring, 1961). Similarly, no significant trends appeared when the intervals to be reproduced contained either simple sounds or symphonic music, and when the subjects reproduced them by activating intervals which themselves contained one of the following: the same sound, the same music, white light that illuminated the darkened experimental room, or a picture projected on a screen (South Africa, European students; prior; Du Preez, 1967a).

A number of investigators have concentrated upon a single factor in the act of intervention, the amount of energy expended. The tasks are listed below again in order of *increasing* temporal judgment, with the objective time of the intervals indicated in parentheses:

1. (60 seconds) resting, lifting a 9-pound barbell. But the difference was even greater when resting was contrasted with producing the interval not by lifting the barbell but by remaining quiet for the specified period of time after lifting it. These differences appeared when the subjects were male students; but no difference could be established when female students rested or lifted a 6-pound barbell (England; prod., prior; Boulter & Appley, 1967).

2. (0.55 seconds) a series of weights from 100 to 250 grams placed upon a small carriage which emitted clicking noises as it was moved by hand to reproduce the interval (U.S., presumably college; presumably prior; Weber, 1927). Whether it makes psychological or physiological sense to describe the differences between these very light weights in terms of "effort" is a question we cannot settle.

3. (15 seconds) unfilled intervals while hearing a buzzer, unfilled intervals while peddling an ergometer (U.S., college; verbal, prior; Schönpflug, 1966). But the reverse effect was also noted: counting letters or numerals while peddling produced shorter estimates than the same task while listening to a buzzer.

4. (24 and 48 seconds) a progression from expending 40 percent of maximum effort to none whatsoever in producing on a dynamometer the interval to be estimated verbally or to be reproduced. It is to be noted (1) that the method here was an unusual one (the subjects used energy to produce the standard and not to reproduce it); (2) that the relation of increase in temporal judgment was non-monotonic, i.e., it did not change directly as more energy was expended; (3) that no psychic stress was involved; and (4) that no difference in judgment accompanied an increased expenditure of energy for intervals of 6 and 12 seconds (U.S., college; Warm, Smith, & Caldwell, 1967).

5. (43 seconds) *decreasing* the amount of muscular tension exerted by the hand upon a dynamometer; the interval's duration was estimated as the subjects exerted pressure varying from zero to two-thirds of their maximum grip and it was found that "muscle tension states affect the accuracy of estimate of short time intervals largely in the direction of reducing the perceived duration of the time span" (U.S., college; verbal, prior; Weybrew, 1963).

6. (1.5 seconds) unfilled intervals, hearing music. This mundane finding, which seems concerned not with output but with input, is noted because the protocols in the classical experiment suggested that the 8 subjects experienced greater feelings of tension in the former than in the latter situation (Germany, university; comparison, prior; Jaensch & Kretz, 1932).

Finally, intervention has been compared not with no intervention but with the perception of activity or events during the interval; with this procedure perceptual input and active output are contrasted. In all of the following studies there was a marked tendency for intervention to give rise to *shorter* temporal judgments than perception alone—once more the experimental manipulations are arranged in order of *increasing* temporal judgment:

1. Performing long division, copying from dictation, listening to a rapid metronome, listening to a slow metronome (U.S., college; verbal, prior; Gulliksen, 1927). As indicated in the previous reference to this study, the change seems to have been from the complex to the simple as judgment increased.

2. Copying pictures, listening to some-

one else recall verbal materials (Poland, children; verbal, presumably prior; Reutt, 1962).

3 Copying dull or interesting material; listening to interesting material during a 10-minute period (U.S., college; verbal, prior; Swift & McGeoch, 1925).

4. Waiting for a mark to pass a vertical line; awaiting the same events but performing "sensori-motor task" while doing so (Czechoslovakia, adults; Bezák & Dornic, 1968).

5. A mild electric shock at the end of the interval; no such shock. The additional input, however, increased only the estimation of the total interval but had no effect upon the rate at which digits were read off during that interval (Sweden, medical students & lab. technicians; PPT, prior; Frankenhaeuser, 1959, pp. 82–6).

The interaction of input and output has been most thoroughly examined in a study employing a 90-second interval and an adequate number of subjects for statistical analysis. The undergraduates viewed a light which flickered at various rates for each experimental group, and one group was not stimulated at all; some of them responded to the flicker by tapping a telegraph key requiring a pressure of ½ pound, others to one requiring 1 pound of pressure, and others did nothing; those receiving no visual input and doing nothing were asked, after their return, simply to indicate how long the investigator had been out of the room. Under these circumstances, there was a general tendency for increased input or output to be accompanied by longer judgments, and the effect of the input was greater than that of the output. Regardless of input conditions, judgments were longest with a maximum pressure of one pound; and regardless of output they were longest with a maximum input of stimulation. No matter what the output, however, all judgments were longer with zero input than they were with a minimum input; but then they regularly increased as the input increased. The effect of the greater motor output was significant only with maximum input (U.S.; verbal, immediate subsequent; Miller, Frauchiger & Kiker, 1967). These findings have been more or less confirmed in a study in which the subjects were simply instructed to produce an interval of 2 seconds. Actually the experimental situation was fiendishly unusual because the investigator sought formally to construct tasks of different complexity: the interval was produced by making 1, 2, 4, or 6 responses after having perceived stimuli whose number fluctuated similarly. Thus in the simplest situation the signal was always the same stimulus which signified the start of the interval and to which the same response was made to end it. The most complicated situation was the presentation of one of six stimuli, to each of which a different response had to be made. The stimuli were visual (pointers on dials), the response consisted of pressing one or more keys. The intervals thus produced tended to be shorter as the number of response keys increased. In contrast, the number of stimuli eliciting the responses did not have such an effect: the produced intervals tended to be longest when a single stimulus signaled their onset, but thereafter the number of stimuli —whether 2, 4, or 6—had no significant effect (Netherlands, young adults; Michon, 1965).

A variant approach to temporal judgment has been to compare input and output not directly but by shifting the attention from one to the other; the set or Einstellung listed first is the one that produced the *longer* judgment:

1. The instruction to pay "attention to the interval between the sounds" and to "put your fingers and your body, breathing and vocal muscles, on [sic] a strain while reproducing the interval" vs. to pay attention "solely" to the sounds and to "pay no attention to your finger movements" during reproduction (U.S., presumably college; prior; Woodrow, 1933).

2. The instruction to pay attention to finger movements in reproducing intervals by tapping a key and to concentrate on "just when the terminal tap should be given" vs. "to reproduce in an automatic manner, to pay no attention to . . . finger movements but to attend primarily to the sounds, and to think of the interval as

'out there' somewhere, i.e., between the sounds" (U.S., college and investigator; prior; Woodrow, 1930).

3. Counting vs. not counting while perceiving intervals varying from 4 to 27 seconds (U.S., college; verbal, prior; Gilliland & Martin, 1940).

ADDENDUM 5.5 "STORAGE SIZE"

As this book was being processed by the publisher, there appeared a small volume (Ornstein, 1970) exclusively devoted to defining the view that many aspects of temporal judgment depend upon what I have called the interval's scope and the investigator its "storage size." His "cognitive metaphor"—the expression is his—refers both to the amount of information conveyed during the interval (the input) and also the way in which that information is immediately or subsequently coded and stored. In a series of nine very ingenious, coordinated experiments he has shown that temporal judgments *lengthened* as a function of an increase in, or the presence of, the following during the interval: the number of tones being heard (but not the regularity with which they were presented), the complexity of the line figures (but "only up to a point"), the difficulty with which a series of sounds or drawings could be coded *while* being perceived, the non-automatic nature of the manual task being performed, the number of segments of a dance film being named, the number of pairs of words and sounds which could be recalled when judgment was passed two weeks later, and non-assistance from the experimenter in coding *after* perceiving line figures. This truly impressive evidence, however, has been collected under the following conditions which inevitably impose limitations upon its generalizability:

1. All subjects were undergraduates, indeed at one university; although there is no reason to believe that other kinds of persons would have reacted differently, the burden of proof remains on the investigator.

2. Only transitory intervals were judged, the range of which was from 30 seconds to over 9 minutes, with the mode between 1 and 2 minutes. These intervals were selected because the investigator had decided to concentrate upon the "experience of duration," which in his opinion is one of the "four major varieties of time experience" (the others are "short-term time," "temporal perspective," and "simultaneity and succession").

3. In all the experiments except one which used Verbal Estimation, the investigator provided subjects with a temporal standard and hence his data stem from the Method of Comparison; with Comparison, he argues, subjects can concentrate upon duration per se without concerning themselves with clock or objective time which is arbitrary.

4. In order to avoid the use of cues other than those being experimentally manipulated, the temporal motive was always subsequently aroused; except in part of one experiment in which the delay was almost two weeks, the arousal occurred immediately after the perception of the interval to be judged.

5. With an exception (to be discussed below), the drive factor was completely excluded, or rather no drive was manipulated experimentally. In the experiment involving line drawings of differing complexity, for example, subjects were given the vague instruction that they were "about to look at two figures" and "we'd like you to look at them carefully as we are going to ask you some questions about them afterwards."

6. In all the experiments only degrees of the variables were compared; by not including an empty interval, the author did not investigate the situation in which, according to our principle in the text (no. 5.2), there is likely to be a decrease in temporal acceleration as the interval moves from virtually zero to some events.

The author refrains from speculating about the reason for his assumption that increased storage space lengthens temporal judgment, rather he seeks to show that many processes involving what he operationally calls that space in fact do just that. He frankly and modestly admits that he has been concerned only with "one particular relationship," "some parametric variations" of which he has been exploring. He believes, however, that his "storage approach alone can integrate the diverse data of duration experience" (p. 109); his list includes:

1. The time-order error: the interval coming first is judged shorter than the one coming second (the so-called negative error) because less information remains in storage from the former than from the latter. The investigator himself demonstrated such an effect in one experiment, but in another he found no difference in order of presentation: "it is difficult to explain this finding on almost any theoretical approach" (p. 67). In addition, positive time-order errors have been reported for ephemeral intervals (see *Addendum 5.2*).

2. The effects of drugs: stimulants, such as LSD and the amphetamines, "increase awareness of and sensitivity to the stimulus array" (p. 38) and hence the greater storage space which is thereby required produces longer temporal judgments, whereas the reverse would be true of depressants. By and large, as the evidence reviewed in *Addendum 9.4* suggests, temporal acceleration is associated with stimulants and deceleration with depressants, but the data are not as "unanimous" (p. 47) as the investigator insists.

3. The phenomenon of the watched pot: expectancy leads to "increased sensitivity to stimuli" and thus "greater amount of awareness of input, and consequently a lengthening of duration experience" (p. 112). I am of the opinion—based on my own pot-watching—that I may be aware of input during the agonizing interval but that I have little to observe; hence other factors must certainly be invoked to explain the phenomenon (see *Addendum 7.4*).

4. Boredom: when bored, "we are forced to attend to more of the stimulus array than we normally would" (p. 112). This, it seems to me, is another debatable point:

when attentive, we pay greater attention to details—and so the only out for the storage theory is to claim that the details under these conditions are so well organized that less space is needed. And again I think the drive factor must be taken into account.

5. Difference between immediate and subsequent judgment: as a vacation ends, we remember all the details and the time seems to have been long; afterwards, we code the material more concisely and the vacation seems to have been brief. I find no fault with this interpretation, except that its scope is so deliberately limited: it omits considering judgments coming from prior or interim arousal of the temporal motive; it does not seek to explain why events are coded differently on the two occasions.

6. Success and failure: successful activity, since the goal is achieved, is better organized than unsuccessful activity; hence its storage size is less and it is judged shorter. *Addendum 6.1* contains a discussion of the problems arising when the effect of the drive factor is explained solely by reference to the event attribute.

In citing previous studies, moreover, the investigator has not considered those in which an increase in storage space has not produced a lengthier temporal judgment (see *Addenda 5.1 and 5.2*). His metaphor, in my opinion, cannot account for the results of experiments involving varying kinds of intervention (*Addendum 5.4*), nor can I immediately deduce from his theory why judgments often become progressively longer in a series of intervals when Production or Reproduction is utilized and no feedback is provided (*Addendum 7.1*). Also in the author's own experiments there is substantial evidence indicating that the principle of storage size admittedly accounts for a limited part of the temporal process. Its vindication in each experiment is obtained only by indicating, of course, that obtained differences between experimental treatments are significant; but individual subjects deviated from the central tendency (as signified by the rather large standard deviations in some of the data) and some of the obtained differences just about reached the level of statistical significance.

In short, I am of the opinion that the theory behind this notable research, though it has been stretched a bit, is in accord with my own principle regarding acceleration as a function of scope; that it neglects the negative acceleration which appears during the transition from an empty interval to one containing few events; and that it accounts for only a limited number of phenomena associated, in words appearing in the monograph's title, with "the experience of time."

6. DRIVES

For a moment or two, let us abandon our friend who is trying to decide whether time is passing slowly or quickly. We leave him suspended after he has decided to pass temporal judgment; after he may or may not have received temporal information; and after he has perceived, and been influenced to some extent by the event or events within the interval. We turn briefly to the general topic of gratification whose antonym, frustration, then again immediately demands attention.

The clearest, the most uncontroversial statement to be made about gratification is that it is temporary. The reasons are legion, but can be grouped under two headings. First, many drives are recurrent. The food is eaten, the sexual act completed, the sonnet written, the greater glory attained—but the satisfaction lasts only until the drive is aroused again and the same or a similar goal must be sought. Then, in the second place, the gratification of one drive leaves other drives ungratified, and that gives rise to new activity. After food or drink and love or poetry, there are yet other goals to be attained—housing, health, and immortality—to which the person then attends.

On the basis of what they remember from their own experiences and from what others have told them, consequently, people believe, and anticipate, that nirvana is almost always a step ahead of where they are, though they may hope otherwise. A normal child in any society learns to anticipate that he will become an adult and that his present joys and miseries are fleeting and will be replaced not only by others but also by responsibilities of varying magnitude. The activities of living and observing, in short, indicate the transitory nature of gratification. If hope also is indeed an eternal human attribute (Cantril, 1964), the explanation must be that each persons's gratifications are incomplete and that in the past some of his anticipations have been realized; hence his temporal orientations move from the past and present into the future. Men would not only reduce the tension of recurrent drives but would also create and satisfy the cravings for novelty and growth (Allport, 1955, pp. 49–51, 65–8).

These demipoetic observations have an important, prosaic implication for our analysis of temporal judgment: every single bit of behavior, including that elicited in the laboratory when subjects, for example, are preoccupied with the task of listening to intervals bounded by clicks, involves striving toward a goal and hence, eventually, its attainment or nonattainment. While ostensibly concentrating upon the event potential of intervals in the last chapter, we have been forced both by the facts and our own intuition to observe that this drive factor cannot be excluded even from classical experiments which generally have concerned themselves with nonmotivational variables, or have tried to do so. Most of what is said in this chapter, moreover, applies only to transitory and extended intervals, for drive attributes probably play a small role in those that are ephemeral.

To link the drive component of an interval to temporal judgment it is necessary to specify when the temporal motive is aroused. You are hungry, but must wait until dinner is served. Your subjective appraisal of the delay varies with your feelings while actually waiting, during the meal itself, or afterwards. Principles are needed for each of these circumstances.

STRENGTH

Obviously drives differ in strength, and we quickly designate two ends of the operational continuum with the adjectives *strong* and *weak*. The man seeking food may be close to starvation or he may simply feel obliged to eat because the noon hour has struck. Almost by definition strong drives, regardless of their overt expression, are more insistent or intolerable than weak ones and therefore their satisfaction is more eagerly sought. Such eagerness has two consequences: every moment of nonattainment is likely to be crowded with events—anticipations, plans, overt activities—which would alleviate the discomfort or restlessness; and there is likely to be wishful thinking concerning the termination of the longing and the frustration. Two Guides thus emerge: one suggests that the reduction of a strong drive is likely to take longer than that of a weak one (although of course more vigorous and quicker action may also be the consequences), whereas the other stems from the hope that the gratification of a strong drive ought to be reduced more quickly than a weak one. These opposing standards can be expressed in a single principle:

> 6.1 *The stronger the salient drive before the attainment or non-attainment of the relevant goal (other than the end of the interval itself), the greater the timing acceleration.*

The word "salient" in the principle serves as a reminder of the probability that temporal judgments are less likely to be affected by repressed or preconscious drives. In one experiment, for example, girls had to wait 13 minutes before the arrival of appetizing food. Those who were hungry tended to overestimate the elapsed interval to a significantly greater extent than those who were not hungry, but only when they had been made conscious of their hunger during the waiting period by being asked to rate food recipes. When during the same period they were instructed to rate clothing, their hunger must have been less salient and hence, in fact, the temporal distortion did not appear (U.S., college; verbal, prior; Schönbach, 1959).

Although the question of whether drive strength can be reduced to number of events is discussed (inconclusively, I admit) in *Addendum 6.1* and also, in passing, in *Addendum 6.4*, it seems useful to illustrate here one facet of the problem, that of pleasantness vs. unpleasantness. After surveying relevant literature, one writer concludes that "relatively small verbal estimates of time intervals tend to occur when emotional interest and preoccupation, particularly of an unpleasant kind, is present" (Orme, 1962b). This view is supported by two investigators who, on the basis of results obtained from psychiatric patients, conclude that "pleasantness is correlated with longer and more accurate time judgments along with an improvement of subjective time sense" (Melges & Fougerousse, 1966). Surely drive strength must play some role: the desire to rid oneself of an unpleasant state of affairs is likely to be stronger than one associated with pleasantness, regardless of the number of actions that are required. Also involved must be the attainment or nonattainment of the goal as such, a factor to be considered later in this chapter.

TEMPORAL HOPE AND FEAR

Smuggled into the principle of the last section is a parenthetical qualification: the goal of the functioning drive does not include "the end of the interval itself." The goal of having an interval terminate or continue can be very powerful; for example, merely waiting for a harmless experiment, involving intellectual activity, to continue produced a reaction sufficiently strong to affect the level of plasma-free fatty acid in the blood stream of the subjects (U.S., college; Back et al., 1967). In another experiment all the undergraduate subjects knew that they would be working on a humdrum intellectual task for 10 minutes. When they were interrupted almost halfway through and asked to estimate the elapsed time, those for whom the ending of the interval was desirable—either because they had been offered the privilege of being able then to leave class or, for those allegedly reaching a certain criterion, the magnificent prize of a box of chocolates—significantly overestimated the interval to a greater degree than the control group, which expected to return to class with no chocolate (U.S.; verbal, immediate subsequent; Filer & Meals, 1949). The subjective estimates of the experimental subjects gave them a push, as it were, in the direction they wished to go in order to be released from the task. Here again is wishful thinking, but in this instance not regarding the achievement of the goal directly but regarding the duration of the interval itself. This variable can be called *duration-dependence* in order to suggest that the achievement of the goal is dependent upon, or related to the duration of the interval. Some goals, for example, must be attained during a clearly specified period of time, others are less dependent on a time factor. The importance of that factor depends on the nature of the goal (whether attractive or unattractive), the portion of the interval being judged (whether elapsed or pending), and the perceived distance from the goal (whether viewed in terms of distance, time, or some other measure) during the interval as judgment is passed. These factors can be placed within the framework of a single principle:

> 6.2 *The stronger the duration-dependent drive accompanying an interim temporal motive concerning the pending portion of an interval, the greater the timing acceleration when that drive involves the interval's ending, and the greater the timing deceleration when it involves the interval's continuation; the reverse is true when judgment is passed concerning the elapsed portion of an interval.*

A corollary is needed to call attention to a factor which increases the strength of a duration-dependent drive, viz., the distance separating the person from goals to which that drive is directed. As originally formulated on the basis of animal experimentation (Miller, 1944), the goal-gradient hypothesis employed the concept of distance literally: the rat ran faster or was willing to endure more punishment as it approached the food at the end of the maze. Here distance is used in a more general sense to refer to perceived progress with reference to the goal, and may be measured in terms of elapsed or pending time, as well as by the degree of success or failure being attained while en route to the main goal.

> 6.2.a *The strength of a duration-dependent drive varies positively with a decrease in the perceived duration of the interval or the perceived progress toward the goal.*

The above corollary occupies a special position in our analysis because it seeks to specify two of the conditions affecting drive strength. Ordinarily we cannot explore this parameter because it would lead us too far afield: how can we possibly indicate the conditions related to drive strength or to the attractiveness or non-attractiveness of goals which catch people's needs or fancies? The corollary assumes that a goal is attractive but it does not, it cannot, indicate why the drive is duration-dependent or how the person knows what progress he is making. Halt, we must halt, and declare another Arbitrary Limitation. The most germane evidence is summarized in *Addendum 6.2*. In the remainder of this section, nevertheless, we shall indicate other special circumstances of a very general kind.

When the individual feels that objective time is passing too quickly—he wishes the interval to continue, not to end; or he would have it last longer—and if upon reflection that primary judgment remains unmodified, then he may be seized with what might well be called temporal fear or, if the fear is unrealistic, temporal anxiety. On occasion he merely feels that time is short, he must hurry if he is not to miss an appointment. But if he is neurotic, the fear may come to dominate his behavior unrealistically and he may be perpetually depressed by the fact that he is growing older every minute: "to be overconcerned with the passage of the years is to numb the freedom with which we live them as we pass" (MacIver, 1962, p. 122). On a primary level the tendency of a person in a state of temporal fear or anxiety is to overestimate the duration of the pending interval, but temporal information conveyed by reality modifies that judgment so that the secondary judgment proves disquieting and augments the fear or anxiety.

The solution to these troubles is either actually to prolong the interval or symbolically to have time stand still. Sometimes the interval may be lengthened: the student asks for more time to answer the examination questions. But when it refers to the interval which is man's lifespan or to an interval whose duration is fixed (as in a game), time cannot be halted, as each man to his sorrow probably notes on some occasion while contemplating his own mortality which is certain, though not precisely scheduled, or the rules of an enterprise which are arbitrary but usually also inflexible. Holding time in abeyance is attained not literally but effectively by legislators, athletes, and others who may cause a clock but not objective time itself to stop, so that they can achieve their ends during an interval which is thus artificially constructed.

The fact that time never in fact does stand still is painfully realized especially in connection with human drives. So much gratification, as we have been forcing ourselves to say from the onset of this chapter, is momentary; if it recurs, it does so only at regular or irregular intervals. The point of course must not be exaggerated: some gratifications, such as the satisfaction and convenience of a home or the reciprocal love of a husband and wife, do indeed endure over time. But otherwise the push forward exists, and people are impatient: they would achieve their goals sooner or they fear that time is too short either to achieve them or, when achieved, fully to appreciate or enjoy them. The hedonistic pill, which brings "everlasting" joy whenever it is ingested, is to be found exclusively in the Elysian Fields, since real-life joy comes not only from achievement but also from the battle against, or with, time to reach the goal of being joyful. In fact, coping not with momentary but enduring temporal fear is one of the cosmic problems requiring a solution by all thinking men. These kinds of problems we find embedded in human beings, no matter where we turn—and I suspect that they give rise to many of the proverbs which express and mold our attitudes (see *Addendum 6.3*). We have previously stressed the fact that a function of society is to regulate and allocate the scarce commodity of time which therefore acquires goal-attracting features of its own. Were time for each of us unlimited, impatience or temporal fear would be pointless: goals could eventually be achieved more or less at our own convenience, and so while experiencing or anticipating we would be sensible to relax. But you know that you will die.

What happens to wishful thinking when punishment looms at the end of the interval being judged? The punishment is avoided only as long as the interval continues, consequently its elapsed portion is likely to be underestimated and its pending portion overestimated—or at least this is what our principle proclaims. But the individual may also wish "to get it over with" when he knows the interval must end and culminate inevitably in punishment. The duration-dependent drive, in short, has as its goal here not the continuation but the ending of the interval. The man condemned to the gallows, we must speculate, may be in conflict *if* he is utterly without hope that his sentence can be commuted: he may want to live and have objective time pass slowly but he may want that time to pass quickly so that he may no longer be compelled to experience the agony of anticipating the unavoidable. Actual evidence on this point comes only from laboratory experiments in which the subjects are punished by mild shocks (*Addendum 6.2*). Under such circumstances the penalty for being sophisticated by hastening the punishment is slight; these young people might have responded quite differently if they had anticipated that they would be executed by pressing a button.

Getting the punishment over with means intervention: the individual controls the interval's duration by shortening it. The fear that time is passing too quickly, moreover, may function as a goad to more vigorous action to attain the goal during the allotted time. In these and other situations the temporal judgment clearly affects overt behavior. The reverse problem being examined here involves ascertaining the effect of that intervening behavior upon subsequent judgments. One point is clear: intervention of any kind affects the event potential and hence changes its scope, as a result of which—we have seen in the last chapter—the judgment may be altered. Otherwise, I see no way to evolve a principle which would indicate the effect of intervention upon drive strength. For on the one hand it can be convincingly argued that intervention must mean a strong drive, otherwise the drive would not spill over into action. On the other hand the very act of intervening may also mean the dissipation of some energy and hence a reduction in strength. Similarly it might be said that any kind of action makes subjective time seem to pass more quickly when you are bored, but some interventions can be more boring than private fantasies.

If there is nothing magical about intervention which invariably affects drive strength, and hence temporal judgment, in one way rather than another, it would be well to challenge at this point an implicit assumption of our analysis so far, viz., that the person passing judgment was, is, or will be the participant in whom the drives have been aroused or the one who intervenes. Instead the judgment may come from an observer, a person who stands idly by or who has only a vicarious experience. It might appear at first glance that there is a sharp distinction between participant and observer; between remembering how long it took you or your friend to reach a goal some months ago; between trying to estimate how long you or he has been trying to reach it; and between anticipating the time it will take you or him to reach it. But suppose he is ill and you are comforting him until the doctor arrives; simultaneously and independently you both may feel that time is passing very slowly and, indeed if you have some reason to know more about the seriousness of his illness than he, your conviction may be stronger than his. Both of you are plagued by inactivity with reference to the enemy, the illness, or the physician whose arrival you cannot hasten. You are not a mere observer; your friendship makes you also a participant.

This homespun illustration suffices, I think, to suggest that the critical difference between a so-called participant and a so-called observer is not only their involvement as such in the interval but also the significance they attach to it and the way in which they perceive its attributes. Obviously the winner of the race is appraised differently by the jockey, whose reward is relatively small, from the way it is greeted by someone in the grandstand who has bet a large sum on the horse; and hence the judgment of subjective time depends not upon their roles as participants in the race or as observers, but upon the meaning of the victory in terms of their own goals. Later the jockey may forget completely this particular race, but the lucky spectator may never forget it. In general, a participant may or may not know his own capabilities and limitations better than a competent observer, so that the two can well have different anticipations concerning the duration of a future interval in which the former intervenes.

ATTAINMENT

We now consider situations in which the impression of the person passing temporal judgment depends not only on the strength but also on the fate of the evoked drive. Does the individual attain his objective, is his temporal judgment affected by pending or actual success or failure? To try to comprehend what transpires in connection with goal attainment, it seems necessary to assume again that men everywhere have common *and* usual experiences. They seek goals connected with the drives that are within them and are evoked; that hungry man looks for food and does not, rather cannot, abandon the search until he finds some. If food is close at hand, the drive is quickly satisfied; if not, more subjective and objective time elapses; in either case, after obtaining food, the activity ceases, or another drive is evoked. Searching for or preparing food normally is more time-consuming than its consumption. In more general terms, nonachievement is likely to take longer than achievement because more responses are involved, especially under conditions of trial and error. Subjects looked at drawings of varying complexity and symmetry for a period of time significantly shorter when they were told to look "as long as the slide is pleasing to you" than when the instruction was to look as long as the figures were "interesting," as long as "you care to," or long enough so that they would be able later to "recognize" what they saw (Canada, nurses; Day, 1968).

In addition, since drives are recurrent, since natural and artificial scarcity exists everywhere and perpetually, and since frustrations are inevitable, everyone probably acquires the Guide that unpleasant intervals are both more numerous and longer than pleasant ones. These alleged facts coupled to a pessimistic conviction lead to a critical principle:

6.3 *The stronger the anticipation, the experience, or the recollection that a non-duration-dependent interval will be, is, or has been pleasant as a result of goal-attainment, the greater the timing deceleration; the reverse is true in connection with the unpleasant.*

Once more it may be argued, as we saw in connection with drive strength, that the significant factor is not the drive as such but the events to which that drive gives rise: fewer trials and errors are likely to be or to have been involved with satisfaction than with nonsatisfaction. In the latter situation the person keeps trying, hence more transpires. According to one view, "stress"—nonachievement—is accompanied "by an increase in neuro-physiological activity per unit of objective time" (Fredericson, 1951). Again, however, I think it important to retain the distinction between events and goals for three reasons. First, while it may be true that stress increases the number of events during the interval (and this may mean an increase in drive strength, as Principle 6.1 suggests), the reverse may also be true: the person who keeps trying may eventually succeed and the one who seems to be failing may abandon the goal—and our principle here would indicate that the former, though experiencing more events than the latter, will subsequently judge the interval to have been shorter. Then, secondly, when attention is directed to the attainment or nonattainment, the emphasis is not upon the discrete events but upon the final result. Finally, the experimental evidence seems to lend some support to the distinction (see *Addendum 6.4*).

Earlier in this chapter the variable of drive strength was considered under the condition of prior or interim arousal of the temporal judgment motive; and now we may well wonder about the interaction of drive strength and goal attainment. According to our two principles, the interval during which a strong drive functions is likely to produce timing deceleration, but not if the drive was, or is, being frustrated. Similarly a weak drive should lead to timing acceleration, but not if that drive was, or is, being satisfied. We are thus confronted with contrary predictions. Let D and d represent a tendency toward deceleration, and A and a a tendency toward acceleration. Further, let the capital letters refer to goal attainment and the lowercase ones to drive strength. Then the four possible combinations may be diagrammed as follows:

	Strength:	
Goal:	*Weak (d)*	*Strong (a)*
Attained (D)	D,d	D,a
Not attained (A)	A,d	A,a

The diagram immediately indicates a congruence of tendencies when the goal of a weak drive is attained (D,d) and that of a strong drive is not attained (A,a):

> 6.3.a *The tendency for goal attainment to be associated with timing deceleration is more probable when the drive is strong rather than weak; and vice versa.*

But what of the other two possibilities in the diagram, the ones going off in opposite directions (D,a and A,d)? They suggest a complicated interaction in which the two factors are probably not equally weighted. I would guess that the factor of goal attainment or nonattainment is likely to be the more powerful guiding standard than drive strength:

6.3.b *The tendency for goal attainment to be associated with timing deceleration is more likely to be the guiding standard than the tendency for strong drives to be associated with timing acceleration; and vice versa.*

This corollary means that, when the goal of a weak drive has not been attained (A,d), the impression will tend to be *long* and not *short;* or when the goal of a strong drive has been attained (D,a), the impression will tend to be *short* and not *long*. All this may sound tedious, but we are seeking precision of analysis which is not always achieved as the problem of time is discussed. Consider a typical laboratory experiment. Blindfolded subjects attempted to solve a stylus maze. The following variables were manipulated: drive strength, which was made either low by telling the subjects they would have to work on dull tasks for two hours regardless of how well they did on the maze, or which was made high by telling them that a good performance on the maze would enable them to be excused from the other dull tasks; progress toward the goal, which, they were informed, was either fast or slow; and alleged distance from the goal, which was reported to them to be 20, 50, or 80 percent of the distance yet to be traversed. Under low motivation, distance from the goal had no effect; under high motivation, however, temporal judgments of the elapsed time increased as they believed they approached the goal and regardless of whether they were told that their progress toward that goal was slow or rapid. The absolute differences were dramatic; for example, the elapsed interval of 6 minutes was judged to be in the neighborhood of 10 minutes under all conditions of motivation and when the distance from the goal was thought to be far; but it dipped to a mean of 3.8 minutes with high motivation, fast progress, and being "near" the goal (U.S., college; verbal, immediate subsequent; Meade, 1959). Thus motivation always played a critical role, but in association with the other variables.

We have repeatedly emphasized that in a book on time we cannot deal adequately with the problem of motivation: we have confined the discussion to drive strength and goal attainment. But another aspect of the subject we dare not ignore because it plays an important role in temporal judgments: the goal of the drive often involves not only the relevant satisfaction from attainment per se but also a decision as to whether that attainment compares favorably or unfavorably with that enjoyed by other persons or groups. From a nutritional standpoint, for example, a diet may be adequate, but the socially mobile person feels frustrated when he cannot have the luxury items enjoyed by neighbors whom he envies. His *deprivation* is not absolute but *relative*. Any aspect of the milieu can give rise to a feeling of relative deprivation, including even the need for air, which becomes a target for envy since some persons cool or warm their homes more efficiently or adequately than others. The feeling may be limited to the drive not being achieved: you are aware that you are not so strong or rich as your neighbor, an awareness that can bring you pain if you also envy him. Or it may extend to many facets of personality: as a person you are aggrieved by whatever achievements are associated with your status; your self, your ego, is easily injured even when you revel in comfort. Relative deprivation may be easier to endure than utter deprivation if you anticipate that with the passing of time your status can be equalized or improved. In brief, therefore, this kind of deprivation is likely to be frequently present in Western society and to affect both the strength of the aroused drive and the definition of achievement; hence temporal judgments so often are overestimations.

UNVERBALIZABLE CUES

The perceptual potential has now been described in terms of the interval's scope, the drives associated with it, and the temporal information supplied by the stimulus potential. But how in the absence of temporal information does the individual make a judgment on the basis of events and drives? The process may be conscious and deliberate: you are asked to judge the duration of a 15-second interval and you note immediately that you are offered no clue to its duration, that the interval is empty, and that you feel indifferent. But then, without further conscious deliberation—the kind more likely to characterize the secondary judgment—you find yourself thinking that a short time has gone by or that less than 20 seconds have elapsed. The mediation has occurred instantly, without much or any verbalization. To describe the transformation of perception into the primary judgment, therefore, I propose a continuum whose end points are verbalizable and unverbalizable cues, with semiverbalizable ones somewhere in between.

The evidence supporting the existence of unverbalizable cues comes from everyday experience and from systematic research. Somehow you attach a temporal tag to most of your experiences and your anticipations without being able to rationalize, that is, to verbalize how and why you have done so. You awaken from a deep sleep, it is still dark, instantly you know—or you think you know—roughly what time it is. Unwittingly you may have utilized cues—the darkness, the quiet, the state of your bladder, the feelings of sleepiness, etc.—but you would be hard put to explain why you think it is 3 (and not 2 or 4) o'clock. Investigations of this phenomenon will be summarized in the next chapter (see *Addendum 7.7*).

The most convincing research evidence suggests the possibility that temporal conditioning of the classical or Pavlovian type may often occur without a corresponding conscious awareness and hence in the absence of appropriate verbalization. In addition, we find investigators reporting again and again instances in which subjects give discrepant temporal judgments without being aware of the difference or the contradiction. The following observation concerns an experiment with schizophrenic patients:

> If an interval of 15 seconds is presented to a subject, he may be able to reproduce it fairly accurately regardless of what his verbal estimation of the interval would be. It could be called 5 seconds or 50 seconds, but the reproduction would still be accurate as long as he is consistent in his judgment from demonstration to reproduction [Clausen, 1950].

In this situation, the experimenter's light, visible and hence perceived for 15 seconds, might have been expected to evoke a verbal response corresponding roughly to the duration of the interval which in turn would guide the reproduction of the interval, provided that the verbal response itself was more or less accurate. Evidently this did not take place: the subjects must have based their reproduction upon non-verbal cues or upon verbal ones different from those giving rise to their verbalization while employing the Method of Verbal Estimation. Non-verbalizable cues we have already mentioned in connection with their role in arousing the temporal motive at the basis of the biochemical clock; now we are saying that both the clock and the cues (whether related to that clock or not) account for some of the primary judgments that are passed. Other relevant evidence can be found in *Addendum 6.5*.

This evidence, though strongly supporting the existence of unverbalizable cues, does not unequivocally suggest when such cues are utilized. I can only venture the following principle:

> 6.4 *Unverbalizable cues are more likely to affect the primary judgment than are verbalized or semi-verbalized ones under one or more of the following conditions: when the interval is ephemeral; when it is not consciously experienced; when it is not deliberately organized as an experience or activity; when it is judged speedily.*

The reasons for each of the conditions are purely ad hoc. Ephemeral intervals slip by so rapidly that there is little or no opportunity to respond to many cues. When you awaken and decide it is time to get up, you are likely to rely upon unverbalizable cues because you have been asleep, and hence more or less unconscious, during most of the interval being judged and because you have little or no material to organize in the few seconds during which you pass judgment. On the other hand, if you try to remember how long ago it was that you had your last swim in the ocean, the interval from then to now is extended, you remember a great deal of what took place in the meantime, you try to be as rational as possible in fixing the date, and presumably you do not make your decision instantly.

ADDENDUM 6.1 THE REDUCTION OF DRIVE STRENGTH TO EVENTS

Obviously I cannot deny that one of the basic assumptions behind Principle 6.1 may often not be true: a strong drive can lead not to more but to less action than a weak one. Or a weak drive may facilitate attention to details: the person is less absorbed in goal-seeking and may distract himself by paying attention to his milieu or his thoughts, and therefore he injects more events into the interval which, as a result, appears longer and not shorter. There is no gainsaying this possibility; I can only suggest that it seems less probable.

Another similar and equally vexing question is whether drive strength can be reduced completely to the number of events perceived within the interval. An earlier writer noted that she herself overestimated the intervals spent traveling by train, whether she was in "a stuffy railway carriage with a screaming baby in it" or "in a compartment which was quiet and sufficiently airy," simply because she was "continually entertained with the scenery or the movements of her fellow travellers" (Sturt, 1923). I rather doubt that the overestimation would be the same in both instances, though she could well have noted internally the same number of events. At any rate, the analysis only mentions the drive factor in passing and instead concentrates upon those events. Another writer raises this approach to what he calls a "law": "any factor which contributes toward an increase or decrease in the number of changes observed has the effect of lengthening or shortening the apparent duration" (Fraisse, 1963, p. 219). But when he describes specific experiments on temporal judgment, he invokes motivational factors which, I feel, do not affect the impression of "the number of changes." For example: "if we increase the difficulty of a task, without fundamentally changing its nature, it will seem shorter" (ibid., p. 222). In order to maintain the law, moreover, he must further assume that "stronger motivation leads to better organization" and that such organization in turn produces an impression of fewer changes or events, which may or may not be the case. Finally, when he reports his own experiments and those of Piaget on children and indicates that on occasion some children use the criterion of work accomplished to pass temporal judgment and on other occasions base their judgments on the quantity of activity they have experienced, he has to squeeze and squeeze in order to embrace both phenomena under a single category. Thus children were asked to judge the time it took two toy cyclists to traverse unequal distances on a table. Actually A and B departed and stopped simultaneously but A went twice as far. The children who thought A moved for a longer period of time based their judgment "on the work accomplished: A achieves more than B." The others who believed it took B longer, the writer suggests, must have been identifying with that cyclist; then "when you fall behind the others and go slower, it is because the task is too difficult, and when this happens you take more notice of every movement which costs an effort." Both groups, therefore, "judge the duration according to the changes experienced, one directly and the other by identification" (Fraisse, 1963, pp. 241–2). It could be argued that more than counting is achieved by identification with the slower cyclist: in the West, the cyclist going a shorter distance loses the race and has not accomplished as much; identification by the child could be said to lead to vicarious frustration and hence to overestimating the temporal interval.

Then, in a previously mentioned experiment in which undergraduates overestimated elapsed time, when they wished the interval to end in order to leave for home or to receive a prize of chocolates, it was also found that these experimental subjects produced more responses than the controls who were not motivated to have that interval end (U.S.; verbal, immediate subsequent; Filer & Meals, 1949). But could the drive factor really be eliminated by emphasizing the difference in the number of events? Apparently not, for in fact there was "no relationship" between the work thus produced and the temporal estimations.

I shall stop the quibble by agreeing that the event attribute is related to the drive factor and that genetically it may be responsible for the temporal impression that factor creates, but the quality of the experience also plays a role in affecting the temporal judgment, especially when the interval is pending.

ADDENDUM 6.2 DURATION-DEPENDENT DRIVES

A frequently cited experiment suggests how temporal judgments depended upon whether or not the continuation of the interval was related to drive satisfaction. Subjects were asked to solve puzzles which actually were insoluble, a fact that was withheld from them. First, they were given a boring "practice" period which, they were told, prepared them for the testing period; throughout the second period, they believed their intelligence, a highly valued attribute, of course, was going to be tested. After each of the two periods they were required to estimate the elapsed time. The result: the "practice" interval appeared longer to more of them than the testing interval (U.S., college; immediate subsequent; Rosenzweig & Koht, 1933). During the first period they wanted the interval to end and they estimated the elapsed time to be relatively long, which brought them closer to its termination; during the second, they wanted the pending interval to be long enough for them to finish the puzzle and thus demonstrate a high I.Q. and, therefore, they presumably estimated the elapsed time to be relatively short. But the experiment has been criticized: a task preformed a second time under different drive conditions is also a more familiar one and for that reason—rather than for one associated with the goal of having the interval end or not end—it might appear shorter (Meade, 1960c).

The critic just cited, and others, have performed a number of experiments which indicate how the strength of the duration-dependent drive can be increased. In one, subjects who were informed concerning their alleged progress in working upon the usual kind of laboratory task tended to judge the elapsed time to be shorter than those who were not informed (U.S., college; verbal, probably immediate subsequent; Hindle, 1951); in similar situations, other subjects needed no help from the experimenter, their own estimates concerning the distance from the goal produced the same sort of effects (U.S., college; verbal, immediate subsequent; Meade, 1959 and 1960c). Alas, this evidence does not enable us to say whether the principle under discussion is validated or not because we are not told how these students felt about the duration of the interval during which they were subjects. If we may assume that they wished the interval to end so that they could escape from the laboratory and go home—a likely assumption—then their judgments concerning the elapsed time should have increased; but if they were enjoying themselves, and wished the exercise to continue, or if they felt challenged to perform well and believed they needed more time to do so after hearing the progress report, their judgments should have decreased. These two studies as they are reported do not enable us to choose between the two alternatives.

Later experiments by the same investigator, however, provide helpful hints concerning the psychological issues that may be involved. With high motivation to achieve a goal, temporal judgments concerning elapsed time decreased if subjects felt they were progressing, and increased if they believed they were not progressing; but with low motivation, rate of progress made no difference (U.S., college; verbal; immediate subsequent; Meade, 1959 and 1960d). This result was obtained with intervals having a duration of 5 or 6 minutes, but then it was impressively replicated with longer intervals ranging from 15 minutes to 1 hour. Subjects worked, as others had in previous experiments, blindfolded on a stylus maze when highly motivated (they were told their intelligence was being tested) or poorly motivated (they were told that their work on the maze was unimportant and probably a waste of effort); while working they were given information which indicated they were progressing slowly or quickly. The judgments for the various intervals under conditions of high motivation and fast progress were always less (significantly so for three intervals, almost significantly for the fourth) than those under high motivation and slow progress or under low motivation and either kind of progress (U.S., college; verbal, immediate subsequent; Meade, 1963). When poorly motivated or making no progress, the investigator suggests, the subjects were bored, satiated, or frustrated; their estimates were longer than clock time, if not always significantly so—they wanted the interval to end, and that end would be closer if elapsed time were considerable. In contrast, those imagining they were making good progress under conditions of high motivation had not yet completed the tasks when they passed judgment; they had an incentive to have the interval continue, and they were being encouraged; hence they wished elapsed time to be short. Both groups must have assumed that eventually the testing interval would end: those motivated wanted the time to be prolonged, those unmotivated to be shortened.

That the anticipation of a punishment at the end of an interval leads not to the under- but to the overestimation of elapsed time is illustrated by an experiment in which subjects were instructed to produce a 5-second interval by activating a motor which moved a platform on which they themselves were standing blindfolded. In all instances, they underestimated the interval (i.e., they stopped the motor before 5 seconds in fact had elapsed); but the degree of underestimation was significantly less when they knew they were moving toward a dangerous stairwell than when they were moving away from it (U.S., adults; Langer, Wapner, & Werner, 1961). Obviously they wanted the interval they were producing to end as soon as possible as they approached the stairwell, for only in that manner could the danger be averted (let me in parentheses indicate that the investigators themselves speak of "overestimation," by which they mean that the produced intervals, being shorter than the designated time, were overestimated by being called equivalent to objectively longer ones; in accordance with my chart on The Length and Speed of Time, I call that underestimating the objective interval of 5 seconds which was supposed to be produced).

In another study the subjects were instructed to produce an interval of 15 seconds by pressing a button after being warned half the time by a signal that they would "feel a slight shock" and on other trials by a signal that they would "hear a bell ring"; they thus anticipated either an unpleasant or a neutral experience at the end of the interval they were producing. These two anticipations did not significantly affect their temporal judgments, but this group continually produced shorter intervals than a control group which received no terminal shock. Apparently an anxious set was acquired after anticipating the shock part of the time, and thereafter all judgments were affected even when the experimenter indicated that no shock was forthcoming (U.S., "mainly" college; prior; Falk & Bindra, 1954).

It may well be that immediate subsequent judgments follow the same principle as those occurring during the interval. One experimenter, for example, made it perfectly clear to his subjects that they would or would not be shocked at the end of an interval of 5 or 20 seconds whose duration they were asked to estimate verbally; he conspicuously attached or detached the electrodes providing the shock. Under these circumstances judgments were significantly longer when the shock was anticipated than when it was not (Canada, adult clerks; Hare, 1963b). It could be argued that the unpleasant rather than the neutral anticipation produced the longer judgments because the subjects in the former condition were highly motivated to have the interval end in order to be rid of the shock—but of course they had already received, or not received, that shock when passing temporal judgment. The investigators themselves venture the opinion that the "expectancy of some motivationally important events," and not stress or anxiety, was involved, so that this kind of overestimation can occur when a pleasant event is anticipated, such as the prize of chocolates in the experiment mentioned in the text (Filer & Meals, 1949); in effect, they place emphasis upon drive strength.

Interaction between the strength of the drive (as inferred from the size of the reward) and the time remaining to reach the goal has been observed in an investigation in which schizophrenic patients were confronted with the task of depressing a switch at a point in time half way between 5.9 and 20 seconds. They first practiced on a series of neutral trials when a red light went on if they pressed for 5.9 seconds or less, a green light if they pressed for 20 seconds or more, and a white light if they pressed within the specified range. They were then told they would be given a reward for cooperating during the next series, $1 or 10 cents, both of which were meaningful in terms of their purchasing power within the institution where they were living. In addition, some were and some were not informed concerning the number of trials in this series and, after each judgment, the number remaining until the end. This latter knowledge which indicated the temporal distance from the goal had no effect upon their performance, but the size of the reward did: those promised $1 pressed for shorter periods of time than those promised 10 cents (U.S.; prod., prior; Simmel, 1963). The knowledge that the interval would end, however, probably was not disturbing since these subjects knew they would be rewarded regardless of their performance.

ADDENDUM 6.3 PROVERBS

What sources should be consulted to discover how time appears in proverbs? There seem to be two kinds: those that have been collected in a particular non-Western society, usually by an anthropologist; and the collections which are found on reference shelves in major libraries. Either has disadvantages: the former concentrates on a single society and hence indicates trends only within a limited group; and the latter is motivated by some nonscientific purpose, such as felicity or economy of expression. I offer examples of each:

1. I have combed through the 416 proverbs and 74 "sayings" (almost, but not quite, proverbs) which have been obtained from the Jabo tribe in Liberia and find, quite arbitrarily but most painstakingly, that only five seem concerned with time.

"No time is set for death" certainly suggests the inevitability of death but offers some consolation. "Rice says: do not weep for me, weep for your dead relative" indicates, according to the compiler, that during the next season—with the passing of time—the rice field which has been damaged by animals will yield another crop, whereas the dead person will not return. The other three seem to deal with the proper allocation of time and may also suggest patience: "The fruit must have a stem before it grows"; "It is the full-grown forest that yields the buffalo"; and "Sheeps says: if you set a date for your man, he will come" (Herzog, 1936, pp. 31, 33, 37, 99, 225). Similarly, an analysis of 113 proverbs prevalent among the Mashona of Rhodesia indicates that only three of them refer to seasons or to time (Irvine, 1970).

2. Western collections are so numerous that I see no way to follow a rational procedure in selecting one. Instead I have let availability guide me—which meant sheer chance—and I dipped into one book which happened to be on the shelves of the only library to which I had access at the time (University College, Dar es Salaam, Tanzania). I examined all the proverbs listed under the heading of "Time" in "A World Treasury of Proverbs from Twenty-Five Languages" (Davidoff, 1953, pp. 413–15). No systematic principles guided the compiler other than the desire to find pungent ones which could be expressed in less than twenty words, and an admitted bias in favor of utilizing English and American sources. I find that proverbial sagacity concerning time can be easily grouped under five headings, each of which I illustrate with a single proverb from the collection. There are 106 in all, and a category is preceded by the number which have been so classified:

35: time is precious; "lost time is never found again"
27: time inevitably conquers; "time and tide wait for no man"
22: time is helpful; "time works wonders"
16: time should be wisely allocated; "there is a time for all things"
6: time can be conquered; "the clock does not strike for the happy"

Noteworthy is the fact that the two most frequently employed categories, accounting for over half of the proverbs, suggest both the scarcity and the irreversibility of time. There is thus fear that time is precious and fleeting.

ADDENDUM 6.4 ATTAINMENT VS. NONATTAINMENT OF GOAL

Experiments in this area, though conducted in the laboratory, seek to arouse genuine emotions. The procedure is necessary when the variable is that of drive, but gives rise to complications, as is to be anticipated, when human behavior is not compartmentalized. Some of the studies require explication. The following *support* the principle that there is a positive relation between pleasantness and timing deceleration:

1. The task: viewing 1-minute dance sequences portraying either dots or girls dancing, respectively, at three or four different tempi. Variable: the tempi and the reaction thereto. Result: the faster the tempo, the more quickly time seemed to pass (U.S., college; immediate subsequent or prior; Goodchilds, Roby, & Ise, 1969). Certainly it seems fair to say that faster tempi meant more events, but the temporal judgment was not correspondingly prolonged. Instead, the critical factor must have been the fact that the faster tempi were more clearly preferred by the subjects and hence presumably brought them more satisfaction. My presumption is that the judgment concerning the speed with which time seemed to pass was either the primary one involving the apparent duration of the interval or the secondary one based upon a knowledge of objective time.

2. The task: superimposing geometrical forms upon pages of zeroes. Variable: "monotony." Result: a non-significant tendency for underestimation to be associated with performing the task before getting bored, in comparison with performing the same task again and again and becoming bored. Overestimation, moreover, increased as the task was repeated and became more monotonous (U.S., college; verbal with immediate subsequent & prior; Burton, 1943).

3. The task: solving mazes by learning to select the correct sequence of letters which indicated the direction to be taken at each turn. The variable: experiencing success or failure as manipulated by the experimenter. The result: the elapsed time was estimated to be shorter, almost always under conditions of success, than under failure (U.S., college; verbal with immediate subsequent which probably became prior; Harton, 1939b). The same investigator points out another variable he has

found to be crucial; viz., the difficulty of the task. He has also shown that not the number of events but the experienced difficulty of the task affected judgment: in one instance, increasing the difficulty of counting, or grouping metronome beats *during* the act of reproducing, led to longer reproductions (so that time "passed more rapidly" with increased difficulty: *longer* intervals had to be used to fill the difficult ones); but increasing the difficulty of the activity performed *during* the intervals being judged decreased the verbal estimates (U.S., not specified; prior; Harton, 1938). How can "difficulty" be interpreted in terms of drive attainment or pleasantness? We could vindicate our principle by maintaining that difficulty during reproduction may have been unpleasant since it was distracting, but that difficulty during the intervals being judged may have increased the pleasure derived from performing essentially dull tasks; however, such reasoning obviously is after the fact and therefore to be mistrusted.

4. The task: becoming ostensibly "adapted" to 2 minutes of darkness and observing the "movement" of a stationary light (an illusion, the so-called autokinetic effect) for the same length of time. The variable: the number of events in the two periods. The result: the duration of both intervals was overestimated, but the overestimation was greater for the period of darkness than for the illusion (U.S., adults; immediate subsequent, verbal; Kafka, 1957). One could say, as the investigator in fact does, that fewer events occurred during the interval and hence the finding contradicts any principle which asserts a positive relation between number of events and magnitude of the primary judgment. But it might also be maintained that waiting in the darkness produced greater unpleasant tension, whereas the actual appearance of the light, harmless as it turned out to be, brought relief and hence underestimation.

5. The task: passive waiting for 6 minutes and solving mathematical problems for the same length of time. The variable: the cultural background of the subjects. The results: American undergraduates and Indian students from three communities estimated that the time spent solving the problems passed more quickly than the period of waiting; but Indian students from four other communities judged the intervals to be approximately equal. The investigator, having previously established that intervals of progress were underestimated, ventures the opinion that the Americans and those Indians responding like them must have had in general stronger drives to achieve, and hence they made note of their progress in solving the problems and considered the passive interval a waste of time; whereas the other Indians, coming from cultures which do not value personal achievement, did not care whether they were progressing or idle (subsequent; Meade, 1968). If this interpretation is correct, then the event attribute during the active period was not overpowering; instead the Americans and some Indians must have been obtaining satisfaction from achieving solutions. This interpretation has been impressively supported by a subsequent study. Again samples were taken from the same Indian communities; an identical procedure was followed, except that the task was changed and the variable of motivation was experimentally manipulated by informing half the subjects that the task had "no particular meaning" and the other half that it was "a test of intelligence." Here exactly the same division between the subcultural groups as in the previous study appeared: those from the cultures valuing achievement responded to the motivational factor by calling the time significantly shorter when highly motivated than when neutrally motivated, but those from the other four communities responded similarly regardless of motivation. One of the latter, however, did decrease its temporal judgment under conditions of high motivation when informed that it was making rapid progress and not when it believed it was making no progress (verbal, immediate subsequent; Meade & Singh, 1970).

6. The task: attempting to solve a stylus maze while blindfolded. The variable: false information provided by the experimenters concerning the progress being

made during or at the end of the task. The result: the subjects gave the longest verbal estimates when they were making or had made "zero progress"; longer estimates were given by those who were told they were making or had made "backward progress" than by those led to believe that "forward progress" characterized their efforts, although the difference was significant for only one of the intervals, that of 30 minutes (U.S., college; immediate subsequent; Meade, 1966b). The tendency for the shortest judgment to be associated with "success" is certainly in line with our principle. That the longest judgments were associated with "zero" rather than with "backward" progress may possibly mean that the belief that they were standing still produced more frustration in the subjects than the conviction that they were sliding backwards. Such an interpretation receives some support from a similar result obtained by the same investigator. This time his subjects were told that they should avoid a particular region in the maze because there they would receive a painful electric shock. Again the "zero progress" group gave the longest estimate, with no significant differences between those allegedly going toward or away from the shock (U.S., college; immediate subsequent; Meade, 1966c).

Negative evidence against Principle 6.3 is also at hand. In one experiment subjects perceived secular and religious words for 0.25 seconds and were asked to compare the length of time they had viewed each. According to the principle, the words with pleasant connotations should have produced shorter judgments than those with unpleasant ones. In fact, the only significant tendency to emerge was for the disbelievers to make shorter judgments for *both* classes of words. It could be, as the investigator suggests, that the subjects unconsciously reduced the time they perceived the religious words which, one might assume, they found disagreeable and that this tendency generalized to the secular words (Australia, presumably students; comp., prior; L. B. Brown, 1965). If this interpretation is correct, then not the pleasantness of the interval but the duration-related drive was the crucial factor; still it is difficult to imagine how that drive could have functioned during such an ephemeral interval. In another study, one half the subjects worked with stylus mazes until they were satiated: they refused to go on, or they reported or exhibited symptoms of satiation. The other half were mercifully halted after achieving a criterion of success. Of the satiated, 87 percent underestimated the interval, in contrast with only 48 percent of the non-satiated (U.S., college; verbal, subsequent; Berman, 1939). This result, which flatly contradicts our principle, however, is equivocal. For the satiated subjects worked much longer than the non-satiated, so that intervals of different length as well as content were being compared. In addition, the manipulation was not valid for all subjects: some of those in the condition of satiation reported not being bored, and vice versa.

In one study it was assumed that students who did not know the examination grade they were about to receive were under greater stress than those who had the information; the task was not to judge the elapsed time as such but to answer scaling-type questions such as "What would seem to you to be a very long time to wait to see your examination?" Thus measured, the temporal judgments of those allegedly under stress tended to be *shorter* than those without stress (U.S., college; Sattler, 1965). I do not think that this finding contradicts Principle 6.3, rather the measuring device required the subjects to think not of their tension but of the end of the interval which would bring relief; hence another principle must have been functioning (and the investigator himself points to the possibility that the subjects without knowledge of their grades must have had a "desire to have time pass more quickly than when they have knowledge of a grade"). The fact that the kind of grades they were told they would receive had no significant effect upon judgments would substantiate the view that the duration of the interval was the critical factor: if they knew their grades, good or bad in comparison with previous examinations, they were not being held in suspense.

Another study I also find difficult to interpret. First, the subjects were exposed to odors that were pleasant, unpleasant, and neutral. Then they were exposed to one of the three and told to wait before they would experience the same odor again, ostensibly to overcome olfactory fatigue. Without prior warning (that is, with immediate subsequent arousal of the temporal motive), they judged the duration of this interval of waiting first by the Method of Reproduction, then by Verbal Estimation; the interval in fact was either 48 or 144 seconds. The result: no relation between the affective tone of the odor that had been anticipated and the temporal judgment. In addition, the tone of those odors also had no effect upon the judgments of other subjects who actually experienced them during intervals of the same length. Most important is the fact that the intervals of anticipation always tended to be judged longer than those of the actual experience; and there was no significant interaction between the affective tones on the one hand and either the anticipation or the actual experience on the other (U.S., college; Schiff & Thayer, 1968). Certainly the failure to find differences within the act of anticipating or experiencing suggests evidence contrary to our principle. It may be, however, that the goal of the interval of anticipation involved not the anticipated odor but the interval's end and that this duration-dependent drive was stronger than the drive experienced while perceiving any one of the odors.

Finally, one study is cited to illustrate the interaction of the variables. The subjects first sorted cards and then, immediately after verbally estimating the time it had taken them to complete the task, some were told that they had done miserably and that this reflected unfavorably upon their IQs; others that they had not responded as well as their peers; and the rest, the controls, were told nothing. Next they repeated the same task and again estimated the interval. The males in the failure group underestimated the interval to a significantly greater degree than all the females, the other males whose stress resulted from comparison with peers, or the control group. In contrast, the experimental manipulation had no effect upon the females (U.S.; Greenberg & Kurz, 1968). The investigators feel that the sex difference resulted from a higher achievement motive among males in our society than among females.

ADDENDUM 6.5 CUES

Conditioning without corresponding verbalizations is reported in two studies. In the first, cortical alpha rhythms, measured electrically, were conditioned to a sound preceding the unconditioned light stimulus by 9.4 seconds; the mean of the conditioned responses which were established was 8.2 seconds, with a standard deviation of 0.7 seconds. But there was "very little relationship" between the physiological responses and the subjects' judgments concerning the interval of delay, which were ascertained by having them press a button when they thought "the conditioned time interval had elapsed": these conscious judgments had a mean duration of 10.5 seconds and hence were longer than the conditioned responses; the standard deviation was 2.5 seconds and hence judgments were more varied than those responses (U.S., adult males; repro., prior; Jasper & Shagass, 1941). In the second the autonomic nervous system, as measured by the galvanic skin response, was successfully conditioned to respond after an interval of 40 seconds had elapsed, during which time the subjects engaged in absorbing and distracting mental activity. Afterwards they could not give correct verbal estimates of the interval which their nervous systems, as it were, judged accurately (U.S., college; Lockhart, 1966).

Variations in the use of Verbal Estimation have produced relevant evidence. One investigator attempted to determine whether intervals of 1 second, 10 seconds, and 1 minute could be judged more accurately on an unconscious or conscious level, as inferred from "voluntary and involuntary responses to questions asked" while producing these intervals; no differences could be detected (England, "normal individuals"; prior; Scott, 1948). Hungry subjects tended to overestimate an interval of delay before being fed to a greater degree than those who were not hungry; but the two groups did not differ with respect to their impressions of whether time during that interval seemed to have been passing slowly or quickly (U.S., female college; subsequent; Schönbach, 1959). Of the 14 subjects (out of a total of 15) who underestimated the duration of an interval of 1.5 hours during which "sensory input" was drastically reduced, 1 gave the subjective report that time had passed more quickly than usual, 8 that it passed more slowly, and 5 that it passed normally—again little relation between subjective feelings and temporal judgment could be ascertained (Canada, unspecified; presumably immediate subsequent; Banks & Cappon, 1962).

In addition, inconsistencies have been revealed again and again between judgments obtained verbally and simultaneously through other methods. No relation, no consistent relation, or significant discrepancies have been reported between judgments obtained through the Method of Verbal Estimation (whether expressed in terms of feelings, speed, distance or clock time) and the Method of Reproduction among subjects under the influence of an anesthetic, nitrous oxide (England, adult, prior; Steinberg, 1955); among psychiatric patients and matched controls (Canada; prior, Cappon & Banks, 1964); among subjects whose body temperature had been raised, on the average, almost 2 degrees Centigrade (England, adults; prior; Bell & Provins, 1963). Similar results have emerged when judgments via Verbal Estimation and Reproduction were compared in the latter two studies just cited as well as among subjects who evaluated their own reproductions (Italy, unspecified; prior, Iacono, 1956), or who were tested most carefully in the laboratory (U.S., college; prior; Woodrow, 1928).

In some investigations we have clear-cut evidence that the subjects themselves recognize the possibility of a discrepancy between their own judgments and objective time. Those in the Italian experiment just mentioned were able to state whether they were satisfied with their own reproductions of intervals in the neighborhood of 1 second and whether they considered the reproductions too long or too short (Iacono, 1956). An adult who had been isolated in a soundproof room for three days stated that "for some reason or other I have a feeling that it must be dark, or nearly so, outside, although according to my reckoning of the time it should still be daylight." Obviously, in the absence of the ordinary cues on which his temporal judgments were dependent, he was trying to "reason" but some other basis for the judgment was present within him: "The important point is that the feeling was there, that it disagreed with his more rational judgments of time, and that it could not be accounted for on the basis of simple sensory cues," as a result of which he always had some sense of time (U.S.; MacLeod & Roff, 1936).

Actual verbalization, moreover, may provide unreliable clues to temporal judgment which therefore must depend on nonverbalizable processes. In one investigation half the subjects—in whom genuine anxiety had been created by the fact that they were about to speak in public before a "critical" audience— reported "a subjectively altered experience of time" when asked to make temporal estimates concerning intervals ranging from 1 second to 19 minutes. Some of them felt time was passing slowly, others more quickly. There was, however, "no relationship between the way in which the passage of time appeared to be altered and the directions of the deviations" from objective time. The medically trained subjects may have tried to ignore their primary judgments while anxious, perhaps because experience had provided an appropriate

Guide. They did so, in comparison with the normal state, however, at a price: their judgments became less accurate as measured by test-retest reliability (England; prod., repro., & verbal with prior; verbal, immediate subsequent; Cohen & Mezey, 1961). In a similar but less well designed situation, it was first found that one group of undergraduates believed that stage fright would increase temporal judgments; a similar group—and presumably with the same belief—tended actually to give shorter estimates of the time spent speaking than the time spent doing nothing beforehand, and there was no relation between self-ratings regarding stage fright and the accuracy of the temporal judgments (U.S.; Henrickson, 1948). When called upon, without prior warning, to state how they were able to judge the duration of a 20-second interval during which they had been engaged in an irrelevant task, 24 percent of a large sample of undergraduates could provide no explanation, 27 percent claimed they possessed this kind of ability as a result of past experience, and the remainder said they had reasoned in various ways, such as recalling their activity during the period itself. There was, however, no significant relation between these replies and the actual judgments, except that those giving longer estimates tended to mention a relevant experience in the past and those giving shorter judgments tended to try to recall the interval in order to judge it (Doob, 1951).

Indirect evidence can be quickly surveyed. No relation was found between the time it took subjects to arrive at a judgment enabling them to reproduce intervals varying from 0.1 to 1.0 seconds and the actual duration of those intervals (India, university; prior; Chatterjea, 1960); perhaps even with such ephemeral intervals it might have been reasonable to expect, if verbal cues had been present, some orderly fluctuation in the decision times. Similarly no relation existed between reaction time and the accuracy with which other ephemeral intervals were reproduced (U.S., boys; prior; Philip & Lyttle, 1945). Diurnal fluctuations in estimating the time of day without consulting a timepiece—such as the reported tendency to overestimate the time between 8 and 10 A.M. and between 6 and 8 P.M., to underestimate it between those two periods, and to be least accurate between 2 and 4 P.M. (U.S., adults; Thor & Baldwin, 1965; Baldwin, Thor, & Wright, 1966)—may well be correlated with fluctuations in fatigue, anxiety, or frustration; but the effects of such fluctuations upon temporal judgments cannot possibly follow a verbal route. Even more nonverbal must have been the tendency for past and future events to appear less distant at noon than when appraised during the same morning and evening hours (U.S., college; Thor, 1962b). And one final note: young children between the ages of five and seven could reproduce a series of sounds as accurately as adults even though they had not yet learned to count (France; prior; Fraisse & Fraisse, 1937).

7. SECONDARY JUDGMENT

Animals as well as men, we have suggested, respond to temporal cues and often demonstrate a keen ability to regulate their behavior with reference to objective time. These cues are probably derived from classical conditioning which always involves a temporal sequence: first the unconditioned and then the conditioned stimulus—and they come to represent particular or general situations. In an anthropomorphic sense, they are the Guides we have mentioned during the discussion of the event and drive attributes of intervals. Probably animals remain on the primary level since they lack sophisticated language and hence presumably a developed proclivity to reason, to anticipate, and to renounce, in a deliberate manner.

What happens after the first impression of duration or succession on the primary level? The primary judgment is likely to be revised or refined: a secondary judgment is passed which in turn may or may not undergo additional changes. This process can be illustrated experimentally. Subjects paid close attention to random digits for 20 minutes as they recorded the digit following each appearance of the number 6; next they were asked to estimate the duration of this interval; and immediately after that they were told to pass judgment upon their initial judgment by means of a 5-point scale indicating whether they considered that first judgment of theirs to be high or low. On a secondary level they tended to make significantly accurate judgments; thus the mean primary judgment of those considering the judgments "pretty high" had been 30.0 minutes, that of those saying "pretty low" 11.7 minutes (U.S., college; Bakan, 1962). This chapter, therefore, addresses itself to subsequent judgments which fold back on each other.

THE STANDARD

Revising or refining a primary judgment involves a series of intervening or mediating responses that influence the secondary judgment, that originate in previous experience or the stimulus potential at hand, and that are aroused by that potential. In addition to the various Guides already mentioned, one response is inevitably evoked: the perception, remembering, anticipation, creation, or utilization of a standard or a standard interval. The ultimate basis for the standard may be a natural phenomenon such as the rising and setting of the sun; but, since such events tend to be unreliable —north and south of the equator the sun rises at a different time each day, and everywhere it is sometimes obscured by clouds—and since communication between people is difficult without a tacit and invariable agreement upon one standard, timepieces and also calendars are preferred as the reference points in most societies. In laboratory experiments, except when the Method of Absolute Judgment is used, the experimenter conveniently supplies the standard. Even when the judgment comes from some kind of unverbalizable cue, including one provided by the biochemical clock, the person in our society who is challenged to justify his judgment usually is able to suggest not necessarily ex post facto reasons for his judgment.

More formally, the standard is some unit providing the comparative basis for passing temporal judgment; metaphorically a yardstick is employed. Perhaps you can decide whether you like or dislike a musical sequence without delving into your past experience or awaiting one in the future, but it is highly unlikely, I believe impossible, for you to say whether the interval has been *long* or *short* except on a comparative basis: *long* or *short* according to what standard? The time seems *long* you say: even if this judgment of yours remains on a primary level, you must have in mind some appropriate or relevant standard of length whether it be the ticking of a clock, the sound of a siren at noon, the gestation period of an elephant, or the erosion of a rock by the side of the sea. You say you think an hour has passed; your standard here is a subjective one having a symbolic relation to the conventional unit of clock time. But if possible, before expressing your secondary judgment, you undoubtedly consult a clock or try to consult one, and then you decide on that basis whether you have under- or overestimated the interval. You have learned to have more confidence in the objective standard offered by an instrument than you have in your own subjective standard.

7.1 *An objective standard for passing temporal judgment provided by the Method of Chronometry is more likely to be employed than any other standard when it is available as a temporal symbol within the stimulus potential, perceived as temporal information within the perceptual potential, and evaluated as accurate.*

I do not think that this principle is ethnocentric from the standpoint of non-Western peoples. For any society through sheer necessity, we have indicated, must objectify standards for passing some temporal judgment.

The principle simply states that these preferred standards are likely to be consulted whatever they happen to be. In fact, if the individual does not rely upon them, then we shall see in a later chapter that his sanity may be seriously askew.

If any objective standard is in fact present—the person is able to look at a clock, a calendar, a star, or a written record—then our fascinating problem largely disappears: the stimulus potential contains this temporal symbol, adequate knowledge is provided, the primary judgment is then confirmed or adjusted, and no further analysis is needed except perhaps to determine whether the person reacts correctly to the symbol. In continuing this analysis, we assume the absence of such an objective standard and, for the moment too, we further assume the absence of the kind of arbitrary standard provided by experimenters for their subjects.

It is clear that the precision provided by the Method of Chronometry, while useful in determining how quickly a motorist can react to danger, or when a chairman should call a meeting to order, does not always provide the basis for the judgment of value to the individual. Chronometry may tell you that only an hour has passed, but your intuitive feeling that the agony made the interval seem like days, or decades, has psychological reality for you, even when you know the fact about objective time and realize you have misjudged it and even though your private feeling may stem from a standard that is very fuzzy and unverbalizable.

The way in which fuzzy and unverbalizable standards come into existence and then serve their intervening function can be observed by glancing at the Method of Absolute Judgment in the laboratory. The task of subjects is to state only whether the interval they perceive is long or short or to rate it on a scale of length. On the first trial they hear two antiseptic clicks which are the interval's boundaries or a noise of short duration which is the interval itself. How can they pass judgment? "At the beginning," the subjects in one experiment are told, "your judgments will, of course, be guesses, but you will soon learn to discriminate the stimuli" (Postman & Miller, 1945), and indeed they are able to do just that. They are given no standard, but they good-naturedly cooperate: an interval appearing to be of 1-second duration will be called short if they use the standard length of time it takes to play the verse of a popular song, long if they think of the exposure times on a camera. What they say on the first trial, consequently, is a matter of chance or arbitrary choice from among their own subjective standards. But on successive trials as they judge other intervals of varying length and similarly presented, they have the experience of the previous intervals to guide them; each interval appears in the context of all other intervals. After a number of trials, they establish a subjective standard that is roughly the mean of the intervals they have experienced. Thereafter they can pass judgment with a high degree of accuracy if those intervals are of discriminatively different durations. Obviously they are aware of the relativity of their judgments, for even during the experiment they know that the interval they call "long" they would consider "very short" in many other contexts.

The introduction of an anchoring stimulus into the series being judged —one that is appreciably above or below the series without the subject being aware of that fact—has been shown to have a marked effect on judgment concerning intervals ranging in duration from 1 to 5 seconds; and this effect also appeared when the anchor was perceived through a sense modality different from that of the series (e.g., it was an auditory stimulus and the series contained verbal stimuli) or when it appeared in the midst of a series composed of both visual and auditory stimuli (U.S., college; Behar & Bevan, 1961). But there is also a "breakpoint" at which an extraneous stimulus has no such anchoring effect. An interval of 0.01 second did not influence judgments concerning a series of intervals whose lowest duration was 2 seconds, but one of 0.05 and 1 second did; and similarly, the 1-second duration of the signal heralding the onset of auditory intervals ranging from 2 to 10 seconds affected the temporal ratings only when its intensity was different from that of the intervals themselves (U.S., college; verbal, prior; Adamson, 1967; Adamson & Everett, 1969). This procedure may be contrasted with any variant of Comparison, e.g., that of Just Noticeable Differences in which the experimenter provides a standard stimulus for purposes of comparison on every trial and naturally that standard is then employed. The Method of Absolute Judgment, with or without an anchoring stimulus, requires subjects to provide their own subjective standards, and I pay tribute to their ability to do so in the following principle:

7.2 *The standard that is employed in passing temporal judgment, if available and relevant, is likely to be the one most frequently or recently reinforced in the past.*

The principle receives support from the general theory that a recently activated response tends either to perseverate or to be easily reactivated. First, we have maintained, most people seek a standard based upon an objective device; for this is the standard that has proved most reliable. In its absence, the recently reinforced and internalized standard is probably compelling because temporal judgments are so intangible: they cannot be anchored to reality, which means that they cannot be placed side by side and resurveyed and appraised the way spatial judgments so often can be referred back again to the lengths being judged. Instead they must leap out of successive rather than simultaneous comparisons; the interval closest in time, consequently, can serve conveniently as the standard. The baffling nature of time is again illustrated by noting that at the moment of comparison both intervals are in the past.

The principle obviously implies that a standard, when it has prestige and is therefore accepted, affects temporal judgments and hence offers one explanation for the fact that such judgments can be easily influenced. I once asked 354 undergraduates to "indicate how recent or distant five events in the past *seem* to you as you look back at them" by rating each of them on a 9-point scale. The events ranged from "Last Christmas" to "The Discovery of America." An inconspicuous parenthetical suggestion was embedded in the mimeographed instructions. "Write the number 5," it was stated, if the event "seems neither recent nor distant but half-way in-between." There followed the parentheses which contained one of the following: "(for example, LAST CHRISTMAS)," "(for example, YOUR OWN BIRTH)," and so on. In comparison with a control group which survived without parentheses, the suggestion tended significantly to raise or lower the ratings in an appropriate manner. The mean rating for "The Civil War" was 7.4 (out of a possible 9) when no suggestion was offered; 6.9 when that event was used as the example in parentheses for "half-way in-between"; 6.5 when "The Discovery of America" was the example; and 7.7 when any one of the three more recent events was the example (Doob, 1951).

But words of caution are needed to understand the principle. First, the standard has to be "relevant": you reach the athletic field on the right day of the week and you arrive punctually, but you do not use calendar or hour-hand standards to time a 100-yard dash which you see there. Frequency of reinforcement, the principle suggests, means that the criterion has been utilized not only often but also successfully: a person who, in the absence of a watch, usually counts to himself in order to judge short intervals, such as the time between a flash of lightning and its clap of thunder, knows or thinks he knows that this measuring device correlates well with intervals measured by clock time. The reinforcing state of affairs, however, is not so easily located as it is in laboratory experiments where subjects are merely trying to please the experimenter or to earn their pay. Circumstances may alter the tendency to cling to the last standard—you return home after seeing the track meet where your standard was in seconds, and there resume reading a book on ancient Greece which makes you think in centuries. Again the importance of objective standards can be appreciated, for they have been most satisfactory in the past and, very often, the ones most recently employed.

The reinforcement for the Method of Absolute Judgment must be based upon a process within the organism, since ordinarily the experimenter who employs this procedure does not indicate whether a given judgment is correct or incorrect. Some kind of totalizing takes place, for obviously the mean about which the judgments fluctuate, though impossible to verbalize, is determined by the range within which the varying intervals fall. I suspect that, in modern society, when no great issue is involved, and in traditional societies under most circumstances, many day by day and minute by minute temporal judgments employ the equivalent of the Method of Absolute Judgment. When you feel that someone is taking longer than usual to complete a task—for example, a child dressing in the morning before having breakfast and going off to school—the judgment stems not from a consideration of clock time, but from a quick intuition cued off by internal criteria. Skilled musicians are likely to judge with great accuracy whether parts of a composition are being played slowly or quickly without counting and without beating time; in passing judgment, they depend upon their sense of time, which means their accumulated musical experience. The evaluation of time on the primary level, if this view is correct, is more likely to use exclusively subjective standards: the person feels inside, or privately, that time is passing quickly or slowly. Lest he make a fool of himself publicly, however, he determines whether he has been over- or underestimating objective time by consulting some outside, objective standard before expressing his judgment—and this is the problem of translation and expression to which we shall turn later in this chapter.

EXPERIENCE

Guides from past experience play an increasing role as primary judgment gives way to secondary judgment. For we note again and again the overwhelming evidence supporting the view that almost every aspect of temporal behavior, perhaps every one, is learned. The confusion of young children regarding time, which gradually gives way to understanding as their experience broadens and deepens and as retention is improved, will be discussed in the next chapter. Subjects in a laboratory who have never used one of the formal methods, such as that of Reproduction, quickly grasp the idea and faithfully follow instructions. And, as has been said repeatedly in these pages, all living organisms must learn to delay many of their responses if they are to survive or to be happy.

7.3 The accuracy of a temporal judgment in terms of objective time varies positively with the amount of reinforced experience.

This principle is true to the point of banality in connection with the Method of Chronometry. Obviously, for example, the young learn gradually how to read clocks and calendars and to interpret various astronomical signs;

their accuracy in this respect improves with experience. American children between the ages of 4 and 6 were asked questions about their usual routine, such as "What time does your school start?" and "What time do you have lunch?" "Initially," the summary of the research states, "descriptive terms are used, or a sequence of activities is cited"; this was followed by "an unreasonable time," and then by "a reasonable but incorrect time" that was designated; lastly and "finally the correct answer is forthcoming" (Springer, 1952). But even reporting clock time can depend upon the kind of clock and the experience of persons therewith. Adults in England responded almost four times as slowly and with ten times as many errors when they noted the time on a conventional clock face than when they perceived it on a digital clock, a timepiece without hands which signifies the time through moving numbers visible in small windows (Zeff, 1965). Clearly they had previously told time more frequently from conventional than from digital clocks, but they had also read the kinds of simple numbers presented unequivocally on a digital clock even more frequently. The information obtained from consulting any kind of standardized instrument, moreover, is likely to be rewarding: you arrive punctually, you write the correct date atop your letter, etc. But sometimes there may be punishment: you arrive punctually, but your companion does not; in an extreme case, the psychotic may feel that it is useless or irrelevant to coordinate his own activities with those of his contemporaries. Under such circumstances, Chronometry may be ignored specifically or generally.

The principle being discussed is not banal when other methods besides Chronometry are employed. And here, before examining the evidence, we must anticipate that the effect of experience upon accuracy is not linear and, indeed, may often be zero, or close to zero. For anecdotally we know that persons in so-called developed countries pass temporal judgments scores of times each day before consulting their watches and that therefore in a sense they practice this activity with reinforcement literally thousands of times in the course of their existence; nevertheless their judgments remain relatively inaccurate. Mature enlisted men in the United States Army, subjects in a vigilance experiment, simply noted and reported when a light was occasionally extinguished for a brief period. While carrying out this task, they could observe a familiar clock which, unknown to them, had been made to operate at twice or one-half the normal rate; they continued at work, respectively, for one or two hours. Afterwards the men "readily admitted to being fooled completely," they did not realize that the clock had been tampered with (McGrath & O'Hanlon, 1967). Similarly, all of us undoubtedly accord the Method of Chronometry more prestige than our hunches.

Experience

The moment Chronometry is unavailable, learning to respond accurately to temporal cues does not proceed with dispatch. Soviet investigators, proving the dictum of Pavlov that time can be considered "a conditioned stimulus" for reflex action in animal and human experimentation, have reported that conditioned responses to time can be established "with much greater difficulty" than to other stimuli, are "less stable" than other reflexes, and can be rapidly extinguished (Dmitriev & Kochigina, 1959). Rats have been taught to discriminate between temporal intervals (beginning with 5 vs. 45 seconds and, in one instance, ending with 5 vs. 10 seconds): they were prevented from running down a maze by being restrained in a compartment for an interval whose duration signified whether they would be rewarded with food by turning right or left. They generally required, however, hundreds of trials before they could learn the discrimination (Heron, 1949).

To test the relation between accuracy and experience, subjects in an experiment, and persons in general, must be given the opportunity not only to repeat the act of passing temporal judgment but also, if the practice is to be effective, their correct judgments must be reinforced and their incorrect ones extinguished by providing them with appropriate temporal information or other rewards and punishments. To learn or to learn efficiently, it would be anticipated, there must be feedback. Most experimenters, especially those in the classical tradition, have deliberately failed to provide feedback because, I would guess, they imagined they could isolate the modes of passing judgment, the process of interest to them, from the contaminating influence of the Guides generated by previous reinforcement.

At first glance, however, it seems curious that blind practice in fact has been shown occasionally to improve accuracy or to have some significant effect upon judgments (for example, a lengthening of judgments with the Methods of Reproduction and Production). The reason for the improvement seems to be a change within the subjects as they repeatedly pass judgment: they grow accustomed to the setting, they are less distracted by extraneous cues, they utilize the cues at hand, they grow more confident, etc. The lengthening of judgments may be a function of sheer boredom. The *if*'s, *but*'s, and other qualifications are relegated to *Addendum 7.1*, but I would emphasize one point here: no matter how novel or bizarre the laboratory setting, the subjects have had some previous experience in judging time in real-life situations; hence somehow with repetition they increasingly utilize that experience as a Guide.

Evidence indicating that accuracy of judgments increases if repetition is accompanied by appropriate rewards or punishments, or by feedback, is very impressive and is collected in *Addendum 7.2*. We need no experimental proof, however, to be convinced that one aspect of temporal behavior usually improves with reinforced practice, viz., the rhythmic production of musical sounds. Playing music means that the musician is able to create intervals of a specific duration and in a specified succession, whether or not he is cued by another musician or a conductor. The novice may have difficulty keeping time or following the designated tempo on the score. But with practice—and the reinforcement may come mechanically from a metronome or a baton or afterwards from verbal criticism by an instructor —he improves, or at least we hope he does. The mystery, I think, involves not the learning process but what probably must be called a genetic component in the sense of musical time which unquestionably plays a significant role in assisting truly gifted musicians.

In most respects, therefore, temporal learning is like any other kind of learning, although less dependable; forgetting in the absence of reinforcement, for example, seems especially rapid. In would appear as if relatively little is to be gained from the experiencing of intervals, even with adequate information or reinforcement, which can be saved and then utilized in the future. The athlete or musician who develops a refined sense of time has unverbalized cues to guide him thereafter, but such cues are not easy to find, at least in psychological experiments—unless counting or feeling one's pulse is permitted, and they seldom are. We crave measuring instruments in the external world, and may even give credence to them when they convey false information—"that clock must be slow"—rather than to our own judgments. Perhaps Guides from past experience are somehow utilized even when they must be followed semi-blindly.

Back to experiments to find more complexities associated with reinforcement. In one, subjects first performed certain tasks, after each of which they estimated the elapsed time and then consulted a normal, accurate clock to verify their judgment. They repeated this procedure except that then, unknown to them, the clock was faulty: for half of them it was running too rapidly, for the other half too slowly. There was a clear-cut tendency for those dependent on the fast clock to increase their overestimations and for those consulting the slow clock to decrease them. The speed with which they worked (as measured by tapping rate and by writing dots in rows of circles) was correspondingly changed: those dependent upon the fast clock worked more rapidly than they had previously, their colleagues more slowly. Finally, when questioned afterwards, only 2 of the 29 subjects suspected that the clocks had been changed; 6 had no idea about the purpose of the investigation; and 21 thought its function was to determine the effect of absorption and interest on temporal judgment. One commented, "I was using my heartbeats [as cues], but it didn't work after a while, so I didn't believe them any more" (U.S., adults; verbal, prior; Craik & Sabin, 1963). Likewise, judgments concerning "an accepted standard unit of time" as the second have been significantly affected by stimulus context, modality, and a host of other factors being considered in this chapter and in others (U.S., college & adults; limits-verbal, prior; Goldstone et al., 1957, 1958b, 1959, 1964a). In all the experiments just cited the subjects previously had had truly countless experiences telling time during their lives, but apparently such experience helped little in these queer situations. What seems to have happened is that in effect they were functioning as if they had been exposed to the Method of Absolute Judgment and hence quickly adopted new standards to fit the problems at hand: the so-called anchoring norms of the present became more efficacious than practice effects from the past.

In the light of such results, some analysts feel that experience as such is not a critical variable for temporal judgments. Through delayed or trace conditioning, according to a Soviet investigator, children may be trained to reach an adult level in reproducing short intervals, but what he calls the "intellectual or rational appreciation of time depending on speech" is likely to be a function not of practice as such but of the number of events perceived and interpreted within the interval (Fress, 1962). The past, in different words, is less influential than the present context and the mode of evaluating it.

I am not convinced, nevertheless, that the basis for remembering time is really so different from other kinds of memory. One writer states that the Method of Verbal Estimation produced results even "after long training" which were "never" perfect because the units of objective time "have no tangible reality and do not give rise to images"; thus, "I can form a representation of a yard, but not of a minute; I can only try to reproduce a similar duration," as a result of which "we try more or less clumsily to put a subjective appreciation into concrete form" (Fraisse, 1963, p. 211). Again, I must ask, is a unit of time any less tangible than any other stimulus attribute—and I have repeatedly referred to color and size, and I could mention volume, temperature, taste, perhaps even beauty—that is not directly perceived?

Then I would quickly add another fact: the experimental work exposing the vagaries of the relation between temporal accuracy and experience has been based in large part either upon repetition without reinforcement or upon ephemeral intervals—or both. I remind the reader and myself, therefore, of the rarified conditions provided by most experimenters. For example, it is said that intervals are judged to be in the psychological present when their duration is probably no more than 2 seconds and when they contain only a small number of elements, perhaps no more than five or six, or when the elements are organized by the person passing judgment into a coherent grouping (Fraisse, 1963, pp. 89–93). Surely more time is usually needed for "rational" thinking to take place, i.e., for the Guides associated with secondary judgments to function. Even so, as *Addenda 7.1* and *7.2* suggest, generalized experience in real life, not all of which involves extended intervals, does lead to accurate judgments. All of us gradually learn during childhood to interpret correctly the position of the hands on a clock and the significance of the figures on a calendar. At a glance you must know whether you have time to cross the street before the car you see approaching will reach the route you plan to follow. You look at a child from your own culture and you know approximately how old he is; you are less certain about one from another area because you have had little or no opportunity to make such judgments.

These commonsense observations provide us with a clue to an important corollary:

7.3.a Reinforced experience is likely to be efficacious in aiding the accuracy of temporal judgments when the individual thereby acquires a Guide which provides insight either into the significance of environmental cues or into his own tendency to commit errors.

The corollary I think is implicit when investigators and analysts draw distinctions between different ways of judging intervals. One of them, for example, suggests that short intervals are directly perceived but long ones must be intellectually evaluated (Pieron, 1936). A statement of this sort, in my opinion, means little more than that judgments concerning ephemeral and transitory intervals stem largely from unverbalizable cues and intuition, and that those concerning prolonged intervals are based upon verbalizable cues and Guides.

Most studies, alas, do not discuss the operational question as to whether with practice students arrive at a Guide for their judgments. The classical investigators, as we shall see later in this chapter, did collect introspections from their highly trained subjects; but the reports provide little insight into the effects of previous experience. In some of the experimental work, however, one has the feeling that the subjects must have consulted or evolved some kind of Guide; or otherwise they could not have passed judgment or perhaps have even survived. One experimenter, for example, in an effort to prove that the experiencing of duration is not "absolutely impalpable" resorted to a fiendish variation of the Method of Reproduction. Instead of simply attempting to reproduce intervals from 1.5 to 3 seconds, he subjected himself and a female student to the task of fractionating what they had just perceived: they sought to reproduce intervals which were supposed to vary from $\frac{1}{2}$ to $\frac{1}{9}$ of the original interval. He himself of course did not rebel, nor did the lady, and their performance turned out to be "rather good" (Germany; prior; Wirth, 1937). It seems unreasonable to believe that fractionating by ninths was completely unverbalizable. Often, too, I have the feeling that, if the subjects had known the investigator's generalization concerning their own performance, they would have been provided with a Guide or that, if they had been told that they needed some sort of Guide, they would have found one; in either case their accuracy would have increased. In one study subjects tended to overestimate the time it would take them to carry out various tasks, such as assembling a jigsaw puzzle, but the experimenter apparently did not suggest that they would probably be in error if they used a Guide they themselves "frequently" expressed: "I had better allow myself enough time, since I do not know just how long it will take me" (U.S., college; verbal; Dudycha & Dudycha, 1938).

How, then, do subjects and human beings in general discount, evaluate, or improve their primary judgments as they pass secondary judgment upon the intervals they experience or anticipate? In one sense this question is a variant of a topic already discussed in *Addendum 7.1*, the improvement in accuracy accompanying repetition without feedback or reinforcement. Getting accustomed to the setting, being less distracted, utilizing the cues at hand, learning to respond to a novel task, becoming more confident—these are different ways of saying that, wittingly or not, the individual emerges with some kind of Guide in the particular situation that is being repeated. Again the experimental data almost require such an explanation. In one investigation, for example, the subjects must have been distracted by the strange task confronting them: they produced intervals by plunging into darkness the entire room in which they were seated. Without any "information about the accuracy of the elapsed time or the direction of the error," however, their judgments became more accurate from trial to trial, from which fact it seems fair to deduce that they must have accustomed themselves to this bizarre situation, so that the degree of underestimation decreased (U.S., college; prior; Eson & Kafka, 1952).

Perhaps everywhere people have a very general Guide: through experience they have learned that primary judgments are so often wrong, there is likely to be a disparity between the subjective time they experience and the objective time communicated by timepieces or their contemporaries. Even the least educated informants in Jamaica did not think it strange when I asked them whether under some circumstances time seems to pass slowly or quickly (Doob, 1960, pp. 192–3); and on an anecdotal level I have found again and again that the same question is a sensible one for Africans living a more or less traditional life in various societies. "I would imagine that my guesses are about twice what they should be, or twice the time," exclaimed one subject in an experiment in which the task was to estimate the duration of a 2-minute interval spent in darkness awaiting a light and of another interval of identical duration during which the stationary light seemed to move. All the subjects engaged in "self-evaluation" or reflection and tried to free themselves from their first impressions; but the investigator thinks that to some extent they remained "stimulus bound" (U.S., adults; verbal, immediate subsequent; Kafka, 1957). Then a subject who had been performing a boring task for about one quarter of an hour, when asked to judge the elapsed time, stated: "It feels like 3 hours, but that can't be right, so I'll say 30 minutes" (Burton, 1943). Obviously he was guiding his judgments by some unverbalizable cues from his environment, and his corrected estimate, the secondary judgment, was much more accurate than what had been either his primary judgment or his way of expressing boredom through a temporal figure of speech.

But awareness of a proclivity to commit errors and the subsequent effort to discount the feelings and other first impressions that are part of the primary judgment do not in themselves promote accuracy; what is needed is a positive Guide. Somewhere in-between a feeling of cautiousness and such a Guide may be a tendency to *compensate* for an erroneous inclination by swinging, as it were, in the opposite direction. "If a person knows that time is underestimated during pleasant happenings and overestimated in boredom," one group of investigators suggests, "he may correct or often overcorrect for these influences" (Gilliland, Hofeld, & Eckstrand, 1946). A laboratory finding that short ephemeral intervals are often overestimated and long ones underestimated has been noted so frequently that it has been given a special title, viz., Vierordt's Law; it may well be that we have here another example of compensating for the initial impulse to judge short intervals as very short and long ones as very long when they occur in the same experimental session (but see *Addendum 7.3*).

The best way to remove the insecurity associated with the primary judgment is to consult a timepiece or its equivalent. In the absence of such a device, various substitutes are sought. With practice and reinforcement, for example, four adults made more accurate guesses concerning the time appearing on the investigator's watch when they tried deliberately to imagine where the hands of that watch were (England; Hall, 1927). Noting one's heartbeat, pulse or respiratory rate or especially counting are ways so frequently used to calculate elapsed time of ephemeral or transitory intervals that subjects in experiments usually have to be warned not to resort to one of those methods; whether counting actually increases accuracy is another sub-problem I have relegated to *Addendum 7.3*. More extended intervals are gauged by perceiving cues that are either internal (such as hunger or drowsiness) or external (such as the position of the sun or the density of pedestrian or vehicular traffic).

Outside the laboratory it seems probable that the belief in the efficacy of a Guide may induce the individual either to improve the accuracy of his temporal judgments or to mistrust them until they have been objectively verified. You have the impression, for example, that the ride home always—or often—seems shorter than when you drove to the home of your new acquaintances. For on the outbound journey you did not know exactly where they lived, you had to stop and make inquiries, you wondered whether you were on the right road, you looked at house numbers until you found the correct one; in contrast, you drove straight home afterwards. Actually you probably took longer to get there than to return but not as long as it "seemed" to you: you did drive more slowly, you did stop, you did experience more separate events, you were a bit anxious, you were stimulated by the unfamiliar, in short, your primary judgment was justifiably a long one. But you tended to discount that judgment by looking at your watch. The fruits of previous experience caused you to mistrust and correct your own judgment. Similarly, were it true that filled intervals are judged "far more accurately" than empty ones (Fraisse, 1963, p. 51)—and I do not think they necessarily are—then people might very well learn through experience to employ the number of events transpiring during an interval as temporal cues; in contrast, they may seldom be called upon either to experience or judge empty intervals and hence they are less efficient or accurate in judging them.

In general, therefore, most persons have learned either to their sorrow or their joy that their primary judgments represent over- or underestimations of time, so that they come almost immediately to distrust those judgments and stand ready to revise them in the light of knowledge of objective time which they acquire by consulting some instrument. And they also realize that an instrument of some kind, or an external cue, is likely to be conveniently at hand. The very passing of a primary judgment, therefore, can be said to arouse yet another temporal motive which stimulates them to gather the necessary information to refine, or eventually to objectify the initial impression. Past experience with time, in short, produces skepticism.

Then the emphasis placed upon timing in modern society probably has a double-edged effect upon the accuracy with which people judge intervals. On the one hand, the presence of timepieces makes it unnecessary to pass judgment: we consult one of them or at least try to do so whenever the temporal motive is evoked. As a result, it has been suggested (Cohen, 1966), p. 257), the sense of time may have atrophied to some extent. On the other hand, the very precision with which so much of life is regulated may instill on an unverbalizable level a feeling for the duration of certain intervals. We intuit, for example, that a commercial announcement on radio or television is exceeding the time normally accorded that unwelcome interruption; we judge that the train or plane is late because its customary punctuality has internalized a set which suggests that something is different. We may prefer and be dependent upon objective time, in short, but privately we continue to make subjective judgments to give meaning to many of our experiences.

Two investigations in Africa make the problem of temporal judgment in our society more concrete. Both used a modified Method of Reproduction: subjects sought to reproduce intervals which they had either perceived by observing a stopwatch or which they themselves had paced off by walking a distance of from 20 to 80 yards. It was assumed that urban Africans living in a city would be compelled to pass temporal judgments more frequently and with greater precision than those living in the bush and hence would be more accurate during the investigation. In fact, however, no significant differences were found between samples of urban and rural Ganda; but the rural group did tend to overestimate the intervals to a greater degree than the urban one (Robbins, Kilbride, & Bukenya, 1968). It could be that the gains from having to consult a watch in order to be punctual, which urban living demands, simply did not transfer to this laboratory-type situation. Then school children and illiterate adults in an African society, the Kpelle of Liberia, were more accurate than a group of American adults in carrying out these same temporal tasks (Gay & Cole, 1967, pp. 72–5). It is to be noted that the techniques here did not enable the Americans to employ the method of estimating time which they have undoubtedly practiced more frequently than the Africans, viz., that of Verbal Estimation, and that therefore the African subjects were not being handicapped. We can only conclude that whatever advantage the Americans might have had from previous experience did not generalize to another method. If it were possible to draw a more sweeping conclusion from these two studies—and of course it is not possible—then it could well be said that the reinforced instrumental response for punctuality in industrial societies involves to a minimum degree reliance upon secondary judgments and to a maximum degree simply consulting clocks and watches.

The effect of reinforced experience upon temporal behavior other than judgment is reasonably straightforward. For here the ordinary principles of learning seem gracefully to apply. Success or failure determines an individual's temporal perspective; if he is punished by anticipating events, his orientation is likely to shift to the present or the past. If he intervenes to achieve goals or to prevent boredom and finds satisfaction in so doing, he is likely to repeat this activity which in turn will have important consequences for temporal judgments.

BOREDOM AND IMPATIENCE

Being bored, another distinctly human capability, raises a host of intriguing problems concerning temporal judgment. Almost all observers of time include the topic among their generalizations. These are the words of a sociologist:

> the hours . . . pass more quickly when we are engrossed or excited or when the emotional tone is smoothly pleasant or when we indulge in quiet reverie. They pass more slowly when we are bored or when waiting impatiently for some anticipated event or when we feel anxiety or when we are in pain or anguish [MacIver, 1962, p. 41].

Very similar is the view of William James that boredom results from being "attentive to the passage of the time itself," for then intervals seem empty and intolerably long: "the *odiousness* of the whole experience comes from its insipidity; for *stimulation* is the indispensable requisite for pleasure in experience, and the feeling of bare time is the least stimulating experience we can have" (James, 1890, pp. 626–7). He also notes that time passes slowly during a night of pain but without boredom, and then approvingly paraphrases a contemporary who said that "what we feel" under these circumstances "is the long time of the suffering, not the suffering of the long time per se."

A very metaphysical philosopher notes that "periods of boredom and periods of intense expectation," in contrast with "periods in which we are deeply absorbed in what we are doing," appear to be "longer than the normal"; and "periods into which many exciting events have been crowded seem in retrospect to be longer than periods of tranquility." He provides an explanation: during boredom or expectation, "we pay more attention to the passage of time than usual, because we are more than usually anxious for it to pass, and because we have little else—or little else on which we can fix our minds—to which to attend. And since we pay as much attention to time in a short period as we should usually pay in a longer period, we judge the period to be longer than it is." In contrast during periods of absorption in activity, we have "little attention to spare for the lapse of time, and so we judge that little time has elapsed" (McTaggart, 1927, p. 277). I think we have here a critical attribute of boredom: a discrepancy between a primary and a secondary judgment.

To explore the nature of that discrepancy, let us first concentrate upon elapsed time. On a primary level, the individual passes judgment in one of two forms: either he estimates or characterizes the interval (in terms of temporal units, length, or speed) or he expresses a wish concerning its duration. Then he perceives a measure of objective time and notes a discrepancy between that fact and his primary judgment: less objective time has passed, time is passing more slowly than he believed or wished. On some basis, therefore, he has overestimated the duration of the elapsed interval. For a pending interval, the same sort of process occurs in reverse, the outcome of which is an underestimation of that interval which is yet to come.

This discrepancy, moreover, is unpleasant: the individual is disturbed that so little objective time has elapsed or that so much objective time is still pending. Invoking our Arbitrary Limitation, we say that the causes of the frustration are too numerous to mention, but they have one feature in common: the person finds himself trapped in a situation which is not satisfying and from which, at least for the moment, he cannot escape. In addition, the frustration is mild and can be easily alleviated: neither pain nor pleasure is ever boring, and the solution to boredom is a change in activity (or a change from no to some activity), no matter how defined.

The analysis suggests that boredom occurs only after a secondary judgment has been made. This means, I suppose, that animals cannot be bored. A child who stops playing with one toy because it no longer interests him is said, in popular language, to be bored; but I am asserting merely that he is satiated or feels dissatisfied with what he has been doing. You find the speaker dull and therefore you wish he would stop speaking; but you are not bored until you discover that too little of the hour to be filled by that speech has elapsed, or until you note that a huge pile of pages from which he is reading still remains to be turned over—and provided of course you have to remain in your seat and wait until the bore stops speaking.

Boredom quickly becomes a very serious symptom when the person experiences the frustration not in specific situations but more generally. For then living becomes extensively unpleasant. Instead of being absorbed in the activities at hand, he notes that time is forever passing too slowly, and he cannot extricate himself.

We may well inquire under what circumstances the individual who finds himself in a mildly unpleasant situation decides to ascertain objective time and thus bring on the possibility of boredom. Here is simply a special case of a duration-related drive: the termination of an interval is anticipated to be related to changes in the activities originally producing the mild unpleasantness. When the boring speaker has finished, you may stretch your legs and go home.

It is not true, I think, that merely paying attention to time produces boredom. You keep looking at your watch as you hurry to meet someone punctually; whether you seem to be early or late, you certainly are not bored. But the assertion does call attention to a number of important points. First, the mild disagreeableness, which is annoyingly unpleasant without being literally painful, may indeed be the reason for the arousal of the temporal motive: how much longer will the metaphorical agony last? Then, according to our definition, the passing of the secondary judgment indicates the overestimation of the elapsed interval and the underestimation of the pending one; hence noting objective time does play a role. This secondary judgment itself *may* actually be the cause of unpleasantness and boredom: the primary judgment possibly contains no affect, the activity seems pleasant or at least neutral until the overestimation or underestimation (that is, the discrepancy) has been perceived; at that point the person becomes discouraged and bored because too little time has passed or too much remains. Also after perceiving the discrepancy, he may seek to escape from boredom by engaging in various activities, one of which may well be to keep noting the passing of objective time, which thus becomes a kind of instrumental response. But he who would thus escape boredom is likely to push himself into a Spiral: mild dissatisfaction makes him look at his watch, looking at his watch makes him bored and increases his dissatisfaction, greater boredom and dissatisfaction makes him look again, etc. Other Spirals, alas, also plague him and us. You know from experience that the man is a bore; when you meet him on the street, you anticipate unexciting behavior, you then are not excited by what he says or does, you wonder about time, and you seem never to be able to extricate yourself.

It is also often claimed that inactivity produces boredom. There may well be a touch of culture in the definition of "nothing." For people within some societies or some persons in any society have the feeling that doing "nothing" is bad and doing something is "good"; and they may then wish the period of nothingness to pass and observe objective time to see whether progress has been made. But more important may be the fact that "nothing" in the waking state must mean few events, little organization of behavior, and no intervention, all of which are the very conditions that associate the interval's scope with a long primary judgment and constitute, in this situation, the foundation for overestimating the elapsed interval. When nothing is replaced with something which represents an effort to escape from the interval and yet to remain within it—the person may fidget, sigh, complain, let reverie have more or less full rein, etc.—there may well be a decrease in the primary judgment: more events, some degree of organization (however loose), and intervention, the very conditions which, in terms of the curvilinear relation we have postulated, give the impression, up to a point, that time is passing more quickly.

To be bored, the individual must pass temporal judgment; no activity is boring unless a temporal motive has been evoked. Two men, one who had been a prisoner of war for seven months and the other for four years, both recalled that the worst threat to sanity seemed to be "the ghost of time"; with no work to perform and nothing in general to do, time soon lost its meaning. For this reason pathetic attempts were made to fill intervals with some kind of activity but, even so, time, as it were, was outlawed (France; Etienne & Kutschbach, 1949). The men, it seems from these accounts, were not bored because they ceased passing temporal judgment. They lacked the incentive to pay attention to time: none of their immediate goals could be duration-dependent because they did not know how long they were going to be interned. Similarly, according to the observations of a psychiatrist who was himself an inmate of various Nazi concentration camps, "the prisoners lived, like children, only in the immediate present; they lost the feeling for the sequence of time, they became unable to plan for the future or to give up immediate pleasure satisfactions to gain greater ones in the near future" (Bettelheim, 1943).

The inevitable spiral relation between boredom and temporal judgment makes it difficult to infer one state of affairs from the presence of the other. In an experiment mentioned on page 123, for example, progressively shorter judgments accompanied an increase in the number of nonsense syllables heard and written down during transitory intervals, but there is no way of knowing whether that outcome resulted from the difference in the way the larger number of syllables had to be organized or from a decrease in boredom. Or, possibly, less boredom could be inferred from the fact that in some instances words produced shorter judgment than nonsense syllables (U.S., adults; repro., prior; Friedl & Lhamon, 1965). ". . . when we are explicitly noting the passing of time, we are more likely to make mistakes than at other times," it has been crisply asserted (Cleugh, 1937, p. 29). Without questioning the truth of this statement, let us assume that one of the temporal mistakes is that of overestimation; then we need to know why the person is paying attention to time. Is he bored, is he awaiting the millennium, is he waiting for the camera to click, is he wondering whether the plane can take off before it reaches the end of the runway, is he testing the accuracy of his watch, or is he just a subject in a psychological experiment trying to earn a little extra pocket money? The accuracy of his judgment is likely to be quite different in each of these situations.

Boredom pertains to the duration of an interval, a series of intervals, or one's very existence, and so does a very similar and, often, coterminous phenomenon, viz., impatience. Here, too, there is a duration-dependent drive and a discrepancy between primary and secondary judgments. The difference between the two relates to the goal: the bored person wants the interval to end so that he may turn to some other activity, the impatient one so that he can attain some positive or significant goal; in addition, the former is in a state of mild discomfort, the latter can be experiencing any degree of frustration as a result of being separated temporally from the goal.

"The watched pot never boils," the folk proverb states. Of course it eventually boils, we must say prosaically, if it is being heated, if it is not too far above sea level, if it is watched long enough. The cook or the cook's friend who has no business being in the kitchen, however, is impatient, he wants the water to boil: elapsed time seems to pass slowly. Why? I dedicate *Addendum 7.4* exclusively to a reply. But in the text here I would pay homage to that writer who, perhaps more than any other in the century, has so greatly concerned himself with the problem of time in the fifteen volumes which he called *Remembrance of Things Past,* that we are truly faced with an embarrassment of riches. I have selected the following passage because therein this sensitive introspectionist explicitly claims generality for his observations concerning impatience:

> The days that preceded my dinner with Mme. de Stermaria were for me by no means delightful, in fact it was all I could do to live through them. For as a general rule, the shorter the interval is that separates us from our planned objective, the longer it seems to us, because we apply to it a more minute scale of measurement, or simply because it occurs to us to measure it at all. The Papacy, we are told, reckons by centuries, and indeed may not think perhaps of reckoning time at all, since its goal is in eternity. Mine was no more than three days off; I counted by seconds, I gave myself up to those imaginings which are the first movements of caresses, of caresses which it maddens us not to be able to make the woman herself reciprocate and complete—those identical caresses, to the exclusion of all others. And, as a matter of fact, it is true that, generally speaking the difficulty of attaining the object of a desire enhances that desire (the difficulty, not the impossibility, for that suppresses it altogether), yet in the case of a desire that is wholly physical the certainty that it will be realised, at a fixed and not distant point in time, is scarcely less exciting than uncertainty. . . . [Proust, 1941, pp. 102–3].

Let us enter into the spirit of Proust and accept his account as a perfectly valid recollection. Then surely, because he imagined himself in love with the other person, there was every reason for him to wish to have the three-day interval of waiting end; he knew that the termination of the interval was "related to the achievement of a significant goal." But he must have consulted his watch; he knew as he went to bed and arose how many days or hours remained; he realized fully that his wishes were making him overestimate the elapsed interval and underestimate the pending one. In the succeeding pages we are told in detail the kinds of preparations he made; this frenetic activity increased the strength of his duration-related drive and the attractiveness of the goal which was to keep the rendezvous. Proust himself proposes "a general rule" which suggests that the strength of the duration-related drive increases as the goal approaches. In addition, he explains this phenomenon by suggesting a change in the standard that is employed ("a more minute scale of measurement") and the influence exerted by the act of measurement. The former makes the pending interval appear all the longer, the latter increases the impatience.

On a less romantic, less charming level, it has been observed that delinquent children became "especially difficult" just before being discharged from reform schools. Their closeness to freedom made their incarceration seem all the more burdensome, and the needs they anticipated satisfying on the outside grew all the stronger (Germany; Lewin, 1931). Apparently the impatience was produced by both the proximity to the goal and the added frustration.

THE TRANSLATION

More often than not, the final step in the secondary judgment is to translate what is being remembered, experienced, renounced, or anticipated into a form which the person passing judgment can express to himself and perhaps publicly to others. The judgment is then also likely to be stored so that it can be fed back through the temporal or total potential into future judgments. Usually the translation must be in socially acceptable currency, which means a verbal formulation. That formulation may be made immediately by the person (Chronometry, Comparison, Verbal Estimation) or it may be derived from him but stem from the experimenter or observer (Reproduction, Production). Actually, a verbal translation may be present even when not required; thus when called upon to reproduce the duration of an elapsed sound by pressing a button, the individual may first translate his impression into verbal units which in turn may mediate, successfully or not, his manual performance.

The first problem would seem to be: how is such a translation made, how does the person arrive at a verbalization of his secondary judgment? The most obvious reply to the question ought to be: well, ask him, ask him how he makes the judgment. On purely a priori grounds, however, it is to be doubted whether a helpful reply can thus be obtained. Surely the subjects in one experiment could not be expected to know how and why increasing the intensity of auditory stimulation produced progressively greater overestimation of intervals having a duration from 0.35 and 0.70 seconds but not of a longer interval with a duration of 1.4 seconds. Did the change in intensity at first make the intervals appear to them to be more crowded, did it cause them to be more active, did it somehow arouse a stronger drive within them; and why did these effects (whatever their nature) cease to function regularly with a longer interval? The investigator herself merely reports the fact, and then makes an "appeal to a more complex perceptual process" as an explanation which is purely speculative in nature and which, I suppose, is equally true and unsatisfying when invoked in connection with creative thinking (France, presumably adults; repro., prior; Oléron, 1952).

Actually many classical experimenters attacked the problem frontally by collecting tons of introspective protocols from their typical or atypical subjects. They were dealing by and large only with ephemeral intervals in their effort to locate upper and lower thresholds, or to verify a psychophysical ratio, but simultaneously they often tried to determine through introspection how the judgments were made. Theirs was an ideal situation, as previously suggested, for unverbalizable cues to serve as the basis for the judgment; most of us just know, for example, whether an interval of 0.25 seconds seems longer or shorter than one of 0.50 seconds but we can provide very little or no substantive explanation for our decision. The subjects in German and American laboratories, however, were trained in the method of introspection; through endless practice and heroic patience they were able to render the unverbalizable into verbalized form, or at least they tried conscientiously to put into words what they honestly imagined they had been previously and spontaneously doing without the use of words. The introspectionist approach has not been very fruitful, though it should not be forever rejected (see *Addendum 7.5*). We do not know how a secondary judgment is reached and therefore all we can do is to follow the procedure customarily invoked when inferences concerning internal responses are made from observable behavior: try to comprehend some of the antecedent conditions which provoke those responses and the consequent responses to which they give rise.

The outcome we do know: a judgment is passed. And so we can state a principle:

7.4 When motivation is adequate and a temporal motive has been aroused, the various responses associated with the perceptual, primary judgment, and other potentials interact in a form most propitious for passing and expressing the secondary judgment.

Motivation, for example, is most "adequate" in the laboratory when subjects agree to judge the duration of intervals and, as in the classical experiments, to report on their mental processes. The organization of the responses occurs, we must remind ourselves again, because there is no separate time sense and hence time is not perceived directly. The limiting case is the ephemeral interval which falls within the span of the psychological present; but even here, when the judgment is not guided by unverbalizable cues there is some kind of psychological organization, so that, for example, discrete events can be included in the psychological present. As the intervals increase in duration, the organization of the events may grow more and more complex, especially when succession is being judged. For then the interval in part or in whole, at the moment of judgment, must be remembered or anticipated, not just experienced; consequently it must be stored and expressed in a form specified by the standard or unverbalizable cues. For this reason minute changes may have marked effects upon the secondary judgment even in the absence of a conscious or deliberate decision.

At any rate, the end result of the secondary judgment is likely to be expressed in verbal form and similarly to be stored as part of memory. As ever, I must quickly add, temporal information is not unique in this respect. What color was the lady's dress? If you are an adult living in the Western world and hence in all probability do not have eidetic images (the apparent ability to produce images, usually visual, in a form resembling the original stimuli), you reply "red," not because you see in your mind that dress and note its color but because you remember the verbal label you attached to it at the time of perception, however fleeting that process was. The label you recall may or may not be accompanied by a weak image or evoke one. I think temporal judgments tend to follow a similar course, except that imagery is likely to be totally absent. You probably cannot replay in your head—or mind—the movement of the sonata as you heard it; instead you resurrect whatever temporal judgment you made verbally at the time, if you made one, or whatever temporal impression that event created as you listened to the music. Fourth-grade pupils in the United States tended to locate hypothetical events in their lives along a symbolic line with marked consistency when retested (Friedman, 1944), perhaps because they made reliably similar judgments a second time, more likely because they recalled their initial judgments. Ninety-five percent of the undergraduates who had been unexpectedly called upon to judge the duration of a 20-second interval were able two days later to recall their judgments correctly; but when asked to indicate how long the same interval now "seemed" to them, only 34 percent gave the same estimates on both occasions (Doob, 1951). This state of affairs is described in the following corollary:

7.4.a *When the temporal motive is delayed and when a previous judgment has been translated into a verbal form, that translation is likely to be remembered and to be utilized again.*

There are other reasons for re-using a previous verbal translation: it may have been verified; it permits consistency, often considered a virtue; and it demands less effort than a brand new judgment. If you know the journey lasted over two hours because you once had reliable information to that effect—you yourself timed it, someone told you, you studied the timetable, etc.—you will be motivated subsequently to recall that information on an appropriate occasion. There will be no reason for you to relive the journey or details therefrom in order to pass judgment afresh. But suppose that you forget the judgment either for harmless or significant reasons, then what can you do? You then must remember what you can: perhaps you noticed when the journey began, you may make a deduction from whatever you did en route, you may remember vaguely how tired you felt when you reached your destination. From this mixture, you derive some sort of impression and pass judgment.

Recalling a sequence of events involves the same process. You are asked whether on that journey of yours you passed through Town A before or after reaching Town B. Again you are dependent upon a series of cues if you did not make note of the sequence at the time. It is these you try to remember; you do not estimate the time between the present moment and your arrival in the past at each of the towns, and then reach your decision by comparing the duration of the two elapsed intervals. The question of whether or why you were originally motivated to learn the sequence is beyond the scope of our inquiry. In passing, though, we note the fact of motivation: the cook worries about whether she first adds flour or salt only when the sequence affects the outcome.

Verbal formulations can be made at every stage of the judgmental process and then revised or preserved. The primary judgment, as indicated in the chart on The Length and Speed of Time, may begin with "short" which refers to an elapsed or pending interval; that in turn, depending upon the anticipation or hope, may lead to the feeling that time is passing quickly or slowly; then as a function whether objective time turns out to be less or more, the secondary judgment may be congruent or at variance with the initial judgment. The choice of temporal vocabulary—the form of the verbal translation demanded always by the Method of Verbal Estimation, and in a sense by the Method of Chronometry, but never by the Method of Comparison—is a function of the language in the society as well as of the cultural forms employed to express temporal judgments. Those forms may be classified in various ways, as has been fruitfully suggested by an anthropologist (Hall, 1959, pp. 87, 170). For example: how long did it take you to finish the job? Adequate translations could be:

1. Informal: not very long; quicker than usual
2. Formal: about three-quarters of an hour; 40 minutes
3. Technical: 42 minutes; 42 minutes, 18 seconds, 23 milliseconds

We need a corollary to suggest which translation is likely to be selected:

7.4.b *The translation expressing a temporal judgment is likely to be one compatible with the standard being employed and/or the one most recently or frequently reinforced.*

Two parts of the corollary demand immediate attention. First, the slippery word "compatible" describes the relation between the judgment and the standard. The solution here must be to give the concept an operational definition in the situation being considered. Subjects in experiments, for example, have little choice except to employ the translations which fit the standard according to the rules of the experimenter: they compare their judgment with a clock or another interval and emerge with formulations which fit his prior decision. Then the reference to recency and frequency in the corollary is an exact reproduction of the principle already enunciated with reference to the choice of standard (*Principle 7.2*). The factor of recency provides an explanation for generalization based upon temporary activity. If you are thinking of the day of your birth and that event seems distant, then certainly your next birthday will appear much closer. Using that first birthday of yours as the reference point will, as long as you do so, make your judgments of shorter intervals seem longer, and vice versa. In the laboratory, however, secondary judgments may be expressed in terms demanded by the investigator, although some of the internal processes leading to those judgments may well deviate from his expectations.

A single experiment well illustrates the corollary. The subjects were instructed to employ either one of two standards, each of which consisted of a 9-point scale: a qualitative one varying from "very, very short" to "very, very long"; and one including the concept of "second" which varied from "very much less than one second" to "very much more than one second." Like good subjects of course they utilized whichever scale the investigator proclaimed they should. Both led to judgments which were affected by the stimulus context in which the interval was embedded, but the qualitative one fairly consistently tended to produce longer judgments than the one involving the "second" (U.S., students; Goldstone & Goldfarb, 1964a). Why this difference should appear is not clear, since we do not know which of the two standards is generally employed more frequently by Americans; but I would guess that the very word "second," even in the context of "very much more than one second," had the connotation of brevity, whereas the qualitative scale fell along a phenomenologically longer continuum.

A critical problem for a translation, temporal or otherwise, is its accuracy or validity. In the present context the terms refer to the agreement between the temporal judgment and objective time, and they are to be distinguished from reliability. Do you agree with yourself when you judge the same interval a second time either through the same or a different method? Your judgment is reliable if on the two occasions you state consistently that an interval lasted 4 seconds, but that judgment is not accurate when in fact the objective duration both times was 5 seconds.

Obviously a thousand and one different factors affect the accuracy of translations and secondary judgments. From where I sit at this moment in an isolated village in Europe, I can scarcely see the clock on the church steeple, so thick is the fog. If I were a stranger here and there were no fog, I might still run into trouble and pass judgment inaccurately: on this particular clock the shorter of the two hands communicates the minutes and the longer ones the hours. Surely you would not have me include fog and unconventional symbols among the variables affecting validity. What I shall do instead is to single out a quartet which seems to possess the greatest generality:

7.5 *The accuracy of a translation expressing a temporal judgment is a function of:*
a. *The modality being stimulated and hence producing perception*
b. *The method of measurement being employed under the specified conditions to judge the interval*
c. *The unverbalizable cues evoked during perception and by the primary judgment*
d. *The duration of the interval being judged*

A word, quite a few words, about each of these four factors. Investigating the modality factor has been a favorite theme for classical and neo-classical investigators (see *Addendum 7.6*). There is a definite tendency for auditory intervals to be judged more accurately than visual ones and also to be judged longer; hence in the latter respect an objectively shorter auditory interval may seem equal to an objectively longer visual interval. Some studies, however, do not reveal this tendency and, moreover, modality clearly interacts with other factors. It may well be, as two investigators point out, after failing to find any difference in accuracy when intervals between 4 and 20 seconds were judged via both modalities (U.S., children; verbal, prior; Crawford & Thor, 1967), that this factor of modality plays a role only when intervals are ephemeral and not when they are transitory or extended: with longer intervals secondary judgments are likely to be affected by mediating responses more compelling than the source of stimulation. In addition, a cultural factor may well be important: all the subjects in the experiments have lived in the West where audition is likely to be the modality for ephemeral intervals more frequently than other sense modalities. Has this tendency, for example, any connection with the kind of temporal discriminations fostered by Western music? The fact that American children tended to be able to judge auditory intervals at an earlier age than visual ones (Goldstone & Goldfarb, 1966) may or may not show "the primacy of audition in the development of temporal judgment," since the subjects could simply have been reflecting emphases in their culture. Certainly it is either ethnocentric or premature to assert that "hearing is the main sense modality in the perception of change and time" (Orme, 1969, p. 9). In general, we are not at all certain how culture affects modalities; all that we know is that the emphasis varies from society to society. In a classical experiment, intervals were bounded by two tones, two lights, or a tone and a light: more precise judgments were made when both stimuli affected the same modality (Germany, university students & staff; comp., prior; Leister, 1933). Can we say that the mixture offered such an unusual experience that the subjects were distracted and hence made less accurate judgments? At any rate, human beings are able to make more or less accurate temporal judgments no matter what modality is stimulated, which is a tribute to their versatility.

The second variable in the principle, method of measurement, has already been discussed in Chapter 1 (see *Addendum 1.3*). We noted there, amid exceptions, a tendency for Reproduction to produce more accurate judgments than Production or Verbal Estimation. Why is this so? One investigator offers the opinion that of all methods Reproduction is "independent of any relation of personal time to objective time": with no need to resort to verbal mediation, the individual may have "personal time units" which "may be long or short, but as long as these remain constant" as he perceives and reproduces they will in effect produce a constant error which will not affect the reproduction (Clausen, 1949). The same reasoning is employed by other investigators who, after finding that most of their subjects counted when using Reproduction, suggest that the biochemical clock "can run at any rate, so long as it runs regularly from the time the tone is presented to the time it is reproduced" (Ochberg, Pollack, & Meyer, 1965). Or one might say that ordinarily with Reproduction the same modality is involved in perception and judgment, whereas with the other two methods different modalities must be utilized. The superiority of Reproduction, however, is likely to be confined to the judgment of ephemeral and transitory intervals soon after they have been perceived, for under these conditions only short-term memory is needed and there can be little forgetting.

Even though Reproduction is not always the method which leads to the most accurate judgments, as *Addendum 7.7* indicates, I propose a promising corollary, the evidence for which can also be found in the same *Addendum*:

7.5.b.1 Accuracy varies positively with the similarity between the method of perceiving either the interval or the standard (or both) and the method of translation or expression.

The third variable affecting accuracy is the presence or absence of unverbalizable cues. We have noted in Chapter 3 a collection of phenomena which suggest the functioning of a biochemical clock: presumably in the absence of external cues, some endogenous stimuli cause the individual to pass temporal judgment or to behave as if he had been supplied with temporal information. The question being considered at this point is whether such judgments are accurate in their own right or whether they are more or less accurate than those stemming from verbalizable cues. William James has written this autobiographical note:

> All my life I have been struck by the accuracy with which I will awake at the same *exact minute* night after night and morning after morning, if only the habit fortuitously begins. The organic registration in me is independent of sleep. After lying in bed a long time awake I suddenly rise without knowing the time, and for days and weeks together will do so at an identical minute by the clock, as if some inward physiological process caused the act by punctually running down [James, 1890, p. 623].

Most persons in our society, however, usually do not, or, because they lack James' ability, cannot trust their biochemical clocks and they therefore have greater faith in mechanical timepieces and their own deliberate judgments. If you wish to be absolutely certain that you will awaken at your usual hour, you undoubtedly prefer to depend upon an alarm clock or another person who agrees to perform its function. Otherwise you fear, regardless of the regularity with which you awaken without outside assistance, that you may oversleep or spend a restless night wondering whether you will do so. Evidence concerning the accuracy of the biochemical clock and endogenous judgments in general is reviewed in *Addendum 7.8*, where it can be seen that the situation is not as clear-cut as that portrayed by James. Judgments stemming from unverbalizable cues can well be accurate —for some persons or under some circumstances—and consequently play a role in the judgmental process. But they themselves are affected in the short or the long run by exogenous processes and they also interact with other processes that are conscious or semi-conscious and hence verbalizable. Have you ever tried to estimate how long you have been asleep after a nap or after your eyes unwittingly have been closed in some kind of a boring situation requiring you to stay awake?

Finally, the duration of the interval itself significantly affects accuracy, as has already been frequently indicated through these pages. Whenever the indifference point or zone is located, for example, we know that some intervals have been judged more accurately than others. And this is the situation in almost any investigation, even though the precise duration of the intervals is seldom the same: one interval always wins the prize, even when those longer than ephemeral are being judged (U.S., children & adults; verbal, prod., and repro. with prior; Gilliland & Humphreys, 1943). I supply no *Addendum* on interval duration because it would be fruitless to do so. No more specific generalization can be made to emerge other than that the prize-winning interval is determined in large part by contextual factors such as the method of measurement and the modality (e.g., U.S., presumably adults; prod. & repro. with prior; Hawkes, Bailey, & Warm, 1960) or that:

> We are more precise in estimating seconds than minutes or hours, presumably because when the period is brief we can keep our attention bound to the interval itself, but when it is long, our minds wander and our judgments are then more erratic, based, as they are, on such indirect cues as the number and kind of activities that have occupied the time [Cohen, 1967, p. 19].

In addition, the interaction between the interval's duration and the method of measurement may not be overlooked; thus in one investigation involving intervals between 2 and 29 seconds there was a more pronounced tendency with Verbal Estimation for the undergraduate subjects to overestimate the shorter than the longer intervals, but the reverse was true with

Production and Reproduction (U.S.; prior; Horstein & Rotter, 1969). It is also true that our opinion concerning the relative accuracy of a judgment varies with its duration and may perhaps be subsumed under some kind of mathematical or Weber function; thus it would be a great blunder to misjudge the duration of a 100-meter dash by more than a second or two, but the estimate of the duration of the entire track meet could err by many minutes and still be considered reasonably accurate.

THE PASSING OF TIME

From the outset of the analysis we have been suggesting implicitly that the accuracy of the translation and the secondary judgment are affected by the time elapsing between the interval being judged and the moment at which the judgment is made. Translations already verbalized in the past are likely to be utilized again when the occasion arises. What happens, we now ask, when the judgment is really delayed and when for the first time judgment is passed upon an interval from the more or less distant past or when the judgment refers to an interval pending in the far future?

One possibility has been pointed out: under certain conditions (when temporal information is not salient, when the drives associated with the interval are not too strong) events are more likely to be salient than drives. But in much more general terms, we now maintain, it appears that:

> 7.6 *With a delayed-subsequent temporal motive, the accuracy of temporal judgment varies inversely, but imperfectly, with the time elapsing between that interval and the judgment.*

The nature of the temporal motive is indicated in order to exclude very brief delays between the termination of the interval and the process of passing judgment. For example: with one exception, the interval of delay, which varied between 1 and 29 seconds, had no effect upon the accuracy with which a group of retarded and normal adults reproduced intervals whose duration was also within the same range. The exception was in accord with the principle: the retardates significantly overestimated the 1-second interval when the delay was longest (U.S.; prior, McNutt & Melvin, 1968). "Remarkable stability" was noted in an experiment in which student nurses reproduced a 2-second light flash and a 1.5 second tone immediately afterwards or at intervals extending from 1 day to 28 days; the intervals were always underestimated and the degree of underestimation, as the principle suggests, increased with the passing of time, though not significantly so (U.S.; prior; King, 1963).

The intervals in the two experiments just cited were neutral ones experienced under laboratory conditions. The principle, I think, is more likely to be valid not only when the intervals in question are distant in either temporal direction, but also when they involve intervention. Distortion concerning past intervals arises in part from wishful thinking and other drives, especially of an unconscious sort, which are likely to exert greater influence with the passing of time. Cruelties come to seem a little less cruel; past events may be idealized; in short, what actually happened tends to be changed into what should have happened. Some of the distortion may also be due to retroactive inhibition: subsequent events interfere with the retention of those learned at an earlier period. Evidence suggests that almost any kind of intervening distraction diminishes accuracy (see *Addendum 7.9*). Similarly distortion may take place regarding intervals anticipated in the future. Fantasy operates in relation to events which have not yet taken place; and the unanticipated may upset the anticipated.

The adverb "imperfectly" appears in the principle to call attention to the possibility that in some instances temporal distance may improve accuracy. It has been suggested, for example, that a recent interval seems longer than one from the more remote past (Fraisse, 1963, pp. 167–8), perhaps because we forget more and more as time passes and hence the latter appears less crowded in retrospect than the former. Still, as one of our previous principles would lead us to suspect (5.1.c) and, as has been suggested, "apart from the disturbing effects of fatigue, an interval in which there are many ideas passes quickly, but it is long in retrospect":

> ... The child who is taken for a treat to the theatre finds something new and exciting each minute—the journey to the town, the marvelous trams, the crowds and shops and traffic, the importance of having a real grown-up lunch, the thrill of getting seated, the tuning-up of the orchestra, the rising of the curtain—and it all passes much too quickly. But afterwards, going home, it seems incredible that so much could have happened in one day, and "this morning" seems ages ago ...
> [Cleugh, 1937, pp. 28–9].

The explicit assumption here is that the child is gratified during the experience ("new," "exciting," "marvellous," "thrill," etc.) and that he wants the pending interval to continue, although he realizes that it cannot do so ("passes much too quickly"); the gratification makes the interval appear short but the wish to lengthen it proves fruitless. "Afterwards, going home," he remembers the events more vividly than the emotions for a number of reasons: the excitement of the day may have produced fatigue; events are perhaps easier to recall than the feeling of pleasure; events when recalled can reinstate some of the previous joy. If an explanation exclusively via events is invoked (cf. Ornstein, 1970, p. 87), it would have to be argued that events during the excursion sweep into one another and as a somewhat undifferentiated whole appear short; but that later they are remembered one by one, hence make the interval appear to have been crowded and long. There is also a complication: the child afterwards must temporarily forget his verbal judgment (which was probably a primary judgment) during the events of the theater, for otherwise he would continue to think the day short. It seems probable, I suppose, that both judgments can be made afterwards: he can think or say that the day passed quickly when he was gratified by the journey, the restaurant, and the theater but—perhaps on a secondary level—that then " 'this morning' seems ages ago." Why, finally, was the motive to pass temporal judgment aroused both during the "treat" and subsequently? Again, if another guess be permitted, I think it unlikely that it was evoked during the "treat" since the child was fully occupied and happy. Later the reappearance of familiar stimuli may have recalled the departure which in turn could evoke the motive.

We may, however, never forget events associated with a very significant interval, in which case the duration of the interval will be judged similarly either immediately afterwards or years later. Let me use a hypothetical illustration. Assume you were bored during the occurrence of an event; time passed slowly; in retrospect, though, you remember little and hence the interval *when judged subsequently* may appear short; but if you remember that you were bored and that time seemed to drag, your memory of the interval may be affected completely by that past judgment. You may also think that the interval seemed long at the time but now appears slow in retrospect. I see no way, moreover, of comparing the initial perception of events when the temporal motive is prior and when it is interim or subsequent. For the argument also can run either way: anticipated intervals may seem to contain either few events because few can be imagined, or many because they may be imagined to be crowded or empty. In short, accuracy is likely to be greater when the temporal motive is prior or interim, inasmuch as the person thus alerted takes advantage of whatever cues are offered in the situation or whatever relevant experience he has had; yet alertness can lead to distraction, self-consciousness, and perhaps the error of overestimation sometimes associated with paying attention to the passing of time.

FEEDBACK

The final, final step in the analysis is the leap from the secondary judgment back to the stimulus, the temporal potential or the total potential. We reach the problem of feedback: past judgment affects future ones. That such effects appear in the short run is clear when the Method of Absolute Judgment is examined: each judgment in the present contributes to a store of information which produces an anchoring effect, with the result that within the experimentally established frame of reference judgments are more or less consistently passed. More generally, any temporal judgment that is experienced is likely to affect what is remembered from the past and anticipated in the future, especially when the judgment also involves an evaluation related to the individual's goals or philosophy.

Forty prisoners in an American penitentiary, for example, were interviewed in an effort to discover the factors associated with suffering. Correlated with those feelings were no objective measures of time—the length of the sentence, the actual time spent in prison—but the men's evaluation of these temporal factors: did they consider the sentence just, did they know definitely when they might be released (Farber, 1944)? The temporal judgment thus gave rise to an attitude which in turn affected their feelings and presumably their behavior. An individual with temporal fear is thrown into panic when he notes the passing of time and, similar to the Spiral effect in boredom and impatience, he frequently observes the passing of time because of the value he attaches to that judgment. Or even simpler is a tendency to speed up or slow down activity after obtaining reliable information from a clock.

Time passes slowly or quickly, but we are more likely to say that it passes too slowly or too quickly, so ever present is some kind of evaluative norm. A person with incorrect temporal information—whether he be Rip van Winkle or a psychopath—is to be pitied and corrected. Most intriguing of all, in my opinion, is the evaluation placed upon an awareness of time which, as we have indicated, is encouraged by Western society. Is it better to be less aware of the passing of time, as presumably traditional peoples are? Since boredom and awareness of time are closely related, does it follow that they are less likely to be bored and hence to demand less diversion and variety than we? But then it can also be said that without awareness of time persons are likely to lack perspective and whatever joy that can bring. Surely we dislike interrupting a pleasurable activity to become aware of the fact that time is passing, but at least with that awareness can come the possibility of extending the activity and of becoming reabsorbed in it.

We have seen, however, that the act of passing temporal judgment does not necessarily improve the accuracy of future judgments unless there is reinforcement or feedback. What, then, is fed back into the past or the future from the present? Here we must refer to the temporal potential:

7.7 *A reinforced temporal judgment produces an appropriate change in the temporal potential.*

Again the tricky concept of "appropriate" appears. A temporal motive is likely to be aroused again whenever a goal which has been previously achieved as a result of a temporal judgment is sought under sufficiently similar circumstances. You discover that the clock on the tower is in error; then you account for your tardiness and you take care not to trust that clock again. If a future orientation leads to anticipations which are frustrated, then the orientation may be altered so that more emphasis is placed upon the present or the past.

The ages of man could be rewritten in terms of this principle, especially for people living in the West. For it is almost impossible to think of situations in which the experiencing of time does not have an effect upon behavior. The schoolboy does not like to creep like a snail to school after he has heard the warning bell: he runs so that he will not be late or too late. "It is later than you think"—the now common observation is employed to call attention to almost any kind of disaster, verging from the personal to the atomic. Presumably being alerted to the lateness of the hour in a literal or metaphorical fashion is reason enough to alter behavior. Ordinarily we do not like to be reminded of our age—at least after adolescence—for the temporal information suggests all kinds of nonaccomplishments and, as ever, the approach of death. An ophthalmologist, on the basis of his clinical experience, suggests that morbid reactions to temporal events symbolized by the holidays of Christmas and Easter or by anniversaries may possibly trigger off serious organic diseases such as cancer (Inman, 1967).

The complicated interaction between temporal judgments and other behavior is well illustrated in connection with dissonance which is defined operationally in terms of conflicting beliefs or values. There seems to be universal agreement that dissonance is a state of affairs which people seek to avoid, to mitigate, or to eliminate. Some of the discomfort involved in many temporal judgments leading to renunciation can be a function of dissonance, for the individual may have two firm, but incompatible, beliefs: he believes that he must renounce achieving a goal but simultaneously that the goal from some standpoint is both necessary and desirable. As a result of a heart attack a man realizes that he must abandon his favorite sport but he also knows that he requires some form of enjoyable exercise. Subjects in a dissonance-type experiment renounce the opportunity to engage in pleasurable activity during the time they are being tested, but they are attracted by the fee they are paid. Among the many ways of reducing such dissonance is to have a future orientation and to anticipate that later on there will be substitute or adequate satisfaction: the gratification is delayed. The man with the weak heart anticipates that, after he recovers, he will be able to engage in a less violent sport; the subjects in the experiment, aside from trying to make believe that the dull activity they must perform during the experiment is really exciting, can also anticipate what they will do with the remuneration after they get it.

On a national level, too, the same phenomenon appears. For the people of a country sometimes make territorial demands, at a minimum for the land they already possess, at a maximum for land belonging to their neighbors or other people. Leaders or followers cannot defend the claims blatantly and nakedly simply in terms of their own needs and desires, for with almost any kind of ethical principles such unabashed assertions would make them feel uncomfortable: there would be dissonance between what they wish and the principles that are supposed to guide their conduct. Under these circumstances, territorial claims must be justified to remove the dissonant tinge. A whole arsenal of justifications is available and utilized (Doob, 1964, pp. 147–55, 189–91), among which temporal factors are likely to be mentioned. The arguments may not be nakedly advanced but, when present, they implicitly point in all three temporal directions. The past may be stressed. "We were here first," which means that we—our ancestors and ourselves—ran risks and made sacrifices to occupy the land and since our arrival we have strengthened our claim by notable achievements. The authenticity of this claim may be established, it is hoped, by consulting the archaeological or historical record. A variant of primacy involves duration: "We have been doing this for a long time and therefore expect to continue"; or, when land is in dispute and absolute primacy cannot be claimed, "we have been here longer than they." Well-established ways of behavior are believed to be satisfying, and their longevity is thought to reflect their goodness: the older the tradition, the more valid it must be. The justification in present terms takes the form of naturalistic ethics: "We are here, we are not going to move, we will not cower before force, we have a right to what we have." The status quo, moreover, derives some of its force from the past ("Why change?") and from the future ("Why run the risk?"). That future itself appears as a justification in connection with posterity: "We must do this—for the sake of our children" or "to keep our culture alive." Especially in this instance, but also in the other two temporal dimensions, the feelings may function not merely on a symbolic level with reference to the country but also quite realistically in family terms: just as provision is made for children to inherit property, so their future is viewed as a continuation of present customs within the kind of nation which now exists.

EPILOGUE

As these three chapters on the judgmental process end, as we complete the discussion of the variables portrayed in The Taxonomy of Time, there remain a few loose ends to be woven into the fabric of the analysis. Time, we have been saying again and again, is a complicated process which has demanded and received a multivariate analysis. There is no simple answer to the question of how people judge duration or succession. You have to begin by considering their total potential, proceed from there to the stimulus potential at hand, and then slowly observe—or try to observe—exactly what is perceived and how the perception may undergo some modifications before judgment is passed. But the pieces must be put together, separate factors cannot be completely isolated from one another. Unless the interaction is emphasized, even a well-designed and elegantly conceived series of experiments produces results which the investigators themselves with disarming honesty call "insufficient for building up a well-rounded hypothesis" (Netherlands, unknown; comp., prior; Roelofs & Zeeman, 1949). When the research in fact concentrates upon many factors and their interaction, we find a necessarily iffy kind of conclusion emerging. That point clearly emerges in a factor analysis of data from over 100 American undergraduates who estimated the duration of intervals of varying length (but generally 2 minutes) during which they engaged in a variety of different activities and after which they also evaluated their temporal responses: "time may seem long during an interval because the activity is boring, because attention is being paid to the passage of time, because the activity is unfamiliar, or because [the subject] is relatively passive" (prior; Loehlin, 1959).

Only a reader with infinite patience will wish to read this paragraph which provides yet another illustration of the complexity produced by interaction. In one experiment drive strength was manipulated by telling half the subjects that their performance (filling in missing letters in words) would simply provide information about the task for future use in other investigations ("task-involved") and the other half that it would reflect their intellectual abilities and personality ("ego-involved"). Goal attainment was manipulated by providing feedback during the task which indicated (falsely, of course) that the individual was succeeding or failing at the task. The temporal judgments themselves were not reinforced. The results, derived from three temporal estimates at the end of 90, 110, and 135 seconds, show that being ego-involved in the task increased temporal judgments when, and only when, no information concerning success or failure was provided:

and it lowered the estimates under conditions of both success or failure; thus our principle concerning drive strength (No. 6.1) is verified but only in the absence of attainment or nonattainment. Then, secondly, the absence of feedback in control groups significantly increased the temporal estimates which is a partial vindication of our principle concerning attainment (No. 6.3); but that principle seems refuted by the fact of no significant differences between the success and failure groups. When the interaction of the variables is considered, the experimental groups can be arranged in the following order of increasing temporal judgments: ego-involved with success, task-involved with failure, ego-involved with failure, task-involved without feedback, task-involved with success, and—by far the longest—ego-involved without feedback. Greatest accuracy tended to be achieved without feedback. In addition, feedback affected the "work" accomplished, hence the number of events which could be perceived during each interval; and the group experiencing failure was more "bored" than the one experiencing failure (U.S., college; verbal, prior; Felix, 1965).

The grim tale is not over: it is possible to conceive of innumerable variables which could be manipulated in experimental situations. One classical experimenter, for example, presented his four subjects with intervals filled with tones of varying intensity and asked them to judge both duration and intensity with instructions indicating to which of the two components they should pay closer attention (Germany, graduate students & staff; comp., prior; Neumann, 1936); the result was inconclusive, but surely someone could justify a repetition of the technique as a doctoral dissertation. In the brave days toward the end of the last century and the beginning of this one, when scholars were not hampered by the thickets created by the epidemic of research, one writer listed no less than 10 factors upon which, in his opinion, temporal judgment depended; they included references to the intensity, number, speed, and position of the "images" that were evoked by the interval as well as "the intensity of our attention to these images or to the emotions of pleasure and pain" accompanying them (Guyau, 1902, pp. 85–6). Progress?

This same sort of complexity faces anyone who seeks to apply our knowledge of time to real-life situations. From Ovid to this day we have testimony to the effect that time seems to stand still and also to flow by very quickly for exiles who long to return to their native land (Solanes, 1948). First, we do not know the facts except through such testimonials: is this actually the view of exiles in general? Then the experience of being an exile must mean a change in way of life: both the event and the drive characters of any interval are different from what they were at home, but the two certainly cannot be blithely disentangled. Another example: one writer has sought to bring to bear on legal testimony what he assumes is common knowledge concerning temporal judgments as a result of experimentation. He isolates three factors which he thinks produce a tendency toward overestimation: "lack of overt activity," "the occurrence of many events," and "an attitude of tense expectancy." We are not interested here in determining whether his appraisal of these factors is correct; rather we would emphasize with him the fact that

> all these last three factors frequently occur together in many situations concerning which witnesses are asked to give testimony. The frightened but motionless bystander watching a bank hold-up; the highway-accident witness waiting, tense and helpless, for the sound of the ambulance siren; the trembling, silent householder, listening to the burglar rummage the downstairs drawers while his telephone call for police aid seems to have been forgotten—these circumstances provide the optimal conditions for overestimation of a short time interval (Culbert, 1954, p. 686).

In effect, the argument would emphasize the context in which any kind of stimulus is perceived. Ordinarily an ephemeral interval is likely to be judged as brief and its emptiness is undoubtedly ignored; but, as William James once suggested by quoting the words of another philosopher-psychologist, a sudden pause in music or speech gives an impression of real emptiness and the interval appears interminable until the flow resumes (James, 1890, p. 626).

Finally, it must be remembered that each of the variables we have considered receives its special weighting as a result of processes within the behavioral and temporal potentials. In this chapter we have singled out experience in judging intervals and feedback. The cultural factors discussed in earlier chapters do play their significant role. If it is true (Hall, 1959, pp. 29–30) that Anglo-Saxons expect an individual to engage only in a single activity during a given interval (you converse exclusively with me), whereas Latin Americans may be more diffuse (you converse with more than one person simultaneously about different matters), then the duration of such an interval may be judged quite differently for reasons associated with its event or drive attribute. Or consider another problem which used to be fashionable: do men and women judge time differently? On a straight empirical basis, the classical experimentalists occasionally broke down their data by sex and, during the early part of this century and largely on the basis of results obtained from Americans, reported a tendency for females to overestimate intervals more than males. This slur on women—supporting the prejudiced view of men that they are never punctual—was not confirmed by later investigators. More recent studies suggest, with exceptions here and there, no differences between the two sexes (see *Addendum 7.10*). Even if the original insult were confirmed, the obtained difference would still have to be explained, and the explanation would have to rest, presumably, upon the varying roles of the two sexes in our society which might lead to diverse ways of appraising the attributes of intervals. Then, too, the divergence might be expected to disappear as the roles of men and women become more similar (cf. Loehlin, 1959). The hypothesis has been advanced that the greater variability among females may reflect a greater susceptibility to external influences (Goldstone, 1968b), again no doubt culturally determined. At any rate, this digression suggests that principles pertaining to time should be employed only in specified psychological and sociological contexts.

ADDENDUM 7.1 EXPERIENCE WITHOUT REINFORCEMENT

Within a given series of trials during the same session, an *improvement in accuracy* in the absence of reinforcement has been reported (U.S., college; verbal prior; Jones & MacLean, 1966) (West Germany, soldiers; prod, and verbal with prior; Spreen, 1963). Similar *improvement* has been noted *from session to session* even when an interval of time intervened (U.S., adult males; prod., prior; Emley, Schuster, & Lucchesi, 1968) (U.S., mostly college; comp., prior; Gridley, 1932). Improvement in reliability (i.e., making consistent judgments) from one session to the next has also occurred (U.S., college; repro. after hearing the standard not once but twice with prior; Doehring, 1961) as well as during a given session (U.S., college; verbal prior; Bakan & Kleba, 1957). Reliability in judging intervals ranging in duration from 15 to 240 seconds was low, but greater within than between sessions which followed each other at intervals of 1 and 2 weeks (U.S., college; verbal, prior; Strunk, 1960). Under similar conditions *no improvement in accuracy* during a given experimental session or from session to session has also been observed (U.S., college; verbal, prior; Schaefer & Gilliland, 1938) (U.S., college; prod., prior; DeWolfe & Duncan, 1959)

(U.S., college; comp., prior; Woodrow, 1928).

Otherwise there is a considerable body of evidence suggesting that repetition in the absence of reinforcement leads to *longer judgments* concerning some intervals but not others—usually the relatively longer ones produced more overestimation than the relatively shorter ones—when measured by the Method of Production (England, various students; Treisman, 1963) (U.S., college; Brown & Hitchcock, 1965) (U.S., "mainly" college; Falk & Bindra, 1954) (U.S., adult males; Emley, Schuster & Lucchesi, 1968) (U.S., college; Shechtman, 1970). Lengthening has also appeared with Reproduction (U.S., normal, schizophrenic, & neurotic adults; Warm, Morris, & Kew, 1963) (Brown & Hitchcock, op. cit.) (Australia, presumably adults; Coltheart & von Sturmer, 1968) (Germany, university; Sixtl, 1963) (Treisman, op. cit.) (U.S., college; Warm, Foulke, & Loeb, 1966). In one investigation, however, the reproduction was longer on successive trials when the subjects during the interval to be reproduced had been learning to sort cards or had been reading aloud but not when they had simply been dealing cards at an increasing speed (U.S., college; Kleiser, 1953). The Method of Verbal Estimation has induced a *shortening* of judgments under these conditions, a result congruent with the tendency toward lengthening when Production is employed (U.S., college; Gilliland & Martin, 1940) (Treisman, op. cit.).

I cannot refrain from citing one other study if only because it was based upon 524 college students and staff members in an American university, who provided a total of 99,480 (*sic*) judgments. The indifference point was shown to be affected markedly by "the preceding experience of the subject in comparing durations," whether that unreinforced experience consisted of having been tested on previous occasions or tested frequently on a single occasion (comp., prior; Stott, 1935). But even such a gargantuan study is difficult to interpret when other investigations are observed: with empty intervals in the neighborhood of 1 second, there was a tendency initially to judge the first of a pair to be shorter, but later in the series that tendency was reversed (U.S., college; comp., prior; Woodrow, 1935), a generalization which in turn was shown not to hold for shorter intervals of about 0.5 second's duration (U.S., presumably college; compar., prior; Woodrow & Stott, 1936).

It must be strongly emphasized that most, though not all results so concisely presented above are in fact less clear-cut than I have made them sound. Thus one investigator has shown that general experience in the laboratory and in the art of comparing the duration of two tones was related to finer and more accurate discriminations, but other factors interacted: what was vaguely called "personality type," the manner of perceiving the two intervals, the presence or absence of a rhythmical sequence before the presentation of the first stimulus (Germany, university; comp., prior; Pankauskas, 1936). Or the evidence of complexity appears when we re-examine one study already cited twice in previous paragraphs (Emley, Schuster, & Lucchesi, 1968). Here 5 subjects produced intervals of 5-, 15-, and 30-second duration 30 times, both in the morning and in the afternoon, on 9 different days. For the group as a whole there was a tendency for judgments to increase, but only for the first of the sessions and only with reference to the two shorter intervals and not for sessions on subsequent days. Those subsequent sessions revealed a tendency to decrease judgments. In addition, the tendency initially to make longer judgments was a statistical one from which some of the subjects clearly deviated. The most important trend over the 9-day period was that these "productions without feedback approached accuracy across trials," with a tendency for the group as a whole to decrease judgments over that period. The investigators do not think the subjects practiced ouside the laboratory—they were asked not to do so—and the greatest improvement occurred on the second session before they had left the laboratory.

Why is there a tendency for repetition without reinforcement to increase accuracy? Either on the basis of experimental manipulation or speculation, the following explanations have been offered:

1. Growing accustomed to the experimental setting or being less distracted by extraneous stimuli. The task in one investigation was to produce an interval of 7 seconds by pushing a button which flashed a picture on a screen. Two of the pictures were of ugly automobile accidents (including a corpse or part of one), the other two were simple ones of moving or parked cars. In all instances the intervals tended to be underestimated; during the first half of the series, the simple ones were underestimated to a significantly greater degree than the dreadful ones; but during the second half there was no difference in the two (Sweden; university; Bokander, 1965). Evidently the subjects learned to pay less attention to the content of the pictures and more to the task at hand which was to judge time. In another study ordinary American undergraduates were able to achieve a high degree of accuracy in fractionating intervals with a duration ranging from 0.4 to 4.8 seconds:

their task was to adjust the duration of an interval (created by a tone) until it seemed to be one-half of a standard interval which was either the first or the second of a pair. They were asked not to count and not to utilize any physiological cues such as breathing or pulse rate. But, the experimenter points out, they were given a preliminary practice session; the intervals were presented in a varied order so that the time-order error could be balanced out; they were told to take as long as they wished in making their judgments and they were permitted to hear each pair again and again until they felt the fractionation had been achieved (Gregg, 1951). Perhaps improvement occurs also because subjects grow more attentive to the temporal task. In one study reliability of estimates concerning intervals ranging from 2 to 10 minutes was greater for those involving a monotonous task than for those containing interesting ones (France, students and laboratory assistants; verbal, prior; Korngold, 1937); presumably interesting tasks were more distracting than dull ones.

2. Coming to utilize cues provided during the interval. On successive trials subjects improved the accuracy with which they produced a 60-second interval when they were simultaneously engaged in solving a problem but not when they were inactive. In the active condition, the investigator avers, "more stimuli" were provided; somehow, therefore, those stimuli were utilized even though they were not directly related to time (U.S., college; Essman, 1958).

3. An "increase in confidence with each successive interval" leading to shorter judgments which in most instances meant greater accuracy (U.S., unspecified; prod. by fractionation, prior; Ross & Katchmar, 1951).

4. Learning to respond to a novel task. There was, for example, a slight and almost significant tendency for intervals with durations in the region of 1 second to be judged more accurately during a second than during a first sitting of 100 trials each, with the elapsed time between the sittings varying from 1 to 38 days; a deliberate attempt was made to eliminate "practice effects"; the slight improvement was slightly greater when the intervals were bounded by tactual stimulation of the finger tips than by auditory clicks (U.S., mostly college; comp., prior; Gridley, 1932). The fact that auditory stimulation produced significantly more accurate judgments at the outset than tactual, suggests that repetition without reinforcement in this instance may have served simply to accustom the American subjects to receiving temporal information through their finger tips, no doubt a rather unusual modality for them. But practice may not always eliminate the effects of novelty. In a classical experiment there was improvement in reproducing intervals only when the model was bounded by stimuli belonging to the same modality rather than by those from different modalities; and I assume that the former situation was more usual than the latter (Germany, adults; prior; Schmidt, 1935).

5. Keeping at a minimum the interval between the model to be reproduced and the actual reproduction. In the German experiment just cited, preliminary investigation revealed that reproduction was difficult and very inaccurate if it occurred more than 5 seconds after the standard interval had been presented. A more systematic attempt to determine whether there was any clear-cut relation between the accuracy of reproducing intervals with durations between 1 and 16 seconds and a delay in reproduction extending from 0 to 60 seconds led to largely negative results: most accurate reproduction occurred with no delay; thereafter inaccuracy increased and tended to be greatest between 5 and 15 seconds but then with longer delays to decrease (South Africa, European university students; Du Preez, 1967b).

The principal explanation of why judgments lengthen with repetition as measured by Production and Reproduction involves a reference to "boredom": judging intervals again and again becomes tiresome and hence the verbal or experienced model seems longer. One difficulty with the theory is that all methods do not always yield the same results. In one study temporal judgments did increase with Reproduction but did not change with Production or Verbal Estimation; why should one method be more boring than the other two? The author merely states that "different processes may be involved" in the latter (U.S., normal, schizophrenic, & neurotic adults; prior, Warm, Morris, & Kew, 1963). Another investigator suggests that, when subjects find the chore of making repeated judgments monotonous, they tend to ignore cues regarding duration and hence their judgments must be "increased in length to match remembered standards." Evidence for this interpretation comes from an experiment in which the tendency to make such judgments was to a great extent halted by providing an alerting stimulus, a bell or a light, before or during the series; these simple additions apparently decreased monotony and so accuracy increased (Australia, university; repro., prior; von Sturmer, 1966). In addition, another series of experiments demonstrated that the tendency to give longer judgments without reinforcement could be inhibited somewhat by manipulating the instructions: in tapping a key at 8-second intervals, the subjects tapped faster and increased the intervals less if they were instructed to tap when "there was the slightest possibility that enough time had elapsed since the last tap"— and similar results were obtained with instructions manipulating hypothetical pay-offs (Australia; von Sturmer, 1968).

Boredom—"flagging attention"—has also been advanced to explain the decrease in accuracy with unreinforced repetition, which may very well be, since in the study giving rise to the hypothesis no less than 100 judgments were made each hour for a period of two hours at each session (U.S., college; comp., prior; Woodrow, 1928). A "judgment drift" toward overestimation, however, has also been noted when boredom may not have been involved: a 30-second interval was produced only 10 times, and the drift was not significantly affected by the reward (money) or punishment (electric shock) anticipated at the conclusion of the series being produced (U.S., college; prior; Shectman, 1970).

Another explanation for an increase of reliability with repetition postulates the advantage of a "warming-up" effect in a given session. Reliability of judgments by subjects who estimated the duration of intervals ranging from 15 to 240 seconds twice during one session, and then again a week later, was high between judgments at the same session but was much lower and generally not significantly above chance between sessions. The relation between the second judgments of one session and another, however, was much higher than that between the first judgments, and for the longer intervals turned out to be significant (U.S., college; verbal, prior; Bakan & Kleba, 1957). Two other investigators indirectly offer evidence supporting the view that repetition may alert subjects to various cues which can aid them in passing judgment: they have demonstrated that the reliability but not the accuracy of Reproduction could be raised and the inter-subject variability increased by utilizing kinesthetic cues which was accomplished by instructing a subject to move his arm in the act of reproduction (South Africa, European students; prior; Danziger & Du Preez, 1963). Blindly but consistently, therefore, it is possible for judgments to be made after repetition. One investigator notes that 10 subjects as a group did not reveal a central tendency to overestimate or underestimate filled or empty intervals but soon each adopted a standard of his own from which there were only minor deviations (France, adults; repro., prior; Orsini, 1958). Results such as these, she suggests, are like those obtained in dealing with loudness where the experimenter commented upon "the paradox of Ss [subjects] exhibiting a high reliability in a judgment for which there appears to be no validity" (Garner, 1959).

ADDENDUM 7.2 EXPERIENCE WITH REINFORCEMENT

One way to test whether the accuracy of temporal judgments improves with reinforced practice is to examine such judgments developmentally, for presumably there is a positive relation between increasing age and the number of temporal judgments which are passed and then reinforced. Supporting evidence comes from a study in which samples of American children made more errors and gave more variable judgments concerning intervals ranging from 9 to 180 seconds than a sample of college students (verbal, production, & reproduction with prior; Gilliland & Humphreys, 1943), as well as from a somewhat comparable study of Soviet children of varying ages (verbal, prior; Elkin, 1928). The American investigators, let me note in passing, invoke a reinforcement hypothesis in trying to account for the fact that the American children performed better than the Soviet children: "It is to be expected," they assert somewhat blandly, "that American children from well-to-do homes would have had more experience in time estimation than Russian children from workers' homes." The relation between frequency of general reinforcement and accuracy, however, is not linear, in part perhaps because of a ceiling effect: it has been shown that samples of American children achieved such a high degree of accuracy in counting up to 30 seconds that no further improvement was possible— and the high degree of accuracy at age 8 or 9 continued through the older groups until individuals over 70 were found to reveal slight deterioration (Goldstone, Boardman, & Lhamon, 1958b).

Does practice with reinforcement improve the functioning of the biochemical clock? One investigator questioned over 1,000 Germans as to whether they could awaken themselves "at a definite time." If we take the replies at their face value, then we might say that practice is helpful in achieving this feat: 79 percent of the adults (psychologists, neurologists, and university students) and 42 percent of the children believed that it was "possible" to do just that (Clauser, 1954, pp. 51–4, 58). More convincing and more damaging, however, is the finding of another German investigator whose 5 adult subjects awakened themselves at times of their own selection on 100 successive and 150 nonsuccessive nights; presumably they ascertained the correct clock time after waking up; improvement after so much practice, he states, was "not demonstrated" (Frobenius, 1927).

What has been demonstrated is the influence of reinforcement, when experimentally manipulated, upon accuracy. A Soviet investigator reports that the accuracy with which intervals of 3, 5, and 10 seconds were reproduced did not improve with non-reinforced practice, but did improve when during intervening trials a conditioned reflex to intervals of that duration had been developed by means of an electric shock; improvement in reproducing the other two intervals was also observed when the intervening training was confined to the 5-second interval; and in other studies verbal instruction concerning mistakes, especially when detailed, led to improved performance (Elkin, 1964). After being told how much time in fact had elapsed as they sought to produce a 30-second interval, high school boys perceptibly improved their performance on a second trial (U.S.; prior; Davids & Sidman, 1962). Similarly, accuracy in producing intervals of 15, 60, and 90 seconds increased as a result of positive feedback, and on the whole decreased with false feedback; under the latter condition, nevertheless, some subjects continued to improve (U.S.S.R.; prod., prior; Dmitriev & Karpov, 1967). The American subjects sought to flash a telephone switchhook for 0.3, 0.5, or 0.7 seconds. Their tendency toward overestimation decreased and hence the accuracy of their judgments increased not when they were given instruction through an appropriate demonstration and not when they used a mnemonic device (saying the word "three," "five," or "seven" while depressing the hook); the effective reinforcement turned out to be the verbal feedback of hearing "too long," "too short," or "good" during the sessions (telephone employees; prod., prior; Schoeffler & Poole, 1967). A group of Australian undergraduates underestimated the duration of one second when counting; but when the correct rate was demonstrated to them, they appreciably improved as determined immediately after the demonstration, an improvement which persisted with slight variations 1 week and 3 weeks later (Kruup, 1968).

Some of the laboratory experiments provide clues concerning the ways in which repetition with reinforcement leads to greater accuracy. In one study, 7-year-old French children were deliberately trained to reproduce empty intervals of 30 seconds by pressing a button: after each trial a light went on if the response had been correct within 4 seconds, a red one if the interval was too short, and a green one if it was too long. Marked improvement was noted, in fact the children after three series of five trials each reached a level about the same as, if not better than that attained by adults before participation in the experiment; and three months later their gains remained. Practice did not help the adults, moreover, because they tended to give correct reproductions during a series, but to perseverate when they were in error, whereas the children seemed able to eliminate such perseverations and thus to increase the frequency of correct responses (Orsini, 1958; Fraisse, 1963, p. 238).

An impressive study suggests that gain from reinforcing judgments about one interval could be transferred to judging others. Without reinforcement the subjects first sought to reproduce intervals of 25 seconds and 2 minutes. Then they were randomly divided into three groups for the next series of trials which involved only the 25-second interval: the control group continued without reinforcement, one experimental group was given valid information concerning their performance, another invalid information (judgments between 15 and 19 seconds were called "correct"). Again without reinforcement, all groups were given a series containing both the 25 second- and the 2-minute interval immediately afterwards and also 24 hours later. First, training was effective: the accuracy of the control group decreased as did the group given invalid information, though slightly less so; and the experimental group with valid information became slightly more accurate and significantly less variable. Then, secondly, the effective training did transfer to some extent to the 2-minute interval during the same session. Finally, whatever was learned on the basis of valid, invalid, or no information tended to persist for both intervals after 24 hours (U.S., college; prior; Bakan, Nangle, & Denny, 1959). Another experiment makes an obvious, but easily forgotten point: the transfer after reinforcement from one situation to another (in this case it involved the production of 2-second intervals by moving the right arm, from more to less resistance or vice versa) did not occur on the first trial but did thereafter (U.S., boys; Ellis, 1969).

Human beings and other animals are capable of learning to respond at roughly the correct time intervals under the following experimental conditions: at the end of a long series of stimuli they are rewarded, provided that in each instance they have not responded prematurely, defined operationally in terms of a constant temporal delay; whenever there is a premature or incorrect response, the series begins all over again so that in a given time period fewer series are completed and fewer rewards obtained; and nothing is gained by waiting to respond for a longer period after each interval—the procedure is know as a DRL (differential reinforcement of low rates) schedule. In one investigation, subjects responded to a light by pressing a telegraph key and were rewarded with nickles: one schedule consisted of 100 stimuli with a delay of 2.5 seconds, the other of 10 stimuli with a delay of 25 seconds. Within the allotted period of time for the repetition of each series—

a full hour—they achieved between 80 and 90 percent of the rewards that were theoretically possible. Such timing behavior was efficient, the investigators point out, for a number of reasons; the subjects knew the nature of the reinforcing schedule (in fact, they witnessed a brief demonstration of it at the outset); they developed a counting sequence to try to conform to that schedule; and they were rewarded with money (U.S., medical students; Dews & Morse, 1958) which at a very minimum may have served the symbolic function of pleasing their egos. In a similar investigation, the reward was even more symbolic ("You will win a point every time you press the button and the light goes on. . . . Your point score is a measure of your ability to learn quickly and you will be compared with other students. . . ."); the Guide for all subjects seems to have been counting to oneself and for one of them counting the oscillations producing the stimulus and swinging his arm like a pendulum (U.S., adults; Carter & MacGrady, 1966). And in yet another series of experiments using essentially the same technique, it was shown that small and even large doses of alcohol had no effect, in comparison with the normal state, upon this kind of timing, even when the subjects engaged in the intellectual activity of subtraction or were compelled to make more complicated discriminations while delaying their responses the optimum period of time (U.S., mostly medical students; Laties & Weiss, 1962)—the reinforcement was evidently more effective than the drink. Thus learning to make temporal judgments has been facilitated, like any other kind of learning, by increasing the size of the reward.

In other ways, too, temporal activity resembles normal activity. In the experiment just mentioned in which the effects of alcohol upon the timing required by DRL were measured, it was also discovered that without the drug subjects tended to make more accurate responses when they concentrated upon the task at hand, rather than when they were distracted by other activities which they were required to carry out simultaneously (Laties & Weiss, 1962). The reinforcement has been shown to be effective even when it is minimal and involves only the presentation of the correct interval. Normal and schizophrenic subjects first judged whether intervals were "more" or "less" than a second in duration. Then the actual duration of 1 second was demonstrated. Both groups improved: the median duration of the normal undergraduate group rose from 0.56 to 0.76 seconds, that of the schizophrenics from 0.31 to 0.67 seconds. In line with our discussion of unverbalizable cues, it is interesting to note that it was this reinforcement rather than the verbalizable cues which apparently were effective. Many in fact measured time by counting —they used "some such phrase as 'one thousand and one'"—but that verbal device did not necessarily increase accuracy; thus some who counted actually judged 1 second to be "less than 0.3 seconds." In addition, "the subject's feeling of certainty with respect to the accuracy of his decision was, if anything, negatively related to the accuracy" of his judgment. Momentary reinforcement, however, could not compete with other reinforcement from the past: persons who ordinarily made temporal judgments in connection with their hobbies —navigation, dancing, photography, horse-training, etc.—tended to overestimate the duration of the interval; they felt "lost without their learned cues" (U.S.; limits-verbal, prior; Lhamon & Goldstone, 1956).

But the learning which occurs as a result of reinforcement in the laboratory, some studies suggest, is likely to be rapidly forgotten. In a series of classical experiments, using largely German university students, errors in judging intervals, especially longer ones and that of 1 minute in particular, could be reduced if the subjects were told how well they performed; this gain, however, "very quickly" disappeared a day later (repro., prior; Koehnlein, 1934). The gain during a given session was much less, however, when variations from the normal procedure of utilizing only a single modality or form of stimulation were introduced; viz., when the reproduction occurred through a sense modality different from that of the interval that had been presented, or when the production of 1 second (with the tutelage being that of demonstration) was accomplished through indicating the start of the production by stimulating one sense modality, and the termination by another modality, or by signaling the start and termination by stimulating different sides of the skin or a different eye or ear. These variations evidently produced even less stable learning since forgetting was still swifter (prior; Hawickhorst, 1934; Grassmück, 1934). An informal report indicates that the accuracy of estimating intervals in the neighborhood of 1 second improved markedly with practice, presumably with reinforcement, but that the improvement "never lasted very long" (Germany, unspecified; presumably repro., prior; Skramlich, 1934). In another experiment subjects were instructed to produce intervals having a duration ranging between 1.5 and 9 seconds. For the first nine trials they were given no feedback: if anything, accuracy slightly decreased. During the next 25, accurate feedback was provided: accuracy improved significantly after the first trial with feedback but thereafter showed no improvement. And during the next 15, feedback again was absent: what had been gained previously with feedback was then rapidly lost, and accuracy dramatically declined, returning in some instances to the pre-feedback level of inaccuracy. The duration of the interval also played a role: there was a tendency for feedback to be more effective with the longer than with the shorter intervals. The investigator suggests that accuracy may have been impaired after the withdrawal of feedback because "motivation flags" and "distractions appear to become more effective in disrupting concentration on the task" (U.S., college; prior; Aiken, 1965).

It may well be that ways of judging time are likely to be generally reinforced and hence to affect accuracy of judgment. Some subjects were first given a preliminary series of 30-second or 10-second tones to judge, others none at all. Then without knowing that the critical test was at hand they judged a series varying from 1-second to 10 seconds. The preliminary series, in comparison with the controls, produced a contrast effect on the subsequent judgments of the 5-second interval within the test series among those who had been instructed to pass judgment in terms of a "situationally relative, novel, arbitrary, and restricted" 7-point rating scale ranging from "very short" to "very long"; it did *not* appear among those whose judgments were in terms of the conventional units of seconds (U.S., college; verbal, prior; Braud & Holborn, 1966). Presumably the brief indoctrination, as it were, resulting from the preliminary series could not affect a heavily reinforced mode of judging time but could influence one not literally reinforced in the past.

Temporal judgments on a highly verbalized secondary level probably improved with reinforcement. Students in the tenth grade who received systematic and deliberate instruction concerning "the construction of time lines, the chronological sequences of events, and historical time concepts" in conjunction with a course on world history improved their performance on a questionnaire measuring that kind of knowledge or ability to a significantly greater degree than those who had only the course but not the instruction. This result is scarcely startling, but of greater interest is the fact that there were no differences on the temporal test between the group which had received no temporal tutelage and yet another comparable one which had no course at all in world history (U.S., Friedman & Marti, 1945). We have here the suggestion that learning about history per se did not automatically transfer to temporal items unless history's temporal attributes were featured,

ADDENDUM 7.3 COMPENSATING AND COUNTING

Evidence supporting Vierordt's Law—relatively short intervals are overestimated and relatively long ones underestimated—can be found in the classical literature (e.g., U.S., graduate & college; repro., prior; Harrell, 1937) and in more modern, sophisticated studies (e.g., U.S., adults; repro., prior; Marum, 1968). When this tendency exists and the intervals are ephemeral, the indifference point or zone can be determined That interval of highest accuracy, as indicated in Chapter 5, is around 0.7 seconds (determined under laboratory conditions), probably as a result of complicated processes involving the capability of sense organs and the nervous system, and perhaps also as a result of an "optimum rhythm" associated with the normal heartbeat and walking gait (Fraisse, 1963, pp. 116–28). In passing it must be added that the location of the point depends, too, upon the duration of the intervals being judged and the order in which they are presented. In effect, we have no proof that the tendency is a compensatory one; we can only speculate that general experience or anchoring effects within the experimental situation itself induce such a Guide.

In addition, the so-called Law does not always hold. One investigator reported among 8 subjects "no universal tendency for long intervals to be underestimated . . . or for short ones to be overestimated" (U.S., college & graduate; repro., prior; Woodrow, 1930); a classical experimenter noted that his five subjects, whose attention was intensified by having them control the onset of the two intervals to be compared, fluctuated somewhat but tended to *under*estimate intervals below 0.5 seconds, to *over*estimate those between 0.5 and 0.72 seconds, and then again to underestimate those above 0.72 seconds (Germany, university; comp., prior; Kahnt, 1914); schizophrenic patients underestimated the longer intervals of 5, 10, and 15 seconds when they used the Methods of Reproduction and Production but not that of Verbal Estimation (U.S.; prior; Clausen, 1950); and when intervals ranging in duration from 0.5 to 8 seconds were estimated through variations in the Method of Reproduction, the only significant difference involved the longest interval which was not under- but overestimated (U.S., college; prior; Doehring, 1961). If our reasoning concerning the possibility of compensation is correct, then these exceptions suggest that the subjects either had not evolved the Guide or did not consistently follow it.

Does counting promote accuracy? One series of experiments, I think, clearly indicates the issues. The duration of intervals ranging from 4 to 27 seconds was estimated verbally, with prior arousal of the temporal motive. Counting turned out to be important under two conditions. First, it contributed significantly to persons who "gave only casual attention to the experiment" but it did not seem to help others who inexplicably were attentive and hence presumably were sensitive to all available cues. Then its effects were even more dramatic when during successive trials subjects were told "the direction and extent of their errors"; this improvement was much greater than the improvement which occurred with reinforcement but without counting. In addition, without reinforcement the error scores derived from the two methods were highly correlated but, in the one instance for which data are reported, the reliability of scores obtained two weeks apart was higher when counting than when not counting (U.S., college; Gilliland & Martin, 1940). It is not surprising, therefore, to uncover experiments in which counting proved significantly helpful (U.S., children & adults; prod., repro., & verbal with prior; & Gilliland & Humphreys, 1943) (Canada, nursing students; verbal, prior; Hare, 1963a) or useless (U.S., college; limits-verbal, prior; Lhamon & Goldstone, 1956). Perhaps counting could always be beneficial if it were practiced; that at least is the informal but convincing evidence provided by time specialists such as referees at boxing matches. But there have been "long" counts.

ADDENDUM 7.4 WATCHING THE POT NEVER BOIL

To discover why the pot never seems to boil, we must observe neither the pot nor a bystander but the cook himself:

1. The cook is obviously passing temporal judgment, and his motive must be connected with his goal, the boiling of the water. Being a cook, he finds boiling very important since, we guess, it enables him to get on with preparing the meal. For all we know, he may be a jittery person who always worries about the passing of time.

2. We are not told whether he is concerned with the events occurring during the interval of watching or with his own drives. If his attention is really riveted upon the pot, he has little to observe: at first he may hear only the noise produced by the fuel (if it makes a noise), he may be watching the source of that heat, he may note after a while the water beginning to simmer and steam becoming visible. But he also is determined to have that water boil. My guess, therefore, is that he is more conscious of his goal than of the events.

3. I cannot imagine what standard he is using when he overestimates the passing of elapsed time. I must guess that he vaguely remembers other occasions when water has come to a boil very fast. He may glance at a clock, but I doubt it.

4. If he is impressed with the few events he observes, then he will have a tendency not to overestimate but to underestimate the interval because it contains some, rather than no or many, events; hence time ought to be passing rapidly and the watched pot ought to come to a boil quickly, which it does not do.

5. And so I assume again that boiling is important, hence his drive is strong, and so the interval must appear subjectively long, which is just what the adage says. As the simmering begins, that strength increase: the goal seems to be approaching.

6. I would further assume that waiting is frustrating in itself; frustration is non-attainment of a goal, another reason for the interval appearing longer.

7. If he is really watching the pot so intently, then he has no opportunity to consult a clock; he must have a subjective impression of a long interval which he translates as "long" or "long time" verbally; if he is semisophisticated, he may even use the adage we are discussing.

8. Here is a situation ripe for distortion: the poor man is powerless to intervene (for presumably the heat cannot be increased, and it just takes time for water to come to a boil); by overestimating the elapsed time, he is underestimating the time which remains before the boiling point, his goal, is reached. I suspect, however, that an experienced cook knows that watching does not help the water and only renders himself miserable, and he therefore appreciates the fact that his overestimation is an illusion.

When the adage is true, therefore, the overwhelming assumption must be that attention is riveted upon the drive attributes of the interval. It is possible to imagine conditions under which the watched pot boils too quickly, and these would be circumstances in which it is thought desirable to delay attaining the boiling point: another ingredient is not ready, a substance like milk will be ruined if allowed to boil, etc. Finally we must say to the cook that the way to get the pot to boil, as he doubtless knows, is to quit watching it; for then temporal judgment is not likely to be passed and hence reaching the boiling point will seem less important, the drive to achieve that point will grow weaker, and the frustration from watching will disappear.

But, it may be asked, how can he manage not to watch the pot? By distracting himself, must be the reply: let him do something else, if only to reflect upon his past sins. The other activity should not be of the kind whose duration he will also judge, for then the increase in the number of events may further strengthen his impression that time is passing slowly. That activity must be satisfying and so absorbing that he forgets the passing of time. The distraction serves another important function: it prevents him from noting again, and hence being frustrated by, the fact that the water is not yet boiling; for the greater the number of frustrations, the longer the interval is likely to appear.

ADDENDUM 7.5 INTROSPECTIVE REPORTS

Below are the conclusions, in full, which one investigator derived from 9 Swiss subjects—2 faculty members, 1 physician, and 6 graduate students—who judged the duration of auditory intervals between 0.4 and 2.4 seconds, Method of Comparison, prior arousal of temporal motive:

> The judgment, based upon observing and comparing two intervals of time, can be traced to two kinds of reaction:
> 1. the interval that is first presented, a, is revived and during its revival the second interval, b, is experienced.
> 2. b produces an independent experience and only then is the experience of the interval a revived. This mode of comparison takes place particularly when a and b are separated by a definite interval [Schmid, 1929].

In my opinion, a summary of protocols such as the above—and be assured that similar ones are available, especially in German—leads us almost nowhere. For the statements are almost as varied as the subjects who made them or at least as the experimenters who sought to categorize them. "It is amazing," one writer notes, "what variety these reports indicate in the internal activities of a subject when he has been asked simply to tell which of two short temporal intervals is the longer" (Woodrow, 1951, p. 1227). In addition, as the investigator just quoted has also noted, and as one of our *Addenda* (No. 6.6) has already sought to demonstrate, such modes of comparisons seem unrelated to the emerging judgments. Let us, however, not be too harsh: hints can be extracted from this painful research. In one investigation, for example, there was some indication that perceiving the quiet of unfilled intervals as a figure rather than the ground may have affected the ensuing temporal judgment (U.S., graduate & college; repro., prior; Triplett, 1931). In another, female college students stated modally that they employed the "amount of work achieved on each task," "guessing," or guessing first and then comparing the work accomplished as cues for passing judgment; in fact, however, the position of the interval and not its content (i.e., the work accomplished) was the only critical variable (U.S.; verbal with possibly immediate subsequent and certainly prior; Postman, 1944). "Almost invariably," the subjects in an experiment who reproduced intervals either normally or while isolated by being suspended in the middle of a tank of water reported, they were unable "to keep the longer duration in their present perspective . . . they lost track of the longer durations." In fact the range of the intervals was not very great: the shortest was 2 and the longest was 32 seconds, but the reproduction was either carried out in the normal manner or the instruction was given to reproduce half or double the interval (U.S., adults; prior; Marum, 1968).

Then direct questioning of 50 subjects who produced intervals ranging from 1 second to 1 minute revealed that all of them had some kind of imagery:

> In the visual field individuals used an imaginary clock or watch, sometimes with and sometimes without a second hand, a pendulum, or a series of numbers which appeared before them one after another. In the auditory field some heard an imaginary ticking or beating or imagined hearing a series of numbers. In the motor field a finger was moved, an arm was swung, the head was nodded, a foot was tapped or the subject counted. In the kinaestheic field some imagined

the feeling of swinging an arm, of nodding the head, of tapping with the foot or of counting. The kinaesthetic imagery was usually accompanied by visual or auditory imagery. The subjects who stated that they judged time automatically, namely that they judged time "without thinking," usually had an image at the beginning or end of the interval. During the interval they were more or less consciously waiting for the task they had set themselves to reach completion [U.K.; Scott, 1948].

In this instance it seems clear that imagery performed an important mediating function.

It is also impossible to dismiss this direct approach to judgment without suggesting that more sophisticated uses of introspection combined with more sensitive apparatus may in the future produce insights we now lack. We infer, for example, that even with very brief intervals, such as a 0.02 second flash of green or red, some "processing" must take place between the initial stimulation and the perception of the stimulus, even though the significance of such a process for longer intervals is admittedly unclear (Efron, 1967). Another fruitful approach is to manipulate experimentally the way in which judgments are passed; thus in one series of experiments subjects tapped a telegraph key at different rates as they followed instructions involving different strategies. For example, tapping was slower when they were told to imagine that they had been given $100 and would lose $50 for every second a judgment was short and 1 cent for every second a judgment was long than when they received the reverse instruction (Australia, presumably adults; von Sturmer, 1968).

Finally, it may very well be that the ingenious, objective methods that are devised in the future will at least reflect the intervening responses which cannot be measured directly. One investigator asked his subjects—Naval personnel, both male and female—to draw three circles of any size and in any position as they responded to the instruction, "Think of the past, present, and future as being in the shape of circles." For these subjects, it is inferred that time was "atomistic" because the three circles almost never touched; and that the future was more "dominant" than the present and the present more so than the past because the largest circle tended to be made to represent the future, etc. (U.S.; Cottle, 1967). Obviously, however, we must know much more about the relation between results obtained from this projective technique and other temporal behavior before we can accept the author's own interpretations.

ADDENDUM 7.6 THE RELATIVE ACCURACY OF SENSE MODALITIES

The very nature of the problem has confined almost all of the studies to brief intervals under laboratory conditions; therefore in all probability we are dealing here with primary judgments. The most frequent comparison has been between audition and vision. Audition has been shown to produce more accurate judgments than either vision (U.S., presumably college; comp. and repro. with prior; Goodfellow, 1934) (Germany, university; repro., prior; Homack, 1935) or tactile stimulation (U.S., college; limits-verbal, prior; Lhamon, Edelberg, & Goldstone, 1962) (U.S., mostly college; comp., prior; Gridley, 1932). Careful experiments have also "shown repeatedly that longer visual than auditory durations are judged equivalent to temporal concepts such as a clock second," which means auditory intervals have been estimated to be longer than visual ones (U.S., college; comp., prior; Goldstone & Goldfarb, 1964b; Goldfarb & Goldstone, 1964) (U.S., college; absolute judgment, prior; Behar & Bevan, 1961) (U.S., adults; prod. & repro. with prior; Goldstone, 1968a).

The same investigators quoted above are of the opinion, as they point out the tendency for auditory intervals to be judged longer than visual ones, that "this curious intersensory difference in time judgment is independent of psychophysical method, stimulus characteristics such as pitch, hue, and intensity, size of and distance from visual target, position in a series and number of response categories; it is present under various background anchor conditions, within very narrow ranges of short duration and in broader ranges up to 5 sec., and with different interval temporal standards." In addition, they have also collected from American subjects impressive data indicating that the same difference appeared among children between the ages of 7 and 13, among schizophrenic patients under most conditions, and under the influence of a stimulating drug (dextroamphetamine) and a depressant one (secobarbital) (limits-verbal; Goldstone, 1964, pp. 127–33, 165–71, 203–13). Even though the work of these investigators is thorough and is based upon careful testing of thousands of subjects, other laboratories report contradictory results. There are instances in which no differences have appeared between judgments based upon auditory and visual stimulation (U.S., presumably adults; repro., prior; Hirsh, Bilger, & Deatherage, 1956; Brown & Hitchcock, 1965) (South Africa, European students; repro., prior; Du Preez, 1967a) (U.S., college; prod., prior; White & Lichtenstein, 1963) (U.S., children; verbal, prior; Crawford & Thorn, 1967). In one experiment—with many more trials than is usually the case—vision actually induced longer judgments than audition (U.S., college; comp., prior; Tanner, Patton, & Atkinson, 1965). In fact, in the very laboratory in which the difference has been repeatedly demonstrated one experiment "failed to produce" that difference; in this instance the intervals being compared were not filled with light or sound but were empty and bounded by "short, discrete sound clicks or short, discrete light flashes" (Goldstone, 1964, pp. 85–8).

This last result, the investigator himself indicates, "focuses attention upon the primacy of stimulating conditions which interact with the procedural factors, or psychophysical method." Judgment, therefore, is "dependent upon the variety of temporal problems posed by the different psychophysical methods, the receptor system stimulated, the arrangements and patterns of stimulation, the physical properties of adequate stimuli, the characteristics of anchoring conditions within and across sense modes and dimensions of experience, the nature of the temporal standard, and the constraints established by the response circumstances" (Goldstone, 1964, pp. 88, 105). This emphasis upon "a fundamental pooling process" is in line with the principle being discussed and calls attention to the way in which contextual or stimulus factors interact with modality. A few additional illustrations of this process follow:

1. Judgments of intervals perceived visually were slightly but significantly more affected by order of presentation and anchoring stimuli than judgments of those perceived via audition (U.S., college; limits-verbal, prior; Goldstone, Boardman, & Lhamon, 1959).

2. A summary of the research aiming to ascertain the minimum interval which must separate two stimuli before they can be perceived as successive rather than simultaneous suggests that "temporal acuity is greater when identical stimuli are applied successively to different receptors of the same sense organ and not to the same receptor cells" (Fraisse, 1963, p. 113).

3. For intervals with a duration from 0.25 to 1 second accuracy was greatest when both stimuli were auditory and least when the first was auditory and the second visual (Germany, adults; repro., prior; Schmidt, 1935).

4. "The degree of discrimination" in a careful psychophysical investigation, in which the subjects were instructed to produce changes in the irregular, intermittent stimulation of the variable interval until it appeared regular, was found to depend not on whether the modality was vision or audition—they appeared equally efficient—but upon "the perceptual task" facing them, such as the nature of the stimulus, the part of the retina being stimulated, etc. (U.S., college; prior; White & Lichtenstein, 1963).

Another study also indicating that within the modality of vision the actual point of stimulation of the retina is of significance is one in which subjects signified whether two lights, each of which was flashed for 0.015 seconds, had appeared simultaneously or at different times; the distance between the two lights was also important (U.S., college; limit-verbal, prior; Sweet, 1953).

5. In another connection, the same investigators who feel convinced that the auditory-visual difference is a general one, have shown with their Verbal-Limits Method that intervals perceived tactually were judged longer than those perceived via audition (U.S., college; prior; Lhamon, Edelberg, & Goldstone, 1962); but with the Method of Comparison no difference between the same two modalities appeared (U.S., medical students; prior; Ehrensing & Lhamon, 1966).

6. Particularly impressive is one very sophisticated experiment in which the subjects were exposed not only to three kinds of stimulation but also to three methods of measurement as they judged four intervals ranging in duration from 0.5 to 4 seconds. The cutaneous modality led to the longest judgments, vision the shortest, and audition in-between. Modality interacted with method for each duration; Verbal Estimation, for example, led to overestimation for all auditory intervals, but to underestimation for all cutaneous ones except that of 4 seconds; or the same method led to underestimation of intervals below 2 seconds and overestimation of 4 seconds when vision was the modality. On the whole, however, the three methods (Verbal Estimation, Production, and Reproduction) yielded similar judgments for each of the intervals, and there was the usual negative relation between Production and Verbal Estimation (rho of $-.78$). We dare not generalize very widely from these findings: they are based upon 9 subjects, presumably adults, presumably Americans—these very simple census attributes are not directly revealed (Hawkes, Bailey, & Warm, 1960).

ADDENDUM 7.7 FACTORS AFFECTING THE ACCURACY OF MEASUREMENT METHODS

The text contains the arguments which suggest that Reproduction is potentially the most accurate method for passing judgment. It should be clear that reference is being made to straight reproduction; for reproduction by fractionation or multiplication (attempting to reproduce as interval that is, respectively, one-half or double the standard) has yielded quite different results (U.S., adults; prior; Marum, 1968). On the other hand, a view contrary to the present one stems from a study which indicated that Verbal Estimation led to "somewhat more accurate" judgments for intervals having a duration of 24 and 48 seconds than did Reproduction, although with intervals of 6 and 12 seconds no difference was evident. The activity taking place while perceiving the interval to be reproduced, it is said, diminished the latter's accuracy. In addition, Verbal Estimation had the advantage of enabling the subjects to pass judgment while experiencing the interval to be judged, whereas Reproduction required them to recall the interval when it was no longer present. With longer intervals, moreover, "the fallibility of short-term memory may increase" (U.S.; college; Warm, Smith, & Caldwell, 1967).

This last statement suggests that the advantages and disadvantages of a method depend not only upon its nature as such but also upon the duration of the interval being judged and virtually any other factor either in the stimulus or the behavioral potentials. When retested either 30 minutes or 2 weeks later, subjects produced more consistent judgments for a 10- than for a 5-second interval (Canada, student nurses; successive time estimation, prior; Kelm, 1962). Likewise in one extremely careful psychophysical experiment with very brief auditory intervals ranging from 0.0004 to 4.0 seconds, temporal discrimination tended to be less accurate as the duration both of the intervals and the intervals of silence separating the intervals became less; and under these particular circumstances there seemed to be no time-order error (U.S., college; contrast, prior; Small & Campbell, 1962). The consistency with which intervals of 30 seconds and 5 minutes were reproduced immediately afterwards or from 15 to 22 days later depended in part, though most irregularly, upon whether the subjects had been doing nothing during that time or working on arithmetic problems (U.S., college; prior & subsequent; Young & Sumner, 1954). For a very small number of patients in an American hospital, the Method of Production was found to be highly reliable (as measured by a retest 6 to 12 days later) for subjects with "constant anxiety levels" (as measured by the same paper-and-pencil test both times) but almost significantly less so for those fluctuating in anxiety levels (Whyman & Moos, 1967). Or the accuracy of any given method may also be affected by details in its execution; thus reproducing intervals of from 2 to 15 seconds was more accurate when the subjects were idle during both perception and reproduction than when they were occupied with a coloring task (France, child. & adults; prior; Wallon, Evart-Chmielniski, & Denjean-Raban, 1957). Concerning one point, however, we may be quite certain: the Method of Chronometry is likely to produce the most accurate translation since only perception and very little judgment are involved.

The corollary in the text receives most clear-cut confirmation in a previously cited experiment in which the subjects themselves first produced an interval whose duration was unknown to them (since it was controlled by the experimenter) and then tried to reproduce one of the same duration. The most accurate reproduction appeared when the activity involved in reproduction was the same as that involved in the original production (U.S., college; prior; DeWolfe & Duncan, 1959). Also in various classical experiments it was found that improvement in the ability to produce or reproduce an interval through reinforcement (by demonstrating the correct interval or by correcting the reproduction) was greater when the limiting stimuli belonged to the same modality rather than to different ones, and also that the gains disappeared more quickly under the latter than the former conditions (Germany, university; prior; Hawickhorst, 1934; Grassmück, 1934; Schmidt, 1935). Similarly with the Method of Comparison, judgments tended to be more accurate when both intervals were in the same modality rather than in different modalities—vision and audition were employed (U.S., college; prior; Tanner, Patton, & Atkinson, 1965). But in two well designed experiments there were no significant differences with respect to accuracy when the interval and the reproductions were in the same or in different modalities (U.S., college; Brown & Hitchcock, 1965) (South Africa, European students; Du Preez, 1967a).

ADDENDUM 7.8 THE ACCURACY OF THE BIOCHEMICAL CLOCK

The following are among the investigations which have sought to determine whether people can awake at a predetermined time and to discover the techniques they employed to attain that objective:

1. A hundred times one person "willed to wake at a certain minute, and I invariably visualized what the face of my watch would look like at the minute resolved on." On 18 of these 100 occasions he was exactly right, 35 times within 15 minutes or less, 22 times between 15 and 30 minutes, and 6 times between 30 and 45 minutes (U.S., Hall, 1927).

2. An individual reported that he was able to awaken himself within 1 to 20 minutes of the time he had specified—at least in two-thirds of the instances—by repeating subvocally to himself, before retiring, "wake me up at x o'clock"; (U.S.; Brush, 1930). Somehow the verbal self-command must have functioned while he was asleep, but in that condition he presumably could not continue to ver-

balize or to respond to the verbalization.

3. Questionnaires and informal experiments with small samples revealed considerable variability in predetermined waking time from person to person—the mean error was around 21 minutes—and greater accuracy was attained when that time was close to the individual's normal hour of awakening (France, adults; Vaschide, 1911, pp. 49–72).

4. Interpersonal variability was pronounced, but also a far higher proportion of the female subjects who claimed beforehand to possess this ability were able to awaken themselves within a half-hour of the predetermined time than was the case among those not claiming to have the ability (U.S., college; Omwake & Loranz, 1933).

5. Over 100 individuals who claimed to be able to awaken themselves at a predetermined time in fact had a tendency to awaken before they intended to do so; on the average, their error was considerably under 10 minutes with a range of less than 20 minutes when too early and slightly over 5 minutes when too late; the duration of sleep and the hour of awakening were not related to accuracy. According to their own testimony, they employed a variety of techniques to ensure that they would not over- or undersleep. For some "the essential aspect of all their techniques appeared to be a translation of their resolution [to awaken] into the pictorial, which could be achieved by concentrating upon the face of a clock, considering the consequences of not awakening, etc., before going to sleep." They did not seem dependent upon external cues; and many of them performed best when the motivation to awaken was very strong or when setting an alarm clock made them feel secure (and we are told that "never" did they fail to awaken "slightly before" the alarm went off). The investigator is of the opinion that, when such conditions are fulfilled from the standpoint of the individual person, then the processes "which we ascribe to awakening via a 'head clock' (Kopfuhr)" are thus "automatically set in motion" and the person "spontaneously" awakens more or less at the correct time without being able to say why or just how he has accomplished the feat (Germany, presumably adults; Clauser, 1954, pp. 62–6).

6. Four university students and an older housewife were tested over a period of 250 nights; they went to sleep at irregular times and were told to awaken a certain number of hours later. On most occasions they were 5 minutes ahead of time, and their errors were within the region of 40 minutes either before or after the appointed time. They generally all awakened "suddenly"; sometimes in the midst of a dream they felt as though they were interrupted with the words, "It's time, you must wake up now" (Germany; Frobenius, 1927).

It should be clear that some persons are able to awaken themselves with a fair degree of accuracy and that such persons are likely to have the impression that the call comes from within and is therefore independent of changes in the external environment. But is it? Evidence on this point is offered in the German investigation just cited. The subjects awakened just as promptly even when suffering from extreme fatigue resulting from having slept only 2.5 hours per night for a whole week as they did under normal circumstances. Equally impressive is the fact that their performance did not fluctuate when a clock in the sleeping quarters struck the hours with a loud sound, when that clock —unknown to them—has been incorrectly set and hence conveyed false information to their sleeping ears, or when there was complete silence (Frobenius, 1927). The same technique, it is claimed without supporting evidence, also had no effect upon another group of subjects (Germany; Clauser, 1954; p. 65).

The ability to state the time upon being awakened was once investigated by waking up four subjects, trained in the classical technique of introspection, on 50 occasions between 12:15 and 4:45 A.M. (and they knew that range beforehand): their judgments were fairly accurate, but they tended to think it was later than it actually was. According to their reports, they depended partially on unverbalizable, "immediate and non-conscious" cues, and partially upon verbalizable, "conscious" ones (such as the state of the bladder, digestion, fatigue, muscular inertness) for their judgments (U.S., adults; verbal, prior; Boring & Boring, 1917).

The functioning of the biochemical clock has also been investigated by deliberately eliminating external cues; this is accomplished by isolating subjects from the outside world and, of course, from all mechanical timepieces. First, there have been results suggesting that the clock is relatively accurate:

1. Two subjects were confined to a soundproof room and were without external cues concerning the passage of time for 48 and 96 hours; they estimated the duration of their incarceration with the small errors, respectively, of 26 and 40 minutes at the end of the experience. Their judgments concerning clock time on the outside tended to be more accurate than the estimates they made concerning specific intervals while in the room; those estimates did not appear to be dependent upon the type of activity (such as making notes, eating, sleeping, etc.) in which they had been engaging, neither were judgments about clock time influenced by bodily cues. At first the errors concerning clock time were large, but gradually they became smaller; the subjects evolved for themselves a subjectively real temporal schedule which, the investigators suggest, may well have been an expression of a biochemical clock (U.S.; university instructors; verbal with prior & immediate subsequent; MacLeod & Roff, 1936).

2. Subjects living in an underground bunker for from 3 to 4 weeks with no clue concerning time tended to maintain a regular circadian rhythm with respect to bodily functions such as temperature and the chemical constitution of the urine, but that rhythm departed from clock time, and temporal judgments were inaccurate. The investigator, for example, reports the following concerning himself:

> On day 8, I got up after only 3 hours of sleep. . . . Shortly after breakfast I wrote in my diary: "Something must be wrong. I feel as if I am on dogwatch." I went to bed again and started the day anew after three more hours of sleep. Judging from the curve of body temperature, my first start happened to coincide with the worst phase of the circadian period, that is, with the low point of temperature. I was mistaken by an effort of will and put to order by my physiological clock [Germany; Aschoff, 1965].

3. Twenty-eight men, women, and children were confined in a simulated shelter for almost two weeks with "no bunks or bedding, bathing water, coffee, or change of clothing" but with a survival diet. They were cut off from all external cues concerning clock time (except on two occasions, the effects of which the investigators believe to have been "negligible"). The manager of the group, at times which he believed to be in the early morning or late at night, asked them to estimate clock time. They were able on the whole to maintain "relative accuracy," with an error of 3 hours in either direction; morning estimates were more accurate than those in the evening, but this effect varied with the period of confinement (U.S.; Thor & Crawford, 1964). Social factors of course affected these results: the relatively slight variability in the estimates reflects the fact that the group adopted a "routine confinement schedule which varied little from day to day."

Under similarly well-controlled conditions, relatively *inaccurate* judgments have also been noted:

1. Graduate students were confined to a dark, soundproof cubicle from 8 to 96 hours but without knowing how long the experience would last; some were instructed to signal at the end of what they believed to be each hour and day; all were asked, as they were released, to state the day and the hour. Their judgments tended to be entirely inaccurate (with a strong tendency to underestimate the time), varied a great deal from person to person, and—according to their testimony—seemed dependent not on a "time

sense" or unverbalizable cues but upon bodily processes, such as hunger, beard growth, defecation, etc. (U.S.; immed. subsequent or prior; Vernon & McGill, 1963).

2. Six U.S. Air Force personnel remained in a soundproof room for 48 hours. Five of their estimates ranged between 45 and 53 hours, but the sixth believed he had been confined for 96.5 hours. Emotional reactions to the isolation may well have upset the biochemical clock, for two other subjects asked to be released before the investigation was terminated; and one of these estimated the 33.5 hours he had been in the room to be 20.5 and the other his 26.4 stay as 64 (Mitchell, 1962).

3. In a simulated shelter, 17 healthy children between the ages of 7 and 12 were confined, together with an adult manager and a nurse, for a week. Confinement had two effects upon their temporal judgments. First, their verbal judgments of intervals between 4 and 20 seconds tended to increase when those intervals were visual but not when they were auditory. Secondly, they overestimated the time of day at first and then underestimated it; and there was also some tendency for these judgments to be slightly more accurate, especially at the outset, in the clock-time morning (U.S.; prior; Crawford & Thor, 1967).

4. Three persons were "completely isolated from objective time" for a period of three weeks; their judgments concerning 3-hour intervals and daily cycles were shorter than objective time and, in general, "deteriorated with the increasing length of isolation" (Czechoslovakia; presumably verbal and prior; Jirka & Valoušek, 1967).

Although the results are somewhat conflicting, it is well to conclude this *Addendum* by noting how deep-seated the endogenous mode of passing judgment is. A man who spent three months underground "was frustrated by a progressive tendency to lie awake to a later hour every night, and to sleep beyond his usual waking time"; he then decided to go to bed when he felt tired, a decision he made later and later every night; as a result he eventually followed a cycle of roughly 24.5 hours, to which his physiological processes adapted (England; Mills, 1964). This man had a watch which indicated to him the correct time above; and it is therefore interesting to note that another man who spent 59 days underground, and 8 out of 9 subjects who were confined in solitude in a deep bunker, also lengthened their daily cycle almost identically even though they had no watches and judged calendar and ojective time most erroneously (France; Siffre, 1963, p. 220) (Germany; Aschoff & Wever, 1962).

ADDENDUM 7.9 DISTORTIONS IN TIME

Classical experiments demonstrate the influence of distortions produced by interposed distractions. According to one summary, "almost any time, apparently, may be judged short or long, depending upon the length of preceding stimuli" (Woodrow, 1951, p. 1229). For example, when one light, whose illumination (0.78, 1.01, or 1.39 seconds) was to be estimated, was followed by a second light, the latter retroactively decreased the accuracy of the estimates; the effect of prior or proactive stimulation under the same conditions was also marked but less consistent (U.S., college; prior; Turchioe, 1938).

One investigator has designed a long series of experiments to test her basic hypothesis that many, if not most, of the phenomena associated with temporal judgments can be accounted for by learning theory. When instructed to read digits at the rate of 1 per second, for example, subjects achieved a high degree of temporal accuracy; but immediately afterwards their estimates concerning the time that had elapsed during the reading and before they were halted by the investigator were significantly shorter than the objective time of the reading or the sum of their subjective judgments during that interval. The difference between the two judgments is ascribed to a loss of retention with the passing of time. This explanation received somewhat striking confirmation from the fact that the discrepancy between the two judgments tended to increase as the objective time of the interval increased from 4 to 53 seconds (Sweden, medical students & laboratory technicians; PPT, prior; Frankenhaeuser, 1959, pp. 41–5).

Other impressive evidence indicating a relation between delays in judgment and inaccuracy comes from experiments in which the accuracy of judgments is compared under conditions in which the arousal of the temporal motive has been varied. When a class of American students was made mildly anxious by being told that some of them would participate in a somewhat painful experiment, and when afterwards they were unexpectedly asked to estimate the duration of the 100-second interval during which they had been held in suspense, the range of their judgments was from 46 seconds to over 20 minutes, with the midpoint (median) at 8 minutes. On another occasion but without the suspense, the mean of their judgments (the author does not provide the median) concerning the interval of identical duration was 2 minutes, 20 seconds; then with prior arousal of the temporal motive the latter mean dropped to 98.2 seconds. The investigator claims to have eliminated "practice effects" as an explanation for these dramatic differences by repeating the last part of the experiment with other groups which had not been previously tested and which apparently were comparably accurate (Culbert, 1954). If not caught unaware or made anxious, in short, the students' judgments tended to be relatively accurate. In another study a comparison was made of the judgments given by normal, schizophrenic, and neurotic subjects who first perceived an interval of 2 minutes without being warned beforehand that they would be asked for an estimate, and who then judged the same interval after being alerted. All the groups tended to overestimate the intervals with immediate subsequent arousal of the temporal motive and to underestimate them with prior arousal, whether or not the intervals were empty or filled with activity allegedly measuring finger dexterity; with the exception of a sub-group of schizophrenics possessing correct temporal information— and for them only with empty intervals— accuracy tended to be slightly or markedly greater when the motive was aroused in advance; and the physical activity had little or no effect with prior arousal but tended to produce less accurate judgments among the abnormal subjects with immediate subsequent arousal (U.S., adults; Dobson, 1954).

But as the text suggests, there is also contrary evidence elicited in some experiments. Delays of from 2.5 to 30 seconds, during which the subjects worked upon an utterly irrelevant task, had no effect whatsoever upon the reproduction of intervals with a range of from 0.48 to 16.2 seconds; regardless of the delay, there was the frequently appearing tendency to overestimate the shorter interval and to underestimate the longer ones (U.S., high school; prior; Kowalski, 1943). In a classical experiment accuracy was adversely affected when the interval immediately followed the one with which it was to be compared, but then was greater when the pause was short rather than long (Germany, university; comp. & verbal with prior; Jaensch & Kretz, 1932); but the reverse of the last finding is reported by another classical experimenter (Germany, adults; repro., prior; Schmidt, 1935).

Germane and damaging evidence, moreover, comes from an experiment in which the subjects searched for specified 2-digit numbers on a printed sheet for a whole hour; the task was intellectually absorbing. Half of them were told prior to the interval that they would be asked to judge its duration, half were not told. The mean of those with prior arousal of the judgment motive was 60.05, that of those with immediate subsequent arousal 54.09; but the difference is not statistically significant (U.S., college; Bakan, 1955).

One experiment reminds us that the content of an interval may serve as a distorting influence even for an ephemeral interval. Undergraduates viewed a series of 24 words which were projected on a screen, one at a time, for a period of 1 second. There was no relation between temporal judgments and the values placed upon the referents signified by the words for the subjects, but there was a slight tendency for the intervals containing the more familiar words, in comparison with those having unfamiliar ones, to be judged—to be judged what? Perhaps the reader thinks those intervals would be judged shorter: a familiar word (e.g., "scientific") could be grasped more quickly than an unfamiliar one (e.g., "percipience"). But no, the reverse effect was obtained: the mean for familiar words was 0.93 seconds, for unfamiliar ones 0.83; the authors think that the familiar ones could indeed be grasped more rapidly, but that then more attention could be directed to time, "resulting in a longer experience of time" (Warm, Greenberg, & Dube, 1964). And all this in one second?

ADDENDUM 7.10 SEX DIFFERENCES

Among the American studies revealing a tendency for females to overestimate temporal intervals more than males are the following, all of which used college students as subjects and the Method of Verbal Estimation with prior arousal of the temporal motive: MacDougall, 1904; Yerkes and Urban, 1906; and Axel, 1924. More recently, samples of males from the same universe were superior to females in passing "concurrent" judgments concerning ephemeral intervals (D. S. Weber, 1965). On the second trial, whether in a lighted or a darkened room, whether claiming to fear the dark or not, both sexes lengthened their estimates of an interval of 5 minutes without increasing accuracy; but the men lengthened theirs less than the women (presumably verbal & prior; Geer, Platt, & Singer, 1964). One study adduced the fact that female undergraduates overestimated intervals varying from 2 to 29 seconds to a greater degree than males with the Method of Verbal Estimation; exactly the reverse tendency appeared for the same interval with the Method of Production (Horstein & Rotter, 1969). Two studies of Americans have shown females to be more variable than males. One concentrated upon children, and found the difference greater with visual than with auditory intervals (verbal, prior; Goldstone & Goldfarb, 1966). The other used an adequate number of adults—120 of each sex—and intervals ranging from 0.15 to 1.95 seconds as well as both auditory and visual stimulation (verbal, prior; Goldstone, 1968b). Finally, the following American investigations revealed no differences between the sexes with respect to accuracy:

1. Swift & McGeoch, 1925; college; verbal prior.
2. Harton, 1939c; unspecified; verbal, uncertain.
3. Gilliland & Humphreys, 1943; children and college; verbal, prod., and repro. with prior.
4. Baldwin, Thorn, & Wright, 1966; college; verbal, prior.
5. Loehlin, 1959; college; verbal, prior.
6. Smythe & Goldstone, 1957; children; limits-verbal, prior.
7. Hawkins & Meyer, 1965; college; verbal, prior.

Also one investigation revealed no differences between male and female Negro college students with respect to temporal information (Johnson, 1964).

The problem has also been attacked outside the United States. During the early period a Soviet study revealed that boys between the ages of 10 and 16 tended to judge 2- to 5-minute intervals more accurately and those between 10 seconds and 1 minute less accurately than girls of the same ages (Elkin, 1928). No differences appeared in a Swiss investigation (normal & adults & epileptics; prod., prior; Pumpian-Mindlin, 1935). More recently, a study in Britain indicated no differences between the two sexes (verbal & prod., with prior; Orme, 1964).

A final suggestion is worth recording: if sex plays a role, it probably does so while interacting with other factors. When Austrian students between the ages of 12 and 15 were asked to indicate on a straight line points signifying their birth, death, the end of the past, and the start of the future, the two sexes did not differ in these respects; but there were a few significant differences when age or social class was taken into account. Thus the older boys showed a greater "increase in the extension of the personal past" (as measured by the length of the segment they used to signify the past on the straight line) than the younger ones, but the girls showed "a slight decrease," very slight in fact, with age. From this limited mode of testing, the investigators extrapolate the generalization that "boys master the future; girls the present" (Cottle & Pleck, 1969), which may or may not be true. In yet another study, first-born and only-children undergraduates tended to have a greater impersonal future time perspective than later borns, "especially among females"; but there was no difference with respect to either birth order or sex when personal perspective was measured (Platt, Eisenman, & DeGross, 1969).

PART II: FUNCTIONING

8. DEVELOPMENT

The analysis so far has been deliberately proceeding on the assumption that anticipation and recollection as well as the resulting patterns of temporal organization and judgments are learned processes. The first principle that was promulgated refers to the acquiring of temporal information as a result of periodic changes in the external milieu (3.1), the last to changes in the temporal potential induced by feedback from reinforced judgments in the past (7.7). That such learning occurs is clear beyond a doubt. At the pinnacle we find in the lives of saints, whether in myth or reality, critical turning points at which they have learned to abandon a life of ease and pleasure, have renounced all, or rather most, worldly pleasures, and have oriented themselves toward an existence in the after life. At the other extreme are infants who, if they could express themselves verbally, would praise only the present but who, little do they know, must start learning another philosophy of living. Here let us consider the circumstances producing such changes in temporal orientation. Also: how does the temporal potential develop and lead to primary judgments which eventually are not automatically accepted, but evaluated, and hence give rise to secondary judgments? We turn first to the infant and then occupy ourselves with age and aging as independent variables. We journey toward the saint, and find him in a later chapter.

THE PROGRESSION

The progression of life appears to be from continuous gratification in the present to anticipation of the future and recollection of the past. During the intra-uterine existence feeding is automatic, or almost so; there is, as it were, no need to remember or anticipate, no need to be guided by temporal orientation or perspective. But immediately after birth the gratification of some impulses is delayed: no matter how attentive and permissive the mother, the infant's need for food and liquid cannot be instantly satisfied; and pain from whatever source persists, even if minimally. Very quickly, moreover, on the basis of an inborn potentiality, the infant learns to adapt to the inevitable requirements of his environment and his body. He begins to extend action beyond what he immediately perceives; he responds to stimuli in large part on the basis of past experience and with reference not only to the present but also to the future. He is able to anticipate that the frustration he is enduring will, as it were, not last forever and that therefore it will give way to some sort of gratification. He can associate the coming of gratification with particular events in his milieu and even take appropriate action to have the events come to pass. "When the infant is hungry, he cries and extends his arms toward his nurse" and so, it has been succinctly remarked, "voilà le germe de l'idée d'avenir" (Guyau, 1902, p. 32). Although we cannot even try to guess whether, and then when, a conscious component is involved, we must infer some sort of anticipation in such behavior from the fact that the child is likely to stop crying when he is picked up. He may enjoy the comfort of being held, but it seems likely, too, that this act by the adult is interpreted—however dimly—as an instrumental response leading to the goal of milk. And again we quote our French observer: "The future is not that which comes to us, but that toward which we go" (ibid., p. 33).

Inasmuch as animals are similarly conditioned, there is no need to postulate elaborate decision-making by the child who cries to be fed. From such a simple beginning, however, must spring the interventions of later life: to be fed, you have to speak up, and eventually go through the elaborate ritual of earning your bread. And to the same source must be traced many of the human virtues, such as patience, that are so highly and universally praised. At the outset the very young child cannot possibly know that this too will pass, and by *this* he—and you and I—must mean the misery and grief he unavoidably experiences. When he acquires an awareness of the present, and therefore simultaneously some feeling for the inevitability of the future, and when he is also able to inject some of his hopes into that future toward which, as a result, he feels himself oriented, the unhappiness of the moment can become more endurable.

Similarly, primitive origins of renunciation can be inferred during these early months and years. This summary of our knowledge concerning animals undoubtedly applies to human infants: "immediate rewards are preferred over delayed rewards," so that "the effectiveness of a reward or punishment diminishes with increasing delays" (Mahrer, 1956). Practices of course vary from culture to culture, and within a culture from mother to mother, yet at some time the child is punished, or at least not rewarded, for crying too much, for failing to control the need to urinate and defecate. Such punishment varies from anguish after being ignored to physical pain after a slap or a cuff. Renunciation is not necessarily a noble, gracious impulse, but an act of necessity motivated by a desire either not to be hurt or not to expend one's energies fruitlessly. The modern master of so many has written that "it is impossible to ignore the extent to which civilization is built up on renunciation of instinctual gratifications, the degree to which the existence of civilization presupposes the non-gratification (suppression, repression, or something else?) of powerful instinctual urgencies" (Freud, 1930, p. 63). His distinction between chaotic, diffuse, instinct-directed "primary" processes on the one hand, and the better-organized, goal-directed, reality-oriented "secondary" processes on the other hand, has proved tremendously useful for a conceptualization of early behavior and as a guide to research, especially in connection with psychologically abnormal and socially aberrant phenomena (Singer, 1955). For Freud and Freudians, primary processes include only a present orientation, and few if any conscious verbal judgments. Orientation can be extended in either direction and judgments become more extended, qualified, and accurate after the appearance of secondary processes which stress the ability to tolerate delay and to plan.

And so as secondary judgments replace primary ones during this distinctively human phase of development, and as the boundaries of events come to be extended both into the past and into the future, many or most aspects of behavior become linked together and an organization, the personality, emerges. Gratification is no longer immediate but often must be renounced or postponed in anticipation of a more satisfactory future; anticipation leads to intervention; and attained gratification, the individual learns to know, is likely to be only temporary. Indeed, it has been suggested (Varga, 1959) that after the person has learned to orient his thought, his language, and his action toward the past and future as well as toward the present, he must then also learn to inhibit the tendency to employ irrelevant or maladaptive impulses in the wrong temporal direction. From these lessons, all the evidence suggests, arises gradually a consciousness of time: the child recognizes that the duration of intervals can, or must be, judged and that choices must be made among temporal orientations. He discovers that he can remember events from the past, he can be conscious of experience at the moment, and he can anticipate events that will occur in the future. A descriptive summary is offered as a Principle:

8.1 *The relevant processes and Guides within the behavior potential which affect the temporal potential, as well as temporal orientation and judgments concerning duration and succession, develop slowly, inevitably, and in culturally determined ways during socialization.*

We cannot halt here to survey the details in the learning process which involve remembering, anticipation, intervention, and renunciation, and which result in the total potential. Before invoking an Arbitrary Limitation, however, I would make note of two points. First, psychoanalysts have a theory that the process can be described in two stages: anticipation is present only in primitive form "during the first year, or oral state"; it really can be said to "appear in man during the second year of life, or anal stage," when "the child learns to give up his tendency to relieve the tension of his bowels, and postpone the act of defecation"; but "in the genital and latency periods, the ability to anticipate the future [develops] enormously" (Arieti, 1947). Then the Freudian view seems to be an oversimplification when we note that any aspect of temporal patterning matures in conjunction with other processes and consequently varies from individual to individual and from group to group. It has been shown, for example, that learning to postpone among normal and maladjusted boys in France (between the ages of 10.5 and 12.5) was associated with the "degree to which personal future events in general appear to be endowed with a sense of reality, as well as the degree of everyday preoccupation with future rather than present events" (Klineberg, 1968). Evidence that this ability to defer gratification is learned gradually and is emphasized differentially appears in *Addendum 8.1*.

During socialization the child acquires the ability to respond to, and with, speech. Again a Spiral may be presupposed since language mediates between the temporal and behavioral potentials: the patterning of existence and language are usually so closely interrelated. The growing child learns to store in language the delays he experiences and the changes in the state of affairs he anticipates, and he soon expresses those experiences—but never quite all of them—through language. And having been provided with the temporal categories inherent within the language of his milieu, he easily utilizes them to describe his experiences, and perhaps he seeks out experiences corresponding to the categories. "You must wait," he is told, "you must wait until . . ."—and the interval then specified may extend from the few seconds before the starter's gun is heard, to his entire life span. "What are you going to be when you grow up?" Such a question may be asked only in societies in which the individual's role is not completely structured by tradition, and therefore significance must be attached to the fact that the question is raised. To reply, the child must project himself into a future he may or may not adequately comprehend, and thus he is induced by the language—or rather the person so employing it—to orient himself toward the future. As a result, the inference must be that, if no one asks him the question, he will be less likely to think about the future and to be oriented toward it. It is, therefore, not surprising that "the use of words denoting temporal concepts shows a parallel development to that of language in general" (Wallace & Rabin, 1960); simultaneously, of course, unverbalizable cues are also learned and serve their important function.

The assistance rendered by language in sorting experience can be indicated in yet another way. The child must learn to distinguish experiences in reality from those in fantasy. Unquestionably dreams present a challenge, for they can easily be confused with the waking state. In some societies, in fact, external reality is attributed to them: the person feels that they represent a part of himself which is wandering away from his body. When the individual is able to say, "I dreamt that," he has begun to grasp the distinction between outside events and his own conscious stream. Usually, no doubt, his family helps him; and therefore he can apply the culturally correct labels to reality out there, and fantasy in here, and prove to himself and others his own alertness.

One final consequence of socialization, of growing up in a society, is the gradual growth of a consciousness of self, of self-awareness, of (if I may drift with the flood and use the fashionable term) an identity. The concept, though difficult to describe and define, points to the fact that each person is not simply a bundle of impulses tied loosely together, but rather, as students of personality—or most of them—have long emphasized, an organized, more or less consistent whole. We have, therefore, some sense of continuity over time. The child learns to believe that it has been "I" who have had the experience in the past, and it will be the same "I" who will have the experience in the future. He may find it almost impossible to realize that what happens to him now will be among the memories to which he can return, to some degree, when he is older; or that if he is naughty or aggressive, he will acquire a reputation, now or in the immediate future, which can bring misery to that self or his family in later life—and thus we glibly say that the "he" of now will affect the "him" of tomorrow. For a person growing old it is often also difficult to imagine that it was the self he now knows who was once a boy: "I was quite different then, and yet I remember clearly that it happened to me. . . ." Portions of the past or the future—or for that matter the present, too—may seem unreal to him; but, unless he is dangerously dishonest or psychotic, he acknowledges their connection with his present state. He may smile at the indiscretions he committed during his youth, but he does not deny that they were "his" indiscretions. Let time or the events occurring in time change his taste and his habits—or some of them—and let his various milieux change, but a core remains continuous. A magisterial semanticist once suggested that dates be attached as subscripts to many nouns and pronouns in order to indicate continuity by the word itself, and discontinuity by the date (Korzybski, 1941, e.g., p. xxix); at any stage of your life, you are you, but you are also quite different.

8.1.a *The effectiveness of the relevant processes and Guides within the behavioral potential affecting the temporal potential as well as judgments concerning duration and succession is increased when they are stored in verbal form and when they are linked to the self.*

I do not think, however, that we know in detail exactly how early experience becomes internalized so that it later affects temporal behavior; we are convinced only that in some way it does. In the last chapter, for example, reference was made to the problem of sex differences in judging time. If there were such differences, could we really account for them in detail by reference to the different roles which men and women in our society must learn, or would we just be rationalizing the ascertained fact? For orthodox psychoanalysts there is little mystery: every aspect of temporal behavior is traced back to early experiences:

> All that is most fundamental to a person's appreciation of time and rhythm originates in a pattern laid down at the breast period, when the body supplies the rhythm . . . where there is gross disharmony between the child's and mother's time, a degree of aggression is aroused which influences all subsequent time relationships, first excretory, then genital, and then passing on to the relationship of work and pleasure to sublimations as a whole [Yates, 1935].

But we need more evidence than the single case history cited by this author or the clinical experience that has given rise to her conclusion, however "tentative." We do have one longitudinal study which shows a significant if not overly impressive relation between the way in which 60 American middle-class children responded temporally to pictures (the so-called TAT) at the age of 14½ and the treatment they had received during the first year of their lives as determined then by observation and by interviewing the mothers: the greater the structuring of time in the stories, especially among the boys, the less the parental indulgence (Zern, 1970). Finding this link between temporal behavior and at least one socialization practice merits wholehearted praise even though the investigator had to examine many possible associations before locating one that exceeded chance expectancy.

THE SENSE OF TIME

Spengler expresses the conviction that, in contrast with space which automatically is part of "the ordinary life of dream, impulse, intuition, and conduct," time must be considered "a *discovery* which is only made by thinking" and which therefore we "create . . . as an idea or notion." Higher cultures alone can have the idea of time, "for primitive man the word 'time' can have no meaning" (Spengler, 1926, p. 122). Valuable here is not the praise of some peoples and the sneer regarding others—for the value judgments are undoubtedly without factual foundation—but the assumption that genetically the sense of time, like any discovery, is difficult to acquire.

It is easy to contend that all the processes of development associated with socialization have a temporal dimension: remembering involves the past, intervention the present, and anticipation and renunciation the future. But a conception of time has to be abstracted from these experiences. "To remember a thing as past," for example, "it is necessary that the notion of 'past' should be one of our 'ideas'" (James, 1890, p. 605); "in order to develop a sense of time, we must be able to realize that what is now was previously not, or was at least less so" (Aaronson, 1968). Even the simplest classical experiment reminds us for the nth time that no single sense organ is assigned to temporal judgment, rather every conceivable modality, potentially at least, can serve to convey external or internal stimuli to the central nervous system and eventually produce such judgment. The intervals between the onset and termination of stimulation by an electric light bulb, a phrase from a Beethoven sonata, a musky perfume, an unpleasant taste, a swift descent in an airplane, a contraction of the stomach or the uterus, etc. —all these intervals can give rise to the same judgment of "short" or of "45 seconds." It is true, I suppose, that a similar point could be made about space which also draws upon many different modalities for stimulation. But space can often be judged at a given instant of time, while the perception of duration demands the passing of time—you may be able to decide instantly which of two lines in front of you is longer, but you must wait to discover whether the piccolo player reaches the last bar before or after the rest of the orchestra. Temporal judgments, finally, are usually uncertain in the absence of adequate temporal symbols and consequent objective information from the external world. It is little wonder, therefore, that a sweeping glance at the literature devoted to the development of the temporal sense suggests to one writer that the first judgments, allegedly like those of nonliterate peoples, are concretely embedded in everyday life and egocentrically oriented toward the satisfaction of needs; abstraction is said to be a later and more sophisticated capability (Werner, 1957, pp. 182–8).

The development of the sense of time can be appreciated only by examining the empirical and, in a few instances, experimental studies which have concentrated upon children. This we shall now do; most of the details can be found in *Addendum 8.2*. But we are faced with a depressing methodological difficulty at the outset, viz., one of sampling. Almost without exception, the children who were the subjects of the investigations have come from a very limited segment of mankind: Britain, France, Switzerland, and the United States. Within these countries, moreover, only small numbers of children have been observed, most of whom seem to have been attending special clinics or schools and hence, I suspect, have come from somewhat privileged classes. The majority of the investigators overgeneralize because they describe their results in universal terms: "the child," they say, "at age 8" does such and such, as if all children in any society performed in the indicated manner. It may well be that temporal development is everywhere the same, but this would be so only if that development depended upon the maturation of other abilities which follow a uniform course. And we know, or think we know, that the development of most abilities, even learning to walk, is also affected to a greater or lesser degree by environmental and hence cultural factors.

Then, too, we are confronted with anarchy so far as techniques are concerned: procedures in these studies have not been standardized with the result that each investigator tends to demonstrate his ingenuity and patience in an original manner, which makes it next to impossible to compare his data with those from another study. Many, but not all, of the studies, moreover, have been confined to what children say about time because, obviously, it is easier and more convenient to operate on this verbal level. Perforce, then, the mediating processes have had to be neglected.

I would illustrate this tantalizing, frustrating state of affairs by quoting here from one study in which children between the ages of 5 and 13 were asked a series of questions pertaining to time, some of which indicated temporal orientation, others temporal information. Samples of the questions are listed below, grouped according to the age at which a minimum of 75 percent could reply correctly (Bradley, 1947):

Age 6: Is it morning or afternoon now?
Age 7: What season is it now?
Age 8: What day of the week is it?
Age 9: How long will it be till you are a grown-up?

Age 10: Does tomorrow come after yesterday or before yesterday?
Age 12: What time is it now by the clock? (*Asked in the absence of a clock*)
Age 13: How long do you think it would take you to walk around this room?

It is impossible to accept uncritically fascinating material of this kind. The investigator did not disclose the number of children questioned and so the data cannot be analyzed statistically. Also, we are told only that the children were British and "urban"; we have no way of knowing how representative of British children they were. We have here only a census report and are left to guess as to why the progression occurred; would the reader, for example, have anticipated that correct knowledge of the seasons would be acquired before the ability to relate "tomorrow" to "after yesterday or before yesterday"? We observe in this study, as we do in all studies which reveal central tendencies, deviations from the modal tendencies, in this instance the 25 percent at each age level who did not conform, from which the obvious conclusion must be drawn that children learn at different rates. Finally, it is necessary at least to question the entire procedure of attempting to establish age norms for the use of time-words: how can uniformity in development be anticipated when the utilization of such words serves many different functions and is also often dependent upon the maturation and modification of a variety of other processes?

Under these circumstances the details in *Addendum 8.2* do not provide us with a complete picture in whose validity and reliability we can have much confidence. There is, however, one general trend in every single study I have ever analyzed, including even a painstaking application of psychophysics and mathematics to two children and two adults (U.S.; repro., prior; Richards, 1964). This I embody in a modest corollary to the principle of the last section:

8.1.b *The ability to pass temporal judgment, to make more or less accurate temporal judgments, to follow a more or less consistent temporal orientation (which is likely to move from the present into the future or past) improves as a function of the rate at which the psychological, linguistic, and social processes affecting the temporal potential improve.*

In some respects the corollary is quite banal. Obviously the child cannot answer questions about clocks or calendars until he has learned the language in which the questions are posed, and he is not ready to learn the language until he has reached a certain stage of maturity. Subtler is the fact that the person must first be sufficiently mature to utilize a method of measurement before the accuracy of his temporal judgments can be ascertained through the use of that method. I yield to the following statement which summarizes a series of experiments on children which the author himself has conducted: "The ability to scale exteroceptive information along continua of temporal magnitude seems to require coincidental development of the conceptual capacity to integrate the durational and sequential attributes of stimuli along with the capacity to scale with yardsticks of increasing complexity" (Goldstone, 1964, p. 147). That should do it.

Eventually, as the individual matures, he acquires an attitude toward time itself which is thus learned in the manner of any other attitude. Adolescents know whether time as such is valuable, whether or not they may put off for tomorrow what they are loath to do today. They grow aware of the fact that there is not only a time, but also a place, for everything, and that therefore they must apportion time in terms of some sort of schedule which then in turn requires them to pass temporal judgment so that they may properly conform. They thus discover, or acquire, a framework in which they can remember, experience, renounce, postpone, or anticipate, depending upon circumstances. This discovery of theirs may well be facilitated, or retarded, by the opportunities in the milieu as well as by their own capacities; thus when subnormal children between the ages of 8 and 11 and also at age 15 were compared with normal children between 5 and 9 on Piaget-type tests involving time, it was found that both the subnormals and the normals followed "roughly the same sequence" but that the retarded simply reached the same stage as their contemporaries at a later chronological age (England; Lovell & Slater, 1960).

Finally, it is well to remind ourselves how highly abstract a sense of time must be. For one thing, concepts are imposed arbitrarily upon astronomical temporal units; thus yesterday it was winter, but today it is spring, even though no dramatic change in the weather or the setting of the sun accompanies the abrupt shift in labels. Children in Western society are taught that people are paid for work rendered during a given interval of time and then discover that some persons receive the same salary each month, although the months differ in length (Sorokin & Merton, 1937, p. 625).

PERSONALITY TRAITS

A person's temporal potential, as already indicated in Chapter 3, is affected by his personality traits either directly or through his behavior potential. Both the temporal potential and the traits simultaneously develop and interact. This relationship, therefore, must be examined in some detail.

In order to relate personality traits and temporal behavior, we must first know whether or not each is organized and leads to consistent behavior. For otherwise we would be pursuing a will-o'-the-wisp: there might be some relation between a particular predisposition and one kind of temporal behavior in one situation and another relation between a different predisposition and other behavior in a second situation, without any interconnection whatsoever between the two.

In everyday life, and in novels and plays, there seems to be little doubt that persons behave more or less consistently and that therefore we ought to be able to postulate with dispatch some central disposition or personality trait responsible for this unity. The difficulty arises in psychology, psychiatry, and in other behavioral sciences when we seek to measure those traits. Systematic investigations, largely confined to people in the West, have been, on the whole, somewhat disappointing. Almost always so-called traits are established or measured by means of questionnaires, tests, or interviews administered in the artificial atmosphere of the clinic or the classroom: either people are asked point-blank questions about themselves ("Are you afraid to walk alone in the dark?") or else they are required to respond to a very limited kind of provocation ("please tell me what you see in this picture"). Then, since human personality is so varied and complex, investigators and clinicians have by no means agreed upon the traits it is important to ascertain, with the result that, almost without exception, there is little standardization of procedures. In short, when we try to relate temporal behavior to personality we have no systematic way either of measuring the traits or of deciding which traits to measure.

The situation with respect to temporal behavior on the surface looks different but is, I think, basically the same. First, it seems evident that we know what it is we wish to measure about time; perhaps, but I am sure that, if this subject had been explored as fully as personality, the research freedom would have produced a long list of categories to be explored. In this section, I shall eventually consider three measurable components of temporal behavior from among the many which might be selected from our Taxonomy of Time: the accuracy of temporal judgments, temporal orientation or perspective, and attitude toward time. And surely our description of research on time up to this point, particularly in the *Addenda,* has revealed virtual anarchy concerning the precise methods employed in measuring these three components.

The task of relating personality traits and temporal behavior, therefore, is like attempting to harness two wild animals from different species to the chariot of truth.

Let us begin by asking whether in fact persons more or less consistently make accurate or inaccurate temporal judgments or under- or overestimate intervals. In one sense, information on the reliability of measurement methods (as outlined very broadly in *Addendum 1.3*) is highly relevant: a reliable method yields similar results on different occasions which must mean, too, that the individual is providing consistent judgments. But of course an unreliable method may conceal the possible consistency of the person. More direct evidence is considered in *Addendum 8.3* where it is shown that actual studies contradict each other by and large and that consistency depends upon many factors, included among which are the duration and nature of the interval being judged, the method of measurement, the temporal relation between the arousal of the temporal motive and the interval, and the momentary state of the person passing judgment. If I may venture a guess, I would say that temporal ability, like the tempo at which people perform various tasks, move their bodies, or even tap (U.S., college; Rimoldi, 1951) is probably more specific than general. That ability, as such, is not easy to acquire, and we have already indicated that it is most difficult to profit from past temporal experiences (*Addenda 7.1, 7.2,* and *7.3*). Contributing to unreality and to nongeneralizability must be the drive factor: during a minute of clock time, almost every conceivable climax can be reached, and he who passes accurate judgment when feeling detached may not be able to do so when emotionally provoked. Under these circumstances no consistent relation between temporal ability and other behavior can emerge when the former is inconsistent or unreliably measured.

For many reasons, nevertheless, the association between temporal ability and intelligence could well be positive. For one thing, intelligence tests themselves are relatively reliable, and certainly their utility in Western society has been demonstrated for more than half a century. To some extent intelligence can be measured only by determining what has been learned in situations to which all persons are supposed to have been exposed, and certainly the various stimuli leading to temporal judgments also belong in the category of actions that are learned. Most intelligence tests, moreover, have a temporal component: usually there are prescribed temporal limits so that subjects must be conscious of time and reply rapidly; and often, too, some of the items on a schedule test a knowledge of time or time reckoning in our society.

Actual evidence is offered in *Addendum 8.4:* intelligence-test scores in fact have been positively related to temporal knowledge and, with exceptions, to the accuracy of temporal judgments. Disappointing are the largely negative findings concerning temporal orientation. It would be useful to be able to maintain that you need to be intelligent to peer beyond the present into the future and, for that matter, to rid yourself of the dead weight of the past; but the available facts do not justify that view. An interesting hypothesis proposes a positive correlation between overestimation via production or reproduction and intelligence: the latter involves an ability to delay action and the former results from delaying temporal judgment too long. Only weak support, however, is at hand: the production of a 15-second interval was *not* significantly related to IQ; and that of a 30- and 60-second, though significantly related to intelligence, was very low (.22 and .32, respectively); obviously these correlations account for only a very minor part of the variance (U.S.; emotionally disturbed boys and adolescents; prior; Levine et al., 1959). Finally, one study suggests a different hypothesis. A slight relation was uncovered between intelligence and judgments of intervals ranging from 15 to 120 seconds for a sample of delinquents using the Method of Production; but no relation was found for the same sample with briefer intervals of 2 and 5 seconds, for the same sample with Verbal Estimation, and for a comparable group of nondelinquent Army recruits when confronted with the same intervals by means of either method (Israel; prior; Siegman, 1966). Could it be that intelligence, even or especially the kind measured by a conventional test, was needed to concentrate upon the temporal task at hand only under certain circumstances (in this instance when a relatively unfamiliar method like Production was used by delinquents who perhaps could not prevent themselves from being distracted by their own generally less controllable impulses)?

Another staple in the clinician's kit is the trait of extroversion-introversion. With notable exceptions, the scanty research summarized in part *a* of *Addendum 8.5* suggests, there has been a tendency for persons considered extroverts on the basis of paper-and-pencil tests to be no more or less accurate in judging temporal interval than introverts, but perhaps to underestimate them to a greater extent. A theory has been proposed to account for the latter trend: extroverts generate inhibition more quickly but dissipate it more slowly than introverts. When two intervals are being compared, it is assumed that the one perceived first evokes both excitatory and inhibitory impulses; that the inhibitory impulses reduce the apparent length of that interval; that extroverts retain this impression longer than introverts because of their general tendency to cast off inhibition more slowly and hence they judge the second interval to be longer. With the Method of Reproduction, extroverts would be expected to cease the act of reproducing sooner than introverts—for their inhibitory impulses presumably build up more rapidly—and hence to give shorter judgments (Eysenck, 1959). Relevant evidence is also outlined in *Addendum 8.5*. Here, it must be added, the same investigator who found a negative correlation between extroversion and length of estimates has also suggested

(a) that, if inhibition plays an important role in passing judgment, then with the Method of Reproduction hysterics should underestimate intervals to a greater extent than dysthymics (extremely anxious and tense persons) since the latter presumably are less inhibited than the former; and (b) that the more extroverted normals should resemble hysterics and the more introverted normals the dysthymics. In fact, he found no overall differences for these groups as well as for a group of schizophrenics; only trends going "roughly" in the predicted directions could be detected (England, adults; Claridge, 1960).

A theory of inhibition, nevertheless, may well have fruitful implications for explaining some aspects of temporal behavior. It has been suggested, for example, that those less capable of the inhibition required by delayed gratification are likely to overestimate intervals: with a more restricted temporal orientation, they use a shorter standard and hence intervals appear longer (Thompson, Spivack, & Levine, 1960). If rural Africans have such a restricted orientation, then this may account for the tendency of one sample of rural Ganda in Africa to reproduce intervals ranging from 15 to 30 seconds not more accurately but with a greater degree of overestimation than an urban sample (prior; Robbins, Kilbride, & Bukenya, 1968). Or, if time can be considered "a conditioned stimulus" which sets up a state of inhibition within the organism before the occurrence of the temporal judgment, if it be assumed that an increase in internal inhibition encourages overestimation, and if it be further assumed that psychological states, mental illness, drugs, in fact "all changes in the state of the subject's health" either increase or decrease that inhibition, then Soviet psychologists may possibly be able to unify many aspects of temporal behavior through such an adaptation of Pavlov's original conditioning paradigm (Dmitriev & Kochigina, 1959).

In the next chapter we shall deal with the relation between temporal behavior and abnormality; but in the present context the question must be raised as to whether some of the traits which, when present in extreme form, constitute abnormality have in normal persons a relation to temporal accuracy. The answer provided by the evidence supplied in parts *b* and *c* of Addendum 8.5 would be—again, no surprise—equivocal and confused, particularly with reference to anxiety. In general, traits considered normal or abnormal have been associated with over- or underestimation or with accuracy or inaccuracy—or frequently no association has been established. It is, therefore, only partially true, as one summary of the literature on time suggests, that "pathology" and overestimation are highly correlated (Zelkind & Spilka, 1965). I have the feeling that, with diverse methods of measurments and diverse subjects, we are dealing here with varying kinds of intervening mechanisms. It could be, for example, that the anxious person tends to cut himself off from external stimuli, included among which may well be those which can assist him in passing temporal judgment. Or, anxiety as a trait is likely to be aroused frequently and hence to increase the drive strength functioning during an interval. On the other hand, it can also be argued (cf. Hare, 1963a) that anxiety ought to lead to greater accuracy: in an effort to perform well, the anxious person may take advantage of all possible cues in situations in which reinforcement is not provided. Anxious subjects might give shorter judgments of enduring intervals than nonanxious ones when the Method of Production is employed, but the two groups could produce similar ephemeral or transitory ones: the longer intervals might be considered more unpleasant than the shorter ones; the only way to reduce the duration of the unpleasantness under these circumstances would be to produce shorter intervals, a solution adopted more readily by relatively anxious subjects, because of their greater need to achieve any goal, than by less anxious ones (Burns & Gifford, 1961).

When we turn abruptly to another aspect of temporal behavior, that of orientation or perspective, we immediately discover again that we are unable to say whether or not this predisposition is a general trait; hence its relation to personality traits or other forms of behavior cannot be constant. In fact, as the evidence in *Addendum 8.6* suggests, relations galore have been established. Many, but not all, American studies seem to suggest a positive connection between a future orientation (measured on the basis of a very limited sample of the individual's behavior) and traits considered desirable, such as sympathy and intelligence. But so many of the associations, though statistically significant, are very low and hence account for so little; exceptions and negative findings also appear; and, obviously, the findings must reflect the American ethos.

It is perhaps not surprising that the relation between orientation and anxiety is not unequivocal. For normal college students it may well be true that "attempts to cope with anxiety . . . involve a deemphasis of the present and a preoccupation with the future" (Rokeach & Bonier, 1960), but the mentally disturbed can attain the same objective by dipping into the past. If this is so, then that particular orientation is adopted which best reduces the anxiety; and from the fact of anxiety no prediction concerning the direction of that orientation is feasible.

The final temporal variable to be considered in this section, the attitude toward time, undoubtedly is heavily intertwined with personality traits. For, as we have seen in Chapters 3 and 4, temporal factors play a role in virtually every human activity and hence appropriate attitudes toward time must come into being. You may be willing to engage in a trivial action, such as a sport or a conversation, provided that not too much time is consumed. If compelled by yourself or social circumstances to do so, you establish "finite boundaries" pertaining to the allocation of time (Moore, 1963, p. 19), as a result of which appropriate actions are regulated: you stop work not when the work is finished but when you look at your watch or hear the whistle blow. If you are patient, you are not disturbed by the fact that the goal you seek is being achieved, in your opinion, too slowly; and if you exhibit such behavior in many different situations, I would posit a trait and call you a patient person. In general, patient individuals know they are powerless to make time pass more quickly; many Africans are said by one of their most distinguished leaders to realize that "you cannot hurry the sunrise, the rains, and the harvest" (Kaunda, 1966, p. 33). The impatient, on the other hand, are likely to intervene: they rush through chores in order to do what they really want to do. Doubtless the temporal attitude leading to the possession of patience or the curbing of impatience is involved in most renunciation: the goal will, or cannot be, attained immediately, you just must wait.

These introductory speculations concerning the attitude toward time, however, do not turn out to be beacons on the road to substantiated truth. For in *Addendum 8.7* we find little hard evidence and only very tenuous data supporting the view, as expressed by one investigator, that for the individual who is "capable of delaying satisfaction of his needs for relatively long periods of time without undue discomfort . . . the rapid passage of time is unimportant . . . and he finds it satisfying to think of time as stable or moving slowly ahead" (Kurz, 1963). In one study the value of time itself was experimentally manipulated. In taking an intelligence test, a group of American undergraduates was told to concentrate upon accuracy and to forget about time, another was given exactly the opposite instructions. Presumably the latter group, at least for the moment, had to place a higher value on time; yet no difference appeared when both groups, immediately afterwards, took a Temporal Metaphor Test (described in *Addendum 8.7*) which seeks to ascertain attitude toward time (Grossman & Hallenbeck, 1965). These authors conclude that "the importance of time is not related to its subjective speed," but they may be wrong: the more enduring evaluation by the aged may be more influential than the momentary set that they experimentally induced.

Before abandoning the quest for personality correlates, we must offer the pièce de résistance, the one body of research which, no doubt because it stems from an enthusiastic theory, has so far offered the most promising and consistent results. The theory, in the words of its progenitor, states simply that persons with a strong need to achieve "tend to be oriented forward in time toward longer-range goals, even when that means foregoing immediate pleasures" (McClelland, 1961, p. 328). Expressed slightly differently, the thesis is that "individuals of high need achievement are aware of time as an energized, expendable, and directional phenomenon—not as an encompassing medium lacking dynamic properties" and therefore time is valuable as it moves swiftly ahead (Knapp & Garbutt, 1965). Implied in the category of the need-for-achievement are the notion of renunciation and impulse control.

The evidence for the connection between achievement and temporal orientation, for once, is somewhat voluminous, and it would be truly impressive were it not for the fact that most of it comes only from the usual American sources, college students. More frequently than not, the founder and his prophets have measured the need for achievement (usually referred to by them reverently as "n Achievement" or just plain n Ach.) by means of the projective technique of the TAT, which has been reliably standardized for this purpose. So measured, that need has been found to be related to what might well be called future orientation, which in turn has been measured in a variety of ways. In addition, an incredible relation has also been found—at least among Americans—between the achievement motive on the one hand and preferences for particular Scottish tartans on the other hand, so that the latter device has been used as a substitute method for measuring achievement. Then, in its own right, the Tartan Test has been related to various aspects of the temporal potential. The reasoning behind the use of Scottish tartans is that those preferring, for example, tartans with "soft," "subdued," "somber" blues are more likely to be achievement-motivated than those preferring the "hard," "brighter," "more commanding" reds because they wish their environment to be soft so that they can manipulate it. In contrast with the laggards, the hustlers "eschew strong, intrusive stimuli" which reds are supposed to be (Knapp, 1958).

I regret to report that sometimes the apostles have failed to uncork a positive relation between achievement and various temporal factors and that many of the positive ones, though significant, are low. Perhaps the association is actually stronger but is obscured by the measures that have been employed; thus the tartan test correlates imperfectly with the projective test, which in turn correlates imperfectly with achievement motivation in real life, and hence it can provide only a twice-removed measure when it is correlated with a temporal measure. Very damaging from a cross-cultural standpoint are the negative findings outside the United States, especially in Tanzania (McClelland, 1961, p. 332; Ostheimer, 1969): either the tests or the relationships may have limited validity and applicability. Indeed it is possible, I think, for the leaders of a society to be achievement-motivated and future-oriented while their followers reveal only the first but not the second attribute with respect to the social policies of the nation. The successes of the n-Ach. researchers, however, have been more numerous than the failures, to which we happily say, "Bravo." In addition, achievement as measured by school attainment has also been related to temporal orientation. The evidence supporting and not supporting the impressive milestone is presented in *Addendum 8.8*.

As ever, this need for achievement interacts with other traits and tendencies, which fact may account for some of the low correlations. In one experiment, in which American college students tried to solve a maze blindfolded, for example, it was again established (as indicated previously in *Addendum 6.2*) that the interval was judged significantly shorter when they were told their progress was rapid rather than slow. Although there was no difference in temporal judgments between those scoring high and those scoring low on the TAT-type achievement test described above, the high scorers were affected more by knowledge of progress than the low scorers: their judgments were shortened to a greater degree (verbal, immediate subsequent; Meade, 1966a).

In conclusion, let me state explicitly a viewpoint I have been implying throughout this drawn-out discussion: I am of two minds concerning the overall significance of the research on personality traits. On the one hand, I am inclined to be pessimistic and to agree with others that the results are too conflicting, the methodologies too diverse, and the subjects too predominantly American for anything solid to emerge (cf. Wallace & Rabin, 1960). And in general it seems to be too true that, no matter what personality trait is measured and no matter what method of measurement is employed, the trait in question will be found to be correlated significantly, though mildly, with some other trait provided enough traits are simultaneously measured, provided the computer is permitted to come up with a host of printouts, and provided the relation and the mode of analysis is established post hoc. Thus almost all the studies yielding positive correlations between an aspect of temporal judgment and a personality trait have failed to find one or more significant relations with other traits simultaneously measured. On the other hand, statistically significant associations do not occur by chance and therefore must have *some* significance. One writer, after a broad survey of the relevant literature, reaches the conclusion that some aspects of temporal behavior may be transient but that others may be "more or less permanently associated with certain personality traits" (Orme, 1962b). In spite of my already expressed misgivings, therefore, I feel impelled to offer a weak principle which extends just a trifle the one (No. 3.6) already offered:

8.2 *Any nonmomentary aspect of temporal behavior is correlated with one or more enduring personality traits.*

The principle leaves as an open question the extent to which any aspect of time is a general trait, for we have seen that generality in these respects has certainly not been established. It is conceivable that generality exists in some persons and not in others, is more prevalent in some societies than in others. We have here one of the explanations for the fact that time remains a baffling, cosmic problem: it may be linked to every conceivable aspect of personality, but need not be.

The principle is stated in correlational form to call attention to the interaction between the temporal potential and other aspects of temporal behavior on the one hand and the behavioral potential on the other, an interaction especially important in the development of both. The orientation of an ambitious, achievement-driven person may be clearly in the future because he can attain his ambitions only then; but it is conceivable that his ambitions are a by-product of his orientation which genetically developed first because he found the past too unpleasant to remember and the present too unpleasant to experience. Or of course both the ambition and the orientation could come into being simultaneously and thus strengthen each other's development. Many of the empirical studies seem to me to be subject to either interpretation. When American school children with retarded reading achievement produced fewer future-oriented themes in stories than those with normal achievement, the investigator suggests that "a degree of self-discipline, ability to postpone gratification, etc., seem to be related to the ability to learn a complex skill such as reading" (Kahn, 1965); he thereby implies that renunciation and consequent future orientation affect motivation to learn to read. It is possible of course that the reading ability affected the measure of temporal orientation or perspective in general, since reading can suggest wider horizons and more distant goals. Another Spiral to which we are forced to make frequent reference.

In spite of that Spiral, however, it may be that knowledge of one factor can provide *some* preliminary insight into the other. Such a successful leap is not unexpected when the starting point is a trait which by definition affects behavior in many ways and in different situations. And here is one wisp of evidence indicating that the reverse is also true. Psychopaths and hysterics who subsequently overestimated an interval of 30 minutes during which they had been interviewed tended also to respond to a maze problem more quickly and less accurately than those who had underestimated that interval, or had estimated it fairly accurately. More than superficial behavior was involved in the temporal judgment: the skin temperature of "high grade" psychopathic patients who had similarly overestimated the interval tended to decrease while performing a frustrating task, whereas the temperature of those underestimating the interval tended to increase. These impressive relations existed in spite of the fact that judgments concerning the 30-minute interval correlated significantly but very minutely (.13) with Verbal Estimates of another interval with a duration of 30 seconds and not at all with the Production of that shorter interval (England, adults; Orme, 1964).

Similarly, sage observations concerning culture may also involve a Spiral: "veracity is only valued where people are in a hurry and set value on quickness," it has been said with reference to the Near East (Paige, 1951, p. 364). The presumption here seems to be that nonveracity takes more time and is, therefore, a luxury which busy people cannot afford: if you are not preoccupied, you have unfilled intervals at your disposal which can be filled by the complications resulting from not telling the truth. Conceivably, however, the custom of telling falsehoods may now in its own right facilitate or demand empty periods of leisure in which the complications are resolved.

Whether in the context of the individual or the society, therefore, the significance of temporal behavior must be apprehended by examining its function for the personality. Two persons may show concern for the future for quite different reasons: one may wish to avoid the present and concentrate upon the future because he knows that his present fantasies about that future cannot be contradicted, whereas another has realistic hopes about the future and would intervene to achieve them. If you refuse to make any kind of public statement concerning your temporal judgments without consulting a clock—you do not trust your primary or even your secondary judgment—you are probably disclosing unsubtly a deep-seated segment of your philosophy which is thus expressed in this trait. But if you distrust all clocks and believe that people have turned the hands forwards or backwards in order to confuse you, and you alone, then this temporal trait is a sign of abnormality requiring psychiatric attention.

On occasion, it seems reasonable to assume not a correlation but a causal relation between temporal behavior and personality. An individual who gradually becomes anxious for reasons not associated with time but with the problems of his existence may change his temporal orientation or his tendency to overestimate intervals as a direct consequence. But does the reverse occur, can changes in some aspect of temporal behavior have repercussions on behavior? Two exceptionally intriguing experiments suggest that possibility. In one, the female teachers who were the subjects were asked at the outset of the investigation to write the first three letters of the alphabet as slowly as possible; they were given no time limit; clearly this was a test of their patience. Then one-third of them, the control group, simply indicated when in their opinion 15, 30, and 60 seconds had gone by, obviously the Method of Production. Another third was told to attend to intervals of the same duration, but they were falsely informed that the duration was 20, 40, and 60 seconds; thus they were indoctrinated in a manner inclining them to overestimate clock time. The remaining third followed the same procedure as the other experimental group except that the lie they heard was in the opposite direction: the intervals were said to have lasted 10, 20, and 40 seconds; hence their indoctrination was in the direction of underestimation. Finally the writing of "abc" as slowly as possible was repeated by all three groups. The control group spent more time at this dull task; evidently practice "improved" performance. The two experimental groups, however, differed significantly from the control group. The first one spent not more but about the same amount of time as before, and the second much more time (U.S.; Thompson, Spivack, & Levine, 1960). Even though the indoctrination had been brief, the experience evidently affected the subjects' way of judging time, and that judgment in turn affected their attitude toward time and the patience they displayed in this situation. Those whose time sense had been speeded up by the falsehood overestimated the time spent on the task and hence grew impatient sooner, whereas the reverse seems to be true of those whose time sense had been artificially slowed down.

In the other study, objective time was again manipulated, first with deceit, and then without it. When a clock was manipulated without the knowledge of the subjects, the obese ones ate more when they believed it was after their dinner hour than when they believed it was before that time; and the reverse was true for those of normal weight (U.S., college; Schachter & Gross, 1968). In this instance knowledge of time affected these persons' reactions to food, but the reactions depended upon body weight which in turn, according to the investigators, reflected characteristic ways of responding to the interval drive of hunger and to the presence of food. In a real-life situation once again the same personality factor interacted with a knowledge of time to elicit different reactions: overweight personnel flying jet planes of Air France claimed they "suffered" less than those of normal weight when, after arriving from Paris in New York at 2 P.M. without having eaten for seven hours, they knew that locally the time was past the lunch hour and long before the dinner hour (Goldman, Jaffe, & Schachter, 1968).

AGING

Passing reference has been made in previous chapters to a conviction that with increasing age time seems to pass more rapidly and that distant events in the past appear to become more recent. Adequate studies, so far as I know, do not exist which would indicate whether temporal judgments of this kind are also made in non-Western societies or, for that matter, how widespread or universal they are in the West. Aging, nevertheless, is a valuable variable to consider in analyzing temporal behavior for a perfectly straightforward reason: wherever we look at the Taxonomy of Time, we find, almost invariably, marked changes developing with increasing age, and these changes are likely to affect behavior. I shall simplify the analysis which follows by attempting to contrast two hypothetical groups at the extreme ends of the age continuum, the young and the old; but I am aware, of course, that the contrasts develop only gradually in most instances.

The most dramatic difference between the two age groups we find within the behavioral potential associated with the biochemical clock: vital processes slow down with increasing age. It takes four times longer, for example, for a wound to heal at age 50 than at age 10 (du Noüy, 1936, pp. 160–1). There must be, it was once argued, corresponding changes in "biological time" within the organism; the biochemical clock functions sluggishly with increasing age; and so time is perceived differently at different ages:

> The days of our childhood seemed very slow, and those of our maturity are disconcertingly rapid. Possibly we experience this feeling because we unconsciously place physical time in the frame of our duration. And, naturally, physical time seems to vary inversely to it. The rhythm of our duration slows down progressively. Physical time glides along at a uniform rate. It is like a large river flowing through a plain. At the dawn of his life, man briskly runs along the bank. And he goes faster than the stream. Toward midday, his pace slackens. The waters now glide as speedily as he walks. When night falls, man is tired. The stream accelerates the swiftness of its flow. Man drops behind. Then he stops, and lies down forever. And the river inexorably continues on its course. In fact, the river never accelerates its flow. Only the progressive slackening of our pace is responsible for this illusion [Carrell, 1935, pp. 176–7].

For the moment, let us agree that the problem raised by this poetic passage does not involve the validity of the contrast which is drawn, but rather the mechanism through which the aging processes of the organism are mediated so that they affect temporal judgment. It could well be that the slowing down of the body—of the biochemical clock—is communicated in subtle ways that are not always salient and hence constitute an extremely important basis for the unverbalizable cues whose significance we have already acknowledged (e.g., *Principle 6.4*). Perhaps, for example, the young and the old use different standards on which to judge a given interval, either because somehow the time it takes for wounds to heal (or hair and nails to grow) affects the standard, or because some process common to both wound-healing and the choice of standard changes with age. Growth curves based upon an index of cicatrization or other physiological processes which change with age and upon the relation between the duration of intervals being judged and the age of the persons making the judgment have been shown to coincide roughly: such an identity implies, it is argued, the existence within the organism of "a subconscious totalizing system which might well be one of the psychological manifestations of the physiological and chemical transformations introduced by age" (du Noüy, 1936, p. 165). More specifically, the flow of blood to the brain and the consumption of cerebral oxygen appear to decline rapidly from childhood through adolescence and then to decline more gradually, but progressively; these changes could lead to a "decreased functional acuity of the aging brain" (Kety, 1956) having repercussions upon temporal behavior.

The ratio between an interval and the person's age may possibly affect his behavior potential and hence his secondary judgment, whether or not that ratio depends upon the biochemical clock. As a hypothesis, this view was modestly expressed years ago (Paul Janet, 1877) and it is often revived, with or without attribution to that earlier source (e.g., Nitardy, 1943). Involved is a kind of temporal calculus: as age increases, an interval of the same duration pointing either toward the past or the future becomes a smaller proportion of the individual total life or of his memory span. Either span, it is implicitly or explicitly assumed, is used as the standard for passing judgment, presumably because it has been frequently reinforced. (*Principle 7.2*). For a child of 10, one year is 10 percent of his existence, but for an adult of 50 the same period is only 2 percent. That interval of a year, therefore, is judged by the child as relatively long and by the adult as relatively short, whether the orientation is backwards or forwards. Similarly, as age increases, a fixed point in objective time, past or future, is, respectively, farther away or closer, but it also becomes a smaller portion of the individual's total life, again assuming that the duration of this life functions as the standard. An event 100 years ago, when a war began or ended, is for the child of 10 at a temporal distance ten times his age; but 40 years later, when he is 50 and although the event is now 140 years back, the temporal distance is only less than three times his age. Or the year 2,000 in 1970 is relatively closer for the man of 50 than for the child of 10, since the span of 30 years is 60 percent of the man's age, but 300 percent of the child's. Obviously, from a psychological standpoint, the figures are not to be interpreted literally.

A theory such as this can only be stated, the different ratios can be acclaimed as correct, but convincing proof that individuals actually function in the indicated manner is difficult or perhaps impossible to gather. The feeling that last Christmas, or your tenth birthday, was a long or a short time ago wells up into consciousness without an awareness of your present age or memory span as the standard of comparison. Two slim, unsatisfactory bits of evidence are at hand which simply illustrate how the ratio may function (see *Addendum 8.9.a*). All that can be done, I think, is to assume that somehow experiences and anticipation most broadly produce and affect the standard.

We shall concentrate, therefore, upon analyzing the different experiences and anticipations of the contrasting age groups, a felicitous introduction to which is offered by the most literate of American psychologists:

> It is certain that, in great part at least, the foreshortening of the years as we grow older is due to the monotony of memory's content, and the consequent simplification of the backward-glancing view. In youth we may have an absolutely new experience, subjective or objective, every hour of the day. Apprehension is vivid, restlessness strong, and our recollections of that time, like those of a time spent in rapid and interesting travel, are of something intricate, multitudinous, and long-drawn-out. But as each passing year converts some of this experience into automatic routine which we hardly note at all, the days and the weeks smooth themselves out in recollection of countless units, and the years grow hollow and collapse [James, 1890, p. 625].

First, then, the event attribute. The young have more physical energy than the old, they tire less easily, they scurry about more relentlessly; if the temporal motive is interim or immediate subsequent, therefore, they are likely to feel that intervals are crowded, full, and so—from one standpoint—long. On the other hand, adults, at least until they retire or grow feeble, usually have more responsible activity to perform (Ennis, 1943), which may also increase the impression of the intervals' scope, a point which James in the quotation above fails to mention. On this score, then, I think the two age groups cannot be differentiated, except perhaps at the extremes: the very young experience a great deal, the very old very little (cf. Benford, 1944); and so, as our principle concerning the curvilinear relation between scope and duration would suggest, time must seem to the former to pass slowly and to the latter quickly.

Most of the subsequent judgments of the young are likely to be closer to the intervals than are those of the old and hence it may be inferred that the memories of the two are probably different. In recalling the content of intervals, the young may suffer less interference from intervening events since they have experienced fewer of them, and so their memories may be clearer or more distinctive. They have, however, less highly developed systems of abstraction and verbal conceptualization, the kind which greatly assist most forms of recall. In contrast, distant events may grow more meaningful with age either because the person has experienced them or because he has acquired more information about them. Thus when a child first hears about a war which was fought before his birth, the event is unreal to him; he may have little more than a sentence or two to describe the episode. Later on, if he lives in the twentieth century, he is likely to experience a war, to know in general the nature of war in all its many horrors and few glories, and to learn in school and through the mass media the details of that particular conflict. The more familiar is likely to have been personally experienced, and what is personally (if vicariously) experienced must have occurred within one's lifetime and cannot be too symbolically distant. With the passing of time we may remember not our actual feelings during an interval, but the way in which we once summarized those feelings when called upon to do so; the summaries can of course become distorted in the interim; and older persons may be more subject to this tendency than younger ones (Allen, 1944), though they may also learn to appraise their recollections more realistically. The old may remember fewer events from a relatively recent period because they have poorer memories in general, because they find the events less impressive or exciting, or because in fact they have experienced fewer of them; yet their ways of storing information may be more orderly, so that they retain more easily than the young what they consider essential. Again, the comparison between the two age groups apparently can be pushed hither and yon; and I can detect no general difference or central tendency likely to affect primary judgments inevitably, in every situation.

The young, in James's words, are more likely to have "an absolutely new experience" very frequently, which means that strong drives are evoked; in contrast, the old find that so much is familiar that their evoked drives are likely to be weaker (Dondlinger, 1943). The young advance from grade to grade in school, are privileged, or required, to change the tasks confronting them as they mature; in contrast, most older persons settle into a routine from which there is little variation and out of which, consequently, distinctive and intense events become less salient. On the score of drive, therefore, the young should clearly overestimate elapsed intervals to a greater degree than the old.

The fact of having lived varying numbers of years must have important implications for the two groups. The young have experienced only relatively short intervals and consequently, no matter what standard they recall from the past, they are likely to tend to overestimate the interval at hand when it is compared with that standard. In addition, with a more restricted temporal orientation, the young are seldom called upon to employ longer standards like the year or the decade; their elders ask them to plan and to recollect in terms of hours, days, or months. Thus the ratio principle again becomes germane.

During early socialization the child's existence is largely controlled by adults. Specifically this includes the allocation of time he is permitted to spend on various activities. Under these circumstances his parents rather than he must pass temporal judgment concerning whether it is time for him to go to bed, or whether too much or too little time is being devoted to a particular task. With less opportunity to practice temporal judgments, children do not judge intervals as accurately as adults, an obvious difference we have already stressed in another section of this chapter.

So far the contrast between the two age groups has involved details concerning elapsed intervals. Another critical difference becomes evident the moment attention is turned to the future and to its varying significance. Here we are dealing with a prior temporal judgment and a duration-dependent drive: by and large—or perhaps only in our culture?—the young are driven to have the interval of their youth end and the old to have the interval of their living prolonged. As a result, the tendency may be for youth to overestimate and for the aged to underestimate the duration of elapsed intervals; hence for the former time seems to pass slowly and for the latter quickly.

What basis do we have for postulating this difference in duration-dependent drives? Frequently by the time secondary judgments can be passed, the period of complete dependence of the child upon the parents is probably at an end in all societies. Youth then becomes, even in the most permissive of societies, a period of some frustration and straining anticipation. The young are not permitted to express many of their impulses. They must renounce for the sake of the future. They are impatient because they realize that critical events—such as independence from parents, marriage, occupation, parenthood, etc.—cannot be momentarily experienced but must be anticipated in the future. Very young children may have a present orientation, but adolescents must begin to be oriented toward the future and must renounce some of their impulses.

In contrast, older people have less to look forward to, unless they are able somehow to view death with detachment or with longing. Some kind of age status is recognized in any society (R. J. Smith, 1961): if the aged occupy an honored, honorable status, then their earlier anticipations have been realized and they have little more to achieve except to obtain whatever satisfaction they can from their position. If they are tolerated with resignation, they can survive psychically only by glancing backwards. The orientation of the aged, in short, is likely to be toward the past (Erdös, 1935) and so with good reason, though with inadequate evidence, a psychiatrist has postulated "the universal occurrence in older people of an inner experience or mental process of reviewing one's life" (Butler, 1963). For them "each added year is a bonus," furthermore, since "all they can hope for is a few more years, and they come to regard the tale of years as itself honorific" (MacIver, 1962, p. 38). The old, consequently, wish to delay the future; to hold back time: they know they have lived a long while, they may feel that they have not accomplished as much as they once wished, and they are compelled to realize that increasingly fewer opportunities remain before they die (Wilen, 1943). And so, especially if they are imbued with the heritage of the Greeks which adored the youthful human form rather than the "Eastern idealization of venerable age," they are less hopeful and expectant regarding the future, they feel that the present is relatively empty, and they may even be dismayed when they try fruitlessly to relive the past (Vischer, 1947). Concentrating upon the past, one writer suggests (Kastenbaum, 1966b), may increase the sense of being isolated from younger persons who cannot really appreciate past experiences which they themselves have not undergone. Actual evidence, reviewed in *Addendum 8.9.b*, comes from American sources and is inadequate; still it lends some support to the thesis of an increase in future perspective with advancing age until perhaps late middle-age is reached. The research also suggests that the critical factor is not age per se but age interacting with other factors such as the sex of the person, social class, personality traits, mode of living, etc.

At any rate, the child must view dying as something unreal and remote. He learns gradually that plants, animals, even human beings can perish and disappear. He probably subscribes to the universal proposition that all men are mortal long before he appreciates his own mortality: they may die but that cannot possibly happen to him. And so he can view the time ahead as almost limitless, whereas the adult who perceives death as a closer reality may even admit that his days or certainly his years are numbered. Ordinarily, therefore, the young find time less precious than the old and hence they worry less about the time needed to achieve a goal—another reason for the feeling of children that time seems to pass slowly and for the conviction of the aged that it is passing quickly. Persons who are oriented to a greater degree toward the past as they grow older probably pay less attention to the future which, as a result, appears to descend upon them more rapidly. They have other good reasons for avoiding temporal judgments; when retired, "particularly in industrial societies," many may be caught "in the highly ambiguous situation in which they have too much time for too short a future" (Moore, 1963, p. 25).

Orientation or even perspective, however, is not an all-or-none matter. The old may prefer to think about the past, but for some purposes they must also contemplate the future. According to one theory of aging, illustrated, and to a certain extent confirmed, by interviewing and testing a sample of individuals in Kansas City between the ages of 50 and 90, older individuals gradually disengage themselves from other persons in their society—their friends die and they have contact with a diminishing number of acquaintances—and the purposes and patterns of the remaining interactions change. At some point, the investigators speculate, each person no longer measures his age by the distance from his birth but by the distance from his death. Some activities then are discarded and it is realized that time is finite. This "anticipation of death" then frees the aging man or woman from "the obligation to participate in the ongoing stream of life" and turns his orientation away from the future and even away from the present (Cumming & Henry, 1961, esp. pp. 14–15, 224–7).

On the basis of clinical experience and his own investigations in the United States and with a heavy dose of commonsense, one writer believes that each age group has indeed a different attitude toward time. Children feel that time is out of their control; adolescents develop a time perspective more likely concerned with the shape of the near rather than the far future; and then old persons again have the feeling that time no longer belongs to them (U.S., Kastenbaum, 1966a & 1966b). This attitude toward time is in part verbal, but as a metaphor "time" thus symbolizes so much of the person's existence. For concomitantly, the same observer argues, each age undergoes so many other changes. The young may not be able to visualize the meaning of old age. Older persons may know that time "is running out," that there is no point in delaying gratification. Occasionally there appears the phenomenon of what has been called the "geriatric delinquent," the individual who like the adolescent is dominated again by "the pleasure principle—orientation toward immediate gratification—frequently at the expense of long-range plans" and who consequently commits petty crimes and asocial acts (U.S.; Wolk, Rustin, & Scotti, 1963). Opposing such social, reckless tendencies may be habits reinforced since childhood which compel continued renunciation even when the individual realizes the imminence of death. This attitude toward time not only reflects but also affects feelings about death: if time becomes less valuable, then death may seem less painful.

Indicating that "time" tends to appear more valuable with increasing age means, I suppose, that duration-dependent drives are more frequently evoked. "Time is a crucial matter for the woman who marries late but wants children"; "the 64-year-old professor who wants to finish a book before retirement finds time short" (Pressey & Kuhlen, 1957, p. 303)—the duration of the pending interval in instances of this kind is of major importance. Perhaps when the interval appears too short or devoid of joy or when the attitude toward a future orientation becomes very unfavorable, the older individual turns toward recollecting satisfactions in the past. Persons who know they have a limited time to live because they suffer from an incurable disease, or those who realize that death may be probable because they are going into battle, may carry on not only by hoping for a miracle which will spare them but also by neglecting the future and the past in favor of the present. *Addendum 8.9c* contains the suggestion that attitudes toward death, as inadequately investigated in the United States, are possibly associated with particular personality traits.

The empirical test for many generalizations concerning youth and the aged ought to be temporal judgments concerning the duration of finite intervals. For if time really does seem to pass more rapidly with increasing age *and* if temporal tendencies generalize, then it must follow that youth will tend to overestimate verbally, and the aged to underestimate verbally, intervals of any duration. More assertions than facts exist with reference to this problem. In the former category is the view that old persons are likely to judge the duration of an hour no differently from the young (Fraisse, 1963, p. 246); or that some intervals can go by slowly, others "with incredible speed," whether one is 7 or 68, and that the critical factor is not the age of the person but his appraisal of the event (Carlson, 1943). On the other hand, a survey of existing studies in *Addendum 8.9.d* does indicate to some extent—with significant exceptions—that increasing age is associated, up to a point, with greater accuracy of judgment and with a tendency to decrease the verbal overestimation of intervals and to increase their underestimation via the Method of Production. We do not know whether these tendencies are as revealing as one writer contends they are on the basis of "preliminary findings" secured from "a series of investigations . . . at an all geriatric institution" and from the reports of "other investigators":

Aged people who underestimate their chronologic age express a desire to live longer than do those who correctly state their age. And those relatively few octogenerians who overestimate their age express no wish to live beyond the age that they suppose themselves to have attained. Furthermore, the overestimators also seem to be dying sooner than patients in either of the other groups [Kastenbaum, 1966a].

Let us gently summarize the discussion with a principle:

8.3 *Temporal behavior varies with age only when changes in the state of the organism and in the total potential themselves vary with age.*

The principle, as we emphasize in connection with every one of our generalizations, is useful only when a particular situation is analyzed in its own right. A father, for example, takes his young son on a three-hour hike; as they return home, how will the duration of the event appear to each? If the child has had more new experiences because he has been on fewer hikes or if he grows weary before the end of the expedition and his father does not, he is likely to pass a judgment that is shorter than his father's; but if his new experiences are thoroughly gratifying and if his aging parent is bored or quickly fatigued, the child's judgment may well be shorter. It is not age per se that creates the differences but the differing experiences.

Likewise if it is true that the young are motivated to have time pass quickly, as suggested above, then on the whole they are likely to be less patient than the old. Impatience here can mean that they wish certain intervals of time to pass quickly and hence are likely to overestimate such intervals when the temporal drive is an interim one. The father and son cannot begin the hike for another hour. Other things being equal, time will drag for the boy more than it does for the man, but only, of course, if he finds the prospect exciting. It is not difficult to imagine that the hike appears to him as a bore; then time may pass all too quickly for him. Or the father may be more highly motivated, and here again the older rather than the younger person may overestimate the interval.

The principle expresses a challenge to search for experiences which fluctuate with age. Often we simply do not know whether what we believe to be changes-with-age are cultural artifacts. How universal, for example, is a tendency to deny that we are growing old, both because "we do not experience the advance of old age either in our vitals or in the unconscious, in short, in anything which forms part of our organic life or of whatever is closely related to this" (Bonaparte, 1940), and because like children we wish to feel that we ourselves may turn out to be immortal or at least that death, being for us inconceivable, must always be far away? Such a tendency, wherever it exists, might make the persons possessing it less inclined to have a future orientation. Or another of the differing experiences which may be postulated certainly involves one's peers: obviously persons who are growing older retain many of their friends who are also changing by growing older. In this circle of associations members are likely to absorb from one another changes in outlook. Even younger persons—in this study, student nurses—tended significantly to extend the range of their own temporal orientation "in the direction of a larger, more complex, and more conflicted domain of psychological time" after intensive and intimate experience over a period of six weeks with aged persons (U.S.; Kastenbaum, 1967).

An important if sour methodological note: whatever correlation exists between age and temporal judgment is rendered imperfect by the obvious fact that changes in behavior themselves are not perfectly correlated with age. Obviously there are relatively young people who engage in little exercise or who retire from their occupation, and there are relatively older ones who exercise violently or who remain occupationally active until they die. An existentialist psychiatrist suggests that "where the past, the lived life, has become overpowering, where the life which is yet to be lived is ruled by the past, we speak of old age"; and then he finds in one of his patients that "existential aging had hurried ahead of biological aging, just as the existential death, the 'being-a-corpse among people,' had hurried ahead of the biological end of life" (Binswanger, 1958, p. 295).

One of the many out-of-bound problems of this book requiring an Arbitrary Limitation must be the explanation for the changes affecting temporal behavior which may fluctuate with age. We need only state the obvious thesis that they are partially biological and partially cultural. The period of dependence of the child upon his mother is the primary biological fact. And the cultural one is the role or roles associated with age status in the society. Often traditional societies have rites of passage to mark the transition from one age to another; and so the young boy after his initiation ceremony is expected to assume some or many of a man's responsibilities. One must suspect that juvenile impulses persist internally, but are carefully squelched behind the new manly facade. Few age-linked events have physiological signals as clear-cut as the menarche; most—such as voting age or the age of compulsory retirement in Western society—have only vague physiological associations and hence are determined markedly by cultural factors. When, and why, does an adolescent in our society decide that he is an adult or an older person that he has reached middle or old age? A portion but not all of the decision is affected by cultural requirements. Men—and women—with gray hair in our society may retain a conception of themselves as young and continue, therefore, to act youthfully until the undeniable infirmities of old age force them to change. And so for us there is what has been called somewhat dramatically "the principle of receding landmarks" (MacIver, 1962, p. 68), a tendency to rearrange criteria for passing temporal judgment so that time can be made to appear to pass less quickly and temporal fear can be diminished. The period labeled "old age," for example, is pushed forward further and further as the person himself grows older. Thus you once thought of your father as a very old person, but now that you have reached the same age, or more, you look back at him as having been very young at the time you recklessly passed judgment.

The fact of age is an important datum of social knowledge. A person may deliberately alter his orientation when he passes from one age to another. And his contemporaries have a way of anticipating to some extent how he will behave both in general and with reference to time when they can place him in an age group. For these reasons, either deliberately or through trial and error, each society evolves ways to communicate age status (Doob, 1961, pp. 218–22). Obviously birthdate is the most certain and accurate medium for conveying this information, provided the date is known and can be ascertained, which is not likely to be the case in many traditional societies lacking formal records. As already indicated, substitutes, such as events in the past or reigns of important kings, can serve as objective reference points. Physical appearance always gives some sort of rough clue. Non-Western peoples have evolved other conspicuous ways of conveying the same information, such as perceptible changes on the human body (circumcision, scarring, the removal of teeth) or the wearing of special ornaments or clothing which symbolize age status.

ADDENDUM 8.1 LEARNING TO POSTPONE

The proclivity to delay or be oriented toward the future has been associated with *age, social class, culture,* and a miscellany of factors, each of which can be briefly examined:

Age. American children between the ages of 4 and 9 were shown two toys, one of which they were told they would receive almost immediately, the other a few minutes or a week later; then they were asked which one they preferred or, in one instance, they were given an opportunity to play with either toy temporarily for a brief spell. The majority always preferred or played with the toy they were about to receive, even when the delay in receiving the second toy was going to be as brief as 3 minutes, signified by the sand running down in an hourglass (Irwin, Armitt, & Simon, 1943). When American college students were confronted with a similar situation—they were offered not toys but prints of famous paintings, or they were questioned concerning their preference for an immediate test or one a week later—the majority again opted for the bird-in-the-hand (Irwin, Orchinik, & Weiss, 1946). In my opinion, this second investigation does not completely "confirm," as the authors maintain, the results obtained with children: although the tasks and other procedural matters were different, higher percentages of the students than the children chose the more distant alternative, which can be interpreted as supporting the view that renunciation in American society is learned gradually. More impressive confirmatory evidence for the gradualness of the process comes from another study employing a similar technique with children between the ages of 5 and 12 (U.S.; Mischel & Metzner, 1962). On the other hand, American children varying in school grade from 5 to 11 were found not to change in temporal orientation as measured by a story-completion test but to decrease in temporal extensity with increasing age. This trend, it is suggested, probably reflected not a decrease in deferred gratification but an increase in realism: the future was being viewed to a lesser degree in terms of personal hopes, wishes, and fears (Lessing, 1968). Likewise there were no significant changes between the ages of 7 and 11 in a group of "normal" and one of "emotionally disturbed" American boys who responded to questions concerning the point in time at which they would spend a gift of 10 cents or one dollar (Davids, 1969).

Social Class. The first hypothesis to be advanced seemed reasonable: middle-class children in America may be reared more permissively than lower-class children but ultimately they are expected to reach a higher level of achievement (Bronfenbrenner, 1958; Lipset & Bendix, 1959, pp. 245–9); hence they go to school longer, defer "heterosexual experience" and marriage until later, save more money, etc. For these reasons there ought to be a corresponding difference in the temporal orientation of the two classes, with the middle-class being more future-oriented than the lower-class. This relation did in fact hold among the maladjusted children in France mentioned in the last section (Klineberg, 1968) and in a number of American studies:

1. Samples of children composed stories quite spontaneously in response to the investigator's request: the time span of those from lower-class subjects was more restricted than those from the middle-class (LeShan, 1952).

2. Among high school students there was a slight tendency for more of those calling themselves working- than middle class to adhere to a "deferred gratification pattern" by claiming that they had engaged in acts of physical violence, that they wished to go to college, that they had high levels of aspiration, that their parents had saved money in their behalf, etc. (Schneider & Lysgaard, 1953).

3. There was a very low relation (.11) between social status and future orientation as measured by the time a group of delinquents and matched and unmatched controls expected events to take place in their own lives (Stein, Sabin, & Kulik, 1968).

This fine, easy relation, however, has not withstood further research in America. First the widely cited study which heads the above list has been validly dissected and the association has been declared unproved on statistical grounds (Greene & Roberts, 1961). Then no relation, or no significant relation, has been found:

1. Between the temporal orientation of middle-class children in a summer camp, which was also measured by means of the spontaneous-story technique, and frustration tolerance as rated by adult counselors (Ellis et al., 1955). This finding can be considered relevant because, presumably, the mediating process differentiating social classes must be related to frustration.

2. Between the social class of a group of adolescents and college students as deduced from their father's occupational status and temporal orientation as inferred from how far in advance each individual claimed to plan details of his existence (Brim & Forer, 1956).

3. Between the social class of children as measured by residence and father's occupation and temporal orientation as measured not only through the story-technique but also—with intelligence held constant—through direct multiple-choice questions (Judson & Tuttle, 1966).

Two studies illustrate how difficult it is to interpret differences when they do appear. In one, the story-telling technique again was employed. Middle-class American children used fewer verbs in the future tense and more in the past tense, and they wrote longer stories with a more extended time span than lower-class children. The differences in the time span of the stories written by both groups may, therefore, be not a function of temporal orientation but of the ability to write a story, inasmuch as the longer stories covered a more extended time period, and the middle-class children were simply more verbal or more familiar with story-telling. The same children were also asked what they would do with $100 if they won the money in a contest; there was a slight but not a significant tendency for more of the middle-class than the lower-class children to claim they would save rather than spend most of it (Kendall & Sibley, 1970). In the other study, upper-class children in Austria allocated less space to their own past, present, and future on a line which was supposed to represent time, than did those from the middle-class; from which fact the investigators draw the inference that the former were "more historiocentric" than the latter; but there were no significant differences with respect to the amount of space on the line allocated to one's personal past, present, and future (Cottle & Pleck, 1969).

My conclusion from this review of the evidence is that in America marked and consistent differences between social class have not been demonstrated, probably because they are in fact very small, conceivably because the variable interacts with others, and possibly because insensitive measures have been employed.

Culture. Once again, as so frequently happens, it is easier to demonstrate differences between cultural groups than to comprehend the significance of the findings. The preference for a delayed and larger rather than for an immediate or smaller reward increased with age among Palestinian-Arab children between the ages of 5 and 10 (Melikian, 1959) and among Indian and Negro children between the ages of 7 and 9 in Trinidad (Mischel, 1958) but not among Australian aborigine children (Bochner & David, 1968). It is tempting to guess that renunciation may have been more rewarding among the relatively sophisticated inhabitants of Palestine and Trinidad than it was among the less "developed" Australians, but it is also possible that the attractiveness of the competing rewards, rather than the temporal factor fluctuated with age. The Trinidad study, moreover, suggests that the mediating process was not age per se but the differing experiences accompanying increased socialization: the delayed reward tended to be preferred by more children (a) from Indian than from Negro families and (b) from Negro families with the father present rather than absent. To explain the first tendency it is presumed that renunciation was fostered to a greater degree in the more ambitious Indian families. Two alternate explanations are supplied for the second tendency. Children may have had greater confidence that the male investigator would actually deliver the promised reward in the future because, having a father on a day-by-day basis, they could consider him to be discharging the role of a father-surrogate. Or the actual presence of a father in the home may have produced socialization in the direction of deferred gratification. These findings have been confirmed, except among Trinidadian children in the age range of 11 to 14, when the same technique was used to compare young Negro and Indian children in Trinidad and young Negro children in Grenada. Again the association between preference for the larger reward and a father-in-the-home appeared, again more of the Indian children preferred to wait, and also the Negro children in Grenada (a culture placing greater stress on saving and in general on independence from non-Negroes) opted more frequently for renunciation (Mischel, 1961b).

Miscellaneous. Among French girls between the ages of 7 and 9, patience in waiting to respond to a signal in a simple laboratory situation was significantly correlated with scores on a sensorimotor test of "emotional stability" (Fraisse & Orsini, 1955). In an American study previously mentioned, toleration of delay as measured by selecting the larger chocolate bar at a later date rather than the smaller one immediately increased not only with age but also with intelligence; yet the number selecting the larger reward significantly decreased, especially among the older children, as the delay in obtaining the reward increased from 1 day to 4 weeks. The sensitive tenuousness of such results is indicated by the fact that the choice of reward could be affected by the person making the offer: one investigator inspired more delayed choice than the other (Mischel & Metzner, 1962). The tendency, on a verbal level, to prefer postponement has been related to scores on a paper-and-pencil test seeking to measure social responsibility (U.S.; Judson & Tuttle, 1966) and, a bit irregularly but rather consistently, to educational status among samples of Luo, Ganda, Zulu, and Jamaica adults (Doob, 1960, p. 284). Conceivably, too, the method of measurement may play a role. In a fractionation experiment, for example, subjects were instructed to reproduce either half or double the length of the interval being presented to them. During the reproduction of "stimulus durations of longer magnitudes" the subjects found it "somewhat boring to wait and therefore they [terminated] their judgments too soon"; as a result, the tendency to underestimate while attempting to double the duration of an interval was more pronounced than when attempting to halve that duration (U.S., college; prior; Chatterjea, 1964).

ADDENDUM 8.2 ACQUIRING A TEMPORAL POTENTIAL

The grounds for criticizing empirical studies of children have been indicated in the text; in this Addendum we shall not forget them while seeking to summarize the imperfect evidence at hand.

The most fruitful and the most stimulating series of studies has been that of Piaget and his collaborators whose subjects have been French-speaking Swiss and whose data are seldom presented in quantitative form (Piaget, 1946; Flavell, 1963, pp. 147–9, 316–22; Piaget, 1966). Their careful observations of ingenious situations of their own devising and their sensitive speculations have sought to establish "stages" through which children go in the development of a temporal sense.

In these Swiss children that sense developed slowly: at the outset they were encased in their own egocentricity and paid relatively little attention to the reality of the external world, except for momentary events having an effect upon their pressing needs. A lost object was looked for briefly, and only later was a prolonged search instituted; it may be inferred, therefore, that the orientation must have been in the present. The first temporal judgment involved the fact of succession, but this the children could accomplish only when they themselves were part of the series of events: they did not begin sucking until they moved their hand toward their mouth. Toward the end of the first year—the norm is not well established and for our purposes largely irrelevant—they continued to relate what they perceived to themselves but without personal participation: they looked for an object they had seen another person hide. Months later they had a generalized conception of time, so that they could indicate where various people had gone even when they were out of sight. This knowledge, however, developed in interaction with other concepts, especially those of movement, velocity, and size. When two objects were moved simultaneously through space so that the onset and termination of the movement coincided, the object moving a greater distance was judged to have taken more time than the one going the shorter distance.

Piaget's children thus revealed again and again a great deal of confusion, the kind that can be expected to occur during trial-and-error learning. Another illustration: when water was drained simultaneously from one vessel into two vessels of identical size and shape, the young children could note that the operation for both began and ended at the same time; under the same conditions but, with vessels of different size and shape receiving the liquid, their temporal judgments were uncertain and they could not always say whether the operation had begun and ended at the same time in both vessels. The larger of two objects tended to be judged the older; thus the taller of two trees was thought to be more ancient, even when the drawings were clearly of different species and hence may have had different rates of growth.

In passing, it must be noted that some of this confusion among Piaget's children has a counterpart among adults. The effect of movement through space upon temporal judgments, for example, has also been observed among adults and, as indicated in *Addendum 5.3*, has been labeled the kappa effect. The Swiss children, however, may have experienced special difficulties because they relied more or less exclusively upon primary judgments: they could not carry out what might appear to adults to be simple reasoning. Young children who were given the task of drawing lines for 15 or 20 seconds concluded that the work lasted longer when they were instructed to draw the lines as quickly as possible than when they were asked to draw them carefully. Obviously the event attribute seems to have been decisive in this instance; they did not employ a Guide which states, in effect, that with increased speed more events can occur in a given period of time.

In a loose sense, it seems fair to say that virtually all studies by other investigators in the West can be included within Piaget's very general framework. This literature can be subsumed under two headings: the learning process and the shift in orientation.

Learning

Critics of Piaget within his own general culture dispute his theory concerning the Guides which young children employ for passing temporal judgment. Piaget himself believes that intervals are judged in terms of the work that is accomplished, whereas it is argued that the Guide is really the activities or events they contain (Fraisse, 1956, pp. 272–7). It has been shown that very young American children were more affected by the kind of activity with which intervals were filled than older children (verbal, prior; Axel, 1924, p. 57), which may mean that with increasing age children were able to avoid judgments based upon either events or accomplishments and to depend upon more reliable criteria. Another study involved two tasks, both with a varying number of discrete events, the one easy and quick (transporting rings from one box to another), the other difficult and slow (transporting counters by means of tweezers, again from one box to another). As measured by temporal judgments and as largely confirmed by an analysis of the reasons given by the subjects themselves, the children tended to be strongly affected by the number of events, i.e., the number of objects transported. This was especially so among the 5- and 9-year olds; but a small group of 13-year olds were more dependent upon other criteria, such as their own interest in the task and its difficulty. Thus at age 5 only 25 percent of the children judged the more difficult task to be longer; by 13 that figure was doubled (France; comp. & repro. with prior; Zuili & Fraisse, 1966). Also whatever is learned at any age must depend in part on available opportunities; though the instruments were admittedly imperfect, sixth-graders from culturally disadvantaged schools performed less well on a test based upon temporal concepts in the curriculum than those from superior schools (U.S.; Foerster, 1969).

These studies certainly do not vindicate Piaget, nor do they conclusively refute him. But they are all in agreement on one point: children learn to pay attention to different cues as they mature. In one investigation employing only children of exceptionally high intelligence, for example, 7-year olds tended to depend less upon the word of other persons and upon "imaginative" and "illogical" thinking and more upon the sequence in which periods of time in general were arranged or upon clocks and calendars than did 5-year olds—or at least that is what they said when questioned about how they managed to supply requests for temporal information such as the day of the week (U.S.; Farrell, 1953). Children must also discover where and how to obtain temporal information. Piaget reports a conversation with a seven-year-old who knew his own age and that of his eight-year-old friend. When asked "Which of you was born first?" he replied:

> "I don't know. I don't know when his birthday is."
> "But come on, think a little. You told me that you are seven years old and that he is eight, now which of you was born first?"
> "You'll have to ask his mother. I can't tell you" [Piaget, 1966].

Piaget concludes that "it is difficult to coordinate the order of events on which a birth date depends with classification of durations which determines age." One of the reasons for the difficulty, however, must have been the previous knowledge of the child that seniority is a function of birthdays which therefore must be ascertained before relative age can be determined; this knowledge that had been reinforced, perhaps too well, prevented him from answering the question with the simpler facts at his disposal, viz., the two ages. He had also learned to depend upon adults—"ask his mother"—for reliable temporal information.

Another point of agreement is the finding that the learning of temporal information and of the Guides to accurate judgment is gradual; and this means of course that increasing age simply produces greater conformity to the temporal norms of the society. Close to 1800 French children between the ages of 10 and 15, for example, were given the following problems:

> In spring the clock is advanced at one fell swoop from 11 o'clock in the evening to midnight. (1) Can you say what has become of the time when one does that? . . . (2) Do you think that you suddenly become older? . . . (3) Do you believe that it is possible to advance the time or on the contrary that it is impossible to do so?

With increasing age, fewer reified time and felt that it disappeared or that they themselves grew older when the clock was put forward (the figure dropped from 37 percent at age 10 to 10 percent at age 15) and increasingly more of them viewed the moving of the time forward as a pure convention and hence seemed to grasp time as an abstraction (the figure changed from 20 to 59 percent) (Michaud, 1949, pp. 56–7, 74, 177; see also Fraisse, 1963, p. 279). The fact that the boys and girls gave different responses either consistently or at different ages (e.g., except at the age of 14, slightly or significantly more boys than girls in each age group reified time) must indicate that a variety of cultural factors had been at work.

Other evidence concerning the gradualness of learning is close at hand. No kindergarten child, but "almost all" third-grade pupils, could name the year correctly; "a short time ago" was understood sooner than "a long time ago"; and, in general, by grade VI almost every child had "a satisfactory comprehension of our conventional time system" (Friedman, 1944). Between the ages of 4 and 6 a sample of American children living in Hawaii revealed a "consistent increase" in the ability to know when various school activities began, to tell time, to set a clock, to explain why clocks have two hands, and to draw a clock (Springer, 1951 & 1952). In another investigation the task was to "write a letter to a friend arranging to go out for a walk with him." At age 8, 33 percent mentioned the time, 92 percent the place of the rendezvous; by 12 the first figure rose to 72, the second to 94; two years later, the first dipped to 59 and the second to 87, though these latter differences are not statistically significant (England; Oakden & Sturt, 1922). Even the somewhat recherché aspects of temporal behavior have been shown to follow a gradual development. One investigator, for example, measured what he called "historical time" in the usual manner by asking questions concerning temporal sequences, temporal absurdities, etc. For measuring "the integral concept of time," he devised a test of 18 miscellaneous questions that "ideally" presented "problems involving time which would be insoluble unless a clear and complete concept of objective time was possessed by the subject" (e.g., "Why do we have leap years?", "What is speed?"); the test, however, did not have very high reliability. Children between the mean ages of 12 and 16 showed gradual improvement on both tests with increasing age, but the most marked improvement on the first test was between 12 and 13, that for the second a year later. Boys did better than girls on the integral-concept test because, the author surmises, his test involves abstract thinking for which boys, as shown by achievement in mathematics, reveal great ability (England; Rogers, 1967).

I conclude this tribute to gradualness by simply mentioning that additional confirmatory evidence appears in a competent survey of the literature (Goldstone & Goldfarb, 1964) and in other investigations: United States (Bromberg, 1938; Gesell & Ilg, 1946, pp. 438–40); Brazil (Rebello, 1937); Soviet Union (Elkin, 1928); England (Cohen, 1967, pp. 27–9); and Japan (Ikeda, 1957).

Some system of rewards and reinforcements in the child's milieu must determine the sequence with which temporal knowledge is acquired. One investigator indicates that the children she studied gave priority, as it were, to learning the labels signifying intervals which were significantly related to their own activities. Being able to tell clock time, for example, lagged behind the application of names to the morning and afternoon, the days of the month behind the day of the week, and bedtime behind supper-time (U.S.; Ames, 1946). Similarly, a superior group of middle-class children in New York City all knew their own age at the age of 3 and could distinguish between past, present, and future; but they did not recognize the day as a "formal unity of 24 hours" until 5 and they were not acquainted with the names and order of the seasons until 6. This sequence which the investigators think moved from the concrete to the abstract seemed to result from an interaction between "more or less private experiences and rhythmic needs" and an "external world"; thus the fact that at age six the children tended to be "vague" about spring and fall "may very well relate in part to the indistinct seasonal nature of New York City" (Schechter, Symonds, & Bernstein, 1955).

Small groups of British children between the ages of 9 and 14 revealed roughly the same trend, for they appeared "to learn first the meaning of time-words in ordinary use"; in addition, like adults, they found it difficult to try to estimate how long it would, and did, take them to walk around the room in which they were being questioned (Oakden & Sturt, 1922). French children at the age of 10—the youngest age at which a "satisfactory response" could be obtained—when asked to estimate the duration of a 20-second auditory interval, sometimes empty and sometimes filled with continuous sound, gave replies ranging from 30 seconds to 5 minutes; this range, however, was drastically reduced when the Method of Reproduction was employed (Fraisse, 1948). Evidently in this instance verbalization lagged behind whatever mediating processes were involved in reproduction. The ability to comprehend succession and duration developed simultaneously but independently of each other, an investigator maintains on the basis of Piaget's data; one of the child's most difficult problems was to relate the two kinds of experiences (Fraisse, 1963, pp. 258–61).

It takes no great perspicacity to suggest that one of the concomitants of increasing age is the discovery that primary judgments must be corrected through the application of some Guide. Virtually every child at the age of 5 was convinced that looking at a picture book, or drawing lines rapidly for 1 minute, took longer than, respectively, simply sitting with eyes closed and arms folded or drawing slowly for 1 minute; but by the age of 9 there was a pronounced tendency for more of the children to judge the two intervals to be equal. This trend was scarcely visible in a group of educationally retarded children (England; verbal, immediate subsequent; Lovell & Slater, 1960). Guides usually imply secondary judgments, the development of which is suggested by one series of studies. Under standardized conditions, children between the ages of 6 and 14 were tested with a variety of methods. In general, they displayed a tendency to underestimate clock time (for example, with a simplified form of the Method of Verbal-Limits, they called an interval 1 second which was in fact shorter than that), but the underestimation decreased with age until a fair degree of accuracy was achieved. Variability also decreased similarly. At all ages counting aloud tended to be more accurate than counting to onself, perhaps because the subjects could take advantage of kinesthetic cues; both methods of counting induced more accurate judgments after the age of 7 than the passive mode of production involved in the Method of Verbal-Limits. The latter, moreover, led to less accurate results than the Rating-Verbal Method which probably was closer to the methods the children ordinarily used. But, as Principle 7.3 wold suggest, improvement with Verbal-Limits also took place, except among the very youngest, when the children were provided with correct information concerning their performance (Smythe & Goldstone, 1957; Goldstone, Boardman, & Lhamon, 1958b; Goldstone & Goldfarb, 1966). For an ephemeral interval—all of the children's judgments involved the unit of 1 second—it looks as though these Americans had to reach the age of 8 before they could profit either from general experience in their milieu or from specific training; and, as we shall see later in this chapter, the ability, once established, apparently remained intact until old age.

Acquiring a temporal potential, like all learning, is linked to general emotional factors. A small group of "normal" American boys, for example, produced an interval of 30 seconds with significantly less error at the age of 11 than at the age of 7 or 9, and the 11-year-olds were markedly superior to comparable "emotionally disturbed" boys of the same age. The latter also showed improvement with age, but the differences were not statistically significant. Most important of all, although at all three ages both the normals and the disturbed improved their performance when they repeated the production of that transitory interval after having received feedback the first time, only the 9- and 11-year-olds among the disturbed showed significant improvement (Davids, 1969). It looks as though the normals had already achieved their potential at a given age, and that the disturbed, perhaps as a symptom of their difficulty, had not done so and required reinforced practice.

Orientation

Studies concerning the development of temporal orientation are not very numerous, but on one point they all agree: young children live in the present, then they gradually develop conceptions of the past and the future. When questioned about their milieu, Polish children between the ages of 7 and 9 revealed an interest only in the present (Bandura, 1936). After observing above-average-intelligent children from upper middle-class American families and after noting particularly at what ages and in what ways they acquired various temporal concepts such as the day and the year, a group of psychiatrists concluded that at the outset there was "emphasis on the realm of immediate personal experience —at first on the physiological, and later on the interpersonal and play activities" but that later "the realm of external factors [in the environment, such as sunlight, clocks, and calendars] becomes more important" (Schechter, Symonds, & Bernstein, 1955).

Do children move first into the past or

the future as they mature? Actual observation of a sample of American children revealed that concepts regarding the present tended to come first and then those pertaining to the future were used ahead of those referring to the past; for example, *today* arrived, on the average, at 24 months, *tomorrow* at 30, and *yesterday* at 36 (Ames, 1946). On the other hand, it is confidently, and almost convincingly, asserted that "ontologically, the past precedes the future"; and so "because we have been able to observe change by holding one state of affairs in memory while observing an ongoing state of affairs, and because successions of such changes can also be held in memory, the inference that such changes will continue to occur beyond the existential *now* becomes possible" (Aaronson, 1968). But no evidence is offered. One astute scholar, generalizing sensitively from his own observations of a single child (and after consulting equally restricted evidence in the journals), advances the reasonable thesis that "the child" must first learn to talk about "an absent situation" in the past or the future before he can be oriented away from the present. This is accomplished as a result of social pressure from adults and the examples they provide. In addition, the child of course has experienced the past, and at the outset each linguistic utterance referring to the future "is *followed* by the event to which it refers: he has had a cup of tea which people talk about and the cup of tea they say is about to appear soon does appear" (Lewis, 1937). When a sample of Austrian children between 12 and 15 years was compared with one between 15 and 18 with respect to performance on four projective or semi-projective tests, there seemed to be a reasonably clear shift from "an orientation of recall to one of expectation coupled with a recognition of the relatedness of the past, present, and future" (Cottle, Howard, & Pleck, 1969). A study of French children between 10 and 17 years, however, suggests the need to consider the function of the orientations as they develop. Although various measures of orientation produced quite different results, one trend seemed clear: more of the younger than the older subjects were oriented toward the future among those who were maladjusted, while the reverse seemed true among those who were normal. The investigator, therefore, concludes that temporal perspective became more restricted among the maladjusted and less so among the adjusted (Klineberg, 1967).

My own inclination is to agree with the conclusion of a competent scholar: available evidence indicates that past and future orientation develop simultaneously (Fraisse, 1963, p. 180). It seems probable that all children eventually remember the past they have experienced and that this past becomes part of the psychological present into which the future must also enter: socialization demands the application of the past and present to anticipations concerning what is about to happen. Indeed the idea of the future must be as difficult to learn as renunciation, and for similar reasons. Whatever positive relation exists between intelligence (as measured by school performance or a test) and future orientation, as we shall see, constitutes evidence on this point. The past must be experienced and remembered, but the content of the future—if not the fact of its inevitability—is always uncertain, whether attractive or repelling. Yesterday is reasonably clear, but tomorrow—we were once told—never comes. Somehow, as indicated in the previous section, the individual must acquire through experience the conviction and confidence that he will be able to attain later that which he does not now have.

ADDENDUM 8.3 CONSISTENCY OF TEMPORAL JUDGMENTS

The following studies have revealed that subjects can make rather consistent judgments concerning intervals of varying length in the absence of reinforcement: (U.S., schizophrenic adults; repro., prod., & verbal with prior;

Clausen, 1950)(U.S., college; verbal, prior; Gilliland & Martin, 1940) (U.S., college; prod., prior; Eson & Kafka, 1952) (France, children & adults; verbal & repro. with prior; Fraisse, 1948) (France, students & laboratory assistants; repro., prior; Korngold, 1937) (France, adults; comp., prior; Gavini, 1959). On the other hand, two different studies report some consistency as well as inconsistency in the same subjects who were also tested under laboratory conditions (U.S., enlisted men; prod. and repro. with prior; McGrath & O'Hanlon, 1967) (U.S., college; prod. & verbal with prior; Geiwitz, 1965). In addition, two factor analyses of laboratory data lead the investigators to the conclusion that there is no reason to postulate a general "time sense" (U.S., college; verbal, prior; Loehlin, 1959, pp. 18–9) (West Germany, soldiers; prod. & verbal with prior; Spreen, 1963). In the words of the former, "people who overestimate minutes do not necessarily overestimate seconds."

Outside the laboratory there is some evidence indicating consistency. A sample of American undergraduates was more or less consistently punctual in various situations on their campus (such as keeping an appointment for a conference, arriving at an 8 A.M. class) as well as in the same situation upon different occasions; but the tardy ones tended to be quite inconsistent (Dudycha, 1936, pp. 26–30). Subjects overestimating temporal intervals in the laboratory, in contrast with those underestimating them or judging them accurately, tended to make less accurate judgments concerning temporal aspects of their performance in two level-of-aspiration situations but not in a third one (U.S., medical students; comp., prior; Baer, Wukasch, & Goldstone, 1963). On a somewhat exotic level the German investigator of the "head clock"—equivalent to what we are calling the biochemical clock—has stated: "I have carried out a whole series of head-clock experiments during the daytime and have determined that these [demonstrations of being able, while awake and without ostensibly the use of external cues, to know when a given and predetermined interval has elapsed] succeed only among such persons who can awaken during the night at the specified time"; and since "not all persons possess a head-clock" (Clauser, 1954, pp. 82, 95), it must follow that those fortunate to own one are likely to be able to employ it generally, i.e., when awake or when asleep. But once again contradictory results can be cited. No significant relations were found among the following three measures: the duration of three brief intervals; paper-and-pencil determination of the rate at which various activities, such as writing a letter, were carried out; and general attitudes concerning the passing of time (U.S., high school students; verbal, prior; Kastenbaum, 1959).

On another aspect of temporal behavior, orientation, almost no consistency appeared in three studies. In one, three measures purporting to measure that process—preference for words pertaining to time, the temporal span of spontaneously composed stories, and verbal readiness to refer one's own interest to the various temporal dimensions—were "uncorrelated" (U.S., college; Matulef, Warman, & Brock, 1964). These results were replicated with three very similar measures of temporal orientation among counseling trainees (U.S.; Sattler, 1964) and with two similar measures among lower-class children (U.S., Brock & Del Giudice, 1963). In another study, however, the tendency for a sample of American undergraduates to be future-oriented appeared "quite stable" when that orientation was ascertained on two occasions about a week apart, when two different modes of measurement (the TAT and Story Completion) were employed, and when the type of instruction and other procedural details were varied (Wohlford, 1968).

Obviously it is not very fruitful simply to count or list studies indicating consistency or inconsistency; instead it seems better to turn to the problem of ascertaining the factors which have been shown to influence consistency or the degree of consistency that can be achieved.

The following have been isolated in one or more studies:

1. The duration and nature of the interval. Consistency in reproducing intervals of 30 seconds and 5 minutes immediately afterwards or 15 or 22 days later fluctuated with the point at which the judgment was evoked, the duration of the interval and—irregularly—with the nature of that interval, viz., whether it had been empty or filled with the activity of solving problems in arithmetic (U.S., college; repro. with prior & subsequent; Young & Sumner, 1954). In an extremely well-designed experiment, four variations of Reproduction were employed: (1) the usual pause occurred between the standard stimulus and the reproduction; (2) the stimulus ending the standard also constituted the beginning of the reproduction (3) the subject was instructed to reproduce one-half of the standard interval as he heard an interval which was twice the duration of that standard; and (4) the last procedure was duplicated except that the subject heard the standard interval twice. By and large, neither accuracy nor consistency of judgment was affected by these variations, although the fourth method produced greater consistency than the third. The critical factor turned out to be the duration of the interval: intervals of short duration (0.5, 1, and 2 seconds) tended to be judged more consistently than longer ones (4 and 8 seconds), and the longest intervals were significantly overestimated (U.S., college; prior; Doehring, 1961).

2. The method of measurement. Contrary to the study just cited, experiments have revealed that the method can be critical, though always new complications appear. The evidence on this point has been reviewed in *Addendum 1.3*, and so here I would only underscore the complexity of the multivariate situation by citing one study. Two investigators employed the Method of Production with intervals varying between 1 and 64 seconds and they also varied, or tried through instructions to vary the way in which the production was achieved: first the subjects "spontaneously" (i.e., without instruction) produced the intervals; on the next series they were told to visualize a clock, then to keep their mind blank and not to count, and finally to count. In addition, they were asked to tap at half their normal "comfortable" rate and also at half the rate at which they had heard a metronome. The four different modes of production resulted in positive intercorrelations which were significant except for the one between the blank mind and counting. Correlations among the judgments for the different intervals within a given method tended also to be positive, with the highest being for counting and the lowest for the blank mind. Production of 1-second intervals with any one of the methods was accomplished most reliably, but those productions correlated, in effect, not at all with productions of 64-second intervals except when counting was employed (U.S., adolescent males; Spivack & Levine, 1964).

3. The relation between the arousal of the temporal motive and the interval. The reliability of judging intervals of 75 and 90 seconds was high (.89 and .69, respectively) when one interval immediately followed the other but low (in the .40s) when a week intervened (U.S., unspecified; verbal, uncertain; Harton, 1939c). Similarly, American college students were asked to estimate the duration of the four parts of an intelligence test immediately after being halted at the end of the recommended time for each part. Correlations between their successive judgments were high and significant, but this consistency was greater for consecutive than for nonconsecutive estimates (Grossman & Hallenbeck, 1965). In yet another study there were relatively high correlations between verbal judgments of the duration of 4-, 14-, and 26-seconds intervals, although they in turn were not significantly related to judgments of other intervals, which lasted 12 minutes or more than an hour. The latter were, of course, much longer than the former, but in addition the temporal motive was aroused at different times: for the shorter intervals, the subjects were asked prior to experiencing the intervals; for the longer

ones, they did not realize that they would be required to pass judgment on how long they would have to wait until the experiment continued or on how long the experiment itself had lasted (U.S., college; Kurz, Cohen, & Starzynski, 1965).

4. The momentary state of the person passing judgment. This factor is more fully treated in *Addenda 8.6* and *8.7* which consider the effect of personality traits on temporal judgment. In passing, a single illustration: the Method of Reproduction by tapping was found to be more reliable than either Production by counting or Verbal Estimation when judgments were passed concerning various intervals ranging from 1 second to 19 minutes, but its reliability was seriously impaired when the subjects were made anxious by knowing that they had to speak in public (England, physicians; prior; Cohen & Mezey, 1961).

ADDENDUM 8.4 INTELLIGENCE

A. *Temporal knowledge.* Evidence supporting the generalization of a positive relation between temporal knowledge and intelligence is reasonably impressive. The list below indicates the temporal items measured in the particular study and, if need be, any variation in the mode of appraising intelligence; all come from the United States:

1. Temporal information, the ability to order historical events in correct verbal sequence and the ability to order events of one's own life in correct graphic sequence (children; Friedman, 1944).

2. The possession of correct temporal information and, as a basis for such information, the use of cues which were related less to other persons and more to sequences or records in the environment. This study was based upon small groups of children whose ages varied from slightly under five to slightly over seven and whose mean IQ was exceptionally high, viz., over 145 (Farrell, 1953).

3. Temporal knowledge measured by a test consisting of items requiring, in effect, the ordering of historical events, with special emphasis upon artifacts. The correlations of mental age with separate portions of this test were slightly higher than they were with either a reading comprehension or a history-achievement test (junior college students & other young adults; Pistor, 1939).

4. Knowledge of common temporal terms. The correlation was higher with mental age than with chronological age; and the relation to intelligence-test scores became more impressive, with school grade held constant, as the grade level increased from kindergarten to the third grade (Harrison, 1934).

5. Performance on a "Time Appreciation Test" which consists of questions pertaining to immediate orientation ("What day of the week is it?"), holidays ("In what month is Christmas?"), definition of temporal terms ("What is a decade?"), etc. (U.S., high school students, mentally deficient or epileptic adults; Buck, 1946). Among Negro college students the same test was found to be more highly correlated with intelligence than with academic grades, .81 vs. .63 (Johnson, 1964).

On the other hand, a measure of the salience of temporal information inferred from TAT protocols was *negatively* related to the IQ of girls (at least it approached the level of statistical significance), but it was unrelated to the IQ of boys; however, the mean of these two samples of middle-class Americans was superior and the range was restricted (Zern, 1970).

B. *Accuracy of temporal judgment.* Intelligence-test scores have been correlated positively with the accurate production of:

1. A 10-second interval, and also with the accuracy of a verbal estimate concerning a 60-second interval (W. Germany, soldiers; prior; Spreen, 1963).

2. A 14-second interval by means of the

differential reinforcement technique (Belgium, adult psychotics; prior; Denys & Richelle, 1965).

3. A 10-second interval. The correlation was slightly higher when the test of mental ability was verbal rather than non-verbal (.37 vs. .25); the temporal ability emerged from a factor analysis as one of the components of a factor designated as "central integrative field" or "organized experience," but was not related to the other factors which involved abstraction, power, and the direction of input and output (U.S., adult head-injury patients; Halstead, 1947, pp. 38–90).

4. A 30-second interval on the first trial but not after feedback on the second trial by a sample of "normal" American boys; but the reverse result was obtained for a sample of "emotionally disturbed" boys (Davids, 1969).

Similarly a group of retardates tended on the whole to underestimate intervals with a duration of from 1 to 29 seconds significantly more than a group of normal adults (U.S.; repro., prior; McNutt & Melvin, 1968).

Intelligence measured on a test was also related to the ability to judge the time of day when confined to a shelter for seven days; perhaps it helped determine how well the subjects "could organize and relate events to the passage of time" (U.S., children; Crawford & Thor, 1967). Intelligence measured by counting the number of movement responses (M) on a modification of the conventional Rorschach Test was positively related to the clock time consumed in producing intervals of 15, 30, and 60 seconds (U.S., emotionally disturbed adolescents; prior; Spivack, Levine, & Springle, 1959).

No significant relation between intelligence and judgments of duration has also emerged in some studies. In a simulated fallout shelter very similar to one mentioned in the preceding paragraph, the accuracy with which the time of day was judged was not affected by IQ (U.S., children & adults; verbal; Thor & Crawford, 1964). Likewise the findings were negative when intelligence-test scores were correlated with judgments of 60 seconds produced by counting (England, adults; Bell & Watts, 1966); the vicarious production of 1-second intervals, at least after practice or reinforcement (U.S., children; prior; Smythe & Goldstone, 1957); estimating the duration of 5- and 20-second intervals as well as ones ranging from 2 to 120 seconds (Israel, students; verbal, prior; Siegman, 1962a & 1966); the production of 85 seconds by reading silently, though there was a small but significant correlation with future orientation (U.S., college; Zurcher et al., 1967); and punctuality in situations on an American college campus (Dudycha, 1936, pp. 31, 42). For a group of delinquent and nondelinquents there was a microscopic relation (.11) between verbal intelligence and the "future age" at which the informants expected events to occur in their personal lives (U.S.; Stein et al., 1968).

C. *Orientation.* This aspect of temporal behavior, it has been ascertained, may or may not be related to intelligence-test scores. On the one hand, those scores have been found to be correlated positively with inclusion of references to the past in projectively produced stories, although there was no relation with reference to the future (U.S., high school students; Kastenbaum, 1965) and with future temporal orientation as measured by a questionnaire (U.S., emotionally disturbed boys and adolescents; Levine et al., 1959). On the other hand, no significant relation has been established between intelligence and:

1. Future orientation of a group of alcoholics and matched controls; yet there was a slight but significant negative correlation between intelligence and orientation toward the past for the former and an almost significant one for the latter (U.S., Roos & Albers, 1965a).

2. Future orientation as measured by the time span of TAT stories; in this instance the intelligence level of the small group was exceptionally high, and that span did have a significant and positive relation to grade averages in colleges. In fact, with the exception of

freshman year, the correlation between TAT span and the averages was higher than that between the scores on the Scholastic Aptitude Test (the measure of intelligence) and the averages (U.S.; Epley & Ricks, 1963).

3. Future and past orientation as measured by the tenses of verbs used to complete incomplete sentences (U.S., psychiatric patients; Krauss & Ruiz, 1967).

4. Temporal orientation as measured by a paper-and-pencil test in a group of educable mental retardates with a mean IQ of 65 and a group of college students of similar ages with an IQ of 100. An overall comparison, however, revealed that the retardates placed less emphasis upon the future and more upon the past; in addition, for them the past was more "negatively toned" and their "pleasant experiences" seemed to a greater degree to be in the present (U.S.; Roos & Albers, 1965b).

5. Ability to delay gratification as measured by asking normal and "emotionally disturbed" American boys how they would spend a small and a large sum of money (Davids, 1969).

ADDENDUM 8.5 TRAITS AND JUDGMENTS

A. *Extroversion-Introversion.* Although there are wisps of evidence suggesting that the equivalent of extroverts may judge intervals more or less accurately than introverts (Japan, high school; presumably verbal, prior; Usizima, 1935) (Germany, university; comp., prior; Schneevoigt, 1934), one study more convincingly suggests no differences between groups possessing these traits in pronounced form (England, students; verbal, immediate subsequent; Orme, 1962a). Extroversion was significantly, though only very slightly, positively correlated with magnitude and/or accuracy of temporal judgment as measured by the Method of Reproduction executed through a linear movement of the hand or by simply terminating variously filled interval; but the relation disappeared when the first mode of reproduction was accomplished by pressing a key or when the Method of Verbal Estimation was utilized (South Africa, European students; prior; Du Preez, 1964 & 1967a).

Next is the question of over- and underestimation, regardless of accuracy. Judgments of extroverted neurotics were shorter than those of introverted neurotics, significantly so for intervals of 5 and 10 seconds, not significantly so, but in the same direction, for intervals of 20 to 30 seconds (England; repro., prior; Eysenck, 1959). These results have been replicated with normal British subjects who were also instructed to use the Method of Reproduction (Claridge, 1960; Lynn, 1961). On the other hand, the judgments of extroverts and introverts did not differ with respect to a 15-minute interval under normal conditions, but the former produced intervals significantly longer after sensory deprivation (England, presumably adults; prior; Reed & Kenna, 1964).

One study reinforces the oft-reiterated point concerning the multivariate nature of temporal judgment. With appropriate reinforcement subjects high in neuroticism and introversion (both measured by paper-and-pencil tests) tended to show less improvement in producing a 1-second interval than those high in neuroticism and extroversion or those low in neuroticism whether introverted or extroverted. Accuracy among the extroverts significantly decreased and among the introverts slightly increased (especially among those also high in neuroticism) under the influence of a placebo pill which apparently was given without any suggestion concerning an alleged specific effect. In this situation the subjects had to restrain their impulse to cease squeezing an ergograph trigger with which they indicated the duration of the interval; it may be that this ability was generally stronger among non-neurotics and extroverts and that it was impaired by the diffuse suggestion conveyed by the placebo among extroverts but not among introverts (U.S., college; prior; Luoto, 1964).

B. *Anxiety.* The one trait related to abnormality which has been most frequently investigated is anxiety. By anxiety is meant in this paragraph not the kind which is experimentally manipulated but the trait as measured by a paper-and-pencil test. A positive relation has been found between that trait and (a) overestimating intervals having a duration of 5 and 20 seconds (Israel, students; verbal, prior; Siegman, 1962a), (b) underestimating those having a duration of 180 and 300 seconds but not ones of 15 and 90 seconds (U.S., technicians & enlisted men; prod., prior; Burns & Gifford, 1961), (c) inaccuracy in judging intervals of 15 and 90 seconds but nonsignificantly for those with a duration of 30 seconds (U.S., hospital patients; prod., prior; Whyman & Moos, 1967), (d) accuracy in judging 20-second but not 5-second intervals, whether counting or not (Canada, student nurses; verbal, prior; Hare, 1962a). No relation, moreover, has been adduced between the same trait similarly measured and (a) judgments concerning intervals having a duration of 10 and 60 seconds (West Germany, soldiers; prod. & verbal with prior; Spreen, 1963) or (b) verbal estimates of a 36-second interval, the production of a 22-second interval, and tapping rate (U.S., college; Cahoon, 1969). Among psychiatric patients there was a curvilinear relation between paper-and-pencil anxiety and temporal perspective—as anxiety level increased, the use of present future tenses first increased and then decreased. In addition, personal items on the orientation test produced more present and future tenses than impersonal ones (U.S.; Ruiz & Krauss, 1968).

The experimental production of anxiety has also led to conflicting results. No relation has been established between temporal judgments and (a) stress as produced by the experimenter's attitude and his reference to "threatening apparatus" or to the importance of doing well (Spreen, op. cit.); (b) real and acknowledged anxiety among British physicians who anticipated that they were about to deliver a speech in public, but subjectively some of them felt that "time seemed to pass faster than normally" and others had the reverse impression (prod., repro., & verbal with prior; Cohen & Mezey, 1961); (c) expressed fear of dark places while passing judgment in a dark room, in the presence or the absence of an experimenter of the same sex, or in a lighted room (U.S., college; presumably verbal and prior; Geer, Platt, & Singer, 1964). In a study cited in the previous paragraph—no difference was found between paper-and-pencil anxiety and temporal judgments (Cahoon, 1969)—a similar nonsignificant result appeared when judgments were obtained concerning intervals in which electric shocks were and were not anticipated, in spite of marked changes in physiological processes; but some significant differences emerged when these two groups were subdivided on the basis of changes in heart and respiratory rates. In contrast, anticipating an electric shock increased the overestimation of 5- and 20-second intervals in comparison with the neutral or control condition (Canada, adults; verbal, prior; Hare, 1963b); and those most anxious while isolated in a soundproof room tended to overestimate intervals with a duration from 1 second to 2 hours (U.S., adults; prod., prior; Mitchell, 1962).

C. *Most miscellaneous.* Here the individualism of the Western world asserts itself: investigators have sought to relate temporal judgments to a heterogeneous selection of traits. My coverage is deliberately incomplete, I quickly present not necessarily representative but symptomatic illustrations.

Overestimating temporal intervals has been positively associated—however slightly—with the following:

1. An "unconscious" concept of the parent of the same sex or of both parents as dominant, leading—it is presumed—to a tendency to overvalue and to overestimate time because its utilization by the child has been previously curbed:

measured by responses to some of the Rorschach plates and a special TAT (U.S., college; prior & subsequent, verbal; Fisher & Fisher, 1953).

2. Anality in the Freudian sense, leading —it is again presumed—to conserving and valuing time: measured by responses to 10 paper-and-pencil items (e.g., "I believe in being thrifty"); the subjects judged the duration of the interval during which they had filled out a longer questionnaire containing the "anality scale" (U.S., students; verbal, subsequent; Campos, 1966).

3. (a) Tolerance of pain as measured by reaction to extreme heat upon the skin in the laboratory and by ratings given slightly ill patients by physicians and nurses and (b) intolerance of sensory deprivation and monotony (U.S., adults; repro., prior; Petrie, Collins, & Solomon, 1960).

4. Dissociation as measured on a paper-and-pencil test (England, various types of abnormal patients and controls; Orme, 1964).

5. Less unpleasant or less disturbed psychiatric states. Patients were examined as they were admitted by the Emergency Department of a psychiatric clinic and then an average of nine days later in the clinic; at the outset they tended to be much more disturbed and agitated than after receiving treatment. Among other tests, they were asked to produce an interval of 30 seconds by telling the examiner when that interval had elapsed after he said "Go." From the initial measure to the later one, the degree of underestimation decreased—that is, judgments increased—and accuracy increased (U.S., adults, Melges & Fougerousse, 1966).

6. The ability to renounce and tolerate delay as shown by behavior involving "good citizenship" when the reward therefor was to be postponed. The emotionally disturbed boys in this study had been punished for misconduct by being deprived of weekend privileges for one week: in contrast with those whose conduct had been bad during this period, those who had been good tended—almost significantly—to overestimate intervals of 20, 40, and 60 seconds to a greater degree (U.S.; verbal, prior; Levine & Spivack, 1959).

7. Low impulsivity-test scores on a paper-and-pencil test; the overestimated interval was one of 8 minutes in which U.S. Navy personnel were seated idly either in a "lighted but small and completely soundproofed room" or alone in a room of their barracks; the effect was more pronounced among those with high than among those with low intelligence (prod., prior; Kipnis, 1968).

8. (Very slightly) Not possessing well organized perceptions and cognitions (as inferred, respectively, from an analysis of Rorschach protocols and free associations (U.S., children; repro., prior; Kahn, 1966).

9. Relative independence of external stimuli in the processes of perception and cognition; the tendency appeared in connection with intervals of 9, 17, and 26 seconds but not 46 seconds (U.S., presumably college; verbal, prior; Friel, 1969).

The *accuracy* with which intervals were judged has been found to be associated with:

1. (1 second) Friendship and prestige status, including being chosen and especially being rejected as measured by sociometric ratings of classmates. There seemed to be an association between two similar kinds of social sensitivity: one involved this "consensual, social unit of time" which requires "the ability to calibrate accurately subjective time with social units of temporal magnitude" and the other one's peers whose judgments depend upon the person's "ability to calibrate private experience with social standards" (U.S., medical; limits-verbal, prior; Goldstone et al., 1963).

2. (9, 17, 26, and 46 seconds) "Field-dependent" rather than "field-independent" as measured by a series of perceptual and cognitive tests standardized by Witkin (U.S., presumably college; verbal, prior; Friel, 1969).

3. (15, 30, and 60 seconds) "Emotional surgency" as measured by Rorschach reactions as well as by ratings concerning

the amount of energy displayed in the wards (U.S., schizophrenics; prod., prior; Singer, Wilensky, & McCraven, 1956).

4. (30 seconds) A measure of motor inhibition before and after being corrected on a first trial; for a measure of motor manipulation the association appeared only on the second trial. These results were obtained from a small group of "normal" American boys. For a comparable group of "emotionally disturbed" boys slightly different findings emerged: slight associations between first-trial judgments as well as improvement on the second trial and motor manipulation; and between improvement and motor inhibition. In neither group was there a relation between the accuracy with which the interval was produced and a proclivity toward delayed gratification inferred from the replies to questions concerning how a relatively small and a relatively large sum of money might be expended (Davids, 1969).

5. (2 hours) A high need for order in general as measured on a paper-and-pencil test (U.S., Air Force; prod., prior; Mitchell, 1962).

In the most-miscellaneous category have been *positive* associations:

1. Between being "non-integrated" and being dependent upon a "head clock" to awaken at a fixed time; in contrast, the "well integrated" were more likely to depend upon external cues. These associations are based upon reports by the persons themselves concerning their ability to awaken and upon the investigator's casual observations and interviewing of persons who patronized a crystal-gazer: those who claimed to see messages in the ball (i.e., they tended to be dissociated) also were more "certain and exact" concerning their alleged ability to awaken at a predetermined time (Germany; Clauser, 1954; pp. 78–9).

2. Between scores on a "Time Scale" (measuring the value placed upon time through items such as "time is money") and two so-called anality scales. Some of the items on the latter skirted the topic (e.g., "You accumulate a great many things because you rarely throw anything away") and others hit it directly (e.g., "I use more toilet paper than is necessary"). Also supporting a Freudian contention concerning time was a negative correlation between the time scale and one measuring impulsivity (U.S., college; Pettit, 1969).

3. Between the span of future and past temporal orientation and the tendency to believe on a paper-and-pencil test in the effectiveness of internal rather than external control of one's destiny. The subjects ascribing destiny to themselves rather than external forces, moreover, tended to be less anxious and better adjusted, again as measured by paper-and-pencil. But the relation between this trait and orientation existed only for some temporal scales (e.g., ones involving personal future and impersonal past) and not for others (e.g., impersonal future, the Time Metaphor Test) (U.S., college; Platt & Eisenman, 1968).

Finally, mirabile dictu, *no relation* has been found between the accuracy of temporal judgment and (a) a variety of "non-clinical personality characteristics" measured by a standard paper-and-pencil test (England, adult & university; prod. & verbal with prior; Bell & Watts, 1966); (b) dependence upon the external environment as measured by two standard laboratory tests (South Africa, European students; reproduction, prior; Du Preez, 1967a); (c) four factors extracted from a battery of tests related to "delaying capacity," such as the M responses on the Rorschach, a motor inhibition test, a measure of frustration, etc. (Singer, Wilensky, & McCraven, 1956); (d) dogmatism measured on a conventional paper-and-pencil test (U.S., college; prod., prior; Zurcher et al., 1967); (e) "Psychosexual role identifications" as measured by a paper-and-pencil femininity scale as well as by the onset of the menstrual cycle (U.S., college; verbal, prior; Baldwin, Thor, & Wright, 1966).

ADDENDUM 8.6 TRAITS AND ORIENTATION

An obvious basic problem confronts the investigator who would relate temporal orientation to personality traits: how shall that orientation be measured? In general, two methods have been employed. The first is direct: the individual is asked to list important events in his life and he thus indicates how far backwards or forwards he reaches. The second is projective: the person is faced with the task of completing sentences or telling a story, and again the time span is analyzed. One group of investigators has shown that, in a sample of American college students, methods such as these yield inconsistent results: subjects were measured in five different ways and none of the interrelations reached a level of significance, from which the conclusion was drawn that temporal orientation is not "unifactorial" (Ruiz, Reivich, & Krauss, 1967). Likewise there were only low or non-significant correlations among three ways of measuring temporal orientation of American children, that of listing future events, completing incomplete sentences, and completing stories (Lessing, 1968). It is thus extremely difficult to compare studies based upon different methods of measurement.

One explanation for the failure to obtain higher associations in the studies just cited may be a function not of the methods but of the persons who may not be consistent in their orientation. For surely, as suggested in Principle 3.1.a, each of us must at some time look in all directions, and so the question of consistency means, I suppose, the modal orientation or perspective, not the exclusive one. A factor analysis produced a cluster involving future orientation in three senses, viz., extensity, organization, and content, from which result the author argues that "there is a relatively generalized tendency to be concerned with future possibilities" (U.S., high school; Kastenbaum, 1961). Similarly American undergraduates revealed a fairly consistent political orientation—"backward," "status quo," or "forward"—regarding eleven public issues such as "the power and influence of labor" and "the decline of the double standard of morality" (Anast, 1965).

Conflicting results have been obtained when the attempt has been made to find an association between orientation and anxiety. On the one hand a future orientation was positively related to the presence of anxiety when the former was ascertained by examining the tenses of stories (U.S., college; Rokeach & Bonier, 1960); on the other hand, exactly the opposite relation appeared when orientation was measured in the same way and anxiety by means of questionnaires, projective techniques, and observations in a stress situation (U.S., college; Epley & Ricks, 1963). And then, as if to increase our confusion, a sex difference appeared when the orientation of American undergraduates was inferred from an inventory of the important events in the past and future which they listed: as paper-and-pencil anxiety among the males increased, future orientation at first constricted, then expanded; with an increase in anxiety among the females, future orientation progressively constricted and past orientation expanded (Albers, 1966).

One team of investigators landed on both sides of the fence. They report a positive association among policemen who, during an interview as part of an evaluation procedure for promotion, responded to the simple question, "Tell me about yourself": anxiety was gauged from speech interruptions (e.g., stuttering, incoherent intrusions) and orientation from the use of the present tense or a combination of the present and the future (U.S.; Krauss et al., 1967). Then they report a negative association for psychiatric patients in a hospital: anxiety was measured in two ways on a standardized paper-and-pencil test and orientation by noting the tenses employed in completing incomplete sentences (Krauss & Ruiz, 1977).

Otherwise it is not easy to summarize research because usually the experiment or investigation has been reported, perforce, with a series of *if*'s and *but*'s. On the whole, however, *positive* associations have been noted between *future orientation* and the following—the method of measuring orientation is noted in parentheses; all the studies except one were conducted in America:

Traits

A. (Tenses of TAT stories) Dogmatism measured by a paper-and-pencil test. The dogmatic and the undogmatic, however, did not differ with respect to future orientation, time span, or degree of extension into the future as measured by the subjects' own ratings concerning the same stories; but more of those with undogmatic, open minds used the present instead of the future tense than those with so-called closed minds (Rokeach & Bonier, op. cit.). The first part of the study regarding the positive relation between future orientation and dogmatism has been subsequently confirmed (college; Zurcher et al., 1967), as has the second part regarding the absence of a relation between extension and dogmatism (high school; Jacoby, 1969).

B. (The time span of events which the individual felt might occur in his future) An expressed interest in activities involving ideas (e.g., collecting books or writing short stories) rather than those involving motors or motion in the literal and extended sense (e.g., operating machinery, playing or watching football). Interest in the past, when similarly measured, however, was not related to this dimension (schizophrenic & neurotic adults; Stein & Craik, 1965).

C. (The time span of events mentioned by the subject referring to the past and the future) Relative independence of the stimulus field in contrast with dependence thereon; the measures of orientation included both extension and coherence, and the independent subjects also made more and better organized references to the past (U.S.; Friel, 1969).

D. (The time span of TAT stories) Higher grade average in college (presumably related to achievement motivation since the group had a uniformly high intelligence) and greater sympathy and involvement with other persons (measured by self-descriptions). Clinical observations and other tests revealed that these American students who had long time spans, thus measured, regarding the future, in comparison with those oriented toward the past, "recalled less sibling conflict, fewer childhood fears, and less solicitousness and anxiety in their parents"; they were "exceptionally 'hard-headed' "; they "emphasized the excitement of change through time"; they were "overly intellectualized, inclined to use fantasy *only* as rehearsal for future action, and insulated too well from immediate emotional experience." On the other hand, a "retrospective span was shown to relate to narcissism, sensitive imaginativeness, and openness to experience" (Epley & Ricks, op. cit.). The investigators themselves emphasize that these far-reaching conclusions and the indicated syndrome were derived only from a very superior group of 17 undergraduates.

E. ("Density" and "identity" regarding the future on a paper-and-pencil test) Length of the time span of stories that were completed and the coherence with which temporal events could be arranged; but the correlations, though significant, were relatively low and in one instance—that between "density" and a paper-and-pencil measure of rigidity—significantly negative (Kastenbaum, 1961, op. cit.).

F. (Content of produced stories) Cognitive organization as measured by the content of words used in free association; but the relation, though significant, was low (.27), and there was no relation between orientation thus measured and perceptual organization inferred from Rorschach responses (children; repro., prior; Kahn, 1966).

G. (The tenses in stories) The tendency to produce movement or M responses on the Rorschach; the relation, though significant, was low (a tau of .28) (children; Kahn, 1967).

H. (Completions of incomplete sentences) "Better personal, social, and total adjustment," "greater sense of responsibility" for one's own destiny, "greater willingness to defer gratification" as measured by standard personality tests, and, when there was a relation—often there was not—"the more favorable psychosocial attributes (e.g., higher intelligence, higher academic achievement, higher socioeconomic status, and healthier personality scores)". But these relations did not appear when other methods of measuring orientation were employed; and they were not particularly stable over time (Lessing, 1968).

Attitude and Interests

A. (Favoring probable future public policies rather than those in the present or past) Optimism concerning the possibility of not having another major war and of people's happiness; satisfaction with the mass media of communication. But there was no relation between this measure and either dogmatism or social values such as approval and ambition (Anast, op. cit.).

B. (A list of future events describing the self) Possessing the interest pattern of high-school counselors; but the relation did not exist between that pattern and either a similar list for the past or the time span of spontaneously composed stories, nor was any one of those three temporal measures related to potential success and ability to be counselors as rated by staff members (counseling trainees; Sattler, 1964 & 1967).

C. (The range in years covered between the present and the anticipated occurrence of events in the future) A preference for metaphors describing time as swift and dynamic rather than the reverse (Israel, students; Siegman, 1962b).

Temporal Judgments

A. (Paper-and-pencil regarding "future time extension," "future time coherence," and "future time valence") The tendency to overestimate intervals ranging from 0.1 to 30 seconds. The relations, though significant and positive, were very low and did not appear when "future time density" and "future time directionality," also ascertained by paper-and-pencil tests, were the measures of orientation (college; verbal, prior; Zelkind & Spilka, 1965).

B. (Paper-and-pencil test) The tendency to overestimate an interval of 12 seconds with the Method of Verbal Estimation and to underestimate one of 20 seconds with the Method of Production. But this tendency occurred only among male undergraduates and not among females; and it did not appear when the estimated interval was 120 seconds and the produced one was 200 seconds (prior; Geiwitz, 1965).

In reporting the above results, exceptions have already been indicated. In addition, one study indicates a negative association between temporal orientation as measured by noting the present and future tenses employed in completing incomplete sentences on the one hand and social introversion and schizophrenic tendencies as inferred from responses to a personality test—the MMPI—on the other hand (Krauss & Ruiz, op. cit.). No relation at all was found between future orientation as measured by the time span between the present and events anticipated in the future and (a) "impulse control" ascertained by noting the slowness with which a circle was traced (Israel, delinquents & Army recruits; Siegman, 1961) and (b) the production of a 64-second interval (U.S., adolescent males; prior; Spivak & Levine, 1964). Similarly negative results appeared when the attempt was made to find an association between the number of dreams graduate students noted in a log book and their future orientation measured by having them imagine events in the distant future and predicting the year of their occurrence (U.S.; Schonbar, 1965).

In the search for correlates between orientation and behavior we might accept the view that "a lack of future time perspective in one's life" is related to the popular syndrome called homesickness. The phenomenon has been investigated among American college students, but what do we find? First, the investigators propounding the theory established such a relation for two but not for three of their paper-and-pencil measures of temporal orientation; and they also report a relation between homesickness and the subjects' own view of the discrepancy between their conception of their present and future selves (Platt & Taylor, 1967). An earlier study established a relation between homesickness or nostalgia and both emotional instability and a tendency to substitute daydreaming for reality. But there were complications: introversion measured with one scale but not with another was similarly related, as were self-consciousness, feelings of inferiority, and lack of self-sufficiency among males but not among females (McCann, 1943). Among freshmen girls there was an association between homesickness and insecurity and the failure to attain a satisfactory adjustment (Rose, 1947). An objective review of the literature, however, suggests that "at one time or another, almost every symptom, both physical and emotional, has been attributed to nostalgia" (Nawas & Platt, 1965). The authors just cited advanced the theory, without supporting data, that "individuals who emphasize futurity in their time perspectives, or others who are characterized as optimistic, goal-oriented, and planful are not likely to fall victims to nostalgia." Perhaps.

ADDENDUM 8.7 ATTITUDE TOWARD TIME

The research problem is to find a way to measure the attitude toward time. In one study note was made of the ability of American children to remain in one standing or sitting position selected by themselves before signaling that they wished to change that position; and it was found that this ability was positively related to a measure of imagination (derived from an interview containing four questions pertaining to "your favorite game," games played alone "by yourself," having "pictures in your head," and having "a make-believe playmate"), but it was not related to future orientation (inferred projectively from the child's responses when asked what he would desire if a "Magic Man" were to grant his fondest wish) (Singer, 1961). On a purely a priori basis, however, it may be surmised that this kind of patience or tolerance involves more than an attitude toward time.

Similarly, punctuality is another related index of an attitude toward time but it is an equivocal one. Arguing from clinical experience, Alfred Adler suggested that neurotics may often demand punctuality in others as an act of aggression and may themselves be tardy in order to enhance their own egos (1916, pp. 361–2). But nonneurotics may make the same demand for the sake of efficiency and may themselves be punctual. At any rate, in one empirical study, punctuality in various campus situations, whether viewed situation by situation or as a general trait, was found to be related very slightly to a tendency to be less "neurotic" and more "self-sufficient" as measured on a paper-and-pencil test; it was unrelated to the traits of introversion and dominance similarly measured or to the ability to predict performance time on laboratory tests. The attitude toward punctuality, which tended to be positive, also was not associated with actual punctuality which was spotty (U.S.; Dudycha, 1936, pp. 31, 38, 42–3).

A direct attack upon the problem has been made by means of what is called the Time Metaphor Test. Subjects rate 25 phrases with respect to their appropriateness as descriptions of time: the greater the number of dynamic metaphors (e.g., "a galloping horseman") and the smaller the number of static ones (e.g., "a quiet, motionless ocean") that are selected, the more time is postulated to be highly valued. So measured, placing a *low* value on time—to view it "passively"— has been found to be positively related to the ability to tolerate delayed gratification as inferred from the number of M (human movement) responses on a group-administered, multiple-choice variant of the Rorschach test (U.S., female college; Kurz, 1963) and to introversion and a concern for "moral" and "rational" discipline in hypothetical situations involving deprivation (U.S., college; Knapp & Lapuc, 1965). Of course these conclusions must be challenged. Aside from the usual charge that they come from atypical samples of human beings, the validity of the Time Metaphor Test as a measure of attitude toward time and of the other measure to which it is related has not been firmly established.

First, the validity of the Test itself. One substantiating bit of evidence comes from an American investigation: a group of older persons (mean age: 71) who were not living in institutions preferred the swift rather than the static metaphors with greater frequency than a group of college students matched roughly with respect to education and intelligence (Wallach & Green, 1961). The investigator assumes that older persons value time more highly, probably a reasonable assumption. Still, in the study relating Test scores and M responses (Kurz, 1963), I have great difficulty comprehending some of the results. A significant negative correlation between those M responses and "wind-driven sand" allegedly substantiates the conclusion; that metaphor, however, does not suggest passivity to me because I think of the very active wind and not the inert sand. Also "a devouring monster," "a large revolving wheel," and "a dashing waterfall" seem mighty swift to me, but the first two have straight zero correlations with the M responses and the third an insignificant positive one. "A massive glacier," strikes me as being the quintessence of slowness but in fact receives another zero. And yet, if I may argue against myself, the devil's critic, there is the possibility that the M responses have some kind of consistent relation to attitude toward time which the Metaphor Test does not reveal; and indeed a factor analysis of the metaphors in the same study does suggest a fairly clear-cut relation with two of the emerging factors called Dynamic-Hasty and Naturalistic-Passive clusters but not with a third, a so-called Humanist cluster.

In addition, another study has revealed a relation between (a) the percentage of M responses and (b) judgments concerning the duration of intervals of 14 and 26 seconds with prior arousal of the temporal motive and of 12 minutes and more than an hour with immediate subsequent arousal: those with many M responses tended to give briefer estimations than those with few M responses. This relation can be interpreted only by noting that the reverse was true for color responses (Sum C) which are thought to indicate lack of impulse control. But even if it were thoroughly established for all time, in the words of these investigators, that "the person who has little M and high Sum C cannot tolerate delays," their own investigation raises serious problems: (1) the relation with M responses did not hold for a 4-second interval; (2) and there was no relation between temporal judgments and movement responses involving animals or inanimate objects (U.S., college; Kurz, Cohen, & Starzynski, 1965). Extremely discouraging in my opinion, moreover, is a study conducted in Tanzania: samples of students from an achievement-oriented society, the Chagga, and from one not so oriented, the Bondei, gave similar results on the Test, in fact the latter tended to prefer not the slow but the fast metaphors (Ostheimer, 1969).

ADDENDUM 8.8 NEED FOR ACHIEVEMENT

The relationships can best be summarized by placing the studies under categories based upon the way in which the need for achievement has been measured:

TAT Protocols (as described in the text)

A. Positive correlations between a high need and:

1. Future orientation inferred from the use of anticipatory tenses (i.e., the future tense itself or any implication of anticipation or expectation) in essays on "What I would Ideally Like to Get Out of an Elementary Course in Psychology" (U.S., college; Zatzkis, 1949).

2. Breadth of orientation inferred from the time span of the same stories which had been written in response to the TAT drawings (U.S., business executives; McClelland, 1961, p. 327) (Germany, students, Heckhausen, 1960). But, as the former investigator himself suggests, (p. 327), these findings are not convincing because the need-for-achievement itself as measured on the TAT is partially scored by taking the time span into account.

3. High evaluaton of time inferred from a preference for swift metaphors in the Time Metaphor Test described in *Addendum 8.7;* but those low in achievement preferred most of all the metaphor of "a devouring monster" which, as the authors admit and as I have already suggested in that previous Addendum, does not readily convey the notion of "very slow motion" (U.S., college; Knapp & Garbutt, 1958).

4. A tendency to push time ahead inferred from their own watches which perchance were incorrectly set: those watches whose owners were high in n achievement were running fast rather than slow, and vice versa (presumably U.S., Cortés, 1961).

5. A tendency not to be distracted by the content of an interval as inferred from the fact that those high in need achievement overestimated intervals filled with Strauss's *Blue Danube Waltz* "played rather loudly," in comparison with empty intervals, to a lesser degree than those low in the same need; they thus paid closer attention to time as such, or else their allegedly stronger egos enabled them to resist the appeal of music (U.S., college; prod., prior; Knapp & Green, 1961).

6. Greater interest in the future, as inferred from superior recall 24 hours later, of tasks that were to be reported in 8 weeks rather than on the following day (cited by McClelland, 1961, p. 327).

B. No significant correlation between a high need and:

1. Reactions to temporal concepts as measured by the semantic differential, a device for ascertaining precisely, though artifically, the connotations of words (U.S., college; Knapp & Garbutt, 1965).

2. The Time Metaphor Test among Chagga students. In passing I must register the fact that the TAT itself, although it failed to distinguish between Chagga students from high- and low-achieving families (as inferred from the number of bags of coffee produced by those families), had some cross-cultural validity: the Chagga students had higher scores than the Bondei students, and the Chagga in general have responded more readily to modernization than the Bondei (Tanzania; Ostheimer, 1969).

3. Two items on a questionnaire pertaining to attitudes toward time which were tested cross-culturally. One, "Do you feel you waste time and spend it uselessly?", was not related to the need for achievement among samples of business men and professionals in the United States, Italy, Poland, and Turkey; and another involving irritation or annoyance when a watch stops or keeps incorrect time (the translation was not exact) was similarly unrelated to that need among samples of boys in Brazil, Germany, India, and Japan (McClelland, 1961, p. 332).

Scottish Tartan Test

A positive correlation for somber colors (blue, green) rather than brighter, nonascetic ones (red, yellow) and:

1. A greater awareness of the rapid passage of time as inferred from the tendencies (a) to underestimate the time it took for a moving point in a piece of apparatus to reach a fixed mark that was invisible (showing thus "the tendency to anticipate future events before they occur"—maybe yes, maybe no) and (b) to consider newsworthy events in the past to be relatively recent rather than remote (U.S., teachers; verbal, prior; Green & Knapp, 1959).

2. "A feeling of harassment with the passage of time together with an effort

to control and manage it," based upon a factor analysis of answers to a questionnaire pertaining to practices and attitudes toward time (U.S., college; Knapp, 1962).

No relation, however, was found between scores on the Tartan and the Time Metaphor Tests among the Chagga students of Tanzania (Ostheimer, op. cit.).

Academic Status

Academic status has been shown in the United States to have a positive relation with future temporal orientation as inferred from the time span of TAT stories (college; Epley & Ricks, 1963); from those protocols as well as recollections concerning the content of conversations (children; Teahan, 1958); and from the themes in spontaneously composed stories (children; Kahn, 1965) (college; Barabasz, 1970b). Similarly, when two groups of very intelligent high school boys were compared, those academically successful inhibited motor impulses while drawing a spiral, composed stories that tended to be future- rather than present-oriented, and claimed they would spend various sums of money in ways suggesting delayed gratification to a greater extent than those who were underachievers from an academic standpoint; but the two groups differed with respect to orientation on only one and not on the other two stories they composed, and their estimations of a 30-second interval were not significantly different (U.S.; prod., prior; Davids & Sidman, 1962).

ADDENDUM 8.9 YOUTH VS. AGE

A. *Ratio Theory of Judgments*

One bit of evidence is somewhat indirect. American children in Grades IV to VI were asked, among other questions, to indicate whether certain events had occurred "a long time ago" or "a short time ago." The numbers in parentheses represent the percentage replying "long" rather than "short": yesterday (1); last summer (44); last Christmas (52); the day I started kindergarten (84); Bible times (89); the time when George Washington lived (93); the time when the Pilgrims lived (96) (Friedman, 1944). We can only guess that many of the children thus judged, on this crudely qualitative level, recent events to be farther back in time than would adults.

Another study includes data on adults. British subjects were told that a 10-inch line represented the interval from birth to the present, and they were asked to indicate on that line "how long" it seemed to them that a specific event in the past had taken place. The results suggest that an event 24 hours ago was relatively more distant than one transpiring a week ago, but that one a month ago did not appear subjectively to be four times more distant than one a week ago. The judgments, when thrown into a mathematical curve, resembled the kind of proportional results obtained in psychophysical experiments devoted to the problem of discriminating between intervals of varying duration, at least in judging intervals up to 6 months; thereafter a more linear relation held. Thus 24 hours ago was placed by all age groups at roughly the same point on the line, but 1 year, 3 weeks, was placed at 7.3 inches by the group slightly over 8 years of age, at 4.3 by the group almost 11, at 3.0 by those in the 20-29 age range, and thereafter at approximately that point except for a slight rise in a small group of old persons over 60 (Cohen, Hansel, & Sylvester, 1954).

A variant of the pseudomathematical approach to an aspect of time involves a simple arithmetical comparison: ". . . when man locates himself in time, he attaches the greater importance to the longer period of his life, taking into account the average expectation of life, that is, the unlived portion when he is young and what he has already experienced when he is old" (Fraisse, 1963, p. 181). This interesting bit of speculation has no concrete, supporting evidence, though commonsense and anecdotes would suggest its validity.

B. Orientation

It is impossible adequately to document the popular view that youth is interested in the present and gradually changes its orientation toward the future; and that the aged in turn change from emphasizing the future and even the present and concentrate upon the past. American studies, using different methods (indicated in parentheses), provide only piecemeal or inconclusive evidence:

1. (Direct questions concerning the way in which they structured their memories, anticipations, and existences) High-school students seemed to be living "in an intense present"; their views on the remote future were less structured than their views of the past (Kastenbaum, 1959). The same tendency to limit oneself to the present and not to think too far ahead has been confirmed among samples of the middle-aged and the aged (Kastenbaum & Durkee, 1964a & 1964b).

2. (Temporal span of spontaneously composed stories) Samples of school children had the shortest orientation regarding the future; the span increased among adolescents, further increased among college students, dipped markedly among business men and less so (though the difference was not significant) among "senior citizens" (LeBlanc, 1969).

3. (Paper-and-pencil questionnaire) No relation was found between age and future orientation for a group of alcoholics and matched controls, both of which had a mean age between 45 and 50; but for the alcoholics there was a significant positive relation between age and an orientation toward the past (U.S., Roos & Albers, 1965a).

4. (Listing of thoughts and conversations allegedly taking place during the past week or two and then responses to interviews concerning the importance, feeling tone, and location in time of each item) Temporal orientation did not fluctuate significantly with age among small samples varying in age from 9 to 69:
all age groups tended to emphasize the future rather than the past, particularly the near rather than the far future; when references were made to the past, again it was the near rather than the far past which tended to be mentioned (Eson & Greenfield, 1962). Although the reliability of these findings, ascertained by interviewing a few of the informants a second time, was reasonably high, polite doubts must be entertained concerning the validity with which such events could be recalled.

5. (The question, "Do you have an unrealized ambition?") Of the 276 centenarians so questioned, 192 were able to reply, and of those about half said they did have and the other half said they did not have unsatisfied ambitions; and about one-half with ambition expected to realize them. More than verbal replies apparently must have been involved, for there turned out to be a statistically significant, positive relation between future orientation so measured and the ability to recall significant events from the past—and the group as a whole, while being questioned, tended to concentrate on memories from the remote rather than from the recent past (Costa & Kastenbaum, 1967). On the other hand, no difference in temporal orientation (ascertained through interviews and projective tests) was found, for two matched groups of American Jews with a mean age close to 80, between those who subsequently died within the year and those who lived at least three years longer, even though the groups differed with respect to other attitudes, such as fear of and preoccupation with death (Lieberman & Copland, 1970).

6. (Direct questions) Adolescents and young adults tended to limit the time span concerning their own future to the next decade or two: most of them thought no further than the age of 35, the highest figure reached was 46 (U.S.; Kastenbaum, 1964b).

Investigations have also isolated factors affecting the relation between age and temporal orientation:

Sex and social class. These two demographic factors affected the central tendency for increasing age, in a group of Austrian adolescents, to be accompanied by "a temporal reorganization" as inferred from the proportion of a line symbolically representing the time allocated by each person to his own birth and death and to the past, present, and future (Cottle & Pleck, 1969).

Personality traits. When young people in the Netherlands between the ages of 14 and 21 wrote essays on their anticipations concerning the future, there were shifts in the content of that future; with increasing age more attention was paid to the future in general rather than to their personal involvement therein as well as to their own fears and apprehensions (Mönks, 1968).

Role. A group of hospitalized patients over the age of 65, when questioned concerning their activities of "yesterday" and "tomorrow," assigned themselves an active role in past events and a more passive or fatalistic one concerning the future; they merely had less to say as the period of institutionalization increased (U.S.; Rosenfelt, Kastenbaum, & Slater, 1964).

Mode of living. Perspective was ascertained directly by asking questions about thoughts and conversations and indirectly by analyzing TAT stories. A group living in infirmaries was much more oriented toward the past and less toward the future than a comparable group living in their own communities. Among the latter those between 50 and 60 years of age were more concerned with the future than those between 61 and 76; but this statistically signficant difference did not appear for those living in institutions. For both groups, moreover, there was a clear-cut, positive relation between future orientation and a satisfying occupation or hobby (U.S.; Fink, 1957).

Impulse control. The hypothesis was tested that renunciation for the future increases internal tension and that older persons can be expected to exhibit less impulse control. A group of elderly Americans, with a mean age of 72, and a group of college students were asked to write a phrase, "New Jersey Chamber of Commerce," as slowly as possible; the number of human movement (M) responses on the Rorschach was recorded because it was believed that an increase in internal tension results in an increase in those responses. The younger group did inhibit writing tempo to a greater degree and they reported more M responses before writing the phrase, as well as a greater increase after writing it, than the older group (U.S.; Pollock & Kastenbaum, 1964).

One investigation, finally, has shown that, on a straight correlational basis, a favorable attitude toward the future, measured by ratings of interview material, among 100 middle-class Germans between the ages of 60 and 75 was significantly and positively associated with men rather than women; with good rather than poor health; with higher rather than lower socioeconomic status, especially among the women; with certain personality traits ("activity," "mood" among men, "general responsiviness," "general adjudgment" among men, "ego control" among younger men, and "feeling of security" among the older women); with intelligence measured on a standardized test; and with satisfaction concerning one's present role (Lehr, 1967).

C. *Attitudes toward Death*

Attitudes toward death, it has been found (Feifel, 1959, p. vii), are difficult to investigate in a Western country such as the United States because apparently death is a subject as taboo as sex once was and a man's income now is. We do know that both college students and older populations have varying attitudes toward death, but a comparison between the two age groups is not fruitful because so far the populations have not been equated in relevant demographic respects and because methods of collecting data have not been standardized (Dickstein & Blatt, 1966). Not surprising but interesting is a study in which American high school students gave their associations to various words under somewhat structured conditions: they seemed to place the word "death" in a frame of reference that was different from that in which the other words—good, real, life, bright, and myself—were all grouped (Kastenbaum, 1959). The undergraduates most fearful of death as determined by their replies to a paper-and-pencil questionnaire wrote stories with a slightly more restricted time range into the future than those showing less fear (Dickstein & Blatt, op. cit.). The unsteady relation between attitudes toward death and personality traits is shown in another American study: the young males scoring very high on a paper-and-pencil test with respect to "outer-oriented needs" tended to estimate their own life-expectancy above the actuarial one for their sex, but this tendency did not appear for those allegedly low in such needs or for young women, whether high or low in that respect (Tolor, 1967). If fear of death increases as death comes closer and if this last result were more generally replicated, then we might expect advancing age to be associated with less frequent arousal of the temporal motive and with congruent distortions regarding future time (or at least the estimated time of death's arrival) in the direction of overestimation.

D. *Temporal Judgments*

Evidence for the view that judgments concerning the duration of ephemeral and transitory intervals, even under laboratory conditions, become more accurate with increasing age and then perhaps decline can be found in the following studies:

1. Ability to judge the duration of a 20-second interval which was either empty or filled with various kinds of low-grade activity; boys between the ages of 9 and 14, and a group of undergraduate and graduate students ranging from 17 to 50: accuracy improved irregularly with age (U.S.; verbal, prior; Axel, 1924).

2. Ability to reproduce intervals ranging from 2 to 15 seconds, empty and with varied contents; children and adolescents between ages of 3 and 16: a general and pronounced tendency for accuracy to improve with age, although there were fluctuations at various age levels (France; prior; Wallon, Evart-Chmielniski, & Denjean-Raban, 1957).

3. Ability to estimate the time of day when confined to a simulated fallout shelter for almost two weeks; children and adults: those between the ages of 21 and 28 made more accurate estimates than the youngest group of school children (7–15) or the older group composed of housewives, the unemployed, the retired, and two factory workers. The young adults, it is pointed out, "were more accustomed to time demands," as a result of the time schedules associated with their normal occupations. Noteworthy is the fact that this age factor was the only significant ecological variable related to the temporal judgment (U.S.; Thor & Crawford, 1964).

4. Ability to reproduce a 5-second interval; really large numbers, viz., 120 children between the ages of 10 and 14, and 286 young adults between 20 and 30 (England; prior; Tejmar, 1962).

5. Ability to make accurate temporal judgments in a variety of situations, such as the reproduction of rhythms and estimating the duration of a conversation; 21 patients between the ages of 64 and 80 suffering from dementia and 6 retired persons between the ages of 74 and 90. The progressive inferiority of the former group, in comparison with the latter, suggests, the investigators state, that under these conditions the aged retrogressed in roughly the same sequence as children progressed and that therefore "temporal disorientation" of this sort is an important symptom of such deterioration (France; Ajuriaguerra et al., 1967).

Conflicting or negative evidence comes from studies measuring:

1. Ability to produce intervals ranging in duration from 30 seconds to 10 minutes; students, hospital attendants, and epileptic patients: the first two, or normal, groups revealed a slight tendency to be more accurate, either when young (between 20 and 30) or old (between 40 and 60), provided they were passive during the act of producing the intervals; the same tendency persisted among males, when the period of production contained the activity of crossing-out numbers or letters, but did not appear among females; the epileptics revealed a decrease in accuracy

with increasing age when passive during production, and the reverse was true for normal males (only slightly so for normal females) when there was activity; and these tendencies interacted with another factor, that of the interval's duration (Switzerland; prior; Pumpian-Mindlin, 1935).

2. Ability to give verbal estimates of intervals varying between 30 seconds and 5 minutes, whether empty or filled with dull or interesting material; college students: no consistent relation to age emerged, but the range of ages was restricted largely to a 10-year period from 17 to 27 and the number in some of the age groups was too small to make meaningful comparisons (U.S., college; verbal, prior; Swift & McGeoch, 1925).

3. Ability to produce an interval of 1 second; children between 8 and 14, young adults with a median age of 24, and healthy adults with a median age of 70: the older group of adults was just about as accurate as the other two groups with the Limits-Verbal Method (they tended to be slightly more accurate than the children without practice or reinforcement, and slightly less accurate after reinforcement), but they were decidedly less accurate than the others in counting to themselves or aloud (U.S.; prior; Smythe & Goldstone, 1957; Goldstone, Boardman, & Lhamon, 1958b).

4. Ability to produce intervals of 16 seconds and 1 minute. Accuracy for the larger interval progressively increased with age among the children (mean age 10.8), adolescents (14.6), and college students (20.2); then it decreased among business men (45.6) and further decreased among "senior citizens." This trend fits the generalization concerning an increase, and later a decrease in accuracy with increasing age; as the author suggests, however, it would have been desirable also to test a group whose age fell between the students and the business men. The perplexing result, moreover, is that the same trend was not evident in the case of the 16-second interval: the adolescent group was more accurate than the children, but the college students and the business men tended to perform like the adolescents; and then so-called senior citizens slumped to the level of the children (U.S.; prior; LeBlanc, 1969).

The following studies indicate—with the irregularities that are noted—that time seems to pass more quickly with increasing age either (a) because of a decrease in the tendency to overestimate intervals verbally or (b) because of an increase in the tendency to underestimate the intervals being produced:

1. Judging verbally the duration of a 20-second interval; the contrasting groups were young boys and a group of undergraduate and graduate students (Axel, op. cit.).

2. Producing intervals of 16 seconds and 1 minute: what has been reported concerning the trends among the different age groups with respect to accuracy applies without modification to underestimation, since in all instances both intervals were underestimated and hence a decrease in underestimation meant an increase in accuracy (LeBlanc, op. cit.). With this method, the trend toward underestimation decreased with increasing age until the mid-forties, and then increased among the senior citizens.

3. Reproducing a 5-second interval; the percentage of adults overestimating the interval was considerably less than that of the children (Tejmar, op. cit.).

4. Verbal estimates of intervals with a duration between 4 and 20 seconds; children between the ages of 7 and 9 and between 10 and 13, while confined for seven days in a simulated shelter: the former group had a stronger tendency to overestimate the intervals than the latter, although the two groups did not differ with respect to estimating the time of day during the period of confinement (U.S.; prior; Crawford & Thor, 1967).

5. Producing intervals with a duration ranging between 30 and 300 seconds: a group of World War I veterans (mean age, 67) produced shorter intervals than Korean War veterans (mean age, 24). Within both groups, those judged to have a "positive" outlook on life on the basis of their replies to questions tended to produce significantly, or almost significantly, longer and more accurate intervals than those with a "negative"

outlook—perhaps the more optimistic the attitude, the more precious time may have seemed and hence the more slowly one sought to have it pass. The investigator himself stresses the possibility that those with an affirmative outlook showed that they were better adjusted because they judged objective time more accurately (prior; Feifel, 1957).

Going counter to the central tendency just mentioned are the results from a study in which the subjects, while taking an intelligence test, were asked at the end of each four sections how much time had elapsed; in spite of the usual restricted age range of U.S. college students, there was a positive and significant correlation (.45) between age and a tendency to overestimate the duration of the intervals, which in fact had a duration of between 2.5 and 4 minutes (Grossman & Hallenbeck, 1965).

For what it is worth, let me end this *Addendum* by noting that 64 percent of a group of psychologists over the age of 65 replied "faster," 4 percent "slower," 26 percent "same rate" to the question, "Do you find that time seems to pass more slowly, more swiftly, or at about the same rate as it did when you were younger?"—the remaining 6 percent fitted into a miscellaneous category or did not reply (Aisenberg, 1964).

9. DEVIATIONS

The statistical sounding title of this chapter would signify its objectivity. In an era when psychedelic drugs have become fashionable for some persons in the West, I propose to examine, not to evaluate, slight and marked deviations from acceptable norms. Under certain conditions, the kinds of renunciations men make appear unusual, their anticipations bizarre, their temporal orientations distorted, their activities just plain deviant.

ORGANIC STATES

The condition of the organism, we have already seen in connection with perception and intervening responses, affects the kinds of temporal judgments people make. The same interval of time is judged differently as moods and needs fluctuate. Current scientific thought is both monistic and materialistic, and therefore seeks to avoid the dualism represented by Spinoza's two parallel clocks which run in preestablished harmony; as a result it is always assumed that changes in behavior reflect and affect physiological changes in the body. If this be so, it may be irrelevant to consider the biochemical processes which accompany the arousal of the temporal motive or the passing of temporal judgment under varying conditions; obviously such processes must exist, we have included a reference to them in our analysis of the behavioral potential, and that ought to be sufficient for present purposes (cf. Doehring, Helmer, & Fuller, 1964). But it is necessary at least to raise the question as to whether there is a specific physiological location for temporal judgment and how such judgments are affected by internal biochemical changes. Clearly some deviations in temporal judgment can be traced to environmental cues (e.g., the content of intervals), but others may have their origin within the central organizing facility of the brain or within other physiological processes such as heart and respiratory rates and body temperature (cf. Eson & Kafka, 1952).

In spite of the fact that the label attached to a part of the cortex, the temporal lobe, might lead us to expect a high degree of localization and in spite of a priori reasoning that "any perceptual event must certainly be associated with cerebral happenings which are extended not only in space but in time" (Longuet-Higgins, 1968), the evidence seems tentatively to suggest that the cortical bases for temporal behavior are varied and difficult to pinpoint (See *Addendum 9.1*). This is not surprising when we remind ourselves of the fact that no single sense organ mediates the temporal potential and that many different psychological functions (such as memory and reasoning) may be involved. When there is brain damage resulting from a wound, a tumor, or an operation (particularly frontal lobotomy), there may be some change, either temporary or permanent, in temporal judgment or orientation, but it is not certain whether the change is caused directly by the brain damage or is one of the indirect consequences of that damage. It is conceivable, for example, that the accuracy of some temporal judgments is reduced after brain damage not because memory or the ability to pass judgment is impaired but simply because patients in a laboratory fail to comprehend adequately the experimenter's instructions. One investigator, who has compared 69 patients with brains damaged from a variety of causes and 50 "average" individuals, feels that under these conditions "there are manifest varying degrees of time agnosia, at least roughly proportional to the degree of deterioration" (presumably without reference to the location of the lesion), with the practical consequence that testing this capability is "useful in helping us to determine the presence and degree of deterioration, both diagnostically and as a post-therapeutic follow-up" (U.S., verbal, prior; Coheen, 1950).

The same kind of Spiral relation may well exist when the effects of the endocrine glands are considered. There are data suggesting that an over-secretion of the thyroid gland is accompanied by an acceleration of the sense of time (see *Addendum 9.2*), but again we cannot say whether the temporal effects resulted from the malfunctioning of the gland or were mediated by some other process. We seem to have, moreover, no adequate information concerning the relation of other glands to temporal behavior.

The existence of individual differences with respect to over- or under-estimating temporal intervals has been discussed in connection with the problem of consistency in personality. Certainly individual differences have been reported in every study that is described in sufficent detail. For decades it has been known (e.g., Klien, 1917) that such differences intrude when any aspect of temporal behavior is studied. In fact, one writer proposes to turn what is ordinarily considered to be a liability in research into an asset: the variations in the same temporal task by the same or different subjects can be factor-analyzed and may yield a useful result (Reuchlin, 1957). These deviations, especially those concerned with sensitivity in connection with establishing thresholds, may well be linked at least in part to genetic factors and hence have an organic basis.

Since patients suffering from coronary artery disease and having a history of myocardial infraction have been shown to make various adaptations which relate to time—one study (Friedman & Roseman, 1959) reported they have "a sense of time urgency"—it was reasoned that their temporal judgments ought to be different. But no differences emerged between a group of patients having one or both of these disabilities and a carefully matched control group of patients having other types of illness when they judged intervals with a duration between 30 and 120 seconds either after working on mazes with immediate subsequent arousal of the temporal motive or when doing nothing except judging the intervals with prior arousal (U.S., adults; verbal; Cohen & Parsons, 1964). Whatever changes took place in their physiological and psychological processes, it would seem, had no influence upon their temporal judgments.

In view of the association between temporal behavior and intelligence (see again *Addendum 8.4*), it is to be anticipated that the feebleminded must reveal some deficiency in temporal behavior. But there are a few nuances worth noting. First, mental age was found to be more highly correlated with temporal knowledge (e.g., naming the day of the week or telling time) than was chronological age in a study of feebleminded children and adults (U.S.; Gothberg, 1949); obviously such information depended more on the ability than the opportunity to learn. The accuracy with which intervals whose duration ranged from 1 to 4 seconds could be compared was related to conventional temporal knowledge or to mental age and IQ only when the feeblemindedness seemed to be inherited or constitutional and not when it could be traced to traumata or other external factors (U.S.; prior, Brower & Brower, 1947). One investigator, on the basis of questioning children in an institution, suggests that the past orientation of mental defectives goes back no farther than between 10 and 20 days (France, De Greeff, 1927); and two others, using a story-completion device, found that institutionalized retarded children tended to be only present-oriented (U.S.; Sternlicht & Siegel, 1968). The association between temporal behavior and intelligence, in fact, is sufficiently close to inspire other investigators to suggest that an available Time Appreciation Test, although it indicated a significantly higher mean mental age than a conventional intelligence test and although the correlation between the two was far from perfect (.65), might be used as a standard way to screen the intelligence of the feebleminded or at least as a supplementary test that takes less time to administer. At least in this investigation, furthermore, there was evidence that psychotics tended to score higher on the Time Appreciation Test than on the standard Stanford-Binet intelligence test (U.S., young adults; Engle & Hamlett, 1950).

A severe disturbance of the nervous system usually has repercussions upon some aspects of temporal behavior. A dramatic illustration is Korsakoff's psychosis which may result from chronic alcoholism and which damages peripheral nerves as well as sections of the brain. Psychologically it is characterized by amnesia and disorientation, accompanied frequently by confabulation as a compensatory mechanism. According to one investigator, "all of the symptoms of the syndrome" can be explained by postulating that the individual's sense of time has been disturbed:

> One of the patients we examined who had been up for an entire day and for the first time had gone for a walk in the garden could explain on the following day exactly where he had gone, which nurse had been with him. He remembered that he had worn his coat, etc., but he did not know *when* the walk had occurred and *how long it had* lasted [Netherlands; Horst, 1932; italics his].

Ordinarily we seldom recall events without attaching thereto a temporal attribute, the same psychiatrist points out; that attribute may not be associated with an algebraic formula we once learned in secondary school and still remember, but it does accompany the recollection of most conversations. When temporal memory of persons with Korsakoff's psychosis fails, the temporal sequence of past events becomes confused; hence events from different contexts may be simultaneously recalled, which then appears to be confabulation. One empirical study demonstrates that on the whole two dozen adults suffering from this disease reproduced filled or unfilled intervals in various ways less accurately than two control groups consisting of out-patients at an alcohol clinic and ordinary hospitalized patients; a similar "deficit" appeared on most of the other perceptual tests that were simultaneously employed (U.S.; Talland, 1958). Our only inference must be, then, that temporal behavior is linked somewhere and somehow to the nervous system, not a particularly sensational finding.

From the brain, the glands, and the nervous system we turn to the entire body and therefore we must once again reconsider the biochemical clock which is the metaphor we are employing to summarize endogenous and hence physiological processes within the body which affect temporal behavior. A physical or even a physiological change in the body, however, need not always rely upon that clock as the mediating mechanism. When the gravitational stress upon the body was manipulated (thus simulating outer space in the laboratory), auditory intervals ranging from 1 to 20 seconds tended to be underestimated. The investigator herself does not invoke the clock: she suggests that the increased stress impaired retention and hence the interval serving as the model for the reproduction tended to be more quickly forgotten than under normal conditions (Sweden, adults; prior; Frankenhaeuser, 1960).

It is usually assumed that the biochemical clock can be affected by, or is really a metaphorical way of referring to variations in body temperature resulting from changes in room temperature, artificial manipulation (dipping feet in hot or cold water), and of course disease. Conceivably, however, such variations could be mediated by less basic processes, such as discomfort or surprise. Many years ago an incident took place which has been widely cited: a competent physiologist reported that "my wife, having fallen ill with influenza, was impressed with the fact that time seemed to pass very slowly." In fact, she seemed to be overestimating objective time. With a temperature of 104° F she asked her husband to go on an errand to a drug store: "although I was gone for only twenty minutes, she insisted that I must have been away much longer." Then her husband asked her to count to 60 at what she believed to be at the rate of one per minute. Though a trained musician, the lady then underestimated clock time: with a temperature of 103° F, she counted to 60 in a mean of 37.5 seconds, whereas with a subnormal temperature of 97.4 F that mean rose to 52 seconds (Hoagland, 1935, p. 108; 1943). Both results are consistent and suggest that her fever accelerated the clock.

On the basis of older evidence in the literature (François, 1927) and additional data which he himself gathered (Hoagland, 1933 & 1966) the investigator postulated the existence of an "internal chemical pacemaker," our biochemical clock, to account for the changes in subjective time and for temporal judgments concerning intervals "with short durations." He and many others have been impressed by the fact that those changes appear to be described mathematically by means of the Arrhenius equation which portrays in very general terms the relation between rates of chemical change and temperature. This postulation has given rise in the intervening years to a large body of research and to considerable controversy.

In this book we have previously considered the postulate in two other contexts: the internal processes which affect temporal behavior by providing unverbalizable cues for arousing the temporal motive and passing judgment and the accuracy of the ensuing judgments. Here we must simply ask whether changes in temporal behavior accompany changes in physiological processes which presumably are indices of the clock. Is there any validity to the original finding of Hoagland that speeding up the clock by raising body temperature will lead to verbal overestimation ond production-induced underestimation?

The evidence, examined at some length in *Addendum 9.3*, by and large supports the original generalization, whether or not the body temperature is raised internally or through external means. But as ever the exceptions are noteworthy. First, a few studies have revealed no relation; and there is the possibility that in some of the investigations the intervals being judged were too short and the subjects were not sufficiently rested to enable the "stabilization of the physiological indices" to take place (Hawkes, Joy, & Evans, 1962). Then other studies have shown that the same temporal effects may be associated not with increasing but with decreasing temperatures as well as with other deviations (sleep deprivation, increase in respiration and heart rate, changes in electro-encephalographic activity). One simple chemical equation apparently does not embrace all these phenomena, although it remains a first approximation.

Certainly there are rhythmic processes within the body—the heart, the blood vessels, the nervous tissues and other parts of the nervous system (including the brain), the lungs, etc. (Gooddy, 1958)—which may have their own oscillations with or without some form of synchronization, as do the separate organs of plants and other animals (Bünning, 1967, pp. 33–44). Their relation to temporal behavior, however, is obscured by three factors. In the first place, the organization or integration of the processes is intricate. A number of years ago, a dozen English men and women deprived themselves of environmental clues concerning astronomical time by living up to 7 weeks in Central Spitzbergen at a latitude of 79 degrees north during the period when the sun did not set. Specially contrived watches regulated their day, one group functioning on a 21-hour cycle and the other on a 27-hour cycle; presumably they were aware of the fact that their day was, respectively, shorter and longer. There was no contact between the two groups. Their body temperatures, with one exception, adjusted to the new day almost immediately, but the situation with respect to the chemical composition of their urine was more complicated: its water and chloride content tended to conform to the new rhythm, while the potassium factor more persistently clung to the normal rhythm, so that the latter was "out of phase" with the former. The investigators feel that their study has offered "conclusive evidence for the existence of an intrinsic 24-hour excretory rhythm in man," but the analysis of the urine suggests that there may be "more than one mechanism controlling physiological diurnal rhythm" (Lewis & Lobban, 1957a & 1957b). We do not know, to be sure, whether there is a relation between temporal judgment and the chemical contents of the urine, but the dogged persistence of the potassium cycle could prove important if it were shown that the mechanism affecting it also mediates processes influencing temporal judgment. In another study, one subject who lived in an underground bunker for more than three weeks with no information concerning objective time displayed desynchronization with respect to rhythms of body temperature and urine secretion, and he also "exhibited two frequencies even in the same organ, the kidney" (Aschoff, 1965). Little wonder, therefore, that there is no one-to-one relation between deviations in physiological processes and temporal behavior.

Then whatever cues, verbalizable or not, which are provided by physiological deviations may not by themselves be the sole determinants of temporal behavior. In the case of the ill wife of the physiologist or of any ill person, we must note that such a person cannot work or play and consequently finds intervals empty and time passing slowly. Or he may be so occupied with his own feelings and anxieties that an interval seems crowded and hence time passes not slowly but quickly; or time may drag because he wishes to get well. The deviations, if they do affect temporal behavior, may be mediated by some intervening process. My candidate here is stress or anxiety, even though I know they affect behavior quite variously (see *Addendum 8.5*) because they could well be evoked by violent changes in temperature, deprivation of oxygen, etc. (cf. Lockhart, 1967). But that is not to say that all effects are thus achieved.

Finally, the ever-present Spiral is illustrated here by an unusually provocative experiment. For a period of one week two subjects were kept in a room having constant light and sounds. Day and night they were fed every 3 hours with varying kinds of food that could not be identified with day or night, and right before each meal 10 physiological measures were taken. On the fourth day they were injected with a narcotic drug (Evipan-Natrium) and, while they were asleep, their watches were set ahead 9 hours. When they awoke, they imagined they were determining the correct time, as ever, by consulting those watches. This effective manipulation of their temporal judgments and information did not immediately lead to a complete shift of their physiological rhythms to subjective time. Without a doubt, however, the bodily temperature of both subjects was affected, as was the urination pattern of one subject; other effects also seemed evident, such as changes in breathing and blood pressure. Least affected were the heart and general circulation. There is a possibility, the investigators admit, that the changes resulted from an alteration not in the time sense but in the sleep rhythms since the men were fed and measured every 3 hours; some of the changes, however, were not the same as those obtained in other investigations when only the sleep pattern was altered. This technique, the investigators tentatively surmise, caused a dissociation of a conscious and unconscious time sense, so that some changes but not all possible ones resulted (Germany, adults; Eiff et al., 1953). In a related study, a manic-depressive patient whose moods had been alternating on a 48-hour cycle for 11 years was placed for 11 days in a soundproof chamber where, unknown to him, the daily routine was compressed to 22 hours: his moods as well as some renal and other physiological processes quickly shifted to a 44-hour cycle (England; Jenner et al., 1968). We see, then, that the basic processes signified by the biochemical clock affect behavior and that behavior in turn may affect them.

And where, then, does this leave the biochemical clock? I think it is a useful metaphor to refer to a host of more or less endogenous processes which, together with many of other processes ranging from objective information to personal experience, affect temporal behavior. We gain nothing by saying that a change produces a deviation in a postulated clock when the facts indicate that so many different physiological processes may be affected and affected somewhat variously. Until we can verify a physiological hypothesis—for example, that temporal behavior is dependent upon brain metabolism which can be affected directly by drugs, body temperature, and the thyroid gland and indirectly by stress and motivation (Fox et al., 1967)—it seems permissible to mention the clock to call attention to the contribution of the body and its numerous, inevitable rhythms.

We dare not end a section concerning organic states on a negative note, simply because our bodies are too important to be so dismissed and because certainly their composition and organization must have a dramatic effect upon the perception and judgment of time. William James and other writers of his generation have speculated dramatically and felicitously on this very point. James makes two assumptions: on the average human beings can perceive—or intuitively include—no more than 10 events in the psychological present (a figure which was obtained from classical experiments upon the span of attention and which need not be debated here) and the number of such events throughout the life of an organism is a constant. If we could perceive not 10 but 10,000 events at a given instant and if the second assumption were to hold, then we would live less than a month and be able to observe only one season and "the motions of organic beings would be so slow to our senses as to be inferred, not seen." Here James cites approvingly Herbert Spencer who suggested that a gnat may appreciate each of the "ten or fifteen thousand strikes a second which its wings make" the way a man does "a quick movement of his arm" and that, if this be so, then "the time occupied by a given external change . . . must seem much longer" to the insect than to the human being. On the other hand, if our span were only one-thousandth as long and if again the same assumption remained valid that "our life were . . . destined to hold the same number of impressions," we would live a thousand times longer and then the seasons would seem "like quarters of an hour" and "the motions of animals [would] be as invisible as are to us the movements of bullets and cannon-balls" (James, 1890, pp. 638–41). I would not dismiss such thinking as fruitless because it stems from a dubious assumption concerning the total number of events the organism experiences, for I think James and his colleagues in this fashion offer insight into the incommunicable nature of subjective time. They are indicating, in effect, that each of us has a normal subjective standard springing somehow from the nature of our bodies.

Shelley also noted the possibility that "the perishing ephemeron enjoys a longer life than the tortoise" because more happens to it during its shorter lifespan. For human beings he observed:

> If a mind be conscious of an hundred ideas during one minute, by the clock, and of two hundred during another, the latter of these spaces would actually occupy so much greater extent in the mind as two exceed one in quantity [Shelley, 1813].

The second interval, he suggests, appears longer because it contains more events; and he further states that "vivid sensation, of either pain or pleasure, makes the time seem long . . . because it renders us more acutely conscious of our ideas." From his standpoint, therefore, not "the actual space between the birth and death of a man will ever be prolonged," rather "his sensibility is perfectible": within the same span of calendar time, he is capable of experiencing more events and hence of obtaining the feeling of longevity.

The poetic excursion enables me to call attention to an aspect of time which, except for this paragraph, I am not considering to fall within the scope of this book: the relation between various natural phenomena, which can be described in temporal terms, and behavior. In another context I have mentioned that births and the onset of menstruation occur more frequently during some times of the day than at others. In addition, attempts have been made to relate menstruation, certain physical and psychiatric diseases, moods, and miscellaneous physiological processes to variations in the week or month; and similarly the seasons, including the time of birth (or the interval of embryonic development), have been linked to the menarche, diseases, suicide, death, intelligence, and other human abilities. All these studies have been most competently evaluated by Orme (1969, pp. 115–27). The mere fact that the day, week, month, and season are temporal units does not bring them within our purview. When and if the relations are more solidly established, however, they will then suggest repercussions upon the functioning of the organism, perhaps on the biochemical clock, and will thus have important implications for temporal judgments. At that point we would remove this particular Arbitrary Limitation.

DRUGS

People may unconsciously inflict pain upon themselves by running the risks leading to illness—and let us never forget, please, that in Butler's *Erewhon* they are punished for becoming sick (1917, pp. 110–7)—but the syndrome of illness itself contains very few elements (other than in some instances the craving for sympathy and attention or the desire to be dependent and hence inactive) that ordinarily are considered desiderata in their own right. Whatever distortion occurs with rising body temperature is an incidental by-product of the illness: nobody probably ever gets sick to change his sense of time. Many persons, however, deliberately affect their organisms, and just as deliberately their temporal orientations and judgments, by taking drugs.

For drugs are almost always used voluntarily to achieve an affective deviation from the normal. The exceptions are patients given drugs to relieve pain and others who take them unwittingly. Various drugs, therefore, may be ingested by persons who have a deliberate, conscious desire to renounce some kind of activity in the everyday world; who in some respects are unable to defer gratification, so bleak is their anticipation of the future; who anticipate satisfaction from the drug-induced experiences; or who may or may not worry about the fact that such experiences ordinarily do not solve realistic problems in the future. These persons would improve what they consider to be, temporarily or not, their well-being. In fact, the use of drugs may well be another universal phenomenon: from recorded time in our own section of civilization there is evidence that drugs have been employed, and wherever we are able to look (possibly with a few exceptions) in non-Western societies, we emerge with the same conclusion. Such universality points everywhere to some kind of discontent with normal existence and, to the extent that some drugs affect temporal orientation, also with the passing of time. Nongratification and renunciation must be so rife that for the moment, and only for the moment, the individual utilizes "intoxicating substances, which make us insensitive" to the fact that "life as we find it is too hard for us; it entails too much pain, too many disappointments, impossible tasks" (Freud, 1930, p. 25).

We do not know whether some aspect of temporal behavior, such as a restricted temporal perspective, plays a role in producing drug addiction. In one study, as a matter of fact, adolescent addicts were as temporally restricted as delinquent peers when both deviant groups were compared with "normals" and controls (U.S.; Einstein, 1965), but we cannot say, of course, whether the restriction was a cause or a consequence.

To assess the effects of a drug, many methodological pitfalls must be avoided. The dose and technique of administration should be standardized (Barber, 1970, p. 85). It seems obvious that the time of measurement after ingestion is important; for example, two hours elapsed before LSD impaired, and epinephrine enhanced, the "understandability" of sentences (U.S., adults; Honigfeld, 1965). Equally important is the fact that many, perhaps most, maybe even all the effects of drugs can be influenced to some extent by a variety of factors that have nothing to do with the drugs themselves. Two writers offer evidence which shows that people's reactions to LSD depended not upon the dosage and the time elapsing since ingestion, but upon the following: personality, somatotype, education, occupation, age, health, previous experience with LSD, previous psychiatric treatment, premedication, time of day, amount of food in the stomach, attitude toward whoever administers the drug, the physical and social settting in which the LSD is given (Hoffer & Osmond, 1967, pp. 104–10). Since this is so, it is to be anticipated that the effect of LSD on temporal judgment, if any, is not likely to be uniform from person to person or from investigation to investigation.

Yet another factor may play a role in generalizing about a drug's effect: the anticipations of the individual who takes it and of the investigator who observes him or hears his report. From a methodological standpoint, here is a source of error which, fortunately, can be eliminated through a double-blind procedure—though one finds that this has seldom been done. During a series of tests, a placebo is used instead of the drug, or some subjects are given the drug, the others a placebo. The fact of whether a subject on a given trial has the drug or the placebo is concealed both from him and the observer.

Persons who use drugs often point to an inability to communicate their inner experiences which, under these circumstances, may seem more private than usual; and so they use the readily available language and figures of speech that are prevalent in their milieu and that may be ambiguous. It is also well to remember that the individual who reports that it appears to take hours or decades to cross from one side of the room to another is convinced of the reality of his judgment. It makes no difference to him whether someone else in that room, with or without the drug, is not having the same experience, or whether scientific investigations have indicated that the modal effects of the drug are at variance with his impressions. Psychic reality on the primary level, moreover, may differ from secondary judgments. In one study the investigators found no relation between what their subjects reported concerning the effect of two drugs upon their time sense and the actual changes in the tested judgments; hence they conclude that the latter "provide a sensitive method for demonstrating changed mental function, even prior to subjective awareness" (England, university; verbal, prior; Goldstone, Boardman, & Lhamon, 1958a).

Some of the available evidence on the relation between drugs and temporal behavior is reviewed in *Addendum 9.4*. There are such marked variations in the effects resulting no doubt from the factors just mentioned, and also the investigations of a specific drug, especially LSD, produce such conflicting results that the most reasonable view seems to be the one expressed by one writer after reviewing part of this literature: "the work on the effect of drugs on time estimation is unsatisfactory" (Orme, 1969, p. 86). Still the trend of the research, I think, does support an "oversimplifying" generalization that "pyretogenic drugs of the amphetamine, mescaline, LSD type" tend to produce excitation and accelerate the time sense, whereas tranquilizing drugs have the reverse effects (Fischer, Griffin, & Liss, 1962; Fischer, 1966). This could well mean that drugs affect basic physiological processes —that is, the biochemical clock—and that their effects vary, even as a rise in body temperature does not always influence temporal judgments uniformly.

The basic importance of that clock in this context seems evident when the effects of secobarbital and dextroamphetamine are asserted to be "independent of psychophysical method, sense mode, subjective standard, and the response scale" (Goldstone & Kirkham, 1968). The senior author just cited, however, is of the opinion that these drugs affect "the event rate" of the person experiencing them (Goldstone, 1964, p. 201). If so, then the explanation would be our Principle 5.2 which proposes a curvilinear relation between timing acceleration and event potential. But more than one principle is probably necessary, for certainly drugs simultaneously affect the drive attribute. If stronger drives are evoked under the influence of stimulants than under depressants—and this seems probable—the postulated and obtained effect on temporal judgment becomes even more compelling. It is possible, nevertheless, that such changes in the event and drive attributes of the intervals are secondary consequences of a primary change in the biochemical clock.

Our discussion of drugs gives rise to an omnibus type of generalization. The tendency for stimulants to accelerate and for depressants to decelerate the time sense reflects changes in the organism's physiological processes but may be influenced by one or more of the following: personality traits, prior anticipations, previous experience, physiological states, the setting, the strength of the drug and the method of ingestion, and the interval elapsing between taking the drug and the observation or method of measurement. But can we always assume that the changes in temporal behavior are mediated by the changes in perception or behavior which are induced by the drug? No, I think, for there is the possibility that the temporal changes result from physiological changes within the organism and hence occur concomitantly with those other changes.

SLEEP AND DREAMS

Normally we seek sleep because we cannot resist fatigue or because we would restore our bodies to a condition favorable for the performance of all kinds of activities. There are individuals who, not just on occasion but frequently, use sleep as an avenue of escape from the realities of living in which they must face all kinds of decisions and make all kinds of judgments, including temporal ones. During sleep decisions do not have to be made, and sleep is the only timeless state known to man which resembles death.

But is sleep timeless? No, the answer must be, we dream while asleep and in dreams the problem of time arises in a new form. Dreaming, moreover, can be considered another universal deviation, in this instance a deviation from the normal waking state. In our society it is scientifically or psychiatrically impossible to approach the subject of dreams without paying tribute to Freud and to utilize his principal assumption and contribution, viz., that dreams are not fortuitous phenomena but reflect (in ways concerning which there is by no means complete agreement, even between the earlier and the later Freud himself) the strivings and current problems of the dreamer. For this reason they constitute a valuable clue to the deeper layers of the personality; and they have also served sociologically and anthropologically to indicate some of the modal traits within a society. Dream analysis is a tricky art in its own right which we shall avoid here after reminding the reader that a valid interpretation of a particular dream undoubtedly is possible, as Freud pointed out, only when the dreamer himself provides the groundwork for the interpretation. This very obvious statement cuts the ground away from the argument concerning the realistic or symbolic nature of dreams: it all depends upon the dreamer. Whether dreams serve the function of providing substitute satisfaction for impulses repressed as a result of compulsion or renunciation or of being an outlet for anxiety concerning anticipations in the future, they reflect the temporal decisions of the waking state.

Are dreams themselves timeless? The most common Freudian view seems to be that the unconscious is timeless and that therefore dreams, being the expression of the unconscious, are also free from the restraints of time. One of the more poetic of Freud's disciples has written that the "flight from the tyranny of time which we are able to renew each night seems to me to represent one of the greatest wish-fulfillments accomplished by the dream for the benefit of human beings who remain harnessed to time's chariot while day lasts." She views most struggles in these terms: we begin life in an unconscious and hence timeless state, and seek continually to return thereto in daydreams, fantasy, love, drink, drugs, mysticism, magic, myths, prayer, and religion. In addition we are always pursuing timelessness to avoid the triumph of time which is death (Bonaparte, 1940). From this standpoint time is an ingredient that comes from repression, it is one of the aspects of civilization that produces discontent. In childhood, time plays a less important role since then more sophisticated responsibilities have not yet been assumed. Dreams which originate in, reflect, or express the unconscious must partake of a timeless quality. The temporal components of dreams, consequently, are artifacts or elaborations of the waking state in which of course all events are likely to have such a component.

And yet time may play a central or significant role within the dream; then the psychoanalytic problem is to adduce its latent meaning from its manifest content. According to Freud, the time of day in a dream may be a reference to the role of time in the dreamer's own life span; dreaming about arriving too late may suggest that the dreamer realizes he can no longer rescue himself from his own difficulties (Gutheil, 1951, pp. 81, 226). Another psychoanalyst asserts, in slightly different language, that "time elements are used—and appear as such in the manifest dream—for the distorted expression of emotionally, highly charged thought processes going on behind the screen" and that "the distortion chosen by the dream work seems to be mainly, but not exclusively, the displacement of emphasis from the personal object to the impersonal element of time" (Gross, 1949). Interpretations of this sort are risky because they assume that dream symbols have universal meaning or can be interpreted out of context. Here it is sufficient to note that time is sufficiently important to require and receive symbolization in dreams.

Perhaps the best way to describe dream time is not to assert that it is absent but that, unlike time in the waking state it is—and the descriptive adjective is not easy to find—more fluid. One of the analysts cited in the last paragraph seems to agree with this view: "In the unconscious the past seems to intermingle with the present, older experiences with current ones, and traces of former events with the happenings of today" (Bonaparte, 1940). The dreamer is not controlled by time because a different set of rules operates from which reality is largely excluded; in fact, according to an analytic view, the elimination of contentional space and time "facilitates the dynamics of the dream process, indeed enables undoubtedly that process to occur" (Simonson, 1928). The sequence of events in a dream does not follow the orderly pattern of objective time and space or any other logical plan, rather it conforms to demands of the drives that are functioning (D. E. Schneider, 1948).

If the temporal motive is evoked in dreams, what kinds of judgments are made? One finds in oneself and in anecdotes that the orientation can be in any direction, but most writers on the subject—even including non-psychoanalytic ones (e.g., MacIver, 1962, p. 43)—seem to agree that a temporal orientation in the present is always characteristic of the dream action. Past events, in whole or in part, as well as fears, hopes, and anticipations concerning the future may affect the content of dreams, but they are transformed almost completely into present actions and present emotions. The significance of such a transformation is not easy to grasp. If dreams are primitive forms of activity, as many theorists contend, then again we have evidence at hand that man's primeval temporal orientation is in the present: when we are asleep, we cannot look backwards or forwards in whatever sense dreams are conscious, we must have our brute experiences here and now. A difficulty arises when the orientation of dreams is investigated because the dreamer when awake reports the contents of the dream either to himself or a second person in the past tense which refers to the event of the dream rather than its time span. "I dreamt I was going to take a ride with a crow," an Ojibwa Indian said. "I sat on its back just below its neck. The crow flew up with me on him . . ." (Lincoln, 1935, p. 283). He and we can only guess that during the dream he was contemplating and then having the ride in some kind of specious present. Or from another standpoint this observation can be made: "It is as if in a dream one would rather experience a pictorially represented cluster of events perceived simultaneously, while the conscious mind when recording these dream events later would automatically put them into an order which we perceive as a logical time sequence" (Franz, 1966).

The principal actor in the dream is the dreamer. Sometimes he sees his own self upon the stage and then he—some other part of himself?—reacts to what he himself is doing or experiencing. At other times, in the same or in different dreams, he observes the play, and experiences the emotions, of the very interested spectator. Or, too, the events are not upon the stage, rather they are real, and they impinge upon him even though he is in the background. Whether the dreamer is the interested or somewhat disinterested spectator in the dream, he has no choice except to feel himself in the present: that is the lot of the spectator who is absorbed in the production.

But more than such concentration may be involved. Many dreams, for example, contain portions of arrested time which, we shall see, is of the essence in art, and more usually the dream events move forward, chaotically or coherently. Some analysts, moreover, report dreams which have the dreamer anticipating the future; and one has discussed the possibility that such dreams, while revealing deep-seated tendencies within the dreamer, may also contain elements of precognition concerning the future (Eisenbud, 1956).

This last view of dreams certainly exists within our own society and, probably too, within most societies. Dreams are thought somehow to portray the future as a premonition rather than as an expression of the dreamer's anxieties or ambitions concerning what is yet to come; what he dreams reflects not what has happened but what will happen to him or to someone else in the real world. The fact that this approach to dreams is widespread does not of course constitute evidence for its validity, since it may reflect man's wish—another universal?—to foresee events which are difficult or impossible to anticipate. We cannot, however, exclude the possibility that the experience of the dreamer is occurring during an interval which, from his own standpoint when awake, and from that of most of mankind, has not yet arrived and which he may later experience—almost any hypothesis is worth investigating. The problem here is an empirical one: does the evidence supplied by scientifically inclined writers (Dunne, 1934) and by humanists (Priestley, 1964) validate the theory or can the "coincidences" between dreams and the future be accounted for by some other means? The question may be unanswerable, but it certainly must be raised—and no sanction should be administered for doing so.

Dreams have the reputation, as we have seen, of distorting time, especially of condensing a long series of events or an interval of long subjective duration into an instant or two of objective time. Subjective duration is appraised only by the report of the dreamer after he has awakened; but how can objective duration be measured? In recent years high and statistically significant correlations have been found between the subjectively experienced dreams of a person on the one hand and objectively measurable rapid eye movements (REM) and characteristic changes in the electrical potential of the brain (via the electroencephalograph or EEG) on the other hand, but then doubts have been raised concerning the significance of such correlations (see *Addendum 9.5*). For the moment, it seems necessary to conclude only that there is no one-to-one correspondence between dream and waking time but that the discrepancy may not be very great.

I find it impossible not to make passing reference to one phenomenon which fits better here than anywhere else in this book: the "panoramic death vision" which, according to folklore, occurs sometimes when a person is close to death, for example by drowning. Allegedly in a second or a fraction of a second he experiences, in visual images, many or most of the important events of his life. It is of course difficult to verify the phenomenon if only because many who may have had the experience do in fact die. The frequency with which the experience has been noted anecdotally and in literature for many centuries, moreover, may or may not be positive evidence. One writer ventures the undocumented opinion that "in this situation the oxygen supply to the brain has been cut off" and "the immaterial brain field is in the process of disentangling itself from its load of incorporated inert matter" so that "normal mental sluggishness disappears" and images can quickly flit by (Stromberg, 1961). If this vision be a real phenomenon, then we have at hand perhaps a very pure illustration of the oft-repeated discrepancy between objective and subjective time.

A priori reasoning, fairly common experience, and the evidence from REM all suggest that most of the time the sleeper, while dreaming, does not awaken: his dream either ends or continues during sleep, and when awake he recollects nothing, at least under normal circumstances. But if he does awaken during the dream or if he does remember the dream or fragments from it, he must first appreciate the fact that he has been dreaming. The capacity to make the distinction between dreams and nondreams must be casually learned—in fact, it may well be that children go through detectable stages in this respect (U.S.; Kohlberg, 1966)—and in some societies adults allegedly do not draw a sharp line between the two levels of psychic reality and, from our standpoint, remain "confused" (Malcolm, 1959, pp. 85–6). What clings to the dreamer in the waking state is some of the emotion of the dream itself: he has had a nightmare or an ecstatic experience, or some event has transpired in-between these two extremes. If he thinks about the dream, he must stretch or shrink its duration to fit into an appropriate temporal interval of waking time, he has no alternative except to use the coordinates of normal clock time. Then the normal Guides from experience—for example, one suggesting that usually many events require a long interval—begin to function again. After awakening the person gives himself the impression that considerable time must have passed during the dream for so much to have transpired or, if an objective check is at hand, he may feel astonished that many dream events could have taken place in so short a time. In addition, whether or not he acts like a psychoanalytic man, he can try to interpret the dream, or have it interpreted, only by fitting it back into his normal temporal span of past and future: the dream has rearranged some aspect of his past experience or it has been pushed into shape by some of his anticipations concerning the future. He has, as it were, atoned for his deviation by assimilating it into his normal existence.

Daydreaming, of course, is a form of dreaming and possesses many of the attributes of dreams with respect to motivation, condensation, and the treatment of time. The master himself expressed the opinion that "fantasy" in this form "at one and the same moment hovers between three periods of time—the three periods of our ideation" (Freud, 1925, p. 177). Obviously the individual can be aware of time within the context of his reverie: his orientation is backwards or forwards, and his absorption can be so great that he experiences physiological changes such as changes in heartbeat, blushing, etc. Objective time or subjective time which refers to the reality at hand, however, cannot be simultaneously experienced; for then the daydream fades and the person realizes that he must return to work or be conscious of the task he is performing while his mind has been wandering.

HYPNOSIS

In spite of a century of casual research and then a half-century of somewhat extensive experimentation, the precise mechanisms underlying hypnosis are not understood. We do not know, for example, how the phenomenon is related to sleep. There is no question, however, that hypnosis is usually classified among the deviant phenomena. For up to a point the subject is highly suggestible: he responds to the hypnotist's commands and apparently, too, he experiences, after suitable suggestions, hallucinations which he would not experience in everyday life. Also solidly established is the fact that in almost all instances the subject must voluntarily permit himself to be hypnotized (that is, he agrees to follow instructions: to relax, to pay attention to the hypnotist's voice, and at the beginning to obey certain simple commands, such as to close his eyes or clinch his fists), he is not likely to be hypnotized "against his will." Here, it would seem, is renunciation writ large: the person ostensibly relinquishes control over his own actions. But such relinquishment occurs only under hospitable circumstances when the subject trusts and respects the hypnotist, or at least feels submissive to him.

It is expected, since subjects in a hypnotic trance are so suggestible, that the impressions concerning the duration of intervals can be easily altered by the hypnotist: they can be told that time is passing slowly or quickly, and then indeed it does. But does it? Certainly they report that it does, and at this metaphysical point we must halt. But one careful investigation has indicated that the same effects upon the subjective impression can be attained through suggestion in the waking state, viz., "It's easy to make time seem to slow down. . . I want you to really feel that time is slowing down. . ." (U.S., students; Barber & Calverley, 1964). All we know, therefore, is that in the absence of information about objective time people are probably suggestible if the suggestion comes from someone with prestige derived from his role as a hypnotist, or in some other way.

"When we hypnotize subjects and ask from them a more correct estimation of time," a psychiatrist once suggested, "we do not address ourselves to a mysterious higher function of time but address to the individual the appeal that he should utilize the perceptual data more carefully" (Schilder, 1936). But is accuracy of temporal judgment in fact increased by hypnosis? The slim evidence we have is inconclusive (see *Addendum 9.6*).

Another question we might well ask is whether the subject in a trance who apparently accepts a suggestion concerning the duration of an interval which is at variance with clock time (he is told that 30 minutes will pass between two signals when in fact only 3 seconds do) behaves as if in fact the equivalent amount of clock time were passing. There seems to be no doubt that the "illusion" of what might be called hypnotic time can be created: the subject really believes that the indicated interval of clock time has elapsed. But is there "an amount of experience under these conditions that is more nearly commensurate with the subjective time involved than with the world time" (Cooper, 1948)? Two American investigators, whose research is also summarized in *Addendum 9.6*, disagree in their replies to this question.

I have been unable to find evidence as to whether temporal judgments can be affected *indirectly* during the hypnotic trance; for example, after the suggestion that an objectively empty interval is crowded with events, does the subject report that the interval is shorter or longer than when he is instructed simply to sit and wait for further commands? Also we know that post-hypnotic suggestions involving temporal judgments are executed even when there is apparently (at least so far as one can tell from direct questions) complete amnesia for the command that was given during the trance; thus the hypnotist tells the subject to scratch his left ear 32 minutes after being awakened and without remembering that he has been told to do that, and he obeys. Without consulting a timepiece, exactly 32 minutes after being restorted to the normal state? Here we have only anecdotes testifying to the accuracy of what must be called, perhaps, an unconscious temporal judgment.

Otherwise it seems certain that any kind of normal temporal behavior can be made to appear during a trance. Three subjects, for example, were told while hypnotized "to take a familiar walk—from house to school"; no time interval was suggested to them; afterwards they located a signal which the investigator sounded during the hallucination at a point in the hallucinated walk which roughly corresponded, with a slight lag, to the point in clock time at which it had been sounded (Cooper & Erickson, 1954, pp. 55–6). Obviously an instance of subsequent arousal of the temporal motive.

That temporal judgments can be affected by post-hypnotic suggestions was shown in one investigation which employed only a single subject who, while hypnotized, was taught to associate three degrees of boredom in his past life with symbolic cues. The classroom situation in which he had been "very, very bored," for example was symbolized with the symbol B3; less boredom was called B2, still less B1, and "not bored, not interested" 0. In the waking state he was ostensibly amnesic concerning the meaning of these cues which were then presented to him on cards one at a time. Two judges observing him rated his behavior without knowledge of the symbols to which he was responding; and their rating corresponded closely to the state of boredom he was supposed to be experiencing. Then he was asked to produce intervals of 5 and 10 seconds. A significant relation emerged between those productions and the suggested state of boredom: the greater the boredom, the shorter the productions and hence the "higher" the "subjective duration." The investigator thus found support for "a common introspective experience" that "the apparently slow passage of time [is] associated with boredom." (U.S., Geiwitz, 1964).

Unquestionably, then, expressed temporal judgments and internal feelings about time are affected by hypnotic suggestion but equally dramatic effects have been validly demonstrated, I think, in connection with basic psychological processes such as pain (Weitzenhofer, 1953, pp. 148–53). More important, it appears as though many or most of the quantitative changes induced in the trance resemble those with which we are familiar in the normal, waking state. You speed up what you are thinking or doing simply when someone tells you or you tell yourself you are pressed for time. Some persons have an unusual temporal talent which they exercise without being hypnotized; Mozart reported he did not always have to "hear in my imagination the parts [of a musical work] successively," instead he could hear them "as it were, *gleich alles zusammen,* all at once" (Fischer, 1967).

Phenomena associated with hypnosis and post-hypnosis resemble those already discussed in connection with the biochemical clock and unverbalized cues. In the opinion of one psychiatrist, the existence of a biological "time sense" among human beings is "demonstrable" only in hypnosis and sleep (Ehrenwald, 1931a). We do not lie awake at night engaging in an activity like counting in order to awaken ourselves at some regular hour. Even when we pay attention to the passing of time—"tell me when you think three minutes have gone by," as in the Method of Production—we have nothing tangible in consciousness that provides us with a clue to say the time is up. Or again when asked to guess the time of day, some of us are able to come up with a judgment almost instantly without consulting a clock, without looking at the position of the sun, and without engaging in elaborate calculations involving the activities in which we have engaged. Telling time post-hypnotically is probably no less or more remarkable than normal phenomena; perhaps the only real difference lies in the fact that the hypnotist rather than the person himself provides the initial stimulus to pass the temporal judgment and that this fact cannot be recalled.

Other phenomena associated with hypnosis can be similarly placed in perspective. Since personality traits are associated with all kinds of behavior, it comes as no surprise to discover that various traits are also found to be correlated with hypnosis-proneness (Hilgard, 1965). Just as persons who seek drugs or success have relevant motives for those goals, so those who wish to receive therapeutic treatment under hypnosis have their own deep-seated wishes. According to the informal observations of a psychiatrist, such individuals desire therapeutic benefits without temporal awareness: during the painful experience of treatment they aspire to achieve "the obliteration of a period of time" which other persons succeed in accomplishing through fuges (Schneck, 1957).

And now, having followed the tradition of trying to debunk hypnosis by arguing that the deviant phenomena associated therewith appear either unsubstantiated or quite normal, I end by admitting that I have not completely convinced myself.

MENTAL DISORDERS

Let me say immediately that bizarre temporal phenomena accompany some mental disturbances. Once a child of eight ran to his mother and said: "Now it starts again. Mother, what is happening, everything is going so quickly! Am I speaking more quickly? Are you speaking more quickly?" (Klien, 1917). In this section we shall try to suggest that elements of such behavior are lawful.

But nasty difficulties are visible on every side of the psychiatric landscape. In the first place, mental disorders are most varied and hence, it may be anticipated, are accompanied by diverse temporal changes and perhaps sometimes by no changes at all. Since psychiatric categories are not completely standardized, it is difficult to establish comparability from study to study. Then there are staggering sampling problems when attempts are made to contrast the normal and the abnormal—and most investigations are based upon two contrasting groups, although of course normality and abnormality fall along a continuum. Throughout this book, we have been emphasizing that the fact of individual differences looms large, and usually significantly so, in almost every study; how, then, can a finite number of so-called normals be used as a criterion group? The problem reappears when abnormal individuals are examined, especially in connection with psychiatric categories as broad as schizophrenia or neurosis. The way out of this difficulty would be to investigate temporal behavior in persons before, during, and after mental illness; but this longitudinal approach is obviously impractical. We must, however, valiantly struggle on.

The temporal phenomena to be observed in connection with any illness are usually diverse and cannot be glibly grasped. If one psychoanalytic view is the guide, viz., that regression plays a role in all or most psychoses, then attention must be paid not only to temporal orientation (which is presumably toward the past—though doubtless unconsciously so) but also to temporal information and judgments (which from the reality viewpoint turn out to be false as a result of the maladjustment). Not unexpectedly, different results are often obtained with the same patients when the method of measurement is altered (e.g., England; Mezey & Knight, 1965). It may be difficult to test those who are mentally distressed: they may be unable to follow instructions as a result of "such psychotic factors as autistic preoccupation or irrelevant affective processes"; and the faulty temporal information they supply may be not a component of the illness but only a reaction to the unchanging routine of the hospital, for "schizophrenic patients are frequently disoriented on arrival at the hospital and become oriented prior to discharge" (De La Garza & Worchel, 1956).

Finally, we are forced to note an almost metaphysical problem: do the errors of psychiatric patients reflect incorrect standards or incorrect judgments? When you are asked simply to estimate how many seconds it takes for an object to fall from a height onto the ground, your error may result either from an incorrect conception of a second or from an incorrect estimate of the duration of the fall. Similarly the schizophrenic may misinterpret the instructions of the investigator or his own sense impressions. But this problem exists for so-called normal persons too; thus it has been suggested that the degree of under- or overestimation with the Method of Reproduction, especially in connection with very short or long intervals, may be due to an error either in perceiving the interval to be reproduced or in the act of reproduction—or to both (Belyaeva-Eksemplyarskaya, 1962). Perhaps, then, the problem of interpreting temporal judgments is no greater but only different when the subjects are psychotics rather than normal persons. At any rate this difficulty makes us enviously appreciate anew the sterilized atmosphere in which classical and psychophysical experiments have been carried out: the danger of responding incorrectly to the instructions is really kept at a minimum.

The basic question to raise, I think, involves the causal relation between mental illness and temporal disturbances. It appears—to me, at any rate—that such disturbances may best be viewed as one of symptoms composing the syndrome of the illness in particular personalities (cf. Clauser, 1954, p. 50). On an a priori basis it can be said that if a person has lost interest in his surroundings for some fundamental reason, he may then no longer take care of his bodily needs, he may not respond to the greetings of other persons, *and* he may not note the passing of time or show concern for the relatively distant past or for the future. If he is in a catatonic state, he may engage in no activity whatsoever and yet not be bored; and if he is not bored, he may not overestimate time. If he is so detached from reality, he may fail to organize his experiences, one result of which can be a discontinuity between the past and the present or between that present and the future; and an occasional awareness of the discrepancy, which may be more or less unconsciously motivated, can serve to increase the severity of his illness. Even though at first glance it may look as if a particular psychiatric case can be described principally in temporal terms, closer examination is likely to reveal the presence of many other related and unrelated symptoms of disorder (e.g., Cohen & Rochlin, 1938). One survey of the literature on temporal disturbances reaches the conclusion that abnormal time perception is not "a disease *sui-generis*" since a "diversity of sensory organs and hence brain centers must be involved in the interpretation of the experience of time" and since temporal disorders accompany a "variety of diseases"; rather, it is suggested, "any disrupting factor leading to either disorganization or discontinuance of higher cortical function seems to disturb orientation in time" (Nettleship & Lair, 1962).

Do temporal disturbances accompany all mental illness? The psychiatric oracles speak in mixed tongues, but the majority appear to reply affirmatively. That this view is taken seriously is shown by the fact that part of the routine interview with a patient suspected of being psychotic involves questions concerning his temporal knowledge: the hour, the day of the week, the month, the season, and the year. It is believed that temporal disorders may be used as symptoms to effect an early diagnosis of psychiatric difficulties (Tscholakow, 1956). It is argued, on the basis of a not competely convincing comparison of adolescents just admitted to, not ready to be discharged from, and about to leave a psychiatric hospital, that being institutionalized enables patients not only to find a sense of self-identity but also to "build a future," so that a study of temporal orientation involving more than the routine of daily living provides a clue to the success of the treatment (U.S.; Rizzo, 1967–8). Then "an investigation of factors prognostic of recovery in hospitalized schizophrenics" has shown that "there was greater predictive value in measures of the patients' attitudes toward the future than in indicators of symptom severity" (Levine, 1960). Temporal factors always play some role in Korsakoff's psychosis, whether direct or symbolic, but "the disturbances in the temporal system of mnemonic organization . . . are only part of a large group of disturbances which result from a basal deficiency" (Talland, 1961). Actually many psychiatrists contrive somewhat careless and sweeping statements when they summarize their experiences; for example, a qualified generalization, "time agnosia *may* be an outstanding symptom in the psychoses and to a lesser degree in the neuroses," is followed by an unqualified one, "*all* neurotics rebel against present time" (DuBois, 1954, italics added) so that one is left confused. Perhaps the following cynical observation summarizes the state of the art: "disorder of time consciousness . . . may be found almost as often as it is looked for in mental disorder" (A. Lewis, 1932). The probability of finding a temporal disturbance, however, is allegedly great: in both phylo- and ontogenesis, it has been asserted, time has appeared after space, and consequently is disturbed first when the individual is in trouble (Davidson, 1941).

From a reverse standpoint, it is possible to comb the psychiatric literature and find therein, as a glance through *Addendum 9.7* might suggest, instances in which every aspect of temporal behavior malfunctions or at least is disturbed. One psychiatrist who has written a very broad, culturally and philosophically oriented book on "the fourth dimension" is able to illustrate virtually every topic with a clinical case, often from his own practice (Meerloo, 1970). A patient may simply lack temporal knowledge of a trivial nature or, as the same writer has also pointed out, at the other extreme his "clinical time" (that is, his "unconscious and partially conscious concept of his personal life span between birth and death") may be so upset that he prefers "sudden death by suicide rather than existence in restless anticipation" (Meerloo, 1966). Under these circumstances it is not surprising that almost any type of conceptual schema has served as the basis for classifying temporal aberrations. A French psychiatrist has utilized Bergson's concepts concerning time for that purpose (Minkowski, 1933). Or Karl Jaspers, a philosophical psychiatrist, divides the "experience of time" along the temporal dimensions of present, past, and future; and then he cites cases to show that somewhere, sometime there has been a patient whose disturbance can be subsumed under each of his subdivisions. In the present, for example, time may seem especially fast or slow; an awareness of time may be lost; the experiencing of time may seem unreal; or time may stand still.

The duration of intervals in the past or the future, he indicates, may be misjudged; patients may be convinced that they have already experienced what they have never experienced (*déjà vu*) or have never experienced what they have in fact experienced (*jamais vu*); or the continuity between past and present may seem to be lost. And there is in such a philosophical system, as might be anticipated in this era (see also part 2 of *Addendum 9.7*), even the existentialist touch: "I cannot see the future," one patient said, "just as if there were none" (Jaspers, 1962, pp. 82–7).

Psychoanalysts have frequently concerned themselves with their analysands' attitudes toward time, illustrations of which appear in part 1 of *Addendum 9.7*. Their precise interpretations are quite diverse and insufficiently validated but, it seems clear, they all agree that the attitude is a component and not a cause of mental illness. The great virtue of this approach, in theory at any rate, is its attempt to uncover the genesis of all mental illness; and so if a temporal disturbance (e.g., obsessive punctuality) is involved in a neurosis, it too is traced to some set of circumstances in the past (e.g., anality) which continues to have functional significance in the present. After assenting to the general principle, however, the analysts again diverge as they outline the sequence of events allegedly leading to temporal disturbances.

Clinical and experimental evidence is reviewed at some length in the *Addendum* just mentioned. Since any aspect of temporal behavior may be disturbed during a mental illness, but need not be, it would be "misleading" to believe that specific temporal disorders are "common to particular diagnostic categories" (Orme, 1969, p. 24). The orientation of psychotics tends to be in the present or the past, rather than in the future: temporal restriction accompanies the more general behavioral restrictions of these disorders, in fact it is difficult to separate the two. But, as one study cited in part 2 of *Addendum 9.7* suggests (Davids & Parenti, 1958), a future orientation in some contexts may reflect a flight from reality with which a present orientation would be concerned.

The temporal information displayed by psychotics is usually spotty and inaccurate in some respects (see part 3 of *Addendum 9.7*), and again the gaps may have compeling psychological significance. It has been noted, for example, that patients may try to conceal from themselves—and maybe from others—the duration of their illness or the length of time they have been institutionalized by stating as their present age at the moment the age they had reached at the onset of the illness or at their entry into the institution. Misinformation may be linked to the entire syndrome of the illness. In one study, patients who denied that they had undergone serious operations, such as a craniotomy, or who were suffering from obvious disabilities, such as hemiplegia, exhibited, among many other symptoms involving confabulation, some disorientation in time. The investigators report that at first such persons were not permitted to look at a wall clock as their time sense was being tested:

> However, it was found that when temporal disorientation was present, it usually made little difference if the patient could see a clock or not. Even though he could read it correctly, he persisted in his error. It appeared that the patient used his own somatic experiences as the index of the time of day and rejected all other data which might have served to correct the error [Weinstein & Kahn, 1955, p. 41].

Doubtless an understanding of why these persons did not utilize the Method of Chronometry, which had previously been so heavily reinforced, would give a significant clue to the nature of their illnesses.

Some psychiatric patients reveal an important symptom when they seldom pass temporal judgment in the course of their daily existence or when they refuse to do so upon being questioned or challenged. What happens, though, when their sense of time can be tested? With many, many exceptions, the mentally ill, especially schizophrenics, tend to judge ephemeral and transitory intervals in a manner suggesting that their sense of time is accelerated: with Production and Verbal-Limits they underestimate intervals, with Verbal Estimation they overestimate them. Their judgments are usually less accurate than those obtained from normal controls. But there is also some evidence suggesting that their sense of time is decelerated in judging extended intervals; again exceptions to the trend have been reported, and again the psychiatrically disturbed tend to be less accurate than the controls (part 4 of *Addendum 9.7*). But I must repeat myself for the nth time and say that investigations involving ephemeral and transitory intervals have been conducted in laboratories under artificial conditions. In a study previously cited (U.S., adults; verbal, immediate subsequent & production, prior; Dobson, 1954) neurotic patients were found to give accurate judgments concerning intervals ranging in duration from 17 seconds to 2 minutes, much to the surprise of the investigator. Since they suffered from "anxiety," he had expected them to overestimate the intervals. His hypothesis, however, may not have been wholly incorrect, for it may apply to intervals which really provoke anxiety and not to the kind of antiseptic ones which he perforce employed. In this situation, he himself adds, the task may not have been an "outlet" for the anxiety, either because the patients' neuroses had previously compelled them to pay such close attention to the passing of time that they had acquired skill in responding to environmental cues, or else it stimulated them to be accurate in order to make a good impression on the examiner. It is, then, important to note here that apparently anxiety did not generalize, rather the schizophrenic ailment did.

Evidence for the basic distinction between primary and secondary judgments appears, it seems to me, frequently in the psychiatric literature. Schizophrenics and others often claim that time seems to be passing slowly, and there are instances in which this primary judgment differed from their secondary judgment concerning the duration of specific intervals (see pp. 328–9, *Addendum 9.7*). Or consider the following excerpts obtained from various parts of an interview with a 59-year-old psychotic:

> How many hours in a day and night?
> Formerly 24 hours; I believe that there are now fewer; there used to be 18 hours. . . .
> How does it happen that day and night now have fewer hours than formerly?
> I have noticed that . . . I do not understand time.
> Are the hours the same for all persons?
> I cannot know in the case of others as well as I can for myself. . . .
> How old is your nephew Klaas? [In fact, 22 years]
> Two years, not a bit more. . . . I do not understand how that child could become so big so quickly . . . I myself grew up slowly, but Klaas looks like a boy of 20—I do not understand that. But it appears to have occurred. . . .

Again and again, therefore, the man kept repeating the phrase, *"ich begreife es nicht"* (I don't understand it) (Netherlands; Bouman & Grünbaum, 1929). He was thus feeling within himself a conflict between his immediate or primary judgment and a secondary one which, he seemed dimly to appreciate, ought to have been more closely connected with reality. Like normal persons, many patients, especially neurotics, must be fully aware, either during or after the interval being judged, that their primary judgment is malfunctioning; otherwise they could not report that time "seems" or "seemed" to be passing slowly or quickly. All of us know that sometimes time passes quickly, sometimes slowly; but if we are normal we are not disturbed by the discrepancies, we recognize them, we reconcile them. The mentally ill may not be disturbed, they may not possess this kind of recognition, or they may be following a pattern which provides little or no continuity to their existence.

The upshot of this discussion, I think, is that in some way temporal disturbances are congruent with the life situation of the sufferer. Hospital patients may lose track of the day of the week or the date simply because they are subject to the routine of an institution and are isolated from their normal existence; but they know when meals are scheduled, in fact an error in this category is considered to be of "diagnostic significance" as is "an error in the year" (Linn, 1967, pp. 559–60). Temporal judgments, therefore, are so important to everybody that, I judge from reading hundreds of descriptions of cases, some part of the universe is appraised temporally quite correctly even when the individual is most seriously disorganized. One woman, age 78, with "considerable" intellectual deterioration, could not supply the most conventional information (age, date of birth, day of week, duration of her stay in the institution), but struggled to be accurate concerning her deceased mother: "my mother has come every day, but today she has not come; she came every day, but I believe she did not come yesterday . . ." (Minkowski, 1933, p. 13; italics omitted). But when there is "very general deterioration," as in the organic psychoses, most contact with the external world may be lost, and then even the order of succession may not be correctly perceived (France, adults; Fraisse, 1952).

One theorist maintains that in schizophrenia a "primitive time" sense is rarely affected; the feeling enabling persons to objectify temporal phenomena is often distorted; but the ability to experience duration as a continuous, unified phenomenon is missing (Horányi-Hechst, 1943). This last view supports our present argument: to survive, the schizophrenic must maintain certain primitive functions but to a lesser degree he frees himself from the need to objectify experience because his psychosis detaches him from the requirements of normal social living. In short, look to the nature of the drives operating during the interval to anticipate whether or not the mentally ill will misjudge time more than comparable normal persons.

It should not be surprising that temporal symptoms may be so closely linked with syndromes of mental disorder, for any illness must mean that the patient has changed his behavior in some respect; and that change is likely to have some temporal implication. It is asserted, without documentation, that inmates of mental institutions, in spite of their inactivity, do not complain about being bored, in fact a symptom of recovery is said to be the appearance of such complaints (Germany; Hoche, 1923). If this is so, the implication is that patients do not pass temporal judgment or, if they do, the intervals are not crowded with events. Similarly, if there is repression, then fewer events can be remembered from the past and therefore distant time may appear shorter. Or phrased in another way, the generalization suggests that changes in the personalities and activities of persons, especially in a serious disease like schizophrenia, must induce some kind of corresponding changes in temporal judgments. The trick is to note the mental disturbances in the first place and then to relate them both genetically and in the present to the temporal potential. Two investigators believe that "loss in conceptualization and lowered attention span involving complex relationships" are the landmarks characteristic of schizophrenia (De La Gaza & Worchel, 1956, p. 195). If this view is correct, then the first loss might be expected to affect the mode of translation: the mentally ill cannot conceptualize and then express their own temporal judgments. The second defect in activity would provide the explanation for the faulty temporal information.

And what conclusion can be drawn from this examination of psychiatric materials? If ever additional proof were needed that temporal behavior is not an epiphenomenon but is most intimately linked to personality and most aspects of behavior, attention could be turned to the examination of a single psychiatric case or to most investigations of the mentally ill.

CRIME

In the West—and this section refers mostly to high-literacy countries—young and old criminals commit crimes for many reasons, but most of them have one attribute in common: they believe that they can escape punishment or that the risk of being punished is slight. There are exceptions of course: those who, consciously or unconsciously, are masochists, those who are unacquainted with the law, and those—such as hoboes and gangsters—who seek the comfort and safety of a cozy cell. In addition, the young, the feeble-minded, the insane, persons under the influence of drugs or propelled by a wild passion, and those acting in self-defense are thought to be unable to distinguish between right and wrong either continually or momentarily; hence they are not considered culpable, or the punishment they receive when convicted is likely to be lighter. The basic assumption seems to be that, if an individual, for whatever reason, is incapable of anticipating the consequences of his actions or of selecting alternative actions, he cannot be charged with any or entire responsibility for his behavior. For similar but opposite reasons, a premeditated murder, if proved, carries a heavier sentence than a spontaneous one—presumably in the first instance the person has appreciated beforehand the consequences of the deed.

From the nonlegal, social-scientific point of view, it is generally argued that perhaps every criminal is unable to anticipate consequences objectively as a result of conditions for which he himself is not fully responsible. A starving man who steals may be able to specify the penalty he will incur if caught; but that expectation is weak in comparison with his hunger and hence cast aside. Of course the potential criminal may simply anticipate incorrectly and underestimate the chances of being caught.

With the exceptions already noted, it seems reasonable to assume that the majority of potential or actual criminals have temporal orientations more toward the present than the future. That starving man is a case in point, and so too is the man who plans the one big robbery to insure his future happiness: his forward-looking fantasy produces excessive gratification in the present. Ordinarily the relation between orientation and crime is probably correlational: the orientation toward the present facilitates the criminal act, and the criminal act reinforces the orientation. A handful of studies unequivocally validating the restricted time span of delinquents and criminals is offered in *Addendum 9.8*.

The shortened time perspective of many criminals undoubtedly reflects deep strains within their motivational systems. The hypothesis advanced by some investigators is that such persons prefer present rather than deferred gratification. This view was once confirmed in a study among children in Trinidad: significantly more delinquents than nondelinquents preferred to receive five cents' worth of candy immediately rather than to wait a week later for twenty-five cents' worth (Mischel, 1961c). Similarly, even in a simple but realistic laboratory situation, those 6th-grade American school boys who in a previous session had indicated they preferred immediate rewards had a stronger tendency to cheat in reporting their scores on a shooting game than those who claimed to prefer delayed rewards; among the cheaters, the petty crime was delayed longer by those inclining toward the delayed rather than the immediate reinforcement (Mischel & Gilligan, 1964). As indicated in an earlier chapter, there is also a bit of evidence suggesting that in American society delay was more likely to be acceptable to middle-class than to lower-class children. Were this really so—and we also showed that the study in question has been challenged—we would have sociological and statistical reasons for expecting a similar difference between criminal and noncriminals: the former tend to come from the lower and therefore the handicapped groups. Delayed gratification, with the anticipation that the future will or can be improved, is a luxury which, perhaps, they simply cannot afford.

Failure to defer gratification means that it is difficult to control one's impulses, another trait frequently attributed to criminals. This point is illustrated by two Israeli studies which, if somewhat ambiguous, at least suggest how such a tendency may have repercussions upon temporal judgments involving transitory and ephemeral intervals: delinquents underestimated intervals with the Method of Production because, perhaps, they impatiently terminated the act of production prematurely (Siegman, 1961, 1966). Then another investigation, also summarized in the same *Addendum 9.8*, suggests that criminally inclined children found it difficult to see the causal connection between the promise of a reward or punishment in the future and its eventual administration (Redl & Wineman, 1957). This latter study reveals the importance of temporal organization most dramatically: unless the individual is able to anticipate the future, the verbal promise or threat in the future becomes meaningless, and hence it may appear senseless to regulate present behavior in appropriate ways. Then, upon the arrival of the future, the fruits of past experience can be gathered only when the connection between the promise or the anticipation and its fulfillment is recalled and experienced.

From the view that the criminal is a person not to be condemned—or, quickly to suggest another philosophy, not exclusively to be condemned—but to be cured of the illness which induced him to commit a crime in the first place comes a philosophy of treatment: while in prison, or otherwise under the custody of authorities, he must be reformed through the use of some form of therapy. After his trouble has been diagnosed—validly, semi-validly, or otherwise—he is placed under medical supervision, taught a new occupation, or provided with psychiatric assistance. Part of that assistance consists of trying to help him change his temporal orientation (although the phrasing may be different): it is hoped that he can find, with or without assistance from authorities and with or without some kind of change among his family and friends on the outside, an incentive to work for goals to be attained in the future and thus to be less of a creature of the moment. At this point we are confronted with the whole problem of change, change in temporal orientation, a discussion of which we postpone until Chapter 13. If the reader is neither criminal nor delinquent, presumably he can tolerate the delay.

SCHOOL DROPOUTS

Are school dropouts (leavers, in British terminology) deviants? In Western society, particularly in America which emphasizes the duration if not the quality of education, they do deviate from the norm that is considered desirable, viz., to finish elementary school or, with increasing frequency, secondary school. From the standpoint of persons in their own milieu, particularly among the lower economic groups, they may be considered, in contrast, conformists when the norm is to get to work as quickly as possible or (if a female) to find a husband as soon as decency and circumstances permit.

The person voluntarily leaving school clearly anticipates that further education will be of little or no benefit, or else he believes that the activities in which he can engage by leaving are, momentarily or semi-permanently, more attractive. The renunciations he feels he must make as a student seem too great. He can identify them because they represent a continuation of those generally imposed on youth and adolescents in the West: homework rather than carefree leisure, dependence upon parents rather than independence, limited spending money rather than a larger sum from his own earnings, sexual abstinence or sporadic sexual gratification rather than marriage. He may hear, and even be able to repeat the oft-repeated argument that the longer he remains in school, the greater will be his annual income or his total income during his working days, but the emotional impact of such figures may be lost upon him when he is intent on gratification at the moment. "The most frequent explanation for leaving college," a 1956 study of a sample of male Negro high-school graduates in the Washington area reported, "was the presence of nagging questions in their own minds about the wisdom of the course they were following . . . they had become unsure of what they really wanted to do with their lives. . ." (Grier, 1967).

In contrast, the individual who does not drop out presumably can tolerate renunciations because he anticipates benefits in the future. But does he? One study of 150 dropouts and 150 matched high-school graduates in six metropolitan areas in the United States suggests that the two groups both favored installment buying and thus did not wish to postpone gratification in this respect. But they differed in two ways: the graduates had been "presocialized" to a much greater degree "into middle class 'deferred gratification' patterns of chastity and monogamy" and the dropouts revealed on the TAT a pronounced tendency to have a "deferred gratification pattern" that was interpreted as being "weak" (Cervantes, 1965, pp. 134, 152, 199). On the other hand, another kind of result emerged from an investigation which concentrated upon two equated groups of lower-class New York boys, one planning to go to college and the other to work. So far as could be determined the concept of postponing gratification did not explain the decision of the group planning to continue their education: "going to college involves the *gratification* of values" already developed "rather than a relinquishing of valued behaviors" (Beilin, 1956). College appeared not as a dreary grind but as a relatively blissful interlude with few responsibilities and with opportunities for exciting pleasures; in addition, of course, there would be benefits in the future. Externally, therefore, the orientation may have appeared to be completely toward the future, but in fact present or near-present gratifications were very compelling.

MARTYRS AND MYSTICS

The culmination of renunciation is the martyr who willingly makes a great sacrifice, including his own life, in behalf of the cause he favors. His anticipations concerning the aspects of living which he thus forgoes or concerning the continuation of his own life may be both clear and accurate, but he also is convinced that those satisfactions will and must pale unless he renounces them. He simply cannot continue his existence no matter how alluring it seems.

The temporal orientation of a martyr, however, is away from a mundane future and toward a much less tangible one: his ostensible concern is not with his own future but with the future of the idea, the ideal, or the persons for whom he makes the sacrifice. It is possible, since all possibilities exist when the many forms of man are squeezed into a generalization, that he may derive comfort from the belief that his own contribution will be long remembered and that thus, like any parent, he will achieve a kind of immortality among future generations. Such a projection, however, may not be dominant and, if dominant, it may remain most spiritual and intangible so far as the person himself is concerned. Still, on a materialistic basis, it is necessary to wonder whether the martyr keeps faith in small or large part for fear of what his fate will be if he were not to do so. The motives may be mixed—and the exact mixture can be ascertained only by examining the individual case—but the existence of martyrs who do not repent or relent in the face of severe physical or spiritual torture is eloquent evidence for a human potentiality. Even so, the motivation to sacrifice oneself must be intense, and hence it is not at all surprising that in the Western world the call so frequently has its origin in religion. The prototype is perhaps Jesus Himself. On a humble level we find a devout Austrian peasant who opposed the Nazis and was eventually beheaded by them: "We are," he wrote, "all in God's hand, who directs and guides our fate, at whose words even the wildest torrents are stilled." Then immediately the realistic call for action: "But experience teaches us that as long as there have been men upon this earth, God has allowed them freedom of will and only seldom has He interfered in the destinies of men and peoples" (Zahn, 1964, p. 216).

In this psychological sense, every sacrifice in behalf of an ideal or another person contains a touch of martyrdom which, like all other behavior, presumably falls along a continuum extending from almost zero to the infinite. Again diverse motives may be operating, but the crucial point remains the same: the individual renounces because or when he is somehow able to propel his sentiments both backwards and forwards and hence to view himself in perspective. But are we becoming sentimental and maudlin when we say this? Years ago female white rats endured the greatest amount of electric shock, crossing a grill (fiendishly designed by an experimenter) not when the lure was to still their own hunger pangs, to quench their own thirst, or even to copulate with a male rat, but to reach and then protect or feed their young (Nissen, 1930). Is this sacrifice or martyrdom? I am not at all sure, for an adequate answer requires the confrontation of the basic biological drive of the rat and the intellectual-emotional decision of the human being or the saint. Maybe ethically we withhold praise from the rat because its constitution gives it no alternative, and we thank the person and venerate the saint because he could have behaved otherwise. Surely, then, martyrdom and sacrifice are not being belittled or even evaluated when the rewards are identified. Everlasting punishment is to be avoided: "It has been shown by the certain testimony of God's word that impenitent dead are to be forever conscious and miserable" (Hovey, 1859, p. 159). Or God Himself becomes angry if the sacrifice is not made, a view which prevails in the Old Testament (De Vaux, 1964, p. 91). The Aztecs, it is estimated, sacrificed 20,000 human beings annually, a deed which must have brought some kind of sorrow to relatives and others who were the witnesses; still "the belief that only with this grim aid could the sun, on which the nation depended for its survival, continue its life-giving functions" (E. O. James, 1962, p. 77) must have been more overpowering. Even a nonliterate person who sacrifices not himself but animals obtains immediate compensation: "he has acquired a religious character which he did not have before, or has rid himself of an unfavorable character with which he was affected; he has raised himself to a state of grace or has emerged from a state of sin" (Hubert & Mauss, 1964, pp. 9–10).

One of the more dramatic aspects of martyrdom is the response to pain. We do know, or think we know, that in general people's reactions depend on cultural and individual elements. Thus in some societies people are supposed to be stoical, in others they are expected to cry out as they experience pain (Zborowski, 1952). Similarly in a given society one group (men in Western society) seldom permit themselves to shed tears, another (women) may weep copiously. Training and discipline must be involved—infants show little or no restraint, though the amplitude and duration of their yowls also vary—as well as temporal orientation: to avoid succumbing to pain, a stronger tendency must be established which is likely to involve anticipating the effect of an outcry upon one's status and reputation. And so a martyr is concerned with matters more important than the pain he endures: being a source of pride to his cause or gaining everlasting salvation is more compelling, for reasons ranging from the masochistic or sadistic to the altruistic or sublime.

The mystic might be considered a special form of martyr who for long or short periods renounces his ordinary existence in order to achieve an experience which he himself and perhaps also his followers consider transcendental. He withdraws into a kind of joyful, peaceful trance in which he either recovers information from the deeper layers of his own being or else, as he himself is likely to believe, obtains extraordinary information, which he considers both compelling and valid as well as quantitatively and qualitatively different from ordinary experience, from some nonmaterial source such as gods or ancestors. He often foregoes his sense of individuality; he may have, for example, a feeling of merging with the external world or with the universe itself. During the trance, as in a dream or under the influence of a drug—in fact a drug may be affecting him—he may lose his sense of time completely, so that he is not driven by a temporal motive and what he experiences may have no temporal reference point; he may, as the glib phrasing would have it, escape from time. When he awakens, that sense is restored, indeed it must be restored if he is to interpret the mystical experience to himself and to others. We deliberately refrain from exploring the metaphysical aspects of this subject, nor do we pass judgment either upon the validity of the mystical experience or its psychological significance or genesis. With proper reverence it is sufficient to say that the person himself believes that during the trance he has undergone an impressive experience and is often able to convince others of its validity, even when he feels he cannot adequately express in words what has transpired. Impressive is not only the content of his experience but also its detachment from reality. Apparently timelessness has mysterious attributes which impress those who have the sensation directly or vicariously.

ADDENDUM 9.1 BRAIN LOCALIZATION

Two broad-gauged studies produced no evidence for localization of temporal behavior:

1. Normal persons and patients were assigned the task of encircling the letter "T" on printed sheets and were told beforehand that they would be interrupted repeatedly and asked to estimate verbally the elapsed time. The patients were suffering from a variety of diseases (menigo-encephalitis, post-traumatic psychosis, postencephalitis, psychosis with organic brain diseases, alcoholic psychosis with deterioration, etc.), but "no relation between the clinical diagnosis and the test performances" could be established, rather the tendency to under- or overestimate the intervals (in comparison with norms from the nonpatients) and the "loss of temporal continuity" seemed to be "roughly proportional to the degree of deterioration" (U.S., adults; Coheen, 1950). One finding in this study is puzzling, especially since the reliability of the judgments from 6 to 10 months later is reported to be "consistent": the patients who overestimated the intervals revealed in their general behavior decidedly more "marked" deterioration than those who underestimated them.

2. A survey of hospital records from 56 and again from 25 British patients revealed no relation between judgments concerning the duration of an interval of testing which lasted over an hour and, respectively, (a) the type of brain lesion and (b) the amount of cerebral atrophy (verbal, immedate subsequent; McFie, 1960).

Other evidence against localization is of

several kinds. First, the effects of an operation may be only temporary, which may indicate that other parts of the brain can undertake the temporal functions. Thus thalamotomies on 30 patients, who had been tested beforehand, produced confused temporal information (e.g., with respect to the day of the week or the patient's age) in 19 instances, a disturbance involving both temporal and spatial information in 4, and no change in 7. With almost no exception, however, the disturbances lasted only a few days or a week. This finding is interpreted as validating the "assumption that multiple circuits participate in the mechanism of temporal orientation, so that a lesion of single thalamic nucleus produces only transitory disturbances" (U.S.; Spiegel, Wycis, Orchnik, & Freed, 1955). Similarly, 10 patients, tested from 45 to 90 minutes after shock therapy, judged the interval quite inaccurately: they inclined toward very marked overestimation. But then their judgments dramatically improved within 2 to 3 weeks (U.S.; verbal, prior; Coheen, 1950).

One investigator has sought to study the effect upon temporal judgments of fluctuations in cortical activity ascertained by measuring brain waves on a continuous electroencelphalographic record. When subjects produced intervals ranging from 0.5 to 8 seconds with eyes opened and closed, the results were largely negative (U.S., adults; prior; Legg, 1968).

The actual experimental alteration of the brain through lobotomy has yielded mixed effects upon temporal behavior:

1. In one careful study, two groups of schizophrenic patients judged, on two occasions, the duration of 5-, 10-, and 15 second intervals by the Methods of Production, Reproduction, and Verbal Estimation; between these two test periods the patients in one group underwent frontal lobotomy and those in the randomly selected control group were not operated upon. The operation on the whole did not affect temporal judgments. In addition, positive correlations between judgments of 15 seconds via the Method of Production on the one hand and subtests of an intelligence test and a sorting test on the other hand decreased after the operation (U.S., Clausen, 1949).

2. After a "standard lobotomy"—as well as "selective ablation of Brodmann's areas 9 and 10" and a cingulectomy (a severing of brain fibers)—patients tapped more rapidly; but after an "open rostral lobotomy" and a "left temporal lobotomy" the reverse effect was noted (France, U.K., U.S.; Petrie, 1952, 1958).

3. Three weeks after a prefrontal lobotomy, 24 chronic schizophrenics revealed, among the most significant improvements in general orientation, a more accurate knowledge of the date; three and six months later they further progressed in being able to tell time from a watch. Another change involved the ratio of immediate-temporary to future-present wants as ascertained by asking them directly what they desired: before the operation, that ratio was 3:2, three weeks afterwards 11:1, and then it decreased to 3:1 three and six months later. Thus better temporal orientation was accompanied by a more pronounced orientation toward the future (U.S., adults; R. E. Jones, 1949).

4. Informal examination of 300 cases of prefrontal lobotomy indicated that a year after the operation all of them were well oriented with respect to temporal information such as the date (U.K.; Partridge, 1950).

Views favoring brain localization include the following:

1. After examining the anatomical evidence and the investigations of the effects of lesions upon temporal behavior (including those just mentioned as well as others), one author cautiously states: "It is supposed . . . that the prefrontal areas of the cortex perform the fine acts of discrimination and judgment in the timing of events, and that the comparison of information about time reaching the cortex from the environment, with the internal standard [e.g., with that provided by the heart, respiration, body temperature, and presumably other processes affecting the biochemical clock], is made at this point" (Dimond, 1964).

2. It is argued didactically that the cortex and the subcortex must be divisible into sections concerned with past and future orientation corresponding to the areas in which, respectively, sensory and motor functions are located, for in this view the former reflects the past and the latter the future (Campbell, 1954).

3. One writer, after exhaustively reviewing the physiological literature, suggests that the biochemical clock "may be placed tentatively in the region of the hypothalamus" inasmuch as "temperature and other eosinophil rhythms" persist even in the absence of cortical activity (Mills, 1966).

4. Another investigator has reanalyzed the data in the hospital study cited above (McFie, 1960) and is convinced that "temporal lesions produced significantly less inaccuracy than frontal or parietal lesions" (Orme, 1969, p. 84).

5. Some clinical histories have revealed an association between alterations in temporal behavior and damage to particular parts of the brain. One patient whose case is frequently cited (because the description is both long and complete) suffered from a wound in an area between the frontal and parietal lobes as a result of an attempt at suicide. Noteworthy is the fact that his judgment of transitory intervals in the normal state and under hypnosis seemed "normal," but his orientation in another respect was disturbed: he revealed "confusion regarding the chronological sequence and arrangement of events, i.e., the 'earlier' and the 'later,' the 'before' and 'after' of time." He also showed similar disorientation regarding space. The investigator advances the hypothesis that the lesion affected not a "primitive" time sense but a "gnostic temporal conception" (Germany; prod., prior; Ehrenwald, 1931b). Another adult, exhibiting motor aphasia after a bullet wound in one temporal lobe, in general was unable to respond abstractly to stimulus conditions or problems in his immediate environment; he seems also to have lost a sense of subjective time and to have become dependent upon external cues. "When he had to be at a definite place at a definite time, he looked at his watch and when he found the hands at the place corresponding to their position at the concerned time [sic], he concluded it was this time . . . All times were moments combined with definite concrete behavior: 12 was lunchtime, 3 o'clock when his mate left the room, etc." (Germany; Goldstein, 1948; p. 213).

ADDENDUM 9.2 THYROID GLAND

An early investigator who pointed out that the thyroid affects "the coarser rhythm of circulation and the more diffuse and intricate rhythms of metabolism" reported no consistent or significant relation between admittedly too small groups representing the various types of thyroid disorder or thyroid surgery on the one hand and temporal estimates on the other hand; thus the group with high metabolism overestimated an interval of 45 seconds to a greater degree than one with low metabolism with the Method of Verbal Estimation, but the two groups gave virtually identical estimates of 1 minute with the Method of Production (U.S., presumably adults; prior; Gardner, 1935). A later study, however, revealed positive results. The task was to reproduce an interval of 15 seconds after a single demonstration and then repeatedly to produce it in a series whose average was ascertained. In comparison with control groups of patients having no hormone imbalance, Air Force candidates, and college students, hyperthyroids tended significantly to underestimate the interval and, in comparison with some but not all of the controls, hypothyroids to overestimate it. The estimates of hypothyroids after thyroid therapy were so reduced that they did not differ significantly from those of the controls (Canada, adults; Stern, 1959). Another study also suggests that hypersecretion accelerated the sense of timing. The median for normal patients in a hospital was 0.80 seconds and for those suffering from hyperthyroidism it was 0.35 seconds when attempting to produce a 1-second interval by means of the limits-verbal method (U.S.; prior; Kleber, Lhamon, & Goldstone, 1963).

ADDENDUM 9.3 TEMPERATURE AND RELATED PHYSIOLOGICAL CHANGES

1. In the following studies, a rise in bodily or room temperature resulted in *overestimation with Verbal Estimation or underestimation with Production* (with counting and tapping considered forms of Production), or vice versa.

a. Counting faster: for two subjects influenza produced the rise, for the third artificial means (U.S., adults; Hoagland, 1933).

b. Tapping faster; there was high inter-subject variability; the relation between the increase in tapping and body temperature was not always linear. Room temperature, however, had a similar though smaller effect (U.K., students; Fox et al., 1967).

c. Slightly, though significantly shorter judgments with the Method of Limits-Verbal (evaluated, as ever, as a form of Production) in comparison with a matched control group which was led to believe that the temperature of the room would be increased (in fact it was not); the body temperature of the experimental group was actually raised from 1.8 to 3.6 degrees F.; the judgments of the control increased slightly (U.S., college; Kleber, Lhamon, & Goldstone, 1963).

d. Faster production of intervals of 8, 24, and 44 seconds when the ambient temperature was raised from 80° to 110° F.; body and skin temperature increased with the rise in room temperature. But Production was also shorter at the low ambient temperature of 40°; with Verbal Estimation, judgments of intervals of the same duration were significantly longer at 40°, though only slightly so at 110°. Also, body temperature did not decrease when ambient temperature fell, but skin temperature did (U.S., soldiers; prior; Lockhart, 1967).

e. Slower production of 60 seconds by counting; amateur scuba divers in Wales were immersed in cold sea water (4° C.) during a week of diving. Their pulses also were lower. The change, the investigator indicates, probably cannot be attributed to anxiety because, as a control he found no change in the speed of counting among the men when they were about to dive into warm water with pieces of ignited gun cotton in their mouths (Braddeley, 1966).

f. Faster production and overestimation of intervals, all of which were 30 seconds or less; body temperature was not raised by a high fever or by artificial means, but fluctuated "normally" during the day. The advantage of this procedure, according to the investigator, is that it factors out the metabolic rate: when bodily temperature rises from a fever or artificial means, metabolic rate also increases, but the "normal" rises during the day are accompanied by a decrease in that metabolic rate (U.S., college; prior, Pfaff, 1968).

g. Faster counting or tapping at a prescribed rate; rise in oral temperature between 0.4° to 1.2° C was accomplished by immersion of legs in warm water. But although the result just mentioned was obtained by comparing the rates when the body temperature was normal and when it was raised, there was not a significant correlation between the two increases; speed of tapping and adjusting the rate of a metronome beat were not related to temperature; and of the eight subjects one or more under varying circumstances showed no change whatsoever as temperature varied (U.K., presumably adults; Bell, 1965, 1966).

2. In the following, *underestimation with Production* has been found to be associated with the indicated physiological processes:

a. Sleep deprivation (for 72 hours);

shorter productions of one-half and 2-minute intervals. This effect, however, was due in part to the fact that on the second and third days the subjects tended to doze and then had "to awaken suddenly" with the result that they felt "more time had elapsed during the 'blank' period than really had passed" (U.S., medical; prior; Bliss, Clark, & West, 1959).

b. Alpha rhythms of electroencephalographic activity; with lower EEG frequency in the resting condition and with eyes closed, shorter productions of 2 and 8 seconds were noted. Constant or intermittent photic stimulation of the eyes did not systematically affect the EEG, although the intermittent stimulation resulted in longer production of the 8-second interval (U.S., college; prod., prior; Werboff, 1962).

c. Respiration and heart beat; with higher rates induced by drugs, shorter intervals were produced. But there was no significant correlation between judgments and blood pressure or skin temperature; and there were no consistent correlations between judgments and these physiological measures with the Method of Reproduction. The design of the experiment required the same subjects on different occasions to estimate four intervals ranging from 0.5 to 5.0 seconds, which were presented, and produced or reproduced, through different modalities (audition, vision, or cutaneous stimulation) under the following conditions: in the normal state, after a placebo, and after ingesting one of five different drugs affecting various parts of the nervous system (U.S., adults; Hawkes, Joy, & Evans, 1962).

With two different techniques a positive correlation between oxygen content and verbal estimation of intervals has been reported. With one, the oxygen content in a tent containing the subjects was lowered and "three hours were estimated by the majority as two, or less" (U.S., medical; presumably subsequent; Barach & Kagan, 1940). With the second, a group of American soldiers judged intervals in the neighborhood of 30 seconds first at sea level, then at an altitude of 12,900 feet.

With Verbal Estimation the mean figure dropped at the high altitude but not quite significantly so; and with Production the mean figure significantly increased (prior; Cahoon, 1967).

A prominent Soviet investigator, in discussing physiological bases for temporal behavior, postulates an internal timepiece which he calls a "visceral clock." That clock exerts its influence directly: "the greatest precision of perception is observed at typically human 72–84 beats of the pulse per minute" and he adds that "the greater the deviation from this pulsation, the greater the error in time reading." "Feedback" also plays an important role, he emphasizes. In perceiving rhythmic stimuli, for example, there are "movements of the head, trunk, and, most frequently, of the extremities" which are fed back to the cerebral hemispheres where they serve "as a source of afferent signals"; any disturbance of these motor components, therefore, leads to temporal errors. The components may be affected by "a corresponding ideation" as well as by "the surrounding conditions" of the organism, both of which forms of feedback can improve temporal performance (Elkin, 1963 & 1966).

Findings not supporting Hoagland's original generalization have already been mentioned in presenting the positive results; in addition, some studies have been negative, viz., with an increase in temperature, pulse rate, etc. there has been a *deceleration of the timing sense:*

1. Body temperature (raised by increasing room temperature) and pulse rate; Production and Verbal Estimation of intervals varying from 45 seconds to 13 minutes (U.K., adults; Bell & Provins, 1963).

2. Mean pulse rate and variability of the rate in a resting state over a 5-day period; Production, Reproduction, and Verbal Estimation of intervals varying from 4 to 32 seconds (U.S., adult hospital patients; presumably prior; Ochberg, Pollack, & Meyer, 1964).

3. "Heart work, pulse rate, blood pressure, lung work, and breathing rate"; Verbal Estimation of intervals ranging

from 4 to 27 seconds (U.S., college; prior; Schaefer & Gilliland, 1938).

This *Addendum* can point up the problem of the biochemical clock by an extended reference to one study employing American undergraduates as subjects. A strange collection of significant and non-significant differences emerged with Verbal Estimation, conventional Production, and tapping as the methods of measurement. Anxiety, whether gauged on a paper-or-pencil test or induced in the laboratory by the anticipation of a shock, turned out to be unimportant. But among those subjects not anticipating a shock, tapping rates were faster and verbal estimates were lower for those with high respiratory rates during the interval than for those with low rates, verbal estimates were lower for those with high heart rates, and there were low but significant correlations between verbal estimates and the alpha frequency of the EEG. Among those anticipating shock, estimates were higher, productions were shorter, and tapping was faster for those with high heart rates than for those with low rates, and there were low but significant correlations between the same EEG frequencies on the one hand and verbal estimates and tapping rates on the other (Cahoon, 1969). All these results are consistent with the relevant principles of this book except for the lower verbal estimates with increased heart rate in the non-shock group; and there, the investigator suggests, "the situation is confused." Otherwise the mixed bag shows, again in the investigator's words, that the biochemical clock cannot be considered a "unitary concept": sometimes a physiological process has an important influence, sometimes not, and in either case the method of measurement plays a role. Whether or not these processes consciously affect judgment, we of course do not know; but the writer assumes that the subject in the shock group "was responding to the perception of his own heart beat in establishing a subjective time rate." If this is so, then experimentally induced temporal behavior is affected by anxiety only when that anxiety is mediated by one of the many components of the biochemical clock.

ADDENDUM 9.4 EFFECTS OF DRUGS

Here the classification of drugs into broad categories is based upon a standard textbook in medical pharmacology (Goth, 1966) and a medical dictionary (Thomson, 1967).

1. Studies supporting the view that stimulants or antidepressants accelerate the sense of time (and hence strengthen tendencies toward verbal overestimation or produced underestimation) and that depressants, tranquilizers, or sedatives have the reverse effect include:

a. Verbal Estimation of intervals ranging from "several seconds" to "a few minutes": overestimation from scopolamine (tranquilizer); 10 subjects, presumably adults. But with the same procedure, the same drug, and the same subjects the reverse effect was obtained when the intervals being judged were relatively long, i.e., over a half hour, and when judgment was passed, possibly immediately afterwards, concerning the duration of the investigation itself which lasted for more than four hours (Switzerland; Heimann, 1952, pp. 34–6).

b. Production of 2.5 second intervals through the procedure of DRL (see above, page 195): more frequent and more rapid response from dextroamphetamine (stimulant) (U.S., medical; Dews & Morse, 1958).

c. Successive, rhythmic productions of intervals of 1 second for 90 seconds and of 10 seconds for 15 minutes: significantly longer productions with secobarbital (depressant) than with a placebo; the effects of the latter, however, could not be differentiated from dextroamphetamine. I have classified the method here as that of Production, although on the first trial a standard provided by the investigator was reproduced. Under these special circumstances the interval of 1 second proved to be "more sensitive" to the drugs; and feedback concerning the accuracy of each series nullified the effect of the depressant and increased intersubject variability but only in connection with the 10-second interval and without affecting consistently the influences of the stimulant, alcohol, or the placebo (U.S., college; Rutschmann & Rubinstein, 1966).

d. Production of 1 second via Method of Limits-Verbal; with secobarbital and a placebo: the judgment was longer than with dextroamphetamine (U.K., students; Goldstone, Boardman, & Lhamon, 1958a). The senior investigator has replicated these results with different methods (two versions of Rating-Verbal). In addition he has shown that the same differences appeared regardless of whether the stimulation was through vision or audition; that the stimulant had its maximum effects 30 minutes after ingestion and continued at that level 60 and 90 minutes later, whereas the maximum effect of the depressant did not appear until 60 minutes had passed; and that secobarbital produced greater intrasubject variability than the dextroamphetamine or the placebo (U.S., college; Goldstone & Kirkham, 1968).

e. Number of digits read during a given interval, the verbal estimate of the interval spent in reading the digits, speed of counting and of tapping at a subjective rate of 1 per second, the production and reproduction of a 10-second interval: in comparison with the effects from pentobarbital (depressant), metamphetamine (anti-depressant) was associated with a larger number of digits read, longer verbal estimates, faster counting and tapping, and shorter production; reproduction produced no significant difference; and on the whole the control condition with a lactose placebo gave results midway between the two drugs. In addition, other related temporal behavior was altered by the drugs: when instructed to tap at their preferred rate, subjects under the influence of the depressant tapped and drew a mirror image more slowly than when affected by the antidepressant; but the drugs had no significant effect upon a task involving word fluency (Sweden, medical & lab. technicians; Frankenhaeuser, 1959, pp. 55–68).

f. Number of digits read during a given interval and verbal estimate of the interval spent in reading the digits: a larger number of digits was read, almost, but not quite, significantly so after caffeine (stimulants), but no difference appeared in verbal estimates. With caffeine significantly more digits were produced than with quinine (presumably neither a depressant nor an antidepressant); and nitrous oxide (presumably a depressant for some, an antidepressant for others in small doses) produced significantly shorter verbal estimates than air or oxygen, but no difference with respect to number of digits (loc. cit., pp. 96–111).

g. Tapping a pencil on table, in a 2-second rhythm, for 2 minutes: in comparison with normal judgments and those obtained from other subjects with a placebo, overestimation after alcohol (depressant) and underestimation after caffeine (Germany; Joerger, 1960).

h. Verbal estimation of transitory and extended intervals; overestimation from marijuana (presumably a stimulant) (Germany, clinical experience; Fraenkel & Jöel, 1927) (U.S., clinical experience; Bromberg, 1934) (South Africa, young adults; Ames, 1958) (U.S., young adults; immed. subseq., Weil, Zinberg, & Nelsen, 1968).

2. Negative or inconclusive results are likely to be obtained when the Method of Reproduction is employed because, as one investigator has suggested, the ingested drug presumably must affect both the perception of the standard interval and the execution of the reproduction. He himself found that meprobamate (tranquilizer) produced greater intersubject variability than a placebo but no significant difference (U.S., adults; prior; Costello, 1961). Similarly, in a study involving the reproduction of intervals ranging from 5 seconds to 5 minutes, in one instance 15 minutes, there was overestimation from tea (stimulant) and

from small doses of quinine and underestimation from alcohol by eight well-practiced subjects. But thyroxine (presumably a stimulant) led to underestimation (Germany, adolescents & students; Sterzinger, 1935). A follow-up investigation with two other subjects, moreover, indicated that the results just mentioned depended also upon the length of the interval and upon the tendency for subjects in general to overestimate or underestimate the passing of time. Under the influence of alcohol, for example, there was a tendency to underestimate intervals between 15 and 25 minutes, to overestimate longer ones (Sterzinger, 1938). Isocarboxazid (antidepressant) induced a slight increase in the number of schizophrenics overestimating their production of a 10-second standard, but left unchanged reproductions of ephemeral intervals and a 30-second interval (France, adults; Guyotat & Burgat, 1965). Negative and inconclusive findings have also emerged with the other methods:

a. Production of 1 second via Method of Limits-Verbal: half of the dozen subjects reported that time seemed to "drag," the other half that it seemed to accelerate after drinking 0.5 gm/kg of alcohol (U.S., college; prior; Kirkham, Goldstone, Boardman, & Goldfarb, 1962). The problem arises here as to whether alcohol in small doses is a depressant or a stimulant.

b. Reports from unspecified individuals; opium (said to produce excitement in small doses) induced the impression that time was passing slowly (May, 1958, p. 104).

c. Verbal estimation of the length of an interview (perhaps with immediate subsequent arousal of the temporal motive), production and reproduction of a 15-second interval; no "consistent trends" appeared when the effects of sodium amytal (depressant), dextroamphetamine, and psilocybin (an hallucinogen) were compared, but with the latter two methods there was "a consistent increase of estimated time intervals in all three subjects after sodium amytal" and greater accuracy was achieved following psilocybin. The last finding, the investigator surmises, may have been due to an awareness by the subjects that the time sense was distorted so that "special efforts" were made to correct it (Canada; Lehmann, 1967).

d. Production of intervals ranging from 1 to 4 seconds: in comparison with a placebo, fencamfamin (antidepressant) had no effect upon producing intervals ranging from 1 to 4 seconds, although in the same experiment another antidepressant, dextroamphetamine, impaired judgment more "severely"; since impairment was least with no drug, the investigator concludes that fencamfamin functioned simply like "a pharmacologically inert compound." But then with the Method of Reproduction subjects under the influence of either of the above drugs made more accurate judgments than with no drug or a placebo (U.S., adults; Goldstone, 1964, pp. 215–9).

3. Research on LSD, as suggested in the text, has not produced a clean-cut generalization concerning temporal behavior; *under the influence of the drug:*

a. The vast majority of subjects reported retrospectively that they had occasionally "lost" their "sense of time" and more than half felt as if time had come to "a standstill or stopped now and then." Both of these impressions did not appear when a placebo was taken, but were present when the same subjects were put in isolation and deprived of sensory stimulation for a period of 8 hours (U.S., adults; Holt & Goldberger, 1960, pp. 13, 16).

b. A "high proportion of patients," as judged by surveying previous investigations and by carefully observing 29 cases suffering from a variety of psychiatric disorders, revealed "a time disorder" but there was "nothing specific" in the produced effects. The investigators believe that the effects cannot be attributed to possible pyretogenic action, especially since they could find no clear-cut relation between mood changes produced by the drug and the temporal disturbances (England, adults; Kenna & Sedman, 1964).

c. On 11 of 17 occasions the "time sense" of 15 normal adults was disturbed and was "characterized by the feeling of time being accelerated or retarded" (U.S., medical personnel; DeShon, Rinkel, & Solomon, 1952).

d. Four subjects in a carefully controlled investigation revealed no consistent change, only an increase in variability; perhaps, therefore, "temporal frames of reference tended to become vague or 'lost' " (U.S., adults; verbal, prior; Boardman, Goldstone, & Lhamon, 1957).

e. 'Subjective awareness of time passing was altered in most" of the 17 subjects; time "stopped, slowed up, or speeded up, and even ran backwards" (presumably U.S.; Stefaniuk & Osmond, 1952).

f. Intervals were underestimated with the Method of Production (France, presumably adults; also Reproduction; Benda & Orsini, 1959) (U.S., adults; Aronson, Silverstein, & Klee, 1959).

g. A "majority" of 21 schizophrenics reported a "slowing of the subjective recognition of the flow of time"—which presumably meant verbal underestimation on the secondary level (U.S.; Hoch, Cattell, & Pennes, 1952).

h. One subject seems to have overestimated time: "My concept of time," she reported afterwards, "was disturbed; the day seemed to cover years." Two years later, again under the influence of the drug, she recorded: "During these two hours, time had no meaning for me— I was bound by neither time nor space. . . . Time was unimportant to me. I seemed to live a lifetime in seconds" (Hoffer & Osmond, 1967, p. 172).

i. No reference whatsoever is made to temporal behavior in a broad survey of the "mental effects": the long list begins with "sensory changes" and ends with "new integrative experiences which may be culturally acceptable" (Blum et al., 1965, pp. 265–6).

j. The effect was that of overestimation, as reported presumably in some kind of interview; actually this is the finding in connection with another drug, DMT, and the investigators simply assume that such an effect is similar to that produced by LSD (Hungary, adults; presumably interviews; Böszörményi & Szára, 1958).

k. Adult men (all of whom were inmates of an institute for chronic alcoholics) tended retrospectively to indicate after LSD that the statement, "hours went by like seconds—or one second seemed to last forever," was "a little like the experience" they had had; they were inclined after chlordiazepoxide (a mild tranquilizer) to claim that the statement was "neither like nor unlike the experience"; and after the drug that was supposed to function as the control, methylphenidate (an antidepressant), they thought the statement was "very much like the experience" (U.S.; Ditman et al., 1969). Obviously the statement being rated offered contradictory alternatives, so that the direction of the effects cannot be ascertained.

4. The effects of other hallucinogens are not uniform. Psilocybin (the active ingredient in certain mushrooms found in Mexico) seems to accelerate the timing sense. After a "mild" dose in one investigation, tapping rates increased and so did handwriting size, eye movements, and gustatory sensitivity. In this and other studies, the author implies, there were large individual differences in response, on the basis of which he classifies subjects into "reactors," "mild reactors," and "non-reactors" (Fischer, 1966). Thus one of his two subjects had a mean tapping rate for an 8-minute period of 1.16 seconds normally and 2.21 seconds after ingestion; the corresponding figures for the other subject were 0.42 and 1.31 seconds. A tranquilizer produced the opposite effects with reference to both temporal judgment and the amount of space used in copying a message on a piece of paper (Fischer et al., 1962). In passing: consistent with this result regarding temporal judgment is the "slowed passage of time" associated with psilocybin which other investigators have found (Hoffer & Osmond, 1967, p. 496). The authors just cited also report miscellaneous protocols from subjects who had ingested the indicated substances:

Ordinary kitchen nutmeg: " . . . when someone speaks to me . . . I can wrench myself back to reality and know what they're saying. The only problem is that it seems like ten minutes between their questions and my answer . . . Time loses all meaning just as in LSD" [p. 54].

Ololiuqui (a seed used by Indians in Southern Mexico): "I'm not disoriented in time but it is extremely difficult to tell where in time I am" [p. 247].

Adrenochrome (a derivative of the secretion from the adrenal gland): "During the day, time passed very quickly and looking back on it, it seems remarkable that I put in a whole 8-hour day" [p. 365].

5. Drugs in other categories sometime affect temporal behavior. An increase in tapping speed has been reported from chlorpromazine, a drug suppressing nausea and vomiting (France, U.K., U.S.; Petrie, 1958). Some subjects under the influence of nitrous oxide, an anaesthetic, indicated that time seemed to be passing quickly, others that it was passing slowly, although—as indicated in a previous chapter—these feelings had no consistent relation to rate of counting (England; Steinberg, 1955). Another investigator found no effect upon temporal judgment with immediate subsequent arousal of the temporal motive, but apparent underestimation with delayed subsequent arousal (Germany; Sterzinger, 1938).

ADDENDUM 9.5 THE DURATION OF DREAMS

The aim of research on this aspect of dreams has been to establish a relation between their reported duration and some objective, measurable phenomenon that can be observed by the investigator. REM and EEG have seemed promising because they can be recorded by means of delicate instruments without disturbing the sleep of the habituated subject. At first it was thought that dreams were confined to periods of REM and were absent during non-REM. But now there is evidence indicating that REM and rapid EEG as well as decreased muscle potentials recur "on a cyclical basis in real time through the 24-hour day irrespective of whether or not the subject is asleep" (Globus, 1966); and also that "reportable mental activity is always present in the sleeping human," whether or not there are REM, and that similar dream content appears in both periods, particularly if the non-REM period takes place late at night (Foulkes, 1962; Pivik & Foulkes, 1968). It seems, however, that the probability of dreaming, or being able to report a dream, is greater in the REM than in the non-REM state. For this latter reason, examples from the older research are worth noting.

Five subjects were awakened 5 or 15 minutes after REM had begun and were asked to estimate whether they had been dreaming for 5 or 15 minutes. Four out of five chose the correct period "with high accuracy" and most of the erroneous judgments by the fifth person resulted from estimating the longer period of REM to be the shorter (U.S.; adults; Dement & Kleitman, 1957). In another investigation the sleeping subject was stimulated with the sound of a tone, the flashing of a light, or a spray of cold water during periods of REM and EEG: "In each of 10 instances, where the stimulus was incorporated [into the dream] and the subsequent interval was precisely timed, the amount of dream action in the interval between the modifying stimulus and the awakening did not vary far from the amount of action that would have been expected to take place during an identical time in reality." These tentative findings, therefore, support "the hypothesis that dream events and real events proceed at about the same rate"; anecdotes to the contrary are either rare or in some instances suspect (U.S., adults; Dement & Wolpert, 1958). A relation was shown to exist between the objective duration of types of REM or EEG and the degree to which that time was overestimated afterwards; it was also noted that the type of EEG during sleep had no effect upon the judgments concerning intervals ranging from 1 to 10 seconds as measured immediately afterwards in the waking state by Verbal Estimation, Production, and

Reproduction (U.S., unspecified; Carlson, Goodenough, & Fineberg, 1968).

Another technique has revealed a discrepancy between the duration of a type of dreaming and objective time: hypnotized subjects were given the instruction, "you will have a dream before you wake up . . . You will remember what you have dreamt and be able to tell me all about it." They were awakened within 1 second or two, and then reported long and very complicated "dreams" (Sweden, adults; Schjelderup, 1960). Whether or not these experimentally induced "dreams" are the same as ordinary ones cannot be stated; but the reports seem most imaginative and suggest at least that in a short time and at the spur of the moment considerable fantasy could be created.

The temporal orientation of dreams has been investigated in one study: the occurrence of noncontemporary dreams among four subjects was positively correlated with the time spent before being awakened to report the dream and negatively with body temperature (U.S., Verdone, 1965).

ADDENDUM 9.6 HYPNOTIC TIME

Accuracy

According to one summary of the literature prior to 1920, subjects in the hypnotic trance were reputed to judge time more accurately than in the waking state (Loomis, 1951). Some of this evidence, I suspect, requires careful analysis before being accepted. In one article, for example, data from two subjects in both the hypnotic and the waking states are given (Czechoslovakia; Ehrenwald, 1923); in spite of the author's claim, his own statistical treatment suggests that hypnosis had no beneficial effect upon temporal estimates which could not have been achieved through repetition without reinforcement in the waking state. The superiority of the hypnotic state, however, has been conclusively demonstrated in the case of a single adult subject who judged intervals having a duration of from 5 to 30 seconds (U.K.; prod., prior; Eysenck, 1941). On the other hand, 8 out of 9 subjects in another experiment in which the investigator expressly suggested that the trance would lead to improvement were not helped by the suggestion in judging intervals of 1, 2, and 3 minutes (U.S., not specified; verbal, prior; Stalnaker & Richardson, 1930).

Effects of hypnotic time upon behavior

Possible congruence between hypnotic time and behavior can be illustrated by a report concerning a single subject who was asked during the trance to count the number of cows in an hallucinated scene and was told that she had 30 minutes in which to do so. In fact only 3 seconds of clock time was allowed to elapse; nevertheless she reported that she had been able to carry out the instruction in detail during the interval: she had counted to 137, she said, and she had "walked around the edge of the field" since the animals "were very close together" (Cooper, 1948). Under the influence of the same kind of suggestion, a number of hypnotized subjects seem to have found creative solutions for problems involving human relations or dress-designing within very brief periods of clock time—from 1 to 10 seconds. These solutions, they either stated or demonstrated, they could not have achieved so quickly in the waking state (Cooper & Erickson, 1954, pp. 83–100). Cases are also described in which time was hypnotically distorted in the opposite direction in order to speed up its passing during the painful experiences of childbirth, of severe headache, and dental treatment; and here on a conscious level the individuals certainly reported that the pain had been diminished because it seemed to last a shorter period (U.S.; Erickson & Erickson, 1958).

There are other instances in which hypnotic time had no appreciable effect upon behavior. Problen.s in mathematics could not be solved more rapidly in terms of clock time. Hallucinated practice at writing with the left or unaccustomed hand for 5 minutes of distorted time— in fact, a few seconds—did not improve performance in the waking state, even though the subjects believed they had engaged in the hallucinated activity which then seemed "easier" to them. One subject learned to associate pairs of nonsense syllables much less efficiently in the normal waking state than in the trance during which he had been told that he would have "ample time" (actually 5 seconds of the clock time) to learn them thoroughly and that during that interval he should print the pairs, repeat them, etc.; but retention after 24 hours was approximately the same for both methods (Cooper & Rodgin, 1962).

In another laboratory, decidedly negative evidence comes from an experiment in which the subjects were first tested in the waking state so that their normal learning efficiency could be determined. Then they learned another set of nonsense syllables under one of three conditions: under hypnosis after being given the suggestion that time would slow down; in the waking state with a similar suggestion; and—the control— in the waking state without suggestion. Apparently the experimental manipulation was successful because 94 percent of the hypnotized subjects, 81 percent of those in the waking state with suggestion, and 12 percent of the controls reported that they had experienced temporal distortion. That experience, however, did not affect their learning ability; if anything the hypnotized subjects learned on the average fewer syllables than either of the other groups (U.S., college; Barber & Calverley, 1964). Two groups in another investigation learned sets of nonsense syllables for 15 seconds and then recalled what they could. They had been matched with respect to hypnotizability as measured on a paper-and-pencil test. But one group had been trained to be hypnotized and, while in the trance, to respond to the suggestion of hypnotic time; during the critical part of the experiment they were told that the 15-second interval would be 3 minutes. The other group, without ever being hypnotized, learned the syllables in the normal waking state. There were no differences in the recall of these two groups (U.S., college; Edmunston & Erbeck, 1967).

The sense of time was altered through the post-hypnotic suggestion that a metronome was going at the rate of 1 beat per second under four conditions; in fact, it was going very fast (180 beats per minute), very slow (40, 20, or 10 beats per minute), at the specified rate, or not at all. Changes in behavior in a small group of students were reported in comparison with the control condition of no discrepancy between the suggestion and the fact; the investigator, for example, thinks that with acceleration behavior had the "characteristics of manic or hypomanic state," whereas with deceleration "a schizophreniform condition" appeared (U.S.; Aaronson, 1968).

It is well to conclude a discussion of hypnosis by emphasizing that temporal judgments, while so often uncertain and unstable, cannot always be affected by suggestions in the waking state. Even the powerful experimenter may be powerless. In one study subjects were instructed to focus upon either a light or a sound which simultaneously stimulated them during an interval of 1 second. This instruction had no significant effect upon their judgments, even though it might have been anticipated that vision would lead to longer estimates with this Limits-Verbal Method than audition; instead, the more intense of the two stimuli in competition turned out to be more influential (U.S., college, Goldstone, Boardman, & Lhamon, 1959).

ADDENDUM 9.7 ABNORMALITY

1. Psychoanalysis

The tendency for psychoanalysts to agree in principle regarding the genesis of mental illness and to differ on the details can be most painlessly suggested by outlining a number of their conceptualizations:

Oberndorf (1941): from clinical observations based largely on long psychoanalyses of three patients, the conclusion is reached that "the sense of time . . . becomes distorted in the absence of the sense of reality and, in cases of profound interference with the perception of reality, time sense may cease to exist." Obviously, according to this view, temporal deviation is a by-product.

Scott (1948): the general thesis is expressed that temporal distortions observable on the surface are "parallelled by variations in the types of unconscious omnipotent fantasies regarding time found in different types of neurosis and psychosis"; thus "an obsessional patient who complained that she was living so slowly and that time dragged lost these complaints in connexion with the analysis of unconscious desires to be finished with life, to reach her death-bed and be with a dead sister so that she could undo harm she had done to this sister."

Bender (1950): this child psychiatrist, generalizing from her clinical experience with more than 7,000 cases, suggests that disturbed children suffer from "defects in time concepts" because, lacking "identification as a continuous temporal process," they are unable to recall the past or to profit from it, as a result of which they "have no future goals and cannot be motivated to control their behavior for future gains."

Jones (1951, p. 257): for a person who reveals "narcissistic exhibitionism . . . the idea of time and its passage is so intimately bound up with such fundamental matters as old age and death, potency, ambitions, hopes, in short with the essence of life itself, that it is necessarily of the greatest importance to anyone who claims omnipotence and omniscience. Like all lesser things it must therefore be under his control, and this belief is revealed in a number of little traits and reactions."

Erickson (1956): "Time diffusion" reflects a "loss of the ego's function of maintaining perspective and expectancy" as a result of some crisis. A normal person is not impatient, for he trusts people and hence waiting for them is tolerable. But a young person, an adolescent who must face a crisis and then tries to find a solution by regression to an earlier phase of his existence, mistrusts everyone and hence time itself comes under suspicion: "every day appears to be a deceit, every wait an experience of impotence, every hope a danger, every plan a catastrophe, every potential provider a traitor. Therefore, time must be made to stand still, if necessary by the magic means of catatonic immobility—or death." Such an attitude toward time, though extreme, is experienced by all of us on occasion.

Meerloo (1954, p. 61): "Perhaps," it is suggested, there is what is called "a typical schizophrenic time experience" consisting "of living in a timeless archaic world without rhythm, without night and day, in a kind of oceanic time such as the fetus experiences in the womb. Many schizophrenics think about their psychotic phase as an eternity. They are aware of their loss of time consciousness. Their weak egos are not able to bring order into their experiences."

Rouart (1962): after arguing that temporal disturbances signify serious psychoses, the writer indicates his belief that the disorders characteristic of obsessional neurotics have their origin in infancy when oral expectations are frustrated, as a result of which growth through the normal phases of development becomes faulty. This hypothesis, to be sure, cannot be easily accepted: it stems from observing patients in one country, France; it assumes the validity of Freudian stages of development, which are not universally recognized, at least in detail; and it leaves unexplained the fact that oral frustration leads to temporal disorders in some persons but not in others.

Alexander (1967): his view seems to be the modest one that aspects of temporal behavior are linked to some specified part of the personality, such as the ego, and it is implied that both develop concurrently.

There are also unorthodox psychoanalysts who follow other stars on the basis of clinical experience which they do not describe in detail. After noting Freud's claim that the unconscious is timeless, two of them point up the obverse: "every conscious or latent fantasy of infantile omnipotence disturbs the normal perception of time." Such "feelings of omnipotence are abundant in dreams and fantasies, evident in games and during coitus," conditions in which unconscious tendencies are dominant. Persons obsessed with time "in our culture" are striving "to achieve mastery of infantile disappointments" and hence in general "the passage of time symbolizes the period of separation" (Bergler & Róheim, 1946).

At least one psychoanalyst seems to believe that some patients, those suffering from temporal phobias, have a syndrome which itself is the factor determining their behavior. They fear they have too little time, they feel as though they are "always being 'cooped up' by their duties"; others are so disturbed by empty time that they try to fill such intervals with activity, almost any kind of activity. He also suggests that a so-called anal personality is concerned not only literally with defecation but also with other actions, included among which may be an attitude toward time or some peculiarity concerning it. These people, he says, "may be stingy or prodigal or both alternately; they may be punctual or unpunctual; they may sometimes be accurate to the fraction of a minute, and at other times grossly unreliable." Whether or not anality is the motivating factor behind such temporal behavior must remain a moot question, but another contention suggests how the two may possibly be related: "How often defecation has to take place, at what intervals it has to be done, how long the process itself should take, how long it may be successfully postponed, and so on, are the situations in which the child acquires the ideas of order and disorder regarding time, and of measurement of time in general" (Fenichel, 1945, pp. 204, 282).

The difficulty with the psychoanalytic approach to time is the same, I think, as with many psychoanalytic theories: there is a paucity of validating data. Any successful analyst can probably produce one or two cases suggesting that punctuality is connected with some anal or oral tendency or has its roots deep in childhood. But this only amounts to asserting a hypothetical connection between time and the personality, and the particular connection thus postulated inspires no great confidence. The earlier analysts, moreover, connected every aspect of time with their classical paradigm. Concerning the hour glass, for example, it has been said:

> It is quite self-evident that this device imitates the process of defecation, that the yellow brown sand corresponds to excrement, the funnel-shaped holder the bowels, the opening the anus [Hárnik, 1925].

Perhaps, though of course the description is unprovable historically; as a matter of fact, the sand in the glass in front of me is pure white—and it could well be green or purple. Still if the above "self-evident" interpretation has meaning for a particular person, then clearly his affection for the device, or his use of it, betrays something about his inner impulses, whether they be normal or abnormal.

2. *Orientation*

Here and there one finds the view that mental illness is not accompanied by a shift in temporal orientation. Temporal disturbances in depressed patients were reported to be relatively rare in clinical practice; but, when present, the orientation was in any one of the three possible directions (Germany; Kloos, 1938). No overall difference between schizophrenics and neurotics could be ascertained with respect to past or future perspective as revealed by the time span of events in their lives which they reported in the two temporal directions. But when these patients were subdivided into those having an interest in ideas, or in automobiles and physical motions, on the basis of replies to a paper-and-pencil list, some differences became evident, such as the tendency for very few of the "motoric" neurotics and very many of the "motoric" schizophrenics to be inclined toward the near, rather than the distant, past (U.S., adults; Stein & Craik, 1965). Similarly, after obtaining no significant differences on some tests and slight differences on others, an investigator concludes that "no easily definable deficits with respect to temporal attitudes, perspectives, and orientations occur in schizophrenia which cannot be attributed to disturbances of a more general nature in the thought processes" (Williams, 1966).

More frequent are assertions, sometimes qualified, sometimes not, to the effect that the mentally ill have a limited time perspective "scarcely extending beyond the present or an attenuated perspective comprising barely a few future possibilities"; from this it follows, one investigator thinks, that normal persons can be differentiated from the abnormal and melancholics from paranoid schizophrenics in terms of their future goals and anticipations (U.S., adults; Israeli, 1936, p. 118). Depressed patients are said not to experience the present as between the past and the future, instead they ruminate and brood while viewing the future as a "shapeless gap" (Eissler, 1952). And a respected French psychiatrist has insisted that "the excited maniac lives only in the present" (Minkowski, 1933, p. 275).

The kind of evidence supporting these latter views tends to be substantiated by investigations of psychotics and neo- or pseudo-psychotics:

a. Thirty-seven paranoid schizophrenics were interviewed and were asked 59—yes, 59—questions concerning their feelings about time. Their replies indicated that they could distinguish between past, present, and future; when confronted with paired alternatives, almost all said they "prefer to look to the past rather than to the future," about half selected the future rather than the present, and about one-third the present rather than the past. The same technique revealed a "vague" future outlook not only among schizophrenics but also among other patients suffering from a variety of psychiatric ailments (U.S. & U.K.; Israeli, 1936, pp. 55, 58, 70–2).

b. When queried whether "yesterday" or "tomorrow" seemed further away, more of the normal adults in an investigation and fewer of the oldest schizophrenics seemed oriented toward the future, with depressed patients and those suffering from a variety of other psychiatric ailments falling in-between (Canada; Lehmann, 1967).

c. A sample of 34 schizophrenics was less oriented toward the future with respect to both time and the coherence of their ideas (as measured directly and projectively) than 34 normal patients in an American hospital (Wallace, 1956), a finding essentially replicated with respect not only to the future but also the past when a similar research design was employed in an Israeli hospital (Schlosberg, 1969).

d. Among the 16 attributes of the protocols produced by 75 schizophrenic patients was a "prevalence of temporal setting in the present perfect tense, and lack of verbs with future reference or with past habitual reference" (U.S.; Balken, 1943).

e. When asked to indicate five events that would occur in the future, normal hospital attendants seemed to have a future orientation of about 3 years, "moderately disturbed" patients 6 months, and "severely disturbed" patients a little over 1 month; but the more severely disturbed had been hospitalized for a mean of 5.5 years and the moderately disturbed ones for 1.3 years. Since years of hospitalization had the same relation to temporal orientation as severity of illness, it seems clear that both variables interacted (U.S.; Shybut, 1968).

f. Children suffering from emotional deprivation revealed a "deficiency in time concepts [which] was without doubt a factor in their limited foresight and in an inability to envision the consequences of their acts [and which] may explain the frequent complaint by foster parents of the inefficiency of punishment and reward" (U.S.; Goldfarb, 1945).

g. Psychotic, epileptic, and hysterical patients were victims of a syndrome of being unable "to evoke the past readily and clearly" or "to distinguish the present and the future." On some occasions time passed slowly, on others it dragged, and there was always "the seeming remoteness of the recent past; the unconfirmed feeling or inability to judge time." They could realize that they were experiencing difficulties regarding the relation of the past to the present: "I wish they would bring me forward in my mind." They might report premonitions concerning the future: "Before a door opens, I know it's going to open" (U.K.; Lewis, 1932).

h. Depressed patients, though aware of their difficulty, are reported to have been unable to abandon the past and advance toward the future (U.S.; Strauss, 1947).

One study suggests how a restricted perspective may affect overt behavior. In comparison with normal males, who happened to be in this instance conscientious objectors, schizophrenics were less confident that, when given the choice of setting up the mechanical details involved in testing their reaction time, they could intervene successfully. Indeed during the experiments they preferred to be controlled by the experimenter and, though slower than the controls, they responded more quickly under those conditions than when they were granted some "autonomy" (U.S.; Cromwell, Rosenthal, Shakow, & Kahn, 1961). The significance of one temporal orientation rather than another, moreover, undoubtedly is affected by a multitude of factors, one of which may be age. Thus, according to the investigators, not present but future orientation had to be considered at an early age to be the neurotic symptom, for such an orientation showed that "the child is dissatisfied in the present and is investing a good deal of his energy in fantasy about the future." Later on, the 17-year-olds who were not emotionally disturbed considered realistic plans for their own future. Among the 11-year-olds there was a tendency for those with present orientations to have more stable friendships than those oriented toward the future (U.S.; Davids & Parenti, 1958). In line with these observations was the failure to find a difference between a group of "maladjusted" children ranging in age from 10.5 to 12.5 years and a matched group of normal children when the subjects were given the choice of a small reward immediately or a larger one later, either in a hypothetical situation or in a real one involving chocolate bars of different sizes (Klineberg, 1968). But these maladjusted children tended to be more future-oriented than the normal ones; whereas the reverse trend appeared when similar groups of older children (between the ages of 13.5 and 17.5) were compared. It must be emphasized, however, that the age differences between the groups appeared only when temporal orientation was measured through the TAT or by asking the children when they expected certain events in their lives to occur; no differences were evident when they recalled conversations and thoughts of the last week or so or listed events they anticipated in the future (Klineberg, 1967).

A psychologist and psychiatrist contend that "anxiety always means insecurity concerning the future"; many psychopaths "are afraid of losing the past and they are afraid of all that the future may bring to them" and schizophrenics "can lose the 'sense of time' almost entirely as if living without time or beyond, giving the impression that they live in another world" (Dodge & Kahn, 1931, pp. 67–8).

Theorists and practitioners who seek to apply the tenets of modern existentialism to psychiatry stress the alterations not only in temporal orientation but also in attitude toward time which characterize mental illness and, perhaps, normal persons in our ill epoch. Although they may employ some fancy heading such as "temporality" or, in German, a complicated neologism, they consider time a crucial factor in their analysis:

> One of the distinctive contributions of the existential analysts to this problem [of time] is that, having placed time in the center of the psychological picture, they then propose that the *future,* in contrast to present or past, is the dominant mode of time for human beings. Personality can be understood only as we see it on a trajectory toward its future; a man can understand himself only as he projects himself forward. This is a corollary of the fact that the person is always becoming, always emerging into the future. The self is to be seen in its potentiality [May, 1958, pp. 68–9].

Existential time, therefore, is "the time it takes for something to become real" (loc. cit., p. 84), and a real experience appears to be one in which the individual relates what is happening to the central values of his existence. Thus one patient revealed "a profound disorder in his general attitude toward the future; that time which we normally integrate into a progressive whole was here split into isolated fragments" (Minkowski, 1958a, p. 132). Another could be described in terms of an "existence" that was "ruled more by the past," "cut off from the future," and hence "robbed of its authentic life-meaning, of its existential ripening, which is always and only determined by the future" (Binswanger, 1958, p. 295). Such temporal symptoms, moreover, always have their own peculiar etiology and interrelationships. "If the depressed patient could become entirely merged with the past without 'knowing' anything about future and present, then he would no longer be depressive!" (Loc. cit., p. 358). Here, evidently, there is continuity between temporal orientations, and the result is not a strengthening of the personality but tension and anxiety. In contrast, schizoid persons and especially schizophrenics are said to "live more in their own, personal time than in the world time" and hence in some instances "lose all awareness" of the latter (May, 1958, p. 108). Conflict or comparison of the two orientations is absent, instead there is isolation within a temporal existence that is incorrect from the standpoint of the patient's normal associates.

If it is true that the mentally disturbed suffer from a restricted temporal perspective, then that perspective ought to expand after successful psychotherapy. One "suggestive" investigation has tested that deduction. For 19 American undergraduates receiving therapy, there tended to be an increase in future and present orientation and a decrease in past orientation when a comparison was made between the tenses of the verbs they used at the outset and at the end of the treatment. This change, however, was not related to measures of the success of the therapy except for those changing most and least; and perspective as ascertained by the span of years mentioned in the same two interviews also was not affected (Smeltzer, 1969).

One final problem: alcoholism. A group of alcoholics and a matched group of controls in Canada had similar temporal orientations as measured on a paper-and-pencil test, and their judgments concerning the duration of intervals ranging from 1.5 seconds to 35 minutes did not differ (verbal, prod., reprod.; Cappon & Tyndel, 1967). Likewise—and also in Canada—there was no significant difference between alcoholics and matched, so-called social drinkers with respect to orientation when it was measured in a structured manner, but it was quite clearly demonstrated that the former had a less developed and less extensive sense of the future at least when confronted with less structured situations (Smart, 1968). When temporal orientation was appraised by means of a paper-and-pencil test which asked the individual to indicate in which direction his activities and feelings tended to turn, a group of alcoholic patients placed significantly less emphasis upon the future than a matched group of nonalcoholics; although the two groups did not differ with respect to emphasis on the past, the alcoholics tended to place more pleasant occurrences there and fewer pleasant ones in the present than the controls (U.S.; Roos & Albers, 1965a). The penultimate investigator suggests that the restricted orientation may have developed as a result of the alcoholism because he found a significant negative correlation between that orientation and age (and age "presumably" reflects years of drinking); but he immediately notes the opposite possibility, viz., that the alcoholics drank heavily because their "lack of a future orientation" prevented "consideration of sanctions or other negative reinforcements contingent upon drunkenness."

3. Temporal Information

The following dialogue took place with a schizophrenic patient:

> "How old are you?"
> "Thirty years."
> "In what year were you born?"
> "In nineteen ten."
> "In what year are we now?"
> "Nineteen fifty-eight."
> "Fifty-eight minus ten, what does that make?"
> "Forty-eight."
> "Well, what is your age?"
> "I am thirty years." [France; Le Guen, 1958]

Material of this sort suggests that such patients may have "a satisfactory orientation in time" concerning the present, but they "negate" or "destroy" the time which has passed since the onset of their illness. It is obvious, another psychiatrist points out, that they can supply correct temporal information from which the fact of their age ought to be calculable; instead—and this is true especially of "regressed schizophrenics" —they seek to avoid "the unpleasant awareness of having spent so many years in mental illness" (Dahl, 1958).

Confirmative evidence for these interpretations is at hand. The age claimed by 50 patients, when directly questioned, turned out to be "approximately equal to their mean age on admission, plus 7 months"; furthermore "nobody gave a chronological age in excess of his true age, as is encountered in the confused organic cases." Many, moreover, named as president of the country the man who had been in office, or they stated as the cost of a car the price prevailing, at the time of their admission. They could show some awareness of their error without wishing or being able to correct it; a man of 60 stated: "I know I should not have white hair since I am only 25 years old" (U.S.; Lanzkron & Wolfson, 1958). Other replications have been cited, particularly among hebephrenics (U.S., Ehrenteil & Jenny, 1960; Ehrenteil, 1964). In one fully reported case, a psychotic patient underestimated not only intervals of recent conversation but also very long periods of time such as the 29 years he had been in the institution which he called 3 years and a year-long period of employment which he called "three or four months" (Netherlands, Bouman & Grünbaum, 1929).

Qualitative theories in the literature of mental illness may well account for such phenomena. It is maintained that "the" schizophrenic lives in a "timeless" world: "he refuses to live by *time* as we know it," to which decision most of his behavior is then attributed (Burton, 1960). According to a psychiatrist, "many patients see themselves as a victim of 'time' and display a wish to control time and turn it back to a pleasant period of life, for example, a desire for rebirth or for regressive experiencing of satisfactions that occurred earlier in life" (Chessnick, 1957).

Similar data spring out of a study of two patients suffering from Korsakoff's psychosis. They represent, the investigator states, "a special case of the entertainment of incompatible propositions in amnesic states," reinforced in this instance by the fact that "sensitivity with regard to advancing age is a common human failing (especially among women)"; the denial of age is "a protective reaction" (England; Zangwill, 1953). Knowledge of age, consequently, is "an isolated fact" not dependent on arithmetic: the majority of a very small group of patients—11 in all—questioned as soon as possible after electroconvulsive therapy first remembered their birth, then their age, and later the current year. Men could give their correct age sooner after the shock than women. If men were still disorganized, they stated they did not know their age, whereas women gave incorrect figures, 10 years younger than the actual fact (England; Mowbray, 1954).

A very special instance of temporal misinformation is the *déja-vu* phenomenon: the individual attaches a wrong label to a new experience, he feels he has had that experience in the past. With our present knowledge it is not possible to specify the accompanying behavioral components; for example, both depersonalization and a "heightened sense of reality" have been singled out as essential attributes (Orme, 1969, pp. 14–15). It seems reasonably clear, however, that the disturbance is likely to accompany some forms of epilepsy (Cole & Zangwill, 1963). The opposite impression, *jamais vu*, has been conceptualized in terms of basic processes within the individual who unconsciously seeks to separate his present perception from his ego by ignoring the temporal cues (Tscholakow, 1956).

4. Judgment

First, I shall offer some evidence indicating how difficult or impossible it sometimes is to induce the mentally ill to pass temporal judgment. Seven patients suffering from various kinds of psychoses with which time-agnosia is associated revealed "confusional states of short duration"; they found verbal units of time "meaningless"; and they could not estimate "short and long intervals of time." These temporal difficulties, however, were symptomatic of the illness which was described as "apathy and unconcern about their condition (the present) and indifference to the past and future" (U.S.; Davidson, 1941), and they clearly reflected deeper layers of the personality. The same sort of interpretation seems plausible even when temporal fear as such is especially critical. Some neurotic patients, for example, have been reported to grow so anxious that they almost threw themselves into a panic whenever they noted that an interval had elapsed without their having been aware of its passing; for them this was "a dangerous break in continuity of consciousness" which produced the fear that objects in the world or parts of the body or the mind would be similarly lost and the person would be left defenseless. "I must never forget myself for a single minute," a young woman with agoraphobia stated, "I watch the clock and keep busy, or else I won't know who I am" (U.S.; Dooley, 1941). Avoiding the arousal of the temporal motive and the inability to pass judgment may be the central feature of some neuroses; "the tendency to resist finality in every form is grotesquely manifest in the procrastinating rituals of the compulsive neurotic . . . " (Cohn, 1957).

EPHEMERAL AND TRANSITORY INTERVALS
When the usual laboratory procedures are followed, sometimes no differences between the mentally ill and normals have been reported. Such a negative finding emerged from a very well-designed and controlled study in which quite comparable groups of normal, neurotic, and schizophrenic persons judged intervals whose duration ranged from 0.5 to 10 seconds by means of the Methods of Production, Reproduction, and Verbal Estimation with prior arousal of the temporal motive (U.S.; Warm, Morris, & Kew, 1963). Likewise small groups of normal adults and adult patients suffering from a variety of disturbances reproduced an interval of 15 seconds with equal accuracy (Canada; Lehmann, 1967).

The studies from the laboratory of Goldstone and his colleagues are compelling because they are based upon adequate numbers of American subjects—"8 years and 5,000 subjects," he reported in 1965—and always have control groups. They have usually demonstrated a markedly pronounced tendency for schizophrenics to underestimate intervals (since they employed their own Limits-Verbal Method, the investigators themselves call this overestimation, by which they mean that an interval shorter than a second of clock time is said to be a second—again the confusion in the use of the terms, and I consistently follow here my own definition rather than theirs). In addition, they have indicated greater variability among schizophrenics than among the controls and greater susceptibility to immediate anchoring stimuli and less to more distant ones (Lhamon & Goldstone, 1956; Weinstein, Goldstone, & Boardman, 1958; Lhamon, Goldstone, Goldfarb, 1965). Their findings had been partially foreshadowed in an earlier study in which both schizophrenic and normal subjects underestimated intervals of 30 seconds and longer, but only the former also underestimated the shorter intervals of 5 and 10 seconds (U.S., prod., prior; Johnston, 1939). Contrary results have also been reported: there were no significant differences between 20 psychiatric patients complaining of "derealization and depersonalization" and 20 nonpatients matched with respect to age, sex, and education, even though the Method of Verbal-Limits was employed (Canada; Banks, Cappon, & Hagen, 1966). As always seems to occur, moreover, Goldstone's own laboratory, although it confirmed the tendency for schizophrenics to underestimate intervals when the method employed was that of counting, reported no differences between schizophrenics and normals when the intervals being judged were separated by steps in a geometric rather than arithmetic ratio (U.S.; Wright, Goldstone, & Boardman, 1962); and similarly there were no differences between groups of "healthy" controls, schizophrenic patients, other psychiatric patients, and the physically disabled, when slight modifications in the judging procedure were introduced to simplify the experimental procedures (U.S.; rating-verbal, prior; Webster, Goldstone, & Webb, 1962). It is not surprising, then, that a complete change in method to that of Comparison produced no difference between schizophrenics and controls (Goldstone, 1964, pp. 157–65). In addition, the tendency to underestimate which appeared with the Methods of Limits-Verbal has been associated only with schizophrenia and not with patients suffering from depression, anxiety, or chronic illness (Goldstone & Goldfarb, 1962).

Consistent with underestimation via Verbal-Limits is the opposite tendency with Verbal Estimation which has been noted among schizophrenics; and here there is also the suggestion that the content of the interval being judged may affect the estimates. Schizophrenic patients viewed tachistoscopically exposed pictures and immediately afterwards were asked to give verbal estimates of the exposure times which varied from 5 to 30 seconds. One-third of the group had "sex" as the major conflict area, another third "aggression," and the remaining third "dependency." One of the pictures, moreover, portrayed each type of conflict. As a whole these schizophrenics had a pronounced tendency to overestimate the intervals to a significantly greater degree than a randomly selected control group of hospital employees, hence their judgments were less accurate. In addition, unlike the normals who responded very similarly to each of the pictures, the patients judged them significantly differently: in the case of those disturbed by "sex" and "aggression," there was a clear-cut tendency for their judgments to be distorted most by the one picture portraying their particular conflict; for those with "dependency" troubles there was a tendency, though not a significant one, in the same direction (U.S.; immediate subsequent & prior; Pearl & Berg, 1963).

Most of the studies just cited indicate that the mentally ill, especially schizophrenics, are less accurate in judging these brief intervals: under-estimation or overestimation means just that. One investigator compared groups of so-called normal individuals, of neurotics with diagnoses of "anxiety reactions," and of schizophrenics, all of whom judged intervals of varying duration under different conditions (unfilled intervals vs. those filled with various kinds of activities; verbal with immediate subsequent arousal vs. production with prior arousal). If anything, the neurotics provided the most accurate judgments and were least variable from person to person; the schizophrenics were the most variable. The latter were also divided into subgroups on the basis of the correctness of their temporal information concerning the day of the week and month, the month of the year, and the year itself. The disoriented schizophrenics displayed greater variability in their judgments than those who were oriented, but on the average were just as accurate; and, with immediate subsequent arousal of the temporal motive, the accuracy of their judgment was affected more adversely by the filling of the intervals (U.S., adults; Dobson, 1954).

Another study, already mentioned in connection with personality traits in *Addendum 8.6,* involved the testing of patients as they were being admitted to a psychiatric clinic and a week or so later after: their temporal judgments were assessed by having them reproduce an interval of 30 seconds. Aside from the fact that as a group they tended to produce longer and more accurate intervals the second time (when presumably they were less agitated than upon admission) as well as when they rated themselves in a more pleasant state of mind, the following differences appeared among the subgroups: psychotic patients were significantly less accurate than the non-psychotics but only while being admitted and not later; those suffering from delusions were likewise significantly less accurate than those not suffering but only during the later examination, a tendency more pronounced during both periods among those with paranoid delusions; and depressed patients produced longer judgments and counted out loud to 30 "at their own pace" more slowly. In short, "patients who are divorced from reality are apt to be divorced" from objective time; they "are not only inaccurate, but also give shorter productions" (U.S., adults; Melges & Fougerousse, 1966). Epileptic patients with slight dementia tended to make somewhat greater errors in producing intervals ranging from 30 seconds to 10 minutes than a group of normal attendants and students; this trend was especially true among the males when they were occupied with crossing out letters or numbers while producing the intervals (Switzerland, adults; Pumpian-Mindlin, 1935). In comparison with norms previously established for normal subjects, schizophrenics revealed a similar tendency to be somewhat consistent in judging both relatively short and relatively longer intervals, but they were much more variable in their responses and hence seemed to reveal "a functional disability in time estimation" (U.S.; verbal, prior; Guertin & Rabin, 1960). Twenty-five paranoid schizophrenics did not differ from a matched group of hospital attendants in reproducing intervals ranging from 1 to 20 seconds; but they were less accurate in producing an interval of 10 seconds. They also suffered from other temporal deficiencies: they lacked adequate temporal information and they were less able to project themselves into the future (France; Géraud, Moron, & Sztulman, 1967).

Again it is necessary to remind ourselves that performance depends in part upon the conditions under which judgments are tested. Schizophrenic patients, for example, judged the duration of 5-, 10-, and 15-second intervals by means of three methods and on two occasions: the only correlations which were clearly significant during the first test and also more than four months later were those between Verbal Estimation and Production (U.S.; prior; Clausen, 1950).

The mentally ill, finally, have been shown to improve less than normals with reinforced practice. A mixed bag of psychotic patients gave fewer correct responses, tended to give more varied judgments, and improved less during a series of trials than "normal" hospital personnel when trying to produce an interval of 14 seconds by means of the differential reinforcement technique of operant conditioning (Belgium, adults; Denys & Richelle, 1965). Subjects low in neuroticism, as measured by a paper-and-pencil test, improved more than those with high scores when the opportunity was given, through feedback, to practice producing an interval of 1 second; but those high in neuroticism, who were also extreme extroverts, performed much better (U.S., students; prior; Luoto, 1964). Apparently, with the qualification noted, the better adjusted could take greater advantage of experience, an observation that has also been made concerning schizophrenic patients in general who are said to be more at the mercy of immediate stimulation than normal persons and hence are less influenced by experience during exposure to a series of stimuli (Weinstein, Goldstone, & Boardman, 1958). These three studies tested adults whose milieu presumably had offered them ample opportunities to practice passing judgment; but, as indicated in *Addendum 8.2*, the estimates of emotionally disturbed children tended to show more improvement after feedback than those of normal children (Davids, 1969), perhaps because their disturbance had been isolating them from such opportunities.

EXTENDED INTERVALS We have already noted a tendency for schizophrenics to underestimate their age or period of confinement as a function of the calendar time at which the illness began. Shorter, but nevertheless extended intervals have not necessarily been similarly underestimated. One technique is to have the patient, without prior warning and hence with immediate subsequent arousal of the temporal motive, estimate the duration of the psychiatric interview or testing session which has just transpired. A small group of schizophrenics and "nonpsychotics" (presumably normal persons and a few psychopathic personalities) estimated the duration of an interview when it was approximately half over and after its completion: the former produced less accurate and more variable judgments than the latter; when in error, both groups tended to underestimate rather than to overestimate its duration (U.S.; verbal; Rabin, 1957). Schizophrenics, melancholics, and anxious and depressed neurotics underestimated a 30-minute psychiatric interview to a lesser degree than hysterics, psychopaths, and manics; among the schizophrenics the underestimation of the paranoid patients was vastly greater than that of the nonparanoids; but in all instances the variability from person to person was great (England, adults; verbal; Orme, 1964, 1966). On the other hand, in another study there was a pronounced tendency for an older group of schizophrenics to *over*estimate the length of an interview in comparison with a small group of normal adults and adult patients suffering from a variety of mental disturbances—that was the only significant difference which emerged (Canada; Lehmann, 1968). The longest estimates of interviews lasting 20 or 30 minutes were given by psychopaths, hysterics, and mental defectives; the shortest by neurotics; and in-between by a comparable group of normal extension and nursing school students (England; verbal; Orme, 1962a). Depressed cases underestimated to a lesser degree than patients during a hypomanic phase or after recovery the duration of an experimental session, but they overestimated to a greater degree the duration of a 3-second interval (U.S., adults; verbal; Mezey & Knight, 1965). Two groups of schizophrenics and depressives judged the duration of an experimental session (31 minutes) less accurately than the normals (U.S., adults; Dilling & Rabin, 1967). The last result was obtained when the subjects may have guessed that they would be called upon to make the estimate (i.e., there was prior arousal of the temporal motive); but all three groups responded with similar accuracy when they were asked to judge the elapsed time half-way through the session (without prior arousal).

QUALITATIVE DIFFERENCES The subjective reports of patients often indicate that they think time is slowing down (probably a primary judgment) when in fact their expressed secondary judgments concerning elapsed intervals do not necessarily reflect that subjective impression. Over three-quarters of a group of patients whose predominant symptom was depression indicated that they had such a temporal impression; their own judgments, however, were "not significantly impaired" (England, adult; prod., repro., & verbal with prior; verbal with immediate subsequent; Mezey & Cohen, 1961)—apparently they retained sufficient contact with reality to discount their primary judgments. In another study, depressed patients revealed a decided tendency to consider that time was passing slowly, whereas patients with a variety of other illnesses suggested medium or rapid speed (Canada; Lehmann, 1967). It is asserted, presumably on the basis of clinical evidence, that time seems to pass slowly for schizophrenics and quickly for persons in the categories of mania and, if not depressed, senility (U.S.; May, 1948, p. 104). A physician states that his patients when suffering from psychoses, epilepsy, hysteria, etc. reported an "increased quickness with which time passes, though it seems also to drag" (U.K.; Lewis, 1932).

There were essentially negative results when a comparison was made between 20 carefully selected psychiatric patients who had been complaining of "derealization and depersonalization" and 20 nonpatients matched with respect to age, sex, and education. The investigation is impressive because three different methods of measuring temporal judgments were employed; because subjects were tested in connection with four different intervals having a duration ranging of from 7.5 seconds to 45 minutes (though not all intervals could be used under all conditions and with all three methods); because comparisons were also made between subjects who reported having and not having experiences of time distortion in the past and, for each of the two groups, between those reporting and not reporting similar experiences during the experiment itself; and finally because judgments were passed not only under the usual "normal" conditions of laboratory experiments but also during experimentally induced experiences which have been known to affect temporal judgments (water striking the ear drum, rotation of the body with eye open and head moving backward and forward, sensory isolation for three hours, and deprivation of sleep for 50 hours). There were a few trends and bits of evidence suggesting that the patients under some conditions, and particularly with an interval of 45 minutes, gave slightly less accurate and more variable judgments than the controls, but virtually none of these differences were large enough to reach the level of being statistically significant. The investigators suggest that here at least there seemed to be little or no relation between the experiences of subjective time, in general or during the experiment, and temporal judgment; but they indicate the possibility that, if differences between patients and non-patients exist, they may arise with reference to ephemeral and enduring intervals which are, respectively, shorter and longer than the ones they employed, and that the act of measurment itself "may bring consciousness and normal judgemental faculties into focus which eliminate or displace for that time the subjective distortions" (Canada, adults; verbal, prod. & repro. with prior; Cappon & Banks, 1964).

ADDENDUM 9.8 CRIMINALS

The high tribute paid to temporal perspective—one writer, after reviewing the literature in various fields, considers it one of the most significant of all psychological variables (Mönks, 1967)—leads one to expect impressive differences between criminals and noncriminals with respect to their orientation regarding the future; to a certain extent the expectation is fulfilled:

1. Delinquents and nondelinquents in the United States were matched with respect to demographic factors (with the mean age being 17) and asked to compose a story by completing a plot suggested by a simple sentence. The mean time span of

the stories supplied by the 26 delinquents was more restricted and more present-oriented than that of the stories by nondelinquent adolescents (Barndt & Johnson, 1955); this finding has been replicated with the same technique and with another group of the same age (Davids, Kidder, & Reich, 1962), but not with younger emotionally disturbed boys (Davids & Parenti, 1958). In the original study, moreover, a higher percentage of the delinquents than the controls wrote stories with unhappy endings; thus the projective mode of measurement suggests that temporal orientation interacts with other impulses.

2. Matched and unmatched samples of delinquents and nondelinquents indicated whether 36 different events could happen to them in the future and, if so, at what age they might take place. Both comparisons yielded very significant differences which suggest that the delinquents were less future-oriented than the controls (U.S.; Stein et al., 1968).

3. During a period varying from 1 to 19 months, a group of 10 whose aggressiveness justified the classification of "predelinquent or delinquent" behavior and who lived together in a special house were observed and treated by a professional staff in the city of Detroit, Michigan. Their temporal disturbances, according to the report on the project, had "a disastrous effect" on their lives and "on the chances for education and therapy to take hold." They tended to be oriented toward the present and hence they "had not developed much of a realistic concept of 'themselves in the future' so that there was little to appeal to, one way or another . . ." They distrusted the future, hence they sought immediate rewards; when one boy asked to go canoeing just before retiring and when he was told that the event had been planned for the next day, he expressed disbelief—"You'll never take us" (Redl & Wineman, 1957, pp. 119–21).

4. In comparison with a normal group, in this instance army recruits, delinquents and criminals tended to be less oriented toward the future as measured by the time span of the events that were anticipated (Israel; Siegman, 1961).

5. Elementary school children in a very poor, lower-class neighborhood were given an opportunity to steal by the experimenter who "inadvertently" left the contents of her purse scattered upon a desk and then, as she went out of the room on a pretext, asked the children to put everything, including scattered change, back into that purse. Those who stole had a more pronounced tendency than those who did not steal (1) to select fewer words involving temporal concepts for telling a story and (2) to tell stories with a shorter time span. The particular words they employed, however, did not clearly differentiate them from the nonstealers with respect to orientation. In addition, there was no correlation between the amount of money stolen and the two temporal measures; among these children no factors—race, sex, age, intelligence, and academic achievement—other than the temporal ones were related to stealing (U.S., Brock & Del Guidice, 1963).

As ever in research of this kind, so much depends upon the method of measurement. One investigator, for example, employed the simple technique of asking "Who are you?" and also a similar question pertaining to the past and the future; and the ideal was tapped with the question, "Who would you like to be?" A group of male prisoners used their name as a way of answering all four questions to a significantly greater degree than groups of undergraduates and students of neuropsychiatric technology; but other differences, such as those involving "negative" or "positive" affect, were less marked and certainly not very consistent (U.S.; Brodsky, 1967). With these data, one can almost pick and choose among a variety of theses.

A group of "youthful offenders" in a Florida prison was tested by a sentence-completion device. They tended to produce items oriented toward the future which had either a positive or neutral tone; in second and third place, respectively, were items concerning the past and the present whose tone was largely negative. No control group of nonoffenders was appraised because the research sought to determine whether temporal distance from parole was related to the responses. No significant differences were found among those who had just been imprisoned, those who would be released in six months, and those whose parole was a week or less away (Megargee et al., 1970). Apparently with this technique the momentary situation did not influence the orientation.

The study of impulse control mentioned in the text and already cited in *Addendum 5.1* involved criminals and Army recruits in Israel: in comparison with the latter, the former tended to overestimate intervals varying from 2 to 120 seconds when the Method of Verbal Estimation was employed and to underestimate them with the Method of Production (Siegman, 1966). With either method, the criminals may have been thus revealing inadequate impulse control: just as they did not curb antisocial tendencies, so they were impatient to have the experimenter's interval end as he employed the first method, and they themselves prematurely ended their own interval with the second method. But a few years earlier and with roughly the same technique, the same investigator had obtained exactly opposite results: delinquents tended to underestimate verbal estimates of 5-, 16-, and 25-second intervals that were empty, and the recruits to overestimate them. He admits "no obvious explanation" for the contradictory findings (Siegman, 1961). In an American investigation, moreover, contradictory results were also obtained: delinquents gave longer verbal estimates than nondelinquents (Barabasz, 1970a).

In the later Israeli experiment, other interesting data were obtained: whereas the nondelinquents tended to find intervals filled with the sound of a buzzer to be shorter than empty ones (as already reported on page 118), the delinquents responded to both the filled and the empty intervals with similar estimates. The investigator himself suggests that in general delinquents and criminals may have a higher threshold for stimulation; if this be so, then perhaps the delinquents in the experiment were less stimulated by the buzzer and hence responded more or less as they did in its absence. There is also the possibility that the delinquents, being perhaps deficient in the ability to produce fantasy (which in turn may be related to their failure to anticipate the consequences of their behavior), were less able to fill the unfilled intervals with ideas and thoughts or to associate such material with the sound of the buzzer; hence both kinds could have appeared equally bland or boring. The author speculates that the same lack of imagination which keeps unfilled intervals empty may account, too, for "the delinquent's phrenetic search for 'thrills' and 'kicks' "—thus he can escape from boredom.

The group of predelinquent or delinquent boys in the Michigan study mentioned above was characterized by a disturbance which was "one of the most plaguing blocks in our treatment plans": they did not distinguish clearly between subjective and objective time. Some of them lacked adequate information about objective time; e.g., they could not tell time from a watch. But, more important, their secondary judgments tended to spring directly from their primary judgments. Equal periods of clock time were judged unequal; thus the children would grow angry and claim they were the victims of discrimination since their judgment concerning intervals spent impatiently awaiting their turns appeared longer than those involving satisfaction. Similarly, when rewarded for behavior, they might not see the causal connection with good behavior and consider that reward "so much good luck." Or the temporal sequence could be misinterpreted: "Punishment is to them not the end of a chain of causation but the beginning of one"—which meant that, after receiving the punishment, they sought revenge for what they thought to be the injustice inflicted upon them (Redl & Wineman, 1957, pp. 119–21, 467–87).

10. GROUPS

Many of the principles and generalizations in previous chapters concern the effects of the social setting upon the renunciations, the anticipations, and the temporal behavior of the persons dwelling therein. Upon analysis social setting is found to be essentially equivalent to the group or groups in which such persons find themselves or with which they feel identified at a given instant of time or over longer periods. And so, in an effort to improve and refine our insights, we turn to groups.

It is a truism of social psychology, and of common sense, that the closer one gets to the particular groups to which an individual belongs, the greater the preliminary insight one has into him, his predispositions, his behavior. If you know only the general culture in which he lives, your insight or predictions must be correspondingly broad: you will be able to designate the language that he employs and some of the general values that guide him. If you know a bit of what there is to know about all the groups to which he has been compelled to belong, such as his family or age group, and also, provided they exist in his society, those which he has voluntarily joined, your wisdom is almost boundless: you can specify more or less accurately the pressures upon him, the beliefs and values to which he undoubtedly subscribes, perhaps also his ambitions, and certainly the way in which he is driven, or chooses, to spend his time. Obviously, however, the best way to understand the degree of enthusiasm he has for these groups, the motives behind his participation therein, or the possibility of his deviating therefrom is, as we have already declaimed in Chapter 3, to concentrate upon his personality. But even if you are concerned with his individuality, you must know something about his groups if you are to appreciate the genesis of some of his traits and behavior or if you are to predict which of those traits and what aspects of that behavior are likely to be activated now or later.

Temporal orientation may be viewed through the lenses of groups. All men must some time look backward, because they remember their childhood, and forward, because they must die. The French by and large have a different view of progress from that of the Bushmen. Within one village in Normandy, moreover, the attitudes toward the past, present, and future of peasants and workers have differed appreciably (Bernot & Blancard, 1953, pp. 323–32). And presumably older peasants there have valued the past more than younger ones, and among the older peasants some have been greater admirers of tradition than others. Eventually the individual peasant, if examined, could turn out to have had a more or less unique perspective. This is the destiny of any trend that is established in connection with groups or variables, such as age, education, or occupation, which reflect group membership: exceptions appear which can be explained by examining either other groups or personality traits.

The casual and the systematic, the subjective and the objective literature on groups is about as staggering as that on human personality. For groups can be approached from so many different standpoints. The historian and the sociologist tend to view particular groups historically and in terms of their current organization: how did they come into existence, how do they now function, what purposes do they serve? The anthropologist may seize upon certain groups, such as the family or the clan, and indicate their role within the structure of society as well as the functions they perform for the society as a whole. Psychologists as usual have sought to make generalizations about all groups, either by classifying them or by studying those which they themselves artificially and purposefully contrive under controlled or laboratory conditions and hence which are usually very small for practical reasons. Here we must select the most relevant generalizations from these ripening fields.

REFERENCE GROUP

Ordinarily every person belongs to numerous groups, starting with the family and ending with a vaguer and sometimes less-commanding group such as the tribe, the nation, and perhaps mankind. In-between are groups arousing different emotional attachments and based upon many different criteria, such as age, sex, occupation, belief system, social status, beauty, strength, etc. It is interesting to note that the temporal factor of age-grading is included among the universal bases for establishing groups (Linton, 1936, p. 118). No matter how values fluctuate, people in every society differ with respect to age, and therefore more or less distinctive values and behavior come to be associated with, and to be anticipated from, the various age groupings. In addition, this chronological designation often supports not a fictive but a real association, such as one's secondary-school or university class in a Western society or the persons who were initiated into the status of manhood at the same time by the elders of the tribe.

At any moment you refer your behavior and judgments to groups or associations in one of two ways. You may be conscious of your membership in one group rather than another. When you are in church, you are most aware of your religious group and not of the profession to which you also belong, though of course you may be simultaneously preoccupied with religion and your own spirituality. Or, in the second place, you may not belong to a group and yet regulate your behavior according to its standards. You are not a member of a clique you know has prestige, but you worry about your reputation among its members for reasons you yourself know best. We are concerned, in this second meaning, with influences upon behavior which do not arise from the situation at hand but which, being within the person, transcend that situation. In this sense a reference group functions like any other predisposition derived from past experience and future anticipation, including the prevailing temporal orientation.

For either type of reference group, the individual has, according to the current phrase, general or specific "role-expectations": he knows or thinks he knows what the group to which he belongs, or aspires to belong, generally expects of him, and he in turn has his own expectations concerning the behavior of other members. These also are anticipations concerning the very next moment (he must be polite to this superior person), or concerning the far-flung future (he will have to own a set of formal clothes if he ever is admitted to the club). The tendency to orient oneself *predominantly* backwards or forwards depends in large part upon the reference group: as a member of a group which prides itself on its achievements, he may emphasize the past; but when considering or participating in one that is struggling to obtain as yet unobtained goals, he must frequently look ahead. The word *predominately* has been italicized to call attention to the exceptions; a reactionary club, for example, must plan its next meeting, a radical one may keep minutes of the previous meeting. Many of the drives, and hence the responses, intervening between the perception of an interval and a primary or secondary judgment concerning its duration likewise spring from the group affecting behavior after the judgment motive has been evoked.

The question of which group is salient at a given moment for a person is another one of those problems which need not be considered here in detail. Certainly the actual setting and the importance of the group are relevant factors. If a citizen sees a flag, he thinks of his country; if he is at a meeting of his club, he most likely conducts himself like a good member. Or if the organization plays an important role in his life, he may be influenced by its standards in many different situations and under a wide variety of circumstances. A monk in a holy order may consider his obligations to the deity most of the hours of his waking day. Members of a group ultimately judge its significance according to the needs it satisfies, one symptom of which may be whether it possesses a distinctive name or title: it has the label because it is important, it is important because it has the label, or the two spiral. Age-grading is likely to increase in influence when each group is clearly recognized as such. If you are one of the elders of the tribe and know that you are thus acclaimed, the role of elder for you will be significant, your age group will often function as a reference group for you and younger members of the tribe. We thus return again to the principle that social importance is correlated with the amount of time spent in the group or with the temporal priority accorded one's membership therein. A reminder: the individual's own personality traits may be formed and function within reference groups, but they may also break loose. For as a person, he is or is not prone to renounce, to anticipate future events, to live in the present or the future.

TRADITIONS AND IDEALS

Groups can be characterized in many ways, but in this context only these two attributes need be singled out. At first it appears as though the concept of tradition could be applied only to groups which survive over time since, being permanent or semipermanent, they have rules of behavior transmitted by one generation to the next and thereafter followed to some extent by all or most members. But has a newly formed crowd or mob a tradition, does a group à deux (such as a psychiatrist and his patient or, for that matter, a hypnotist and his subject) have traditions? Certainly they have no rigid set at the outset, the way a formal association complies with parliamentary procedure or celebrates anniversaries. They do obtain from the society at large certain prescriptions which they follow, modify, and thus convert into their own traditions, which regulate behavior as long as they interact in the temporary group. The members of the crowd or mob, for example, have a general attitude toward law and order or toward particular kinds of leaders; the patient knows that he must accord the psychiatrist a certain amount of respect, and the latter follows a code of ethics; and the subject places some confidence in the hypnotist.

The existence of newly- or well-established traditions in a group means that members behave in a particular way with relation to one another and to outsiders. That behavior may or may not involve from their standpoint some degree of renunciation. If they have been raised as children to revere the tradition, they will find only gratification in conforming; thus to wear traditional garb on certain occasions or to dance and sing in traditional ways can bring pure joy. At the other extreme, conformity can be frustrating and hence involve renunciation. The younger generation, for example, may seek to change its clothing or its mode of dancing and singing and be constrained from doing so by older members of the group.

Members of a group may or may not believe that there are alternative ways of behaving besides adhering to tradition. "We do this because our ancestors did it that way" or ". . . because that is our custom"—such a view is likely to be correlated with resistance to change. Even though traditions may vary within a group, if only as a function of age and sex, traditionalists may be unable to conceive of appropriate or satisfactory behavior that is different from what prevails. On the other hand, the possibility of nonadherence can arise from an interpretation of the group's history which does not sanctify what has come to be (probably a very infrequent occurrence), or from direct or symbolic contact with other cultures or representatives therefrom.

An awareness of tradition, when it does exist, stimulates a temporal orientation toward the past as a guide to the most efficient or best activities of the present. All persons, however, are tradition-bound to some extent: during the first years of their lives they have been members of the family group, and family practices tend to be hallowed by tradition. Here perhaps is one of the reasons it is so easy to look backwards, and why social change is slow. The way of all flesh is difficult to disrupt until the family of origin no longer functions as a dominant reference group.

It may very well be that the tendency to place a high value upon traditions just because they are old—a practice that is certainly extant in Western societies (except for revolutionary and evolutionary minorities) and also, available evidence suggests, in other societies—is another cultural universal. For age as such is likely to have value, since it is obvious that older people have wisdom as a result of experience (whatever defects they may simultaneously possess) and that all persons, unless they die, grow old. Perhaps then it is easy to generalize from what one knows, as a result of personal experience, about aging generations, to customs and beliefs whose value likewise increases when they reflect the wisdom and experience not of a few but of many generations. Let us continue this speculation for one more sentence and express a really wild fantasy: men everywhere may unwittingly be Darwinians in their judgment, viz., the fact of a tradition's survival may indicate to them its adaptability and hence its utility and holiness.

The references to tradition, however, need not be confined to the past but may also include the future. "We do this because only by following the ways of our ancestors will our group survive"—or "live peacefully" or "be happy." When traditions are so viewed, they are really strong: they have become hallowed in all temporal directions. Sanctions can then be applied by invoking the displeasure of ancestors, the discomfort of nonconforming here and now, and the hostility of decendants.

Traditions in the contemporary sense give rise to social interaction that reflects and enforces the regulations of the group and that makes use of appropriate communication channels. Such interaction requires conformity and coordination, which in turn must rely to some degree upon timing and hence temporal judgments. An interesting question concerns the amount of temporal precision necessary for a society to function effectively. We know only that if people are to work together or to conduct meetings and ceremonies, some sort of temporal patterning is needed.

Certainly the most tradition-saturated group seems to be the nation, especially the modern nation. Efforts are deliberately made, as emphasized in a previous chapter, to glorify the past, and people come to feel that they share a common culture inherited from their ancestors and therefore worth defending and preserving. Equally important in nationalism is a forward surge: the future must be even more glorious (Doob, 1964, pp. 235–7). This utopian tinge takes the form of ideals which may visualize the future in terms of a brave, new world or the restoration of a golden age or some point in-between.

Many groups within a nation, especially in Western society, have as their ideal the breaking of at least some tradition. In our time it is necessary only to mention words like peace, revolution, civil rights, profits, radicalism, conservatism, the establishment, alienation, etc. to remind oneself of tradition-fighting groups by the score. And there has been a vocabulary of change also in the emerging nonliterate societies outside the West: independence, unity, industrialization, modernization, self-reliance are the words frequently employed to suggest the need for change away from former colonial regimes and toward contemporary nationhood. These are the goals, the positive ideals, of group members.

It seems fair to add, however, that tradition-bound groups also have an ideal concerned with preserving what they have. No doubt every group pursues both positive and negative ideals, the proportion of which varies with the particular group and with circumstances. Today, as most parts of the world keep changing at least externally, even a very conservative group is likely to seek change, perhaps in the direction of achieving in the future more of the practices of the past.

The existence of an ideal signifies that members of the group antipicate the achievement of different goals through intervention; hence some sort of future orientation becomes obligatory (cf. Lippitt, 1942). Obviously effort is required, and effort in turn undoubtedly demands some renunciation. You work for the party because you are convinced that, if it is elected, your welfare or that of your country—and you are part of the country—will be improved.

The generalization is risky, but there seems to be some intuitive evidence supporting the view that greater sacrifices are made in behalf of highly abstract goals than for very concrete ones. In fact, in the early days of communism in the Soviet Union, it has often been pointed out, Marxism and communism had such a strong appeal because they had not been adequately tested in practice; therefore people were able to have almost whatever fantasies they wished concerning the utopia which would, perhaps inevitably, appear after the revolution. In contrast, the opponents of communism at the time had a concrete, meaningful model to offer, and that model, being imperfect as workaday models have to be, seemed much less attractive.

Thus it appears that a utopian element, as it has been called, can really stir men to action. Again in our time the cry of "freedom" has been heard, and millions have fought or made sacrifices to attain an imagined bliss; without being able to specify in detail their understanding of freedom's meaning, they have been induced to take action and cooperate by slogans and symbols which appear to them to be attractive and compelling. In the West, too, there have been utopian groups in the literal sense, men and women, who after renouncing the society in which they were raised, have established communities operating on the basis of principles derived from their faith and beliefs. Such persons have anticipated that society at large could not be appreciably changed and that therefore they could attain their precious principles only in a somewhat isolated enclave. Often, perhaps always, they have had the secret or public belief that their communities would also function as models or at least stimulate others to adopt some of their reforms.

Messianic cults provide an extreme illustration of utopianism. Their members believe so ardently in the immediate or eventual coming of a savior that they eagerly renounce many significant activities in the present and await his coming. Or, if they do not or cannot renounce a great deal, they find a somewhat intolerable existence more tolerable as a result of their anticipations. My misery will end soon, they say in effect, because the whole earth is going to be destroyed; time passes slowly, the great day is so impatiently awaited that it often appears imminent.

The cults and, to a lesser degree, many of the utopian groups believe primarily that certain changes are inevitable, as a result of which they intervene, or at least behave, in such a way that their anticipations are more or less realized. The phenomenon of the self-fulfilling prophecy reappears: you do what you anticipate will happen and thus you make it happen. For more than a century almost every country has produced witnesses who have forcefully illustrated this process, viz., members of communist parties whose patron, Karl Marx, believed in the almost complete inevitability of revolution and who simultaneously, except in the most philosophical of his writings, advocated the organization of the working classes to achieve, or at least to hasten, the inexorable achievement of this goal. A reasonable conclusion seems to be that people are more strongly motivated to renounce and to intervene when they anticipate success with a high degree of certainty.

The question arises as to whether groups can continue to exist when their aim is only that of perpetuating traditions they now enjoy. Do they not also need ideals which are not being fully realized (whether they refer to a golden past now vanished or a glittering future never before contemplated) and which conceivably could be attained in the future? According to a functional view in anthropology, traditions are not mechanically perpetuated; they are abandoned unless they provide some satisfaction in the present. In addition, morale and esprit improve when there are goals to be achieved and when the ensuing struggle produces at least partial success. The most powerful incentive of all, as a well-established truism in social science affirms, is the presence of a threatening outgroup. For then tradition in the face of the enemy appears especially precious, and the ideal becomes that of defense which can quickly be transformed into offense. Thus the real bravery and heroism in time of war—when men are not simply trapped by a compulsory draft or motivated by the situational requirements of their unit or of the battlefield, but when they nobly and voluntarily renounce careers and comforts and sometimes their own lives—stem in large part from the anticipation that defeat for their side will really mean disaster for one's countrymen, one's friends, and oneself.

Aside from affecting temporal orientation, a reference group may influence or provide a standard for passing temporal judgment when the appropriate motive is aroused within the frame of reference of the group. Well-established associations are likely to have longer standards than recently established ones, and hence temporal judgments about group activities will be shorter in the first instance than in the second. The decade or so during which newly established nations have been independent appears to have begun only yesterday to outsiders in the West but to be in the distant past to the nationals of those countries.

Small groups, like large nations, also conceptualize an ideal person, included among whose attributes are traits pertaining to all three temporal dimensions. The perfect member must clearly have reverence for traditions; he must work for the benefit of the group in the present; he must make personal sacrifices or renunciations in behalf of the future; and to some extent he must have so much confidence in the group and in himself that he can anticipate a bright future with overwhelming assurance. Any one or more of these attributes are likely to be associated with the actual or ideal leader, to whom attention is now turned.

LEADERS AND PROPHETS

In very general terms we know a great deal, and pitifully little, about leaders. On the one hand they are so obviously important that they have been both admired and investigated perhaps for as long as men have had the wit and leisure to think and draw conclusions; and in modern society leaders are so precious that they are deliberately cultivated, or at least an attempt is made to do so through so-called scientific procedures. But on the other hand our knowledge tends to be concentrated on specific individuals or upon common denominators expressed in statistical form. From biographies, autobiographies, and scraps of information we sometimes think we comprehend particular leaders; we believe we know why they had unusual talents and how those talents fitted into the needs of their time and their group. Research indicates the kind of traits leaders ostensibly need in order to function effectively in particular groups; yet the information thus obtained, valuable as it may be for purposes of recruiting and training masses of leaders, leaves us some distance away from the individual who tends to be more or less unique. It also generally is information applicable to leaders in one kind of group situation and not in others.

For present purposes, however, it suffices to call attention to one central problem, which is the relation between a leader and his followers. Most broadly, the question is: who is leading whom? The leader can dominate his followers, not completely but almost so. Whether issuing orders or making suggestions, he is limited by some of their capacities or drives; at the very least, for example, he must communicate with them in a verbal or nonverbal language they can understand. He can also give them what he believes they want or what they say to him they want. The difference is cleanly and clearly illustrated in a study carried out among squads of cadets at the United States Military Academy: their actual performance as scored by experienced observers was unrelated to the group's morale as measured by the mutual esteem of members toward one another; but the esteem of leaders for followers and vice versa, a feeling which in fact tended to be reciprocated by each status group, was *negatively* correlated with squad performance. A well-liked leader, therefore, was not necessarily an effective one (Gottheil & Lauterbach, 1969).

At the opposite ends of the continuum, then, we find different principles at work. The leader who pleases without dramatically leading is not demanding very great renunciation from his followers: he caters to them, he gives them rewards more or less immediately. In contrast, the leader who pushes himself and his followers to new or different goals demands much greater renunciation: he is not doing what people wish, he requires sacrifices. But often the tougher leader really must obtain support too, perhaps by compromising with his principles but more characteristically by raising people's anticipations concerning the better future, the ideal or even utopian state of affairs which his leadership will bring into being. The temporal orientation is toward the future.

Some men, whether leaders or not at the outset, become convinced that they can foretell what is to come when ordinary men are silent. They speak up with deep conviction because—like the prophets of biblical Israel (Lindblom, 1962, p. 6)—they have such a conviction completely and sincerely, a conviction likely to be strengthened by the belief that not they themselves but some outside force like a deity or an ancestor (or, in modern times, the spirit of a nation or of science) has chosen them as an instrument. If they are believed, and if their message involves action, they may become leaders: their followers have faith in them and therefore behave in ways consonant with the prophecies. Immediately or ultimately such prophets demand some form of renunciation, if not of goods of goals consider desirable at the moment, then of the ideas or beliefs to which the followers have previously adhered. In Western society, perhaps in other societies too, otherworldly or metaphysical goals may have little appeal at first—you believe that Man of God must be crazy or subversive—because time must pass before novelty is understood or transformed by those seeking or repelling understanding. But then people may respond, perhaps because the prophet expresses his vision about the past, "the hidden past," or the future in a poetic form which suggests divine inspiration (Chadwick, 1942, p. 14). Inspired prophets of doom may also seek to evoke anxiety and then to offer their solutions (Hyatt, 1947, pp. 113–17), or else they may be able to assuage the guilt people already possess concerning their own shortcomings. No doubt almost any man is a potential follower of an individual who can convincingly reveal the nature of a future concerning which human beings are curious, uncertain, or anxious. If the vision of the prophet takes hold, the behavioral attributes of conversion become apparent: men can drive themselves to extreme behavior, for the Lord as it were is on their side and hence presumably allows them to do no wrong from His and their own standpoint. When the actual history of prophets in the West is surveyed, it would appear that the record as judged by subsequent events has been far from perfect—no self-fulfilling prophecies here—but their significance lies in the fact that their followers have been credulous (Forman, 1936). During the Middle Ages in Europe, minor and major messiahs arose again and again, for "the dream of a New Era of absolute justice, to last a thousand years, was not so much stamped out as driven underground, and survived as a continuing influence on the popular imagination" (Toulmin & Goodfield, 1965, p. 71).

We do not know under what circumstances prophets can attract followers and then lead or try to lead them out of the wilderness. At the least we can say that truly happy people are not eager to be guided down dubious or untried paths and that therefore prophets are likely to come into existence and then be effective among people who have been suffering adversity. What is not so immediately self-evident is the success which prophets of gloom sometimes have: they foretell disaster or even the end of the earth, and yet attract people to their cause. One can only guess that the converted do not entirely renounce the present in favor of the calamity, rather they derive some satisfaction from knowing that others share the pain and from associating with those facing the same fate. On the whole, however, I imagine that the prophet, the astrologer, or the oracle has a greater appeal when the future is envisioned brightly rather than darkly, and that, wittingly or not, the choice of hue depends not only upon the temperament of the forecaster but also upon his skill in interpreting the predispositions of his audience.

Similarly, individuals who gamble may or may not have accurate knowledge of the odds for and against them, but they are attracted by the possibility of gain resulting from a small or great risk involving the actual investment or their own welfare. Leaders here are the people who control the game or the mechanism. Through publicity (in the case of a lottery, by shifting the odds) or by changing the rewards (the share of the winner), they can affect the number of their followers and hence their own profits. The skilled pitch man knows who the suckers are.

In connection with most leaders and prophets the question must be raised: how long can people continue to be motivated by unfulfilled promises or prophecies? The question is intriguing, but it is much too broad to permit any kind of a sensible answer. If you are hungry and I promise you food within an hour and then no food arrives, you grow impatient and irritable. If you are unhappy with the present regime and I promise you a better life after a revolution in which you participate, and if that better life does not come, you may also be impatient. And you may or may not be convinced by what I subsequently tell you concerning the need to postpone personal benefits until the enemies of the state abroad have been vanquished. Or I may offer you everlasting salvation as balm, and then I shall never know, if I live longer than you, whether my promise has been kept. So much seems to depend upon the immediacy and insistency of the need; so much is affected by whether the unfulfilled promise exists in the midst of other needs which are bringing satisfaction.

RELIGION

The one institution in any society which regulates remembering, experiencing, anticipating, renouncing, and temporal patterning more than any other in an ultimate sense is religion. For religion would provide sets of answers to the eternal questions of man's existence: how could he have originated, what is the purpose of his mortal life, and what will happen to him after death? Accompanying these answers are rules of conduct to which people are expected to adhere if, in some sense or other, they are not to incur metaphysical wrath either here and now or for eternity—or for both. Religion as a separate institution probably is an abstraction of Western thought, for in most societies outside that orbit the separation of what we call religious practices from economic ones, for example, is probably not clear-cut: you do not simply worry about the gods on feast days or when you are worshiping them, rather aspects of your entire existence—when you cultivate fields, when you eat food, when you kill an enemy—are bound up with your view of them.

The explanation of the whence, why, and whither results in ethical codes requiring some form of renunciation. Human impulses are curbed: murder may not be committed, neighbors' wives may not be coveted, property may not be stolen, and so on through ten commandments or their equivalent. No doubt these regulations have evolved historically through a Darwinian process of trial and error whose dimly perceived goal has been the production of more satisfactory modes of social existence. At every given moment, regardless of the explanation, people having the religious beliefs, whether gracefully or grudgingly (and usually the former), make the sacrifices they consider necessary. Necessary for what? Necessary to avoid punishment in the present or the future, necessary also to obtain some kind of eternal salvation. The religious pattern of temporal reinforcement, in short, is oriented toward the present or the future, and—how difficult it is to be prevented by an Arbitrary Limitation from trying to specify the economic and psychological circumstances affecting the orientation—sometimes toward both.

The omniscient god or gods—or some substitute, such as ancestors or benign or evil spirits—are watchful and mindful of transgressions. They are in effect alive and ever ready to pounce upon those who violate the regulations and to reward those who adhere to them. The divine, ancestral, or spiritual presence is usually made meaningful through a series of concrete symbols which can be perceived by believers. Thus the world's great religions are symbolized by single objects such as the cross or the crescent, to which are attached all manner of associations. There are buildings or shrines for worship, prayer, and receiving admonitions. On special holidays some aspect of the belief system is strengthened by being reenacted or celebrated.

Perhaps the strongest appeal of most religions, the principal reinforcement for the sacrifices that are made, is its promise for the future after death. For death is always an insoluble challenge for men who soon come to appreciate their own morality and, unless they seek desperately to avoid dwelling upon this unavoidable part of their destiny or unless ghastly misery has made them seek out annihilation, they are miserable and must eventually long for some kind of eternal life. That life may resemble, they hope, not the kind of existence they have either led toward the end of their own lives if they are old, or have seen old people lead if they are young, but the kind they enjoyed at whatever they consider the peak of their powers or of someone else's. Eternity in an inferno may also loom as an alternative to be avoided. Or there may be neither heaven nor hell but a return to earth in another form. There is no gainsaying the fact of death, and religion offers the hope of escaping the cruelest prospect of all, perishing into nothingness.

Religion, in brief, may enable us to anticipate an existence without end, which means that temporal motives are never aroused, temporal judgments never passed. Most, perhaps all, of the heavens or hereafters that have been portrayed in holy works or in the poetry of the West suggest an existence in which time is unlimited and gratification certain. A doctrine of reincarnation makes time appear reversible or infinite since other existences are in the offing. In an ultimate sense, therefore, the temporal orientation of the religious person is toward the future or it transcends time. Within the social setting of each faith—certainly in the West, possibly almost everywhere—everlasting bliss, pie in the sky, salvation, or whatever form heaven or nirvana takes, is not automatically attainable: sacrifice, renunciation, and discipline are required on earth, in this mundane, sad existence.

Even when religion points toward a state of timelessness in another existence, it must also emphasize temporal conformity here and now. Religious observances, especially in the large organized churches, are closely keyed to the calendar and the clock. The most religious of persons, those in monasteries and convents, are the very ones whose daily routine is carefully regulated by timepieces. Lay members of the church are thus reminded by clocks, bells, or chants not only of "the eternal order represented by religious authority" in broad symbolic terms as well as of the brief time that men may live on earth (Moore, 1963, p. 25), but also of particular religious symbols and responsibilities. These temporal mechanisms likewise serve very practical ends. The devout hear the summons and so the divine service, communal in nature, can begin punctually; coordination is thus achieved. In addition, believers and nonbelievers alike are made aware of the time of day or night and, especially in rural areas before watches and clocks and radios become prevalent, they can plan details of their lives in unison and with certainty. The combination of religious and world functions thus discharged is well illustrated in the South Tyrol where the custom of ringing church bells as severe thunderstorms approach still persists. The sound, according to tradition with which most people seem acquainted, reminds men that in an emergency their destiny as ever is in God's hands; it conveys the message that other persons, the bell ringers or those in authority, perhaps more competent than they, are impressed by the danger of the menacing clouds and the flashes of lightning and so all must seek shelter; and formerly it was believed to produce vibrations which would break up the clouds and hinder the formation of damaging hail. The last function more recently has been discharged by shooting cannon balls into the thickest clouds while the bells ring.

The conception of time to which a religion subscribes, though basically metaphysical, is likely to have impact upon all true believers. An influential and controversial Swiss theologian, for example, has contrasted the views of the "Primitive Christians" and the Greeks in this respect:

> [For those Christians] salvation is bound to a *continuous time process* which embraces past, present, and future. Revelation and salvation take place along the course of an ascending time line . . . all points of this redemptive line are related to the *one historical fact* at the midpoint, a fact which precisely in its unrepeatable character, which marks all historical events, is decisive for salvation. This fact is the death and resurrection of Jesus Christ. . . .
>
> Because in Greek thought time is not conceived as a progressing line with beginning and end, but rather as a circle, the fact that man is bound to time must here be experienced as an enslavement, as a curse. Time moves about in the eternal circular course in which everything keeps recurring. That is why philosophical thinking of the Greek world labors with the problem of time. But that is also why all Greek striving for redemption seeks as its goal to be freed from this eternal circular course and thus to be freed from time itself [Cullmann, pp. 32-3, 52].

In effect, according to this particular interpretation, the New Testament teaches that Jesus' crucifixion has occurred at the midpoint of time: the creation, development, and expansion of the world before His life among men was a foreshadowed prelude to His coming, and after His resurrection men eventually and inevitably—at the Parousia—will be redeemed by Him and granted salvation as a result of His sacrifice. Those subscribing to this "linear conception of time" and believing eternity to be "endlessly extended time" (ibid., p. 62) are likely to view their own existence in a manner quite different from those perceiving their destiny as part of an endless cycle and eternity as "timelessness." Theological differences like these, it seems to me, affect not the kind of temporal judgments reached in psychological experiments or in ordinary social life but the central, core value placed upon time and man's attitude toward it, both of which must have either ephemeral or vast behavioral consequences. Whether an individual is truly religious or not, consequently, is reflected in the salience and content of his ultimate appreciation and evaluation of time.

11. WORK AND RISK-TAKING

Men seldom obtain fruits by waiting to gather those that tumble down from trees, rather they plant the trees and cultivate and fertilize the ground, and may spray the leaves before they can eat. They must work, and work—the effort necessary to reach what is considered to be a productive goal—usually requires the renunciation of other activities, including of course nonactivity; in short, it "claims time that could otherwise be spent in other ways" (Linder, 1970, p. 13). Work, moreover, means intervention, the carrying out of actions which, it is anticipated, lead to the attainment of desired goals. Even the slave, who has no alternative other than to work for his master, thereby achieves goals, the negative one of avoiding punishment, the positive one of being fed and housed. The anticipation of goal attainment through work further demands some sort of future temporal orientation: if you work, you are deferring gratification in behalf of an objective you anticipate in the future, and you are unquestionably aware of the location of that objective in time. In modern society, most people, whether they work in a factory or in a law office, must conform to a rigid time schedule concerned with the hours of work, the length of the work week and holidays, the amount of work produced in a given time period, the mode of coordinating their own efforts with those of others, etc. If you work in the West or, for that matter, almost anywhere, you must frequently look at a watch and a calendar if you are to achieve success. The material rewards from working in modern society, moreover, are saturated with temporal factors. The amount that is earned is usually a function of the objective time spent on the job (and the rate of pay for most piece work eventually is based at least in part upon a temporal criterion in the mind of the employer). Then the frequency with which wages or salaries are paid and the temporal unit for which payment is made (hour, day, week, month, year) is negatively correlated, at least in the United States, with the prestige of the occupation, for it is thus unsubtly suggested that those in the "better" positions are capable of (a) renouncing for a longer time the money they earn and (b) planning their budgets or the equivalent in order to anticipate the delay.

PRODUCTIVITY

The statements just made must be challenged. For the moment a descent is made from those abstract clouds onto the reality of the work situation it becomes clear that present gratifications as well as anticipated rewards determine the quality and quantity of work. Factory workers in Western countries have been studied systematically for more than a century, and their productivity has been found to be affected by a host of factors. There are conditions outside the work situation itself: the worker's physical and mental health; the amount of time necessary to reach the job; the status of the factory or of his particular position within it as viewed in the larger community; the day of the week or the month of the year; the status of a labor union; etc. And on the job other influences can be critical: illumination, ventilation, toilet facilities; fatigue and rest periods; the location of the men on the factory floor with relation to one another; the possibility of social contact while on the job; the intrinsic interest of the work; the coordination and organization of the work process; the time of the work day; etc. Even the remuneration, the reward for working, cannot suggest the actual incentives that operate, for it involves not only the absolute wage or salary and its purchasing power, but also the method of payment (e.g., fixed rate of pay vs. piece work), the relative amount in comparison with that received by other persons or a different reference group, the opportunities for advancement, etc. The list of variables is by no means exhausted, but enough has been sketched to underscore the complexity of the situation.

The large number of factors associated with industrial productivity does not mean that the phenomenon cannot be understood or controlled. We are simply confronted again, as we said we would be in Chapter 1, with another multivariate situation involving temporal and other kinds of behavior which interact. An investigator, when he approaches a new factory to determine whether productivity can be increased, for example, cannot know in advance just which factors will turn out to be crucial; but he has in his intellectual kit the knowledge of such variables which in a very valuable way can guide his empirical investigation of the situation at hand. He can quickly check illumination (which may have been crucial in the last factory he visited) and discover that it is satisfactory; then he can turn to union recognition, and go down the list of factors until he discovers the relevant ones. Relevant to what? Relevant to changing the situation, that is, to increasing productivity. But they are all probably important to some degree if the men are to work; thus the expert discards illumination only as a possible factor to be changed, but obviously the workers must continue to have adequate lighting regardless of what other changes are introduced or not introduced. The conclusion must be, therefore, that the anticipation of reward from work and the concomitant renunciation of present goals in favor of deferred gratification do not, by and large, provide sufficient incentive when productivity as a general phenomenon is considered; they are essential undoubtedly, though only in conjunction with other incentives.

I would make passing but relevant reference to the problem of efficiency because time is one of the three factors involved in appraising productivity from this standpoint—the other two are the work or goal achieved and the energy expended (or its equivalent, such as money). An operation is efficient when it is accomplished with the smallest expenditure of energy and time. Information, therefore, must be available concerning objective time, and then the work and the amount of energy expended can be determined. If you are both lazy and inefficient, you either do not care how long it takes you to attain a goal or else you do not seek the goal, since what you wish to do is not to expend much energy. In modern society the attainment of operational efficiency involves a paradox in the utilization of time. To attain such efficiency, it is usually necessary to expend extra time or energy initially: a machine must be designed, perfected, constructed, installed—and the same criteria of efficiency can be applied to that machine. Thus the person who uses the machine, or who has it built for his employee, values time or energy so much in the long run that he conserves it by expending vast quantities in the short run, which I suppose is what must be meant in part when people pass judgment on modern life by calling it a rat race. For machines do break down, so that servicing becomes important, demands in its own right an expenditure of somebody's time, and hence—as it becomes scarcer with an accompanying decrease in quality—contributes significantly to the "harried" nature of modern living (Linder, 1970, pp. 40–6).

MONOTONY

Monotonous work is disagreeable, of course, but it all depends—and here we succumb again to the compulsion of adding the *if*'s and *but*'s to what might appear to be a perfectly good generalization. The dictionary definition of monotony includes a reference to sameness: repetition with little or no variation. The feeling tone of such repetition is negative: the individual dislikes what he is doing not because he experiences difficulties but because he is unable to express other feelings or engage in other activity. Immediately we see that the mere fact of repetition, with little or no variation by itself, may not be disagreeable. Monotony is frustrating only when it becomes boring. Day after day your doubtless endure the same routine when you get out of bed in the morning, but you do not complain, for the very repetitiousness enables you to waste as little time as possible —and you accept the fact that you are powerless to escape the details. You are so pleased to have a job that, even though you must keep doing the same thing over and over again, you are not disturbed: the thought of the pay you will receive or of your fate if you were not employed prevents you from faltering or from feeling bored. Similarly ardent amateur fishermen are not likely to be bored when they repair their lines or nets since such repetitious work contributes to a joyful end; and nonboredom can be almost guaranteed if they distract themselves by humming, listening to music, conversing with friends, or daydreaming. In another situation the individual likes the work he must perform without variations because it is within his capacity; if he had options or if he were compelled to make decisions, he fears he might fail; he knows what he is supposed to do and derives pleasure from doing it. Or after a hectic day overflowing with a variety of activities, it can be soothing and restful to perform a routine chore such as painting a wall or cultivating a garden.

Each of these hypothetical anecdotes might not stand closer analysis, but together they perhaps suggest that the unpleasant feeling of monotony springs from the individual's psychological state at the time he is performing the work, and not necessarily from the repetition as such. By watching a group of workers perform endlessly repetitive work you cannot single out the particular ones who are bored; you need to know more about them than the nature of their activity; at the very least you would guess that the more intelligent or alert workers are the ones more likely to be bored. Nonsense, some devil's advocate ought to exclaim at this point: even if you know that the reward for repetitive work is a guaranteed entrance into the kingdom of heaven, you can pretty well guess that at times all the men must be bored, must be finding the work monotonous. But why?

Part of the reply, at least in the West, must stem from the fact that the constant and effortless achievement of similar goals, especially when absolutely basic drives are not involved, is by itself not satisfying: the thrill of achievement disappears. If you beat your opponent easily in a game, you are not particularly elated—unless you want to humiliate him or unless you find some reward from winning other than the fact of winning. For a while, the effortless achievement of a goal can be gratifying if that achievement has involved great skill or another admirable quality; later, then, automatic success again and again is devoid of challenge.

In addition, the consciousness of time is involved when tasks are felt to be monotonous and then a duration-dependent drive is evoked. A challenging, nonmonotonous task demands the full attention of the person if he is to succeed. He need not concentrate, however, upon what he is doing when the goal is relatively effortlessly achieved; he may think of other matters, including the passing of time. For time is important to him because after an interval—when the noon or the five-o'clock whistle literally or figuratively blows—he will be released from performing the task. He wants time to pass quickly to obtain release, but instead it passes slowly. The kettle does not seem to come to a boil: he may believe that a long interval has elapsed and he finds that he has overestimated the interval, he is frustrated, and the feeling increases even more than the apparent duration of the interval.

Monotony, however, must involve even more than the anticipated release from the work and the degree of frustration that work and the passing of time entail. I suspect that the reaction to the content of the work interval as such plays a role. An interval devoted to a monotonous task must seem less filled than one in which the individual exercises initiative and hence is more active. One study cited in *Addendum 11.1* (Kerr & Keil, 1963) suggests that the overestimation of time while working was not related to the men's expressed interest in the work but, among other variables, to the variety of activity.

The remedy for boredom and monotony, when the feeling creeps in upon consciousness, seems to be to forget the passing, the slow passing of time and to become absorbed either in the task at hand or some substitute activity; evidence for this rather obvious point has been collected in the *Addendum* just mentioned. Sometimes piecework may help, for then the worker is motivated to perform the same task again and again just as frequently as he can in order to attain the maximum reward. Or, detachment from the task is useful: you reminisce or daydream as the rest of your body goes through the required motion. It has also been shown (again see the *Addendum*) that the time span of the work—how long it takes to complete a given task, the amount of planning that is required, etc.—may affect morale. But unemployment, when it is involuntary, also affects temporal orientation and judgment. It is, moreover, the fear of unemployment or, in positive terms, the craving for security (e.g., U.S.; Walker, 1957, p. 176), whose drive strength is so great that the restrictions of a job can be seen in the perspective of life goals and appear necessary and hence not monotonous.

I return again to the existence of individual differences: one individual becomes bored more quickly than another. Differences of this sort obviously have varied origins. Some persons are bored because they are negativistic: they reject responsibilities in general or the task facing them by invoking the complaint of monotony which they then ascribe to circumstances outside themselves. Others may be so skillful that they quickly master the problems at hand, and then they immediately begin to feel dissatisfied when they cannot be faced with new challenges.

DECISIONS

Economic goods, economists not surprisingly also note, reflect human preference and choices (Alchian & Allen, 1964, pp. 19–28). Natural resources, for example, acquire economic relevance when they are scarce and when they satisfy the needs of individuals. A simple market price—beans are selling for 27 cents a pound—represents the outcome of scores and scores of decisions by the sellers and buyers involved in the transaction. Some sellers might have been willing and able to sell at a lower price, some buyers might have been willing to buy at a higher price; other sellers and buyers would have traded if the prices had been, respectively, higher or lower. At the moment of sale, both groups are presumed to be maximizing their present preference; perhaps they anticipate that a delay in the transaction will not provide more favorable terms or, if they do so anticipate, they find their present preferences more compelling. They are thus confronted with the perennial problem of allocating time.

Usually, unless the buyer has unlimited means at his disposal, he is confronted with a number of alternate ways of spending his money. In a free or semi-free market he decides whether to purchase one product rather than another (cheese vs. fruit for dessert) or, among competing products, one type rather than another (camembert vs. gorgonzola). The money to be spent has no real properties of its own, although paper bills could be admired as aesthetic creations or used to kindle a fire; it symbolizes the reward which the individual himself, or someone remotely or intimately connected with him (an ancestor, a husband), has obtained from working in the past or will obtain from work in the future. When a person makes a decision concerning even the simplest expenditure, he wittingly or unwittingly engages in an economic act of double renunciation (or, as economists are more likely to say, sacrifice): he renounces the alternate way or ways of spending the same money, and he renounces the money (however he has earned it) in anticipation of the satisfaction to be gained from the purchase or from utilizing the service he buys. The temporal orientation can be in all three directions: toward the past, as he remembers or is dimly affected by whatever experience he has had with the product or service; toward the present in terms of his present needs; and especially toward the future as he anticipates the utility to himself. The future orientation, moreover, may include the contemplation of intervention: "the universal form of conscious behavior," not a psychologist, but a classical economist, once stated, is "action designed to change a future situation inferred from a present one" and hence "we must infer what the future situation would have been without our interference, and what change will be wrought in it by our action" (Knight, 1921, pp. 201–2).

The renunciations of alternatives and of money or time may be trivial or serious, depending upon the magnitude of the transaction and its significance to the buyer. Let me use capitalistic, Western, homespun illustrations. If that buyer selects chocolate ice cream rather than strawberry, or if he prefers ice cream rather than chewing gum, his investment is small and he probably will have the opportunity soon again to make a different choice. But if he is not wealthy, and buys a car, then he knows he has had to work a long time to earn that much money and also that he will be unable to purchase another model the day after tomorrow. Indeed, the long-time renunciations to which a person commits himself, as in the purchase of a home either outright or with a mortgage, is usually a sensitive index of his anticipations concerning the relevant phases of his existence. It is psychological factors of this kind which determine the choice of goods and services and which then are indistinctly but in some way reflected in the shape of the demand curve.

There are various techniques to diminish the pains of economic renunciation. Sellers in a free economy can affect renunciation by changing the price; the lower the cost to the customer, the less of a competing good or service the latter will be compelled to renounce. But the matter is not so simple since there are other psychological considerations to take into account. In many strata of Western society, a high-priced article has prestige, or high prices are equated with quality; raising the cost, consequently, must mean the expenditure of more money by purchasers and hence greater renunciation of other goods or services. And yet a compensation for that renunciation may be an increase in the psychological utility of the article because as a more expensive item it acquires greater prestige and allegedly brings deeper satisfaction. The buyer himself may seek to increase the subjective utility of the product after purchasing it. One provocative but not completely conclusive investigation suggests that the purchasers of cars tried to reduce some of the dissonance arising from this momentous decision by noticing and reading advertisements inviting them to own the car they had just bought (Festinger, 1957, pp. 50–4); for where else could they find such high praise for the very decision they had made? The insecure buyer, moreover, may diminish his uncertainty by displaying to his friends the product he has purchased and by soliciting their admiration.

Mortgaging and installment buying are devices used in competitive societies to achieve several important economic and psychological functions. Most of all, they obviously enable purchasers with insufficient money to obtain immediately what they might never have been able to obtain or what they could have obtained only by waiting and accumulating the resources. But the consumers must anticipate a series of renunciations in the future as they pay off interest and pay back principal—and, bien sur, the producers or distributors must anticipate that they will do so. The buyer's temporal orientation, however, may be less significant in determining his decision than his interest in the present. Indeed that is the psychological rationale behind such buying and selling; you buy intervals of gratification in the present and pay for them with intervals of work (or its equivalent) in the future. The renunciation at the outset is small—the papers are signed, a down payment is made (and sometimes the latter may be skipped)—and you move into the new home or begin immediately to enjoy the new car or suit of clothes. The future payments of course are always small in comparison with the principal that has in fact been borrowed; and we have, fortunately or unfortunately, no hedonistic, temporal calculus enabling us to determine whether the sum of renunciations from small payments plus interest is less than the single renunciation of the total amount without the interest. Certainly if the decision turns out to be a happy one—people like their new home, the car runs well—the gratifications being continuously obtained may help to offset the pain of paying the monthly bill. The present is easier and more tempting to enjoy than the future, and the future may seem an unknown distance away when no temporal motive concerning the pending interval is evoked or when that interval is judged most vaguely or casually.

In commercial spheres in the West all the subtle weapons in the arsenal of advertising and public relations are turned loose upon the consumer in order to increase the apparent utility of goods and services from his standpoint and thus to have him change his preferences. Through the use of the mass media, it is almost impossible to avoid at least perceiving the content of advertisements: to do so would mean being cut off from the news columns and other features of newspapers, from the entertainment and education provided by radio and television, from the highways which remain cluttered with billboards, etc. Then those who prepare advertising copy or messages are continually making true or, to employ a euphemism, partially true claims concerning the benefits to be derived from using the product or service. When the purchaser finally purchases and consumes the product, therefore, it is highly likely that his satisfaction depends in some part not only upon the objective attributes of the product when appraised from a technical or engineering viewpoint but also upon the qualities associated with it as a result of advertising. You prefer and select the product because it tastes good, and it tastes good partially because advertising has given you a set of anticipations concerning its nature. Finally, a miscellany of commercial devices may add still further to the psychological utility of the product or service after the decision has been made: if he has been told that he is getting a bargain or that only a limited supply is available (as if the product were an antique or an objet d'art or the service one that could be rendered only by Zeus), the consumer may then enjoy the alleged advantage he has gained and consider himself the envy of all.

DIMINISHING RISKS

In Western society, perhaps elsewhere too, there are some individuals who approach their work or the spending of money like gamblers: they are thrilled when they must undergo risks, often because they like the danger, but perhaps more frequently because risky enterprises are more challenging, may produce greater returns, and if successful can bring the satisfaction of having emerged the victor in a dubious enterprise. Most people, it appears, however, do not like risks or very great risks, at least in reference to the basic activity of work and the essential goods and services which working enables people to command; they would diminish them. While it may be true from a strictly economic viewpoint that insurance has "much in common" with gambling and speculation—"all consist of the transfer of wealth from one person to another which is contingent upon an unknown, usually, a future event" (Hardy, 1923, p. 329)—the motivation behind the anticipations is quite different.

Within recent years risk-taking has been investigated in the laboratory by observing how subjects, largely American undergraduates, react to relatively simple gambling games: do they run a larger risk in order to gain a bigger reward (tokens, dimes, meaningful money, symbols, etc.), or do they cushion their possible losses (i.e., take out insurance) by risking less at the cost of eliminating the possibility of gaining a great deal? The results so far, though often phrased in precise mathematical form, are not unexpected: the decision depends upon the way in which the alternatives are presented, upon the odds and payoffs, and upon various factors present in the situation (e.g., other students who, being confederates of the investigator, behave in predetermined ways), and upon the personalities of the players (U.S.; Rapoport & Chammah, 1965). Another approach is to ask subjects how they themselves would react or how others should react in a hypothetical situation. For example: should a scientist work on a difficult problem which might bring him great honors if he solves it but at the risk of retarding his professional career if he does not, or should he concentrate on a series of minor, less important problems which are easier to solve and thus virtually ensure no damage to his career? Not unexpectedly this kind of behavior has been related to personality traits in a multivariate manner; thus, in one study there was no relation between "impulsiveness" and risk-taking for "the male sample taken as a whole," but that relation emerged "in striking fashion for strategy measures in males low in test anxiety and defensiveness" (Kogan & Wallach, 1964). A phenomenon given the flaming title of "the risky shift effect" has also been noted: for some persons and under some conditions greater risks are considered more desirable after the issues have been discussed in a group than when a decision is made on an individual basis without advice or criticism from anyone else. Investigations of this sort offer us no great insight into the broader problems of interest here, other than dramatizing some of the variables and significantly reaffirming the principle of determinism: the temporal decisions apparently are not made willy-nilly but spring from the structure of the situation and the people.

Diminishing risks may require two kinds of renunciation. First, the individual anticipates the possibility of failure in the future, to avoid all or part of which he makes a disproportionately small sacrifice in the present. Or else he agrees to diminish his literal or figurative return in the future which becomes less than what it would have been if he had been willing to expose himself to dangers. One or both these renunciations can be found in most aspects of economic life. At a simple level, the well-satisfied consumer in the West is likely to be loyal to his brand if it is proving satisfactory. Why, he tells himself, should he run the risk of trying something new which may turn out to be less gratifying? He thus renounces the possibility of an improvement and he may even pay more for his preferred brand than he would for its competitors. At the other extreme of complexity are insurance policies which cover almost every conceivable kind of risk by means of a simple principle: the greater the risk, the higher the premiums, that is, the larger the renunciation in the present and the correspondingly less the return in the future. You drive your car, you know there is the possibility of an accident, you realize you can ill afford to pay the damages which might result, but you do not therefore stop driving to avoid the possible disaster; rather you make the relatively small sacrifice of paying premiums on an insurance policy which guarantees to compensate you for damages incurred under specified circumstances when and if you have an accident. All very simple: and the complicated calculations of the insurance companies which set the insurance rates stem from the principle that their income from a large number of policy holders will be greater than the payments they must make to the unlucky few.

Most insurance is of a similar type. You anticipate the possibility that your house will burn or that you may require hospitalization; those anticipations become less painful—or they can be forgotten completely—if you renounce a bit of your income by taking out insurance. Or you plan to stage an outdoor festival which could be ruined by rain: an insurance policy guarantees some return on your investment if it does rain: you will have paid out for the policy only a small sum in comparison with your receipts if it does not rain. Similarly, while perfectly well, the individual renounces part of his income in order to contribute to a health plan which thus enables him to diminish or almost eliminate the risk of paying large bills if or when he becomes ill. As a young man, he likewise contributes a small portion of his current income to a fund which he will not utilize until he is retired.

Slightly subtler is life insurance in Western society. Here the beneficiary is not the person who makes the sacrifice in the present by paying the premiums, but his survivors whose gratitude when they receive the benefits he will never enjoy. The policy holder derives momentary gratification from the conviction that he is providing for his descendants and thus diminishes the anxiety he may be experiencing for them then and now; the latter in fact may praise him while he is still alive.

A psychological problem involving insurance broadly defined is the determination of the circumstances under which certain kinds of persons run risks without insurance or with insurance of varying amounts. Only some passengers pay for accident or life insurance before boarding a plane; only some Europeans and Americans are attracted to gambling tables or games. It is noteworthy that, where private enterprise prevails in Western society, insurance companies are large, important, and generally prosperous. Here is impressive evidence that Western adults are willing and able to renounce small sums in order to avert future losses or the anxiety from anticipating them. In these instances the risks of individuals are diminished by being widely distributed. In effect, those remaining healthy pay the medical bills of those becoming ill. If employers contribute part of the premium or if government supplies money for whatever deficit a national scheme incurs, again renunciations are made by other persons, the employers or the taxpayers. Inevitably, therefore, we are involved fiscally in one another's lives.

The act of purchasing insurance does not always signify that the individual is only attempting to diminish risks. Surveys among Americans have shown that some policies are bought "because it is the thing to do." Or the reverse may also be true: what conceivably can be anticipated from insurance is not completely understood. Again a generalization from the same American surveys: "Life insurance is undoubtedly the most common form of saving and the most common form of investment—but most people do not consider it as either saving or investment" (Katona, 1951, pp. 104, 110). In these instances, therefore, behavior that appears to have a future orientation is mediated by nontemporal impulses.

In very critical spheres of activity, notably health and old age, private insurance schemes in the West by and large turn out to be inadequate for perhaps the majority of people and hence the state must intervene and force all persons to promote their own welfare through renunciation in the present. Many are realistically sugar-coated since they include the fringe benefit of requiring the employer to contribute to the policy. The element of compulsion is again very prominent: people do not easily look to the future. Some of the difficulty here, to be sure, is realistic: many persons cannot afford to pay for insurance. But there may be psychological components: they may be unwilling or unable to appraise their present situation, to look ahead, or to renounce for the sake of the future.

One explanation of the conservatism of most non-Western societies is the absence there of the equivalent of insurance plans. The radical, the aberrant, the nonconformist under most circumstances is likely to be punished, sometimes even when he succeeds in making some sort of innovation which, however, is then rejected by most people or by those holding the power. Conspiring against him may be not only his contemporaries but also the unseen yet effective power of ancestors or tradition or magic or whatever it is which symbolically enforces sanctions within the society. And if he fails, so much the worse for him. He has no policy to protect himself against the risks of the enterprise. It may be, therefore, that the way to achieve rapid social change in a society, as one anthropologist has implied (Firth, 1969), is somehow to provide insurance in behalf of innovation; in this fashion the potential innovators are likely suddenly and quickly to restructure their own temporal orientation as well as the activities in which they engage to achieve new goals, or at least old goals in new ways. In our own society the individual innovator may have no form of insurance to diminish his risk, but he can distribute the risk—and the potential gain—by forming a company with limited liability, or, in some instances, by obtaining support from government. Much of our industrial innovation occurs within companies whose innovators receive a regular salary.

Producers, too, would scale down their own risks, for they cannot always be certain that what they produce will be purchased by consumers or at least purchased at a price profitable to them. For this reason they seek to ascertain consumers' preferences and demands. Market research assumes either that individual buyers reveal their future preference by what they do or say at the moment or that, when questioned directly, they themselves can anticipate their own future behavior. Either or both assumptions can turn out to be incorrect. Sometimes, therefore, all that is needed is a knowledge of relevant behavior from the past to provide the clue to the future; thus no survey is needed to anticipate that more warm clothing will be purchased as cold weather approaches (provided money is available), but the manufacturer's risks can be reduced if somehow he can ascertain consumer preferences and purchasing power in greater detail. In fact, any type of promotion or service, including, certainly, satisfaction from actual consumption in the past, strengthens the bond between producers and consumers and hence diminishes the producer's risk.

Acquiring additional relevant information enables producers to plan and hence to decrease uncertainties. Farmers, for example, cannot organize and operate their farms effectively unless they are acquainted with prevailing prices; with production trends; with persons in their immediate vicinity, such as hired workers or county agents; and with institutional regulations ranging from tax rates to conservation programs (U.S.; Johnson et al., 1961). Some of this information is unwittingly absorbed as part of daily living, some must be deliberately pursued (Hart, 1940).

Producers and workers alike form organizations in order to eliminate some of the hazards of functioning alone. Groups of producers have various objectives: public relations for an entire industry, research, trade agreements, and cooperation with reference to the demands of labor and strikes. And workers join unions to be assured that their rights can be defended, to secure for themselves improvements in wages and working conditions, and to receive benefits during strikes when they are cut off from their normal earnings. Contributing dues to the central organization represents yet another sacrifice in the present for the sake of future gain. A strike requires renunciations too, as each side engages in the power struggle whose outcome depends in no small part upon the ability to renounce for a longer period of time than the opposing side: workers are willing to lose wages since they anticipate that ultimately a victory will bring them better terms of employment; employers lose the profit from what would have been produced in the absence of the strike since they prefer not to suffer the losses resulting from the demands of the unions. During strikes, moreover, risks may be reduced for the two antagonists: related companies may close down in sympathy or offer compensation to those against which the strike is directed; and workers who have not struck may refuse to cross picket lines. In unity there is strength, or at least less risk.

Risk-taking is really not an all-or-none matter since each person is generally willing to run risks in some areas and not in others. One man who has bet on a long-shot at the races does not pass another car as he approaches a curve on the way home; another is very cautious in his business or profession but, being unconcerned about his social reputation, he makes blunt, honest statements at cocktail parties. We have here little more than another instance of the generality or specificity of traits and behavior: risk-taking for each person may be largely situational, but it is more likely to be a general disposition or tendency. The big gambler at the races, I would be willing to bet, will in fact run the risk of passing a car near a curve.

CAPITAL

No matter what the approach to economics, orthodox or Marxist; no matter how underdeveloped, developing, or developed the society; in fact, no matter what kind of work is being appraised, the need for capital must be recognized. For almost without exception capital is required in order to produce: ordinarily the farmer must first build or buy a plow before he can plant his seed, the fisherman needs a hook or a net before he can catch fish, the industrial capitalist needs a building equipped with machinery before he can begin manufacturing. Somebody must supply the capital, whether it be the producer himself, his family or acquaintances, stockholders, moneylenders, or the state. At least one person must have a future temporal orientation which enables him to defer gratification: he must be willing to renounce something in the present for the sake of a future reward for so doing. Thus the cultivator who needs only a simple stick to turn over the soil must find that stick and make it into a tool, his labor in itself is economically (if not physically) nonrewarding but is expended because of the crops that he hopes to reap. The state which invests in a new enterprise is buying equipment and paying the wages of those who build the factory with money it has obtained from taxes or loans or from some other industry already functioning. Human beings have often been characterized as toolmaking animals: though they may not have an absolute monopoly in this respect, they do have a pronounced capability of anticipating the future and taking appropriate, productive action to reach future goals.

Immediate consumption by everyone is impossible if there is to be capital: there must be renunciation in the present for the sake of anticipated gratification in the future. That future, obviously, may be close at hand or far away. In the Western world people who deposit money in savings banks or who purchase stocks or bonds are renouncing immediate spending, but all or part of their reward for doing so may be close at hand, viz., the interest or the dividend paid every fraction of a year. Some investors in private enterprise receive no immediate return but await dividends or profit in the future. Again, as in gambling or many games, the return is proportionate, to some extent, to the risk; so-called gilt-edge stocks pay relatively large dividends regularly and do not appreciate or depreciate much in value, whereas "glamor" stocks which pay small dividends less regularly, and sometimes none at all, fluctuate markedly up and down so that there is always the chance of a large profit—or a large loss. Why are some persons within a society more eager to take risks than others? Stop—Arbitrary Limitation.

Still it must be said that the motivation in the West to renounce and defer gratification through savings or investment—in economic terms, to substitute future for present consumption—is undoubtedly a complex process. In the first place, the magnitude of the renunciation fluctuates as a function not only of the time, money, or energy that is invested as capital but also of the quantity of each at the individual's disposal. Obviously the value or utility of a given sum of money to a wealthy man is different from that of the same amount to a poor man. The motive for renouncing also varies. In the area of savings, it can range from a desire not to be tempted to spend all one's income (and hence Christmas Clubs flourish in American banks because they compel depositors, as it were, to save for 10 or 11 months in order to splurge once a year) to the conviction that only in this fashion can one have enough income from savings or annuities after retirement. Many years ago the provacative thesis was advanced that a Protestant ethic —the principle that it is sinful to expend recklessly in this world and that only through good deeds here and now can eternal life be attained—encouraged people not to spend but to save and hence created in advance the renunciatory motives necessary for industrialization or capitalism to come into existence (Weber, 1930). This view swings the pendulum too far away from economic determinism, but it does call attention to the important relation between the values in a society and the motive to renounce some of the fruits of one's labor.

The complex machinery of a more or less open economy in the West has ways to encourage or discourage investment and the use of capital, some of which have already been suggested in the previous section of this chapter. When the interest rate on savings and on loans changes, the motivation of savers and borrowers is affected. When the rate goes up, savers receive more for renouncing the use of their money, and borrowers must renounce more, or run greater risks, in order to make capital expenditures or to buy goods and services in the present; hence saving presumably is stimulated and borrowing discouraged. The reverse of course is true when the rate goes down. Some banks in America offer eye-catching gifts, as suggested on pages 95–96, to encourage people to open savings accounts; immediate consumption sweetens the pill of renunciation and initiates renunciatory activity which thereafter becomes somewhat automatic or at least less painful. The use of capital, consequently, provides a complex index of people's judgments concerning their allocation of time and concerning the future; they must make decisions involving their temporal orientation and they must also pass judgment concerning the length of the interval which must pass before they can repay or regain the loan.

The question of how much reward a saver or investor must be offered to induce him to renounce and how much a producer or consumer must be charged if he is to utilize or enjoy the effects of that renunciation has been discussed here only in psychological terms; but that frame of reference is closely related to the economic one of determining how much capital is actually available within a system at a given moment. There are, in addition, grave ethical issues involved in the use of capital which are sometimes overriding in the modern world. Thus the radical, and sometimes also the conservative, maintains that the interest or dividends paid on capital investment are unjust: the rate is too high for the alleged sacrifice that is made, or else the source of capital should be not private individuals but a group of peers (like a consumers' or producers' cooperative) or the state itself. Eventually the same kind of searching questions have been and are being raised concerning property, land, and labor which in this sense are forms of capital and therefore demand renunciation if they are to be acquired and utilized. These psychological and economic issues have had in our time grave political consequences, resulting probably everywhere in changing the role of the state in men's affairs. In fact, it is important to note that when their own cultural and personal standards convince them that the amount of renunciation required to obtain a just portion of earthly goods, security, and spiritual gratification is too high, men may no longer postpone satisfaction but may resort to the kind of violence characterizing the present era.

CONSERVATION

Often work can be avoided by squandering or exploiting the resources at hand: you eat and are merry and are unconcerned about the morrow. Conservation, on the other hand, demands renunciation, and the price paid is a sacrifice in the present, and indeed additional work. Consider again an illustration previously mentioned: children in the South Tyrol—whose farming population has a proverb which states that "the forest is the savings bank of the peasant"—are given a holiday each spring to picnic in the mountains and there to plant the seedlings which much later in their life or after their death will grow into lucrative trees. The objective is thus not only to have the trees planted, but also to orient the children toward the future; and the orientation occurs on the basis of gratification obtained in the present from the outing and other pleasures associated with the day itself. This is almost an ideal device to achieve conservation: postponement of gratification is obtained almost effortlessly because of the rewards in the present, while simultaneously a general orientation is, hopefully, established. The element that is lacking is renunciation: the children at the time of planting are themselves making no sacrifice, it is all so very jolly. Then in the future they are likely to see the trees mature, remember their role at the outset, and thus be reinforced with respect to some type of deferment pattern.

Ordinarily renunciation looms large for somebody: he could reap a profit now from mining the forest, but he is asked or required to restrain himself for the sake of the future. But whose future? If he is a young man, then the future may include benefits for himself: the trees will mature in the meantime and his profit be all the greater. The pyschological problem for him, however, is quite different if he will be dead at that time. Then how can he be motivated to conserve the forest?

The simplest answer is to force him to cooperate. But that answer may be too simple, for it sidesteps the problem of why the society adopts the policy that forces him to do so. Many who endorse conservation in a democratic society, like the individual himself, will not experience the benefits; for those who will be the beneficiaries the incentive is clear. The question, therefore, has to be broadened: under what conditions do those in power adopt conservation measures, what rewards can be offered them when they themselves will not directly enjoy the fruits therefrom?

There seem to be three, all somewhat intangible and indeed symbolic and all quite similar to those functioning in connection with life insurance. First, an appeal is made to "future generations," and this comes closer to being concrete and meaningful when those hearing it think of their own children. It is pointless, here at any rate, to push back the problem and to inquire why parents would provide for their children. The bases may be purely altruistic or almost so: one loves them and wishes them to be eternally happy; or one feels that immortality can be achieved only in their memories. Then, secondly, the reinforcement may be a bit more tangible: as with life insurance, the gratitude of the beneficiaries and one's contemporaries will be expressed in one's old age or now; the sacrifice, in short, is acclaimed. The importance of such symbolic rewards depends upon the modal temporal orientation of the society. If that orientation is directed largely toward the present, as it probably must be among nomads or food-gathering peoples at least so far as sustenance is concerned, then the incentive to conserve will be small. But of course other factors affect this incentive, such as the prevailing attitude toward nature: will God provide or will He only help those who help themselves? Most societies, however, seem to have an almost biological urge to perpetuate themselves, a feeling which carries over to individuals who seek their own perpetuation and that of their group through their children. There is a certain amount of ego involved in the thought that one's name or deeds will be remembered after one's death by grateful descendants.

The third appeal stems from the second but is less easy to capture: the expression of gratitude to past generations, to one's ancestors, for their accomplishments and for their legacies. One has, as it were, a debt to them which is discharged by bestowing gifts upon future generations. That debt may be concrete and meaningful when the elders who have conserved in the past still exercise influence in the society, or at least are at hand to remind their offspring of the sacrifices for the future they once made, but on the whole, it is discharged symbolically. The ancestors' deeds are celebrated in word and pageant, and thus the need and feeling for continuity from generation to generation is reinforced.

Once more, the difficult nature of renunciation and of future temporal orientation has become apparent. Man has these capacities, and the ability to think of future generations, but he exercises them reluctantly and then only when some benefits, frequently only of a spiritual nature, occur to him in the present.

OCCUPATION

A person's occupation determines many of his earthly rewards. His earnings enable him to command goods and services. The time spent working brings him varying degrees of satisfaction. The prestige associated with his work may affect the evaluation he places upon himself. At every turn, therefore, an occupation in the West may require a critical choice which is likely to involve renunciation in some form. The simple experience of having been counseled briefly with respect to vocational opportunities produced in a group of American college students a stronger orientation toward the future as indicated by a projective device having no direct connection with occupation as such: in comparison with a comparable group of controls not given such advice, they wrote stories with longer time spans. This particular experience, however, did not significantly affect their interest in words pertaining to time nor their verbal readiness to refer to themselves in varying temporal contexts (Matulef, Warman, & Brock, 1964).

In most traditional societies, the problem of occupation is not complicated by the need to make a decision: the individual has few or no alternatives, and therefore he does what is appropriate to his sex, his age, his family, his clan, etc. But even then some choice may exist, for example, concerning the amount of land he cultivates or concerning whether or not he tries to become the type of artisan or worker his group favors. Nowadays such societies are likely to be affected by the modern world, and many men and, less frequently, women decide whether to remain within the tribe or to migrate to cities, there to learn and then follow another occupation. In Western countries the number of choices that are open to subgroups, such as the children of workers, may in fact also be very limited—miners are sons of miners—but here the belief in the possibility of mobility or change may make the restriction seem not natural or desirable but frustrating and unjust.

The selection of an occupation, when that is feasible, has implications for ensuing renunciations. An interval of time ordinarily is required to learn the relevant skills or to secure the necessary information. Even the most elementary form of labor demands some training, though it may take place within a few minutes and the actions can be perfected on the first trial. At the other extreme would be the modern surgeon who, after the usual education, must then begin his medical training; in very basic ways, consequently, until he approaches middle age, he must renounce many of the customary rewards of this society whose realization he can in the meantime only anticipate. The more specialized the training, moreover, the less the likelihood that the individual will be able to change his occupation in any radical fashion later on: his investment may make him less eager to consider an alternative, he may be too old to seek one, and he may find his work habits maladapted to most other occupations.

Even within each occupation, a person may be confronted again and again with choices between what he would have and what he would renounce. In Western society he may have to decide between being an employer and an employee. In the former role, he runs greater risks but may achieve more material rewards and great distinction, whereas in the latter he may have more security and contentment but less reward and distinction. Or within the professions of modern society he will certainly have to choose between the security of being employed by an institution and the risky adventure of private practice. Similar choices may exist within a given role; as an employee, for example, does he prefer pleasant working conditions and less pay or less pleasant conditions and more pay? It almost looks as though modern man is faced with two factors whose product equals a hypothetical constant, viz., security and momentary reward: an increase in one means a decrease in the other.

ADDENDUM 11.1 REPETITION AND MONOTONY

An example of the empirical approach to monotony on a laboratory level is a study in which American children and college sophomores were closely observed as they drew the same "moon-face" again and again. They sought to vary the actual drawing they produced and, in addition, they also talked, whistled, sang, and found a substitute in "an active inner fantasy"—any activity other than straight repetition seemed preferable (Burton, 1943). A psychologist speculates, on the basis of experiments in academic settings, that time in a factory can be made to pass quickly, with a consequent improvement in morale, if the worker is aware of some subgoal in the task, if he is motivated to reach a final goal, and if he obtains information concerning his progress (Meade, 1960b).

After being deliberately misinformed that the company clocks were incorrect, workers in an American factory were asked unexpectedly to give the correct time. No relation was found between their estimates and expressed feelings concerning their interest in the job; overestimation, however, was found to be positively related to (a) a variety-type rather than a monotony-type job as classified by the investigators (.43), (b) long-cycle rather than short-cycle jobs (.27), and (c) absolute error in temporal estimation (.42); and it was negatively related (−.46) to length of service on the particular job (Kerr & Keil, 1963).

Workers in a British factory, mostly girls and women, did not complain of boredom when they were fatigued if the incentive of overtime pay was provided; otherwise they spontaneously sought ways to relieve the boredom of repetitive work by not looking at their watches too early in the day (lest they discover that time was not passing quickly), by eating sweets (especially those which had to be distributed surreptitiously), by changing their bodily position, by singing, by wishing for something to happen (during World War II even an air raid could serve that function), by going to the toilet, by adjusting their make-up, etc. (Jahoda, 1941).

One study has addressed itself to the question as to whether there is a relation between temporal orientation or evaluation of time and the type of position men occupy, as well as their productivity and attitude toward their work. By speculating on the basis of a thorough analysis of one British factory, the writer suggests that the tasks there involved increasingly longer time spans and hence higher levels of abstraction as one moved from the worker dealing with the piece of metal or paper at hand to a foreman planning a future production schedule; and he suggests that this "time-span capacity" somehow be assessed as an individual is considered for appointment or promotion (Jacques, 1965). If this generalization is valid, then the time span of the task confronting any worker involves anticipating future events, delaying gratification until the task is completed, and tolerance of uncertainty. Guided by this frame of reference, a study was made of managerial personnel in an American plant with "time extension" measured by having the individual indicate how certain he felt about events in the future, and "time value orientation" by having him indicate his attitude toward future events; in addition, he was asked how far ahead his job permitted him to plan, the extent to which he allocated his time on the job to planning, and his satisfaction with the work. Correlations were found between the temporal measures and type of work, but so many were not significant and all of the significant ones were so very low that the investigator quite rightly concludes that, in the factory he studied and with the measures he used, there was little support for the original British theory. Evidence was uncovered, however, that the workers were more satisfied with those jobs which were more or less congruent with their "time span propensities": dissatisfaction resulted in large part when the requirements of the jobs were *less* and not more than the men's own time spans (Goodman, 1967).

Worse of course than boredom while working is to be unemployed and to have no work at all when a job is essential to maintain one's standard of living and self-respect. The unemployed males in an Austrian community during the depression of the thirties soon became unable to think in terms of a future since their outlook was so bleak; they seemed to lose track of time and to forget about punctuality because, with no schedule to guide them other than a few responsibilities at home and with long periods of empty time at their disposal, they had no incentive to pass temporal judgment (Lazarsfeld-Jahoda & Zeisl, 1933, pp. 59–69). The inability of the unemployed to peer very far into the future has been shown to characterize similar groups in Scotland and England. In fact, the unemployed who were interviewed there seemed about as "downcast" about their future as psychotic patients, particularly those suffering from anxiety (Israeli, 1935).

PART III: MANIPULATION

12. THE ARTS

The arts, the graceful, the lively, the sublime arts are tantalizingly elusive. I do not enjoy the delusion that they can be captured glibly. In fact, I shall simply pursue the ways in which they treat time from the standpoint of the audience, the reader, the listener, the spectator. By and large the artists who create them will be deliberately neglected. Here the arts are considered only to suggest dimly how they embrace us as we are and how they alter us temporally before releasing us for normal time-bound living. The focus for better or worse is upon the aesthetic reaction.

It is tempting to begin by seeking to define the arts very broadly in behavioral terms, and I succumb: they are the stimulators of contemplation. And what is contemplation? It is quiet, thoughtful activity pursued for its own sake. Basic drives and ulterior ends are cast aside for the moment when that is feasible. Temporal satisfaction is derived, and some insight of a cosmic or transcending nature is experienced. Such a conception of contemplation, with its fancy words and phrases, though fashionable and maybe even intelligible, may well be ethnocentric. For in the West we have a tendency to place the arts in a separate category, to detach them from everyday life, and to view them as "exquisite" and "pure." The moment we turn away from ourselves and observe non-Western peoples, we see that what, from our standpoint, is unmistakably art—a wood carving, a story, a poem, a song, a dance—has a more basic or at least a clear-cut purpose for its indigenous audience. But, I assert, the decoration of the hunting spear of the nomad, while it may serve the mundane function of identifying the owner of the weapon, or the more ethereal, though utilitarian one of placating the gods, can also be an aesthetic object for him if it reminds him of the stability or instability of his own existence or if, however briefly, he manages to become absorbed in its design and coloring. It looks as though we are thus quickly gliding into the problem of beauty and utility. This we must not do: so much has been spoken and written about the moot problem that somewhere, some time, some one has already made the brilliant observation, drawn the convincing conclusion which inevitably has been more or less successfully challenged by an equally penetrating successor. I shall not disturb the windmills, I hope. Instead I seek peaceful passage by agreeing to accept any other definition of the arts that is proposed, for I confidently suspect that even in their non-Western implications such conceptualizations must contain a touch of contemplation.

In my innocence, too, I propose denotatively to welcome as art any object or composition so designated, whether it be a contrived sonnet or a simple sea shell in its natural state. It matters not, therefore, whether the stimulator of contemplation is called a scientific treatise, an advertisement, a woven basket, a piece of shiny machinery, or a human body—or it may well be a painting, a song, a poem, or an oboe solo. What matters is that contemplation is the consequence of perceiving the pattern of stimuli. And yet we in the West are more likely to experience this aesthetic reaction when we are confronted with what is conventionally called art rather than non-art. Why? There is nothing mysterious here other than the mystery of art itself: the person striving for the artistic effect has had that experience himself; among other motives he is striving to communicate it; and for innumerable reasons he may have the talent and the training to do so.

This broad view enables us to note the universality of the arts: apparently in some form they coexist within human society everywhere. They do not float about invisibly: they are appreciated by those perceiving them. Their universality suggests that some human impulses can find satisfaction in no other way. Regardless of how arduous and time-consuming ordinary existence is, there is extra energy and time for contemplation, however fleetingly. The arts are not being placed upon a pedestal: just as all men and women must sneeze when adequate conditions are at hand, so they wish for and enjoy contemplation. But of course we know that irritation of the membranes leads to sneezing, whereas in the case of contemplation we believe we know that the conditions are deeper within the organism and involve a larger variety of states such as loneliness, frustration, joy, or the striving for the unattainable and the unknowable.

ORIENTATION

Whether displayed in the market place or hidden in a museum, whether communicated in a mass medium or a love letter, whether stimulating the eyes or the ears, art contains components which, when perceived, remind us of the past and which therefore help to conserve memories associated with the history of a group or ourselves. The madonna points to our faith, her halo to its significance; the music comes from another era of which we are proud, or it recalls our youth when associations more easily attached themselves to harmonies and discords; and that sea shell elicits again canons of aesthetic taste or bewilderment concerning its almost perfect symmetry which have always pleasantly stimulated us. This interest in form, a philosopher of art suggests, is "a pleasure shared by artists, collectors, and historians alike" and derives from "the discovery that an old and interesting work of art is not unique, but that its type exists in a variety of examples spread early and late in time, as well as high and low upon a scale of quality, in versions which are antetypes and derivatives, originals and copies, transformations and variants" (Kubler, 1962, p. 45).

The past in art, however, is not contemplated for its own sake but as an ingredient of the future. The artist may call upon his audience to act in behalf of any conceivable cause, although until recently in the West artists and their followers have been rather inclined to call artistic causes pure and lofty, and they did not want them confused with propaganda or preaching. Both directly and indirectly art advocates some kind of value involving judgment and conduct, whether conventional or iconoclastic—and values always mean a judgment based somehow upon past experience or indoctrination that now serves to anticipate the future. Indirectly the joy from contemplation, the sheer fun which art can offer, provides satisfactions too numerous to categorize: they range from catharsis to comedy. The satisfactions in turn add mightily or minutely to a person's resources; for just a moment or more they remove the feelings of frustration and futility likely to be pervading him.

The arts push persons in all three temporal directions and therefore, it may be supposed, the orientation dominating them or being vividly or dimly expressed both reflects and affects the one prevalent within the society. Are the paintings, the carvings, the songs, or the legends pointing toward the ancestors who must be honored or appeased, or toward the future life with possible joy everlasting? Or do they simultaneously push us in both directions? Contemplation has these temporal implications.

Anyone who has ever had the slightest philosophical whim may know that there are levels of satisfaction and that allegedly some levels are higher or better than others. This kind of evaluative thinking a psychologist may avoid without apology or regret. We need only agree that there are differences in behavior without pouring praise upon some actions and abuse upon others. And certainly there is a difference between the contemplation evoked by art and the actions unrelated to art. The actual taste of the prune is not the same as the portrait of the dried fruit, and we care not which is more soul stirring or nutritious. When you contemplate the drawing of a prune, you are not eating what you see, no matter how hungry you are. You are in a state of suspension which may be gratifying in its own right—

never before have you appreciated the sweep of the prune's wrinkles until this artist displayed them in his masterpiece—but which surely demands that you remember delights from the past or anticipate some in the future. The arts gratify in one fashion while demanding either recollection or anticipation—and the anticipation suggests deferment. While under the influence of art, or of good or effective art, reality for the moment blissfully steps aside; and the pieces of existence fall into place before tumbling apart.

THE ARRESTING OF TIME

Ordinarily events move past successively and inevitably, and from the viewpoint either of science or the naive observer no two of them are ever exactly the same. But when an event is witnessed by a person—someone hears the tree fall in the forest, and so spares the metaphysician his verbal dilemma as to whether a sound heard by no one exists—it may be recorded in some fashion. The witness remembers, or tries to remember what has happened; but with the passing of time distortions creep in, or he may forget or die. In literate societies some kind of written message may be kept to aid memory; and if there are recording instruments, parts of the event may be seized directly, as when a photograph is taken or a voice put upon tape. In nonliterate societies the event, or its actual or desired consequences, can be preserved by some variant of the spoken word or—according to sophisticated speculation concerning paleolithic society (Brandon, 1951)—through one of the visual arts such as carving or painting. In any case, there is a difference between an event perceived and experienced by a participant or observer and one recalled by a person or an instrument. Some media transmit messages which instantly perish, such as the spoken word, a gesture, a telephone circuit; others preserve the message over time, such as the written word, a tattoo on the skin, a typewriter (Doob, 1961, pp. 99–110).

To the extent that recollecting results in contemplation which includes experience similar to that which the individual has had if he was a participant, which he would have had if he had been one, or which he could have if ever he is one, to that extent the communication can be considered artistic for him. So-called participant art—dancing, group singing, some forms of the modern theater—includes activity from which similar emotions are derived directly, whereas the more passive kind of behavior associated with other art makes the experience relatively indirect. Clearly the aesthetic response is a matter of degree: a hurried report by a journalist, if it satisfies the criterion just stated, may be judged by the recipient to be just as artistic as a painting in oil. The spark that is struck in the audience by a work of art, then, is this emotion, feeling, or experience.

But "spark" obviously is a metaphor, I submit, for the phenomenon at the core of the psychological contemplation: the arresting of time. The event no longer has moved on but has been captured, slowed down to a standstill, suspended, contemplated. Any event, whether it be a riot or a simple human face, is so complicated, contains so many different elements and shades, that it can be never completely recorded. But its essence, the critical pattern or combination, can be captured, and that is art: time is overcome by being rather than becoming.

The objection may be muttered that human beings change like events, and that therefore the identical communication is perceived differently by the same person at different times—because his mood is different, because he has had different experiences in the interim, etc. Of course, but the event itself would likewise have been perceived differently by the same person in a different mood, with different experiences in the meantime, and—I choose the next word again with conscious deliberation—etc. In fact, reaction to great art probably does change with the audience's enduring mood, and so it should if it is to arrest time in the way people would have originally perceived and experienced it, or could do so in the future. When it does not transcend its own era, art is likely to be shoddy, no matter how impressive its impact upon the generation for which it was created.

The event whose temporal element is arrested may not be so concrete as a riot or a human face; it may be inside the artist, an intangible emotion which he then expresses in his medium. None of the renaissance artists who painted a nativity or a crucifixion had been present during the occurrence of those events. What they did was to express in their paintings not only the emotions they themselves felt as they considered what they believed to have once transpired, but also their own faith from the Christian standpoint; they sought to arrest forever their inspired internal experiences and in addition, no doubt, to convey them to others. Even if the author of a novel or a story has not wished to touch upon cosmic themes but merely to write a good tale, he may succeed in halting that period of time as a result of the insights or values his characters or his plot, intentionally or unintentionally, expresses and communicates.

The carving, the painting, the novel, after they have been created, exist and with virtually no change—other than one involving a mechanical factor, such as lighting, or the type size, or layout—stand ready to excite an audience. There are, however, some forms of art which must always be re-created again and again if they are to be effective media of communication. Most notable is any kind of music, whether written down in conventional notation or simply transmitted orally from person to person or from generation to generation. The original composer leaves behind the structure he has created, with or without explicit or implicit directions concerning technical features such as tempo, loudness, or instrumentation; but then the talent and temperament of the performer, the artist, markedly affect the recital of the composition. Similarly the playwright is dependent upon a host of technicians and actors for the impression that the production in fact elicits from the stage. There is no need here to go into the question of whether the performer should try to be true to what he believes to be the original intention of the composer or author, or for that matter whether such fidelity is ever possible. I am concerned only with the fact that the performer unquestionably affects the way in which time is arrested and hence also influences the emotions which are evoked and the truth which is conveyed. But I am tempted to argue that the role of audience is much more important than that of the performer in appreciating art and in arresting time: a bad headache can ruin the finest production of a Shakespeare play, just as a sensitive, sophisticated student of Shakespeare can inspire himself as he employs the miserable performance he witnesses as a way of remembering the greatness of the play which is already interiorized within him.

From the standpoint of arresting time there is no absolute difference between the arts. One can certainly say, as Lessing did, that "painting, in her coexisting compositions, can use only one single moment of the action, and must therefore choose the most pregnant, from which what precedes and follows will be most easily apprehended"; whereas "in the same manner poetry also can use, in her continuous imitations, only one single property of the bodies, and must therefore choose that one which calls up the most living picture of the body on that side from which she is regarding it" (G. E. Lessing, 1766, p. 55). But if painting must choose a moment of time and poetry a single property of space, one could argue the reverse with almost equal cogency: a poem occupies space on a page, a painting takes time to apprehend. By being just a trifle perverse and violating the unwritten agreement which the reader or listener makes with the artist when he joins the audience, all "principles of time" at the basis of literature and music (Mendilow, 1952, pp. 23–8, 236) may be violated. The member of the audience may not follow the artist's composition and perceive parts of it in an order different from what was intended (he skips ahead to see how the story ends; on a phonograph he changes the order of the symphony's movements) or he may return to sections he does not at first comprehend or appreciate. But of course from the standpoint of the creator there is frequently "a predetermined order" if the optimal effect is to be achieved: "Buildings in their settings are a sequence of spaces best seen in an order intended by the architect; the sculptured faces and separate parts of a public fountain or monument also should be approached as planned; and many paintings were originally meant each to have a fixed position in a sequence, from which a total narrative effect might arise" (Kubler, 1962, p. 97).

I fear further to tred upon the sacred soil of what is called the aesthetic experience, for in this delicate realm the fact of individual differences is all compelling and therefore armies of sensitive scholars and poets—I obtain the word "sensitive" from them in a most objective manner and hence do not employ it derisively—have failed to agree on what transpires when art is ingested by its audience. I claim no special sensitivity, but it seems reasonable to assert that, if it is to arrest time, a work of art must demand and receive complete attention or absorption. At the spectator responds, his own temporal orientation is fixed, fixed upon the art affecting him, fixed upon the present. This effect is seen, is felt most clearly in the case of music which, being a pattern in time, demands attentiveness lest the bar or the phrase or the motif be passed over and, at least for that rendition, be forever missed. Before, and certainly afterwards, the result of the experience of being absorbed in the moment can be related to the past or the future, but those are effects and not attributes of the experience as such. Perhaps music soothes more than any of the other arts because it may be perceived so effortlessly (though less effort means a great loss) but, more importantly, because the absorption must be so complete and the distracting associations with other experiences and other temporal orientations so few.

Of course from one standpoint the arts do not have a monopoly on the arresting of time. An influential semanticist, a remarkable jack-of-many-intellectual-disciplines, has maintained that the uniqueness of man stems from the fact that he possesses not only the characteristic of plants, chemistry-binding, and of animals, space-binding, but also one peculiar to himself which is timebinding (Korzybski, 1941). This is a somewhat flamboyant way of suggesting that people have a social heritage which transmits the wisdom and folly of previous generations. You know more than Aristotle not because you are brighter but because you have acquired some of the knowledge, perhaps a little of the wisdom accumulated during the centuries since he lived. Thus any medium which preserved knowledge or records deeds, whether a piece of marble or magnetic tape, binds or arrests time. But the reference here is to information gathered at a particular period, and this is science. Such a process is different from the re-creating of the emotional, internal state during the experience, the unique province of art.

THE SUSPENSION

Time in any work of art, as an almost brilliant analysis of this factor in the novel suggests (Mendilow, 1952), may be considered in three different frames of reference. There is the duration of the interval during which the artist created the art. Ordinarily this aspect of time is of interest to no one other than the artist himself—it presumably has no effect upon the aesthetic experience—but some writers do feel impelled to include a description of the circumstances under which they put down their thoughts and feelings. And here I can most appropriately quote an endearing passage from Laurence Sterne whose biography of Tristram Shandy is in large part the autobiography of an eighteenth-century gentleman who is writing his autobiography; at about the midpoint of the account Tristram tells us:

> I am this month one whole year older than I was this time twelvemonth; and have got, as you perceive, almost into the middle of my fourth volume—and no farther than to my first day's life—'tis demonstrative that I have three hundred and sixty-four days more life to write just now, than when I first set out; so that instead of advancing, as a common writer, in my work with what I have been doing at it—on the contrary, I am just thrown so many volumes back—was every day of my life to be as busy a day as this—And why not?—and the transactions and opinions of it to take up as much description—And for what reason should they be cut short? as at this rate I should live 364 times faster than I should write—It must follow, an' please your worships, that the more I write, the more I shall have to write—and consequently, the more your worships read, the more your worships will have to read [Sterne, 1759–66, chap. 13].

The second kind of time in art, the actual duration of the interval as the audience perceives, understands, and reacts to the artistic communication, Sterne has also indicated in the final sentence above, and indeed throughout the account he similarly teases his audience: "Stay," Chapter 8 of Volume 5 begins, "I have a small account to settle with the reader before Trim [a character from the preceding chapter] can go on with his harangue—it shall be done in two minutes." In the nontemporal arts like painting and architecture, this temporal feature is largely out of control of the artist after the product has been created; but of course their simplicity and complexity have some effect upon the quantity of time the audience devotes to the aesthetic experience. No doubt topics worthy of scholarly research would be the determination of the factors in these arts as well as in the audience which affect the decision concerning the allocation of time. In the temporal arts, especially music, drama, and literature, the artist retains much greater power over the expenditure of the audience's time: obviously a sonnet takes less time to read than an epic, a song less time to hear than a symphony.

Of greatest significance is the third temporal factor, the interval actually portrayed in the art itself. In a single painting—as distinguished from a series of paintings and some murals or frescoes—only elements of time can be represented: the historical period, the season of the year, the approximate hour of the day, the age of the character or object, etc. In drama and fiction, on the other hand, any length of interval and any succession of intervals can be offered, and these need not reflect temporal conditions in the real world. Thus an event of importance which transpires quickly according to the criterion of clock time may occupy many more pages than one of little significance which in fact takes longer to transpire. And the succession need not follow the chronological order of real life; suddenly, if the author chooses, he may flash back to an earlier period in the character's existence or leap ahead. Similarly in the theater, the intermission between Act I and Act II we know lasts 10 minutes because it says so on the program and we have verified that fact as we returned to our seats. When the curtain rises, however, we are both willing and able to accept the fantasy that ten years have elapsed; hence it seems reasonable to see characters whose makeup reveals that they are older and who carry within themselves the consequences of the events of the last decade. Or without questioning our own sanity we are undisturbed when Act II takes place before Act I from the standpoint of calendar time. Also within a single scene, without the dropping of a curtain and even in a drama ostensibly naturalistic and neither impressionistic or expressionistic, we are undisturbed by the equating of a long period of clock time and a short one of stage time (cf. Priestley, 1964, p. 120).

The arts, then, offer an opportunity to those with relevant predispositions and in a suitable mood to escape from the psychological present. Usually, we have been reminding ourselves again and again, we experience only that present, though within it the past is brought back in memory and the future appears in anticipation. Simple words and objects can elicit these internal responses and can cause the temporal motive to be active in some subsequent or prior form. These mundane signals, however, are likely but not certain to lack the vividness of art. Through literature the individual may again feel the touching agony of youth or love, or he may anticipate vividly the reactions associated with being old and feeble. His attention may be similarly arrested by a carving in wood or stone: he views more than a likeness, he feels within himself something akin to the emotions he would have in another incarnation, at another time. A photograph of himself as an adolescent elicits an aesthetic response not by reminding him that he was once young but by rearousing in a mature context some of his feelings from that era and by causing him to reflect upon himself. And in wordless music he experiences timelessness which can orient him in all directions simultaneously without causing him to feel insane or to distrust his temporal judgments.

Art cannot achieve such miracles, unless members of the audience agree to make the kind of sacrifice which the normal responsibilities of living cause them to abhor or avoid: they must tacitly, but fully, renounce virtually all other impulses during the aesthetic experience. "We enter into the author's world"—or the world of any creator, I would add—"by existing within the time relations which largely define it" (Lynen, 1969, p. 19). The medium must be all-engrossing, it alone provides the content of the psychological present. There can be no purer experience, as we have been using the term in this book, than a sublime love of man or God which is perhaps also aesthetic in large part. The renunciation involves both the past and the future; for the moment the past is forgotten and the future is not anticipated. Certainly the past and future as ever continue to exert their influence upon the aesthetic present; at the center of the consciousness produced by art, however, they are suspended. To some degree the schizophrenic or hypnotic patient also turns away from normal reality, but the motivation and the experiences are on the whole quite different.

It seems self-evident that the artistic media which are extended in time—music, opera, drama, literature in general—require the renunciation of other activities during the clock time when they are transmitted. The more static arts, furthermore, do not really differ in this respect. The paintings on the ceiling of the Sistine Chapel which have been there for centuries and which are not changing, at least as they are viewed, clearly cannot be apprehended at a glance. Conceivably, however, a quick glimpse of them can induce a perseverating feeling as profound as hours of concentration, even as a few bars of a symphony may release a series of internal associations that have little or no relation to the remainder of the composition but that are basically aesthetic and hence detached temporally. Absorption in art also evokes impulses that have extensions in time. Will the hero win? Do I not know someone like that character? But here we are dealing with a play within a play: the spectators are participating in, or identifying with parts of the plot and thereby during the very moment, when normal temporal orientation is arrested and suspended, they are transcending time by re-experiencing it in a timeless context.

Since reactions to drugs may well include aesthetic components, does that mean that drugs belong in the category of art? No, of course not. For in a nonmetaphysical sense the reactions they produce are involuntary and hence do not depend upon a decision to renounce the usual mode of behaving. But here, as in any state of euphoria produced by the body (whether through physical well being or sexual activity), the distinction can become blurred.

The tacit agreement to renounce reality during the experiencing of art means, among many other matters, that the principles ordinarily employed in judging the duration of intervals are suspended. Given sufficient patience and erudition, a critic could pounce upon every principle formulated in Chapters 3–8 and show how on occasion each may be transgressed in the domain of art. I have proposed, for example, that "the stronger the drive saliently functioning before the attainment or nonattainment of the relevant goal . . . the greater the timing acceleration" (6.1). Surely you can think of a character in fiction (e.g., one created by Edgar Allen Poe) who functions in exactly the opposite way because he is not driven by wishful thinking or some other nonperversity; you find the portrayal reasonable. Or apply the following to yourself after viewing an abstract painting: "the stronger . . . the recollection . . . that a non-duration-dependent interval . . . has been pleasant as a result of goal-attainment, the greater the timing deceleration" (6.3). What goal were you achieving as you observed the swirls and angles? Whenever you have a profound feeling, does time seem to pass slowly or quickly? If these assertions concerning the effect of the aesthetic agreement are only half-true, then I must enter two caveats in behalf of the principles.

First, it may well be that many of the exceptions I have never been loath to note in connection with the principles—the constantly appearing *"but's"* —have their genesis in situations or under circumstances similar to those associated with art. Certainly some reflect individual differences, and it is conceivable that, as the artist affects his audience, so the numerous dispositions from the past within a person produce more or less unique reactions. But this is just an instance of multivariance and should offer no excuse to cry that scientific analysis is hopeless. It is difficult, that is all, and it is not likely to be very successful in dealing with the nuances of the individual or the event. Certainly, too, many of the deviations suggested in Chapter 9, such as the tendency for schizophrenics to underestimate the duration of extended intervals or such as some of the phenomena associated with hypnosis, bear some resemblance to the aesthetic experience. Indeed, it seems reasonable to assert that systematic knowledge of time, whether derived from a personal Guide or from scientific investigation, helps the individual appreciate the different kinds of experience art can afford.

Then, secondly, the agreement to step aside from reality for the moment, as art is experienced, demands analysis in its own right. When this is done, many, maybe even all the apparent exceptions to the general principles turn out to be alterations in the state of the person's goals or drives induced by the artistry affecting him. You cannot eat the prunes in the picture, true enough, but your inspection of the drawing may evoke a strong drive within you involving your effort to comprehend what the artist sought to achieve or the symbolic significance of what you see in front of you. If so, then the strength of that artistically induced drive and the degree to which it is satisfied—I mean, of course, the degree to which *you* satisfy it—becomes the independent variable which affects your judgment concerning the passing of time while contemplating what you see. Thus approached, the miracle of the agreement to renounce and of the experiences induced by art becomes, or should become, explicable in terms of the more miraculous changes which can be quickly and succinctly wrought in the spectator's ongoing processes. This assertion by no means minimizes art's accomplishments by reducing them to psychological processes; if anything it increases the mystery by suggesting what must be accomplished.

Sometimes the arresting of time in art occurs because the audience is given the impression that it can master temporal intervals metaphorically and vicariously. The past is not lost; you turn the page back, you replay the symphony. The future comes sooner: you look ahead in the book, you skip the slow movement on the phonograph record. Or the author, blatantly or subtly, outlines the destiny of his characters before the characters themselves know their fate or—like Macbeth warned by the witches—when they are powerless to avert the tragedy they realize will come. The crudest form of this phenomenon occurs in detective stories, I am told (Priestley, 1964, p. 142), many of whose characters never change from tale to tale. Chronos may eventually devour us all, but art provides an escape from him, temporarily at least.

In all the arts, including even music, I would contend, the artistry of the creator involves a delicate interplay between the clock time spent in perceiving the art and the fictitious time portrayed therein. At first the interplay is made possible by the willingness of the audience to submit to the work of art, but thereafter—unless there is some irrelevant distraction—the successful artist must be able to give the impression that vicariously the listeners or the viewers are experiencing or comprehending the time he himself would designate. I suppose, as no doubt every critic and analyst of art in some place has noted, the most telling criterion of success to be achieved by a work or art is its ability to evoke such complete absorption that people forget clock time completely. Then, when the aesthetic experience is over, they either cannot estimate the duration of the interval or—as an index of satisfaction—they say that time has passed quickly or seemed to. The concern of the scientific investigator for "the ismorphism, such as it is, between private and public time-keeping" can vanish and, as in myth, time in the arts may have "neither a fixed metric nor a uniform flow" (Cohen, 1967, pp. 72, 81).

THE PORTRAYAL OF TIME

The agreement to renounce reality when reacting to art leads to experiences, we have said, which enable the spectator to pass temporal judgment within the framework of the medium. Two generations are described within a single novel, and the reader has the impression that he has absorbed the salient facts of their existence in a manner no different from what would have occurred if he had been present during their joys and sorrows; he virtually feels as though a half century has gone by. "For it is a difficult business—this time-keeping; nothing more quickly disorders it than contact with any of the arts" (Woolf, 1928, p. 275).

In addition to the temporal experiences created by its content, art may also portray time directly. In the visual media the effect can be obtained in a single production—as when a group of persons of different ages appear in a painting—but portrayal as such very probably demands a series of presentations, as in a motion picture. Without a doubt, however, time as a topic can be found most explicitly in literature. Here we are confronted with fictitious characters in novels and plays who philosophize about time and with philosophers who vent their own imaginations in connection with matters temporal. Writers in this category enable their readers to comprehend many of the problems associated with time and thus to form their own opinion concerning its nature and importance. Their views are not summarized in this book for reasons not completely clear to me: either I have the conceit to think that some place in these many pages I have in truth mentioned the issues they propound (not because I am more perspicacious than they, indeed not, but because a psychological approach is bound to subsume at least the outline of what they have had to say), or else I willingly suggest that the simple task of identifying and summarizing their views with attribution to them demands many a volume in its own right. And such volumes exist (e.g., Poulet, 1956). Perhaps, then, I am paying highest tribute to Mann and Proust and even to Kant and Bergson by not daring to simplify their accounts.

Surely I cannot beat a complete retreat from the portrayal of time in literature without at least making a few bows. For temporal generalizations in literature may be no different with respect to content and scope from those considered to be scientific, but they are presented in a different and often more challenging form. Except for discussions on the mountain they appear not as principles or premises but as insights attached to characters. On the surface they make no claim to be universals because they are embedded in the life or thought of particular persons, but they would transcend their locus. They are conveyed, moreover, in a manner more compelling and, therefore, likely to be remembered more easily and more pleasantly than any precisely stated formula. For that is the nature of literature. From countless illustrations, I select only two, both from the same author. Why quote again Virginia Woolf whose fame certainly has no connection with philosophizing about time? Mere chance: I have had to be arbitrary and, while writing one of the nastier chapters in this book during an otherwise happy sojourn in Africa (during which interval, needless to say, time sped by like a cheetah in full chase), I thought I would escape from time by reading a book of hers I found in a university library. Instead of obtaining relief I discovered that the first chapter tossed me right back into temporal perplexity;

> Since he belonged, even at the age of six, to that great clan which cannot keep this feeling separate from that, but must let future prospects, with their joys and sorrows, cloud what is actually at hand, since to such people even in earliest childhood any turn in the wheel of sensation has the power to crystallise and transfix the moment upon which its gloom or radiance rests, James Ramsey, sitting on the floor cutting out pictures from the illustrated catalogue of the Army and Navy Stores, endowed the picture of a refrigerator as his mother spoke with heavenly bliss. . . . Why, she asked, pressing her chin on James's head, should they grow up so fast. Why should they go to school? She would have liked always to have had a baby. She was happiest carrying one in her arms. . . . And so she went down and said to her husband, why must they grow up and lose it all? Never will they be so happy again. And he was angry. Why take such a gloomy view of life? he said. It is not sensible. For it was odd; and she believed it to be true; that with all his gloom and desperation he was happier, more hopeful on the whole, than she was. Less exposed to human worries—perhaps that was it. He had always his work to fall back on. . . . [Woolf, 1927, pp. 1, 94, 95].

Unquestionably these passages, though ripped out of context, poignantly convey intuitive wisdom concerning temporal orientation and fear.

With what aspects of the temporal problem does literature deal? What a silly question—all themes must have preoccupied the world's writers ever since their thoughts could be recorded. A basic observation must be the sociological one, viz., that the society's prevalent conception of time is likely to be expressed in its literature. Well and good, but the observation is convincing only when the conception is both distinctive and ascertainable. It may be that in nonliterate societies "time itself is regarded as a recurring cycle, in which events repeat themselves in a definite regular sequence, as the seasons of the years with their appropriate growth of animals and plants" and hence that a principal task of oral literature is to delineate the powers or forces responsible for these recurrences (Bowra, 1962, p. 235). Similarly, as a tour de force suggests, the literature of Greece and Rome may well have embodied the view "in the early phase of classical thought," viz., that time as such is "not yet an abstract frame of reference within which events take place at definable points, but is so closely tied to events that, apart from them, it does not appear to exist" (Thornton & Thornton, 1962, p. 119). Thus the Hebrew prophets, the Greek dramatists, and the writers of the Elizabethan stage (particularly Shakespeare) had quite different concepts of time which permeated thir creations since they lived in different eras and wrote for different audiences (Driver, 1960). "The Victorians, at least as their verse and prose reveal them," it is stated at the outset of another study, "were preoccupied almost obsessively with time and all the devices that measure time's flight"; they expressed the conception of time then promulgated by influential philosophers and particularly scientists (Buckley, 1966, pp. 1–13).

Whether the matching of literary treatment and *Zeitgeist* is more than post hoc hindsight is a challenging, if not impertinent question. Indeed, if we shift to a less explored relation, we may feel less confident in our judgment. What, for example, is the conception of time in the modern age which literature or the arts portray? The answer is not simple, I think, because our society is too heterogeneous to give rise to a unified characterization other than the one employed by scientists in their own theorizing. In addition, it can be argued that the conception of time in the arts has a Spiral effect upon audiences and hence in turn becomes a component of the prevailing conceptualization of, and attitude toward time. I venture an opinion, impossible to prove, that the great work of art reflects the artist's age and its temporal norms with greater accuracy and expresses them with greater effectiveness than the tawdry and the sleazy because the better artist is more sensitive to what has stirred, is stirring, and will stir in his own milieu.

If forced at dagger point to give a substantive reply to the question two paragraphs ago concerning the treatment of time in some or maybe all of the arts, I can only say that my impressions agree with those of others more competent to judge (for example, Meyerhoff, 1955, p. 72): any great writer or poet must concern himself with a sliver of the problem of time's irreversibility. For man's mortality is the ultimate mystery which the arts acknowledge and about which they offer balm or bane. A scapegoat must be found; and so, for example, Father Time himself is portrayed as being so old that *he,* the somewhat lovable villain, looks as if he were about to perish together with his scythe. After being given a glimpse of that drooping figure and his weapon, the reader is often then given the advice to gather rosebuds while he may.

TRUTH AND WISDOM

Other than idiosyncratic motives such as sublimation or displacement and self-seeking ones such as fortune and fame—and these are irrelevant unless they interfere with productivity, since we judge fruits and not roots—the genuine artist must have some basic reason to be creative. And the audience must likewise have a compelling motive to renounce reality and to pay attention to the art rather than to the host of competing and simpler attractions within the society which also offer entertainment and relaxation. Here, I feel, the creator and the audience have a common link and another reason for joining together, as the phrase of the moment would have it, into a communications network: the truth and wisdom which the artist feels compelled to express and which the audience would share or experience.

Consider once again, for example, the easy but subtle medium of proverbs. Usually their author is unknown, but the function they serve for those who use them is almost consciously clear-cut: they embody many of the traditional rules of behavior within the society and are cited, in teaching or in conversation, to enforce and justify those rules or to demonstrate that the user himself is conforming or at least knows what conformity means. In some African societies men stage contests by hurling proverbs at one another, and you win an argument by finding a suitable proverb embodying your point of view. Similarly many but not all the verbal media of art embody a value considered by those in power to be important, if not now, then in the future, and hence they transmit the communication. If the audience pays attention to the message, with or without an agreement to renounce reality, and with or without aesthetic reactions, they are thus offered a coating of sugar in the form of entertainment and even, on occasion, enlightenment.

But what truth and wisdom are contained in the dance, in music, or in a carving? That question could well be asked concerning great verbal media too. For surely the significant men in a literary tradition, such as Shakespeare, Molière, Dante, Goethe, Vergil, or Homer do more than transmit and emphasize the mores in the manner of proverbs. Their wisdom is less culture-bound, their truth transcends their milieu, the evidence for which is the fact that their works endure and inspire audiences in different times and cultures. And so while the function of a dance or a song or a bust may be to perpetuate the status quo by allowing people to enjoy themselves, by giving them pride in their group, etc., its greater value more often is the momentary insight it affords into some overpowering aspect of human affairs. The words sound fancy, but they can become meaningful the moment the artistic product itself is contemplated. For great art raises cosmic questions concerning man's fate and the basic problems of his existence and then provides answers which manage, if only slightly, to transcend the artist's own prejudices as a human being living in a particular age. Pure music (always the baffling art), divorced utterly from verse and dance, makes its contribution too, I think, by transforming the listener—and before him the performer—into an individual who fleetingly can undergo such meaningful, liberating experiences. Here the transformation is fairly direct, without the use of too many past associations utilized by other artistic media. But in any communication, there is always a transformation mediated by words and other associations from the past.

The wisdom and truth in art are different from those transmitted in everyday or scientific propositions. The latter, whether deduced or induced, are, or pretend to be, statements of fact which can be verified by competent people in the society, verified through an appeal to public sense data. In contrast, the artist expresses or provokes positions that seem intuitively true or wise without an appeal to such sense data. The absence of the appeal, however, does not preclude consensus since human beings, who are more or less similarly constructed and hence responsive in similar ways, can reach agreement in spite of the sollipsism in which their private feelings are encased. One dictum, however, the artist may not transgress or transgress too often if he is to be effective and retain his audience: he cannot communicate propositions that are blantantly false, for then disagreement interferes with the appreciation of his art. If you as the consumer of art keep saying, "what nonsense," or if you become more interested in reality than your own subjective feelings, the sweet receptiveness of the aesthetic mood soon vanishes.

In what direction does art influence people? Especially in traditional societies but also to some extent in ours, proverbs are likely to perform the function of reinforcing social rules by stressing the virtues of renouncing whatever it is that custom wishes people to renounce as well as the benefits which accrue from doing just that. Many, like rituals, give directions concerning the future, and thus assist people in anticipating what they will sooner or later experience. Any collection of proverbs, my own dipping suggests, contains a host of observations and predictions concerning the weather: no man can escape bad weather completely or enjoy good weather eternally, and so he seeks and obtains guides on the basis of empirical experience which seldom turn out to be entirely incorrect. All this is perfectly obvious.

What may not be so obvious is the important role which transformation through art can play. For here the individual catches a glimpse of the intangible, whether it be a smile on the face of a painted madonna, the lilt of a scherzo, or the metaphor in an ode. It is, perhaps, the same sort of glimpse which religion also seeks to provide, but it is effected less blatantly. In religion men are taught that their eternal future will be good if they follow certain rules and practices. But in art they are offered no heaven, instead they are given the feeling that somehow they have the potentiality of drawing a little closer to what they dimly seek; they are granted, however, only the feeling of closeness, they are not offered a definite and certain route. Paradoxically, therefore, by arresting time for the audience, art pushes the temporal perspective forward: these are the values you seek to realize and will never quite realize, it says in effect, but keep pushing, keep trying, for now after this glimpse you may possibly know forever after that the struggle is not completely futile. You cannot experience great art without being somewhat mystically stirred *and* changed. If this interpretation is valid, then one elusive fact may have been effortlessly explained: it is little wonder that art and religion are usually so closely intertwined, as they were once, and often still are in the West.

ON MATTERS OF STYLE

It is hopeless, rather it is suicidal to attempt to pontificate concerning artistic style or, for that matter, to delude oneself into believing that any uttered proposition on that subject can achieve assent or acclaim. For there exists in our society a special breed of persons dedicated to discussing matters of style and, in passing judgment, they are convinced they demonstrate their own superiority over the artists whose work they judge. One does not have to bow to critics: being alive and awake, I assume I have the privilege of expressing a few thoughts concerning style without asking their permission. For this embattled subject cannot be avoided, some attempt must be made to comprehend how artists are able to produce aesthetic reactions that are temporally detached.

First, from a psychological viewpoint it would seem that art which would persuade the audience to renounce reality, to experience arrested time, and hence to change can probably achieve these mighty objectives in a variety of ways, all of which I think share one attribute: they demand that spectators engage in intellectual and emotional activity not only to comprehend but also to react at some length to what they perceive. A comic strip may arouse and command the interest of a child, but the flimsy plot almost always can be followed effortlessly and then can be soon forgotten. In contrast, consider this line from Goethe which I give first in German because, as ever, a bit of the flavor disappears in translation: *Es ist so gut als wär' es nicht gewesen* (it is so good as if it had never been). The thought here cannot be instantly grasped and, when grasped, gives rise to many philosophical perplexities which in turn suggest a breathtaking insight into the nature of human existence and experience. The sadness, though certainly not new, is expressed so concisely and precisely that it strikes you forcefully—or perhaps you rejoice and do not moan. The style of Goethe is impressive, that of a comic strip ephemeral. The difference between the two is more than a matter of content, for a character on the strip might express Goethe's thought ("too good to be true, eh, bud"); no, Goethe makes you work, the silly strip encourages you to relax.

Then it follows from this distinction that a great deal, but not all art achieves its effect through economy of expression. Poetry here is the example par excellence, for in good verse each word, phrase, or line adds a delicate tone or color. The reader or the listener dares not miss any part if he is to comprehend the total. Sometimes for people in a given culture—or in a culture at a given time—the changing of a single word can have vast consequences, one of the reasons poetry has been defined as "what is lost in translation." A carving or a painting may also be dependent upon a single line or coloring or patterning, and consequently more than a fleeting glance is necessary if its meaning is to be experienced.

Such economy, such efficiency is possible in media whose transmission is controlled, as it were, by the audience. The poem is in front of you, and you read it as slowly or as quickly as you wish; if you want to see a line again, you turn back. Similarly you look at the painting for a short or a long spell; you get an impression of the total pattern or you concentrate upon parts; you keep your eyes open or you squint. But during the performance of art transmitted through aural or aural-visual media, such as music, a dance, or a play, you cannot glance backwards except in your mind; and if you glance too long, you miss what is happening. Here there can be no economy in pure form; some repetition, some redundancy is needed for the sake of comprehension. The main theme in the sonata returns later, sometimes unchanged, sometimes in a minor key, sometimes accompanied by another theme, sometimes taken over by a different instrument; thus there is pure repetition, or repetition with variation.

One dancer approaches the other, he withdraws, he approaches again, and the movements may be repeated a number of times until the climax is reached or a new sequence introduced. Our written poetry may strive for economy of expression, but it must be noted that peoples who are nonliterate and therefore cannot turn back to read sentences a second or third time possess poetry which is recited (and, more often or not, chanted or sung) and which echoes and reechoes the same thought and feeling within the same stanza and also from stanza to stanza (cf. Bowra, 1962, p. 71). Conceivably, too, repetition, pure and simple, can be attractive in its own right: we want to be surprised, but we also prize the familiar. A golden mean is often but not always the solution.

Redundancy in art, though uneconomical, therefore, is necessary: it can be useful not only to encourage understanding, as has been said, but also to achieve some effect, such as emphasis, in the most economical way. Paradoxically, one kind of redundancy may be the best way to avoid another kind.

In like manner repetition occurs in those media whose perception is regulated by the audience. The refrain in a poem is repeated; the character in a novel reveals the same trait again and again in situations which vary very little; the curves on a baroque altar do not sweep the eye upwards as in a Gothic structure but undulate in planes with little or no change and thus may retard all or most movement in the minds of the viewers. Such styles in fact are criticized by those who seek only the pure beauty of the nonredundant. But they can be defended on the same grounds of economy, if they are truly artistic: they also arrest time in a different manner, they demand and receive active participation of their audiences. The example should repel you because it is so trite, but I cannot prevent myself from resurrecting a youthful impression that the climax and immediate collapse achieved by Ravel in his Bolero depend solely on repetition with obvious but effective variations.

In short, there is no legislating concerning style. The variables are too numerous: the art, the medium, the audience, the period, the fashion, all these play some role. But one can say that good style, whether it be efficient or redundant, conveys the artist's profound feelings and hence has more of a chance to transcend many of those factors.

Finally, it cannot be too strongly emphasized that good style places a strain upon the audience. In contrast, much or ordinary communication is in bad style: there the aim is to inform and to inform unequivocally, to which end redundancy becomes essential in order to avoid misunderstanding. Unless you have a completely captive *and* attentive audience, you repeat your message, often many times, in order to avoid misunderstanding—

and this sentence is an illustration of what is meant, for it adds little to what has already been stated in the preceding one. Scientific communication, however, often conveys information with the same or greater precision as any impressive work of art. For knowledge that can be expressed in a mathematical formula (e.g., chemistry) is derived from, and embodies, a whole series of controlled observations of experiments briefly and accurately. The same standard is achieved, if not so elegantly, by any abstract principle, generalization, hypothesis which condenses into a limited number of words the essential aspects of the phenomena in question. Thus on the level of style good art and good science meet, though they part with respect to content and the effects they have upon people.

But do they with respect to the effects? Both summarize experience in different ways and therefore serve to shape people's anticipations concerning the future. Science, however, points to the attainable and the verifiable; whereas art, when it does point, suggests subjectively satisfying goals that in many senses are unattainable. Science provides procedures for replicating what has been declared to be true. In contrast, art may arrest time but, when individuals no longer perceive the communication, they are again alone and must grope to experience once more the harmony of the orchestra now silent, or the pang of love suggested by the sonnet but never bestowed, or the detachment from gravity apparently attained by the dancers before withdrawing from view. To the extent that the artistic performance affects you profoundly, to that extent the experience cannot be repeated, though you may undergo a more than adequate and satisfying experience upon repetition.

The style of successful art, consequently, is like the experiencing of time: it is irreversible and can be appreciated only in passing. Let the point, however, not be exaggerated. The person who fails to perceive the subtlety of rhythm in a musical passage can ask to have the passage repeated; and the excitement of a filled interval of time does occur. And yet, without being puristic, it must be said again, the repetition is never exactly like the original, though the difference in many cases may be quite unimportant or—when one thinks of a musical interlude upon high-fidelity, stereophonic tape—virtually nonexistent. But then the person himself has changed. At any rate, a good style which produces good art enables us to experience the finality of time while reminding us again and again of the fleetingness of our existence and enabling us simultaneously to forget that this is so.

13. CHANGING

In an earlier chapter we pursued people down their lifelines and noted how they become socialized and how their temporal judgments and orientation vary as they move toward death. Now we consider how and why they change as a function of innovation and planning, rather than of their biological constitution and the normal expectancies of their society associated with traditional roles. Obviously we shall emphasize the role of temporal factors in these changes. Planning comes first.

PLANNING

At the outset it seems evident that all planning demands a future orientation and that anticipating future activity is a form of planning. An animal learns the quickest route through the maze in order to get food: after previous experience he anticipates the reward and then goes through a series of trial-and-error movements to find it. Or an ambitious person knows that he dare not express his aggression in front of a superior if he is to be promoted: here there is renunciation for the sake of an anticipated reward, and the renunciation is part of the intervention which must also include other activities related to the promotion.

It is a long leap from such simple planning to the strategy adopted by a general to win a battle or the blueprint promulgated by a leader to diminish poverty. The difference is one of degree: many more factors and uncertainties are involved in planning campaigns against an enemy or a social disorder. Then, too, not one but many people are involved and interact. The time spent planning a future activity may be greater or less than the time it will actually take to carry out that plan; and it has been argued that one can speak of a decision-making process (and hence of a plan) only when there is a "temporal disparity" between the processes involved in planning in the present and executing the plan in the future, for otherwise the future is just rehearsed in the present as in practicing a play (Kolaja, 1968). There is one other factor which distinguishes most planning from individual anticipation and intervention: the novelty of the behavior that is anticipated. Little wonder that planning is considered to be the highest of the "functional dimensions" of the mind's temporal activities (Wallis, 1967).

What will you be doing tomorrow? You are going to arise at a particular hour, have breakfast, set to work, etc. That is what you anticipate, and to realize the anticipation you may have to renounce some activity at the moment, so that you can have enough sleep and rise on time; and you may have to engage in an instrumental activity, such as setting an alarm clock. This habitual carrying out of a routine is not a plan, nothing takes place. Naturally no one day is exactly like the next, but the degree of novelty can be fairly close to zero. In contrast, if you intend tomorrow to break the routine, to do something you have never or seldom done, then you have a plan; and what you do now and tomorrow requires the anticipation of novelty as well as intervention.

In this sense, therefore, planning involves action which, it is anticipated, will achieve a future goal in a manner only partially or incompletely known from past experience. A very young child does not plan: he may anticipate reward or punishment from his actions, but at first he does not or cannot intervene to attain his objective. As he grows older and his experience increases, he gradually adds elements of novelty to his existence. Later on, as an adult, after he has achieved a number of his plans—in connection, for example, with his occupation and marriage—he may grow conservative and feel that novelty is undesirable in comparison with the security and certainty provided by habitual ways. To plan you need hope and confidence, and therefore planning is likely to be linked to various personality traits and predispositions. In one American study, for example, a sample of white, married women who began using contraceptives before their second pregnancy was compared with another who adopted this form of family planning only after the birth of two or more children; the scoio-economic status of the two groups was held constant. More of those planning earlier than those planning later (a) were oriented toward the future as measured by a paper-and-pencil test which included items concerning willingness to sacrifice present goals to achieve long-term ones both for themselves and their children; (b) had an activistic rather than a passive philosophy; and (c) appeared to feel generally optimistic rather than pessimistic (Kar, 1971).

But you may also need luck to carry out a plan: circumstances over which you truly have no control can upset the best laid schemes. A follow-up study of thousands of American consumers who anticipated that they would be purchasing a durable product, such as an automobile or a garbage disposal unit, showed clearly that those intending to buy had "higher purchase rates than nonintenders" (Juster, 1964, p. 9). Anticipations thus turned out to be a reliable, if not certain, guide to future behavior as a result in part, it may be assumed, of both unforeseen and unforeseeable events in the interim.

The motives behind overt behavior which obviously demands planning are seldom easy to discern. In our society, for example, the less affluent may be compelled by their meager incomes to keep budgets; otherwise they spend too much before the arrival of the next pay check. And the wealthy may also have budgets not because they fear they will ever be without money to spend but because they believe in budgeting as a principle of living or as a way to distribute expenditures according to a philosophy. In either case, future needs and actions must be anticipated, but for different reasons; the temporal orientation must be away from the present, although clearly the present must be the starting point for that future. "Why should we plan?" The answer of a book which raises that question and then urges American communities to plan their own development is: "cite examples of what happens when we do not plan" (Kyle, 1955, p. 11). In more general terms, it may well be that anticipation of future gratification, to be effective as a motive, must be accompanied by some immediate reinforcement. "Saving has its best chance," notes a psychologist who has analyzed data obtained directly from representative samples of Americans, "if future needs acquire some degree of immediacy." A person whose present wage is satisfactory but who fears that he may lose his job and be unemployed in the future has within himself at the present moment a perceived, an experienced drive to begin saving (Katona, 1951, p. 72). Here anticipation evokes a real response.

For the individual, some planning is obligatory, some voluntary. In modern society various forms of social security are compulsory so that the persons involved cannot be said really to plan for the future. But anyone who of his own accord takes out an annuity makes certain assumptions about his own future or that of his family. I suppose those who own property, especially land, must plan for the future of their descendants, at least within the varying limits permitted by custom and law. If primogeniture is mandatory, then a father does not have to indicate that his land shall be divided among his sons; otherwise he must somehow anticipate their needs or follow some principle of just division. According to one analyst (Moore, 1963, pp. 80–1), inheritance in American society "tends to be off-phase:" when young parents need property or money most, their parents are still alive; instead their legacy is likely to arrive after their children have grown up and after their own earnings have almost reached a maximum.

The use of the term voluntary with reference to planning immediately suggests that people in our society have varying attitudes toward intervention. As a result of the freedom with which ideas and philosophies circulate, it seems highly likely that these attitudes reflect not only many of the diverse views observable in history and other cultures but also to some degree the social status of their possessors. Their underlying dynamics are also important to observe. A hobo and a man of wealth may both subscribe, consciously or not, to a philosophy of fatalism, the hobo because the absence of wealth and of a fixed abode gives him no incentive to plan, the man of wealth because his accumulation of wealth and property makes personal planning for himself superfluous. Whether or not there is a relation between attitude toward planning for oneself and attitude toward planning by government is a question I leave to Cassandra to consider.

All planning involves risk. The calculations at the basis of the anticipated gains may turn out to be incorrect, or an extraneous factor may suddenly be introduced into the situation. It is probably true that some people are better planners than others as a result of innumerable factors, such as previous experience, access to relevant information, personality traits, intelligence (by which is meant the ability—native and acquired—to put together relevant bits of information into a coherent whole), motivation, and, of course, temporal orientation. The feebleminded child lacks the essential facts as well as the self-confidence to plan extensively; and many neurotics and psychotics are too impulsive to be able to apply themselves to the future (Werner, 1957, pp. 197–8). Even if you are are normal, can you predict precisely what you will be doing tomorrow, next month, next year, next decade?

The risk involved in planning is especially great when multitudes of people are to be affected, since each of them has somewhat unique attributes and since they themselves are likely to interact with others in unknown, perhaps unknowable, ways. To be a social planner, no matter what the size of the region or the nation, demands a special kind of skill. For somehow it is necessary to extrapolate present trends within the society (e.g., the increase in population, the frequency of floods) and then to estimate the maximum form of intervention and the consequences therefrom. "Voir pour prévoir," as Comte once said of science, but in addition foresight must include the novel elements introduced by the plan itself. The self-fulfilling prophecy may invalidate so-called social laws—you make come true what you say is going to be—but in connection with planning this is exactly the goal that is sought.

Here it is necessary only to allude to some of the major problems involved in social planning. For how long a period can or should a plan be formulated? This is a way of inquiring about the interval in the future toward which the planner is oriented. There can be no easy answer, though at least two factors seem to be of critical importance. First, the goals of the plan's beneficiaries may set certain limits. Plans for education, for example, must extend over a long period since education is not a quick process. Then, secondly, the complexity of the activity must influence the temporal limits. In industry and agriculture, the number of discrete elements is so great that often plans must be short-run so that the effects of intervention can be empirically ascertained. Perhaps it is true, as common sense and most social analysts suggest, that difficulties and mistakes increase as the time period of a plan lengthens: there is more to anticipate, there is a greater possibility of relevant change in the interim that can disrupt the plan. But over longer periods of time, it may also be argued (Moore, 1963, p. 99), expertness can be acquired and opposition overcome. In addition, if adherents are needed, plans with immediate consequences may provoke anxiety, but those with far-flung benefits may appear to a greater degree to be fantasies and hence less threatening.

The skill required to formulate social plans and the time span of the plans are both affected by the degree of coordination to be achieved. In the planning stage, it is necessary to bring together people and data from a wide variety of departments, agencies, or strata; and likewise in executing a plan, activities must dovetail. On every level judgments must be passed concerning the time required to obtain information or to carry out details of the plan.

The very existence of a plan and of information concerning its progress may offer an incentive for the planners as well as for the relevant population. Psychological issues appear: should the specific goal (e.g., number of tractors or secondary school students) be realistically set so that people will not be disappointed, or should it be set too high so that they can be induced to expend more effort? Since some coordination and cooperation are always necessary in executing a plan, how can such activity best be achieved and then later how can the fruits of joint activity be utilized in connection with future plans? Indeed, the acceptance of a plan in the first place requires that relevant persons be willing to carry on an activity in the present and in the near future. Usually some renunciation or postponement of immediate satisfaction is demanded—taxes are increased, consumer goods are not available, another project is deferred—in return for the greater gains that can be anticipated. The challenge is best summarized by saying that the temporal orientation must be toward the future, away from the past or the present: if you do this now, you will be happier later. And one question becomes highly relevant: under what conditions does a shift in temporal orientation take place? For reasons which I think will become perfectly clear, it is useful and necessary to distinguish between gradual and rapid changes; but the referent will continue to be planning.

GRADUAL CHANGE

The one basis for all learning and hence for all change is some kind of dissatisfaction with the present. The term "dissatisfaction" is vague, deliberately so, since it would subsume all the discontent of which people are capable. It is difficult, or perhaps impossible, to conceive of a person totally satisfied except at a given instant; usually he is likely to desire some kind of change in his present circumstances. The change, however, may involve the simple repetition of a previously learned response: he is happy by and large, but he is a bit thirsty, hence he does what he usually does under those circumstances, viz., he drinks water. Problems of planning arise when momentary dissatisfaction cannot be relieved in the usual manner: there is a drought, and so the individual must change his orientation, he must think ahead and dig another well. Or his community must plan to build a reservoir or lead new water to an existing reservoir from a distant place.

Our problem thus becomes: what kinds of plans requiring renunciation and postponement are people willing to accept when they are dissatisfied? We need a set of propositions to summarize our knowledge concerning a problem so vitally connected with temporal behavior. On the assumption that the reader is now surfeited with temporal principles, I shall refer to the generalizations of this chapter as hypotheses and change the format from indented italics to italics modestly embedded in the text: plus ça change, plus c'est la même chose. All these statements are assembled in the Index of Principles and Hypotheses.

The inspiration for the hypotheses comes generally from what is known about learning and social change; previous formulations (Doob, 1960 & 1968a) are twisted to fit our present concern. *Hypothesis 1: People are likely to accept a plan when it is in accord (or not in conflict) with the goals they seek, their predispositions, and the modal personality traits of their society.* The predispositions can be variously conceptualized, but for plans involving many persons they include especially the traditional beliefs and values of the society. The likes and dislikes to which tradition gives rise cannot be blithely ascertained: the path to wisdom here is cluttered with unfounded generalizations and bright intuitions concerning the nature and content of modal personality traits in a society. The degree of control people think they can exercise over their own destiny is, as suggested previously, important, and their views may come in large part from those prevailing in their society. Thus fatalism increases the difficulty of planning —what will come will come, and why bestir yourself to change? The word "conflict" in the proposition is a relative matter since all plans by definition require change, which means some degree of incompatibility between the new and the old. An important practical predisposition at the center of planning is the temporal orientation: do people look toward the present and the past and thus wish to perpetuate old ways, or toward the future and thus seek new ones? There is evidence that in societies given to innovation for reasons that we dare not go into here—perhaps because we really cannot specify what those reasons are—they are more likely to accept additional changes when they are not antagonized by the idea of change as such (Barnett, 1953, pp. 56–64).

From the standpoint of the dispassionate observer or planner, people may possess the relevant predispositions, but they themselves must be convinced that they do. Corollaries are needed to suggest the conditions which will facilitate such a conviction. For the plan itself, this much can be said: *Hypothesis 1a: People are likely to accept a plan whose demonstrable advantages they can comprehend or anticipate.* The advantages revolve around alleviating the dissatisfactions which in the first place have made people receptive to the changes being promulgated by the plan. Planning as such may be attractive, prestige is associated with the proposed innovation (Barnett, 1953, p. 503). When, however, are the advantages of a plan "demonstrable"? The answer is elusive because demonstrability can mean something different to each society or person. Consider for a moment the ancient, artificial distinction between material and nonmaterial traits in a society, such as, for example, the wearing of shoes and a system of religious beliefs: it has often been maintained that material traits change more rapidly than nonmaterial ones since the advantages of the former can be more easily demonstrated than those of the latter. But the statement is not always true, men's ideas may on occasion change more rapidly than their beliefs; many converted Christians in contemporary Africa, for example, go barefoot even when they might afford shoes. The strength of the need plays a role, and a spiritual urge to find one's place in the universe can be as pressing as the desire to avoid bruised toes. So much also depends upon the initiative of the innovator who offers the demonstration or upon the channels of communication available to him and his audience. Often the missionary preceded the trader in Africa not, as critics of missions have charged, to prepare the way for profit but really to achieve only his stated objective, which was to save souls. From another standpoint, it may be asked, when can advantages be meaningfully demonstrated simply through verbal symbols? Or, it may be that the feasibility of a change is more likely to be appreciated when it is first demonstrated "on a small scale" or "on a trial basis" (Rogers, 1962, p. 84); if that be so, then do the persons with such experience generalize what they have learned and thus acquire a favorable attitude toward planning or toward a future orientation?

Finally, each society and each person has certain limits within which planning is considered feasible (Mannheim, 1941, pp. 155–62). How far into the future are you able and willing to peer, how extensive is your future orientation? In part the answer here must depend upon the value of foresight which has been demonstrated in the past, in part upon the knowledge that is available. Sometimes the distant future may appear more certain than the immediate future: you may not be able to anticipate and plan for the day after tomorrow, but you have greater confidence in your predictions about what you will be doing next Christmas. Then the advantages and disadvantages to be anticipated are influenced by the conception of the area over which the individual feels he has more or less direct control. If he ascribes a great deal of the future to destiny, he is less likely to respond favorably to the promises within a plan.

Just as the reaction, at least the immediate reaction to a communication depends not only upon its content and the medium of transmission, but also upon the attitude of the audience toward the communicator, so the success of a plan is related to the reputation of the demonstrator. This parameter can be stated in a more general form: *Hypothesis 1b: People are likely to accept a plan suggested or demonstrated in what they consider a favorable social context.* This corollary would signal a whole complex of factors besides the attitude toward the innovator: the symbols in which the plan is clothed, the timing, and the role the beneficiaries play in formulating and working out the details. In the last section the educational value of planning has been mentioned. Here we can add what is perhaps a prejudice of a democratic society: people may wish to be consulted and will refuse to participate or cooperate unless their desires have in fact been ascertained. In some societies, however, it may be sufficient only to have the assent of the elite, or of those so considered: others feel incompetent to pass judgment or they believe that it is not their function to do so. In either case, however, the beneficiaries are likely to have confidence in what they are supposed to anticipate if a plan reaches them through leaders with prestige.

One more corollary is required in order to suggest the factors which induce good intentions and convictions to spill over into action: *Hypothesis 1c: People are likely to accept and execute a plan when they have confidence in their own ability to carry it out.* This confidence springs from two main sources. First, some of the activities demanded by the plan may have been previously learned; they need only be utilized in another context or combined in different ways. Cultivators, for example, know how to plow and plant; to realize an objective of conservation, they must modify only the direction of the furrows and the type of crop they plant. Or people can come to appreciate their own capability after trying or exercising it. They try a new crop on a small plot; if the returns are satisfactory, their confidence in the method and themselves leads them to follow the same procedure on the remaining plots. What happens, however, when individuals have had no relevant experience in the past and when they are not in a position to assess their own abilities? This seems to be the situation in developing countries when families move from rural into urban areas; they can only dimly realize that they will be called upon, as one observer has theorized (Apter, 1965, p. 60), not only to acquire new skills but also to fulfill quite different roles. Under these circumstances they utilize what they consider relevant from their past, and then they anticipate their own performance, realistically or not.

RAPID CHANGE

The hypotheses just considered assume that change occurs slowly. And indeed, when mankind is viewed from almost any standpoint, that seems to be a valid assumption: people tend to be very conservative. Anthropologists stress the many customs from the past which survive in traditional societies ostensibly undergoing rapid change. The therapy provided by the psychoanalytic forms of psychiatry is based in large part on the belief and experience that patients cannot easily change traits and behavior acquired during childhood and persisting tenaciously in adult life. The major religions of the world have been asking men for centuries to renounce some or many of the pleasures of this world, in order to lead better lives and to obtain final salvation but, no matter what criteria of wordliness are employed, the conclusion must be that they have not had appreciable success with the vast majority. For centuries, too, profound educators in the West have agreed that one of the major aims of a liberal education is to broaden the temporal perspective of students: they must come to appreciate the past, to experience the subtleties and complexities of the present, and also to comprehend the relevance of the past and the present for future anticipations; yet today there is certainly little agreement as to how this ideal can be achieved in or outside the lecture hall, the seminar room, or the library.

At the same time it is also evident that sometimes men change relatively rapidly. On an anecdotal level every person can recall experiences of such change. Think of the lives of many saints, or of the changes following more or less permanently after a traumatic experience such as an unhappy love affair or a dreadful accident.

Relatively little of a systematic nature, however, is known about rapid change. But we are able to view such change through three descriptive categories, and a classification at a minimum is an important first step in the direction of real explanation: *Hypothesis 2: Rapid acceptance of a plan results from learning induced by conversion, transformation, or transplantation.* Each category requires a word of explanation.

Conversion has a definitely religious connotation: the individual who is leading a secular life is suddenly converted, and thereafter innumerable aspects of his behavior are different. What happens is that one of his central traits is changed, as a result of which the equilibrium of his personality, in a metaphorical or literal sense, is upset, and so adjustments and changes are facilitated. Thus the convert whose temporal orientation shifts from the enjoyments of the flesh at the moment to the contemplation of cosmic values views himself in quite a different light; he must revise his daily routine in many respects. Conversion is not confined to religion, it can be detected in connection with any sudden shift in behalf of a dedicated cause. In our time, the person becomes a communist, a pacifist, or a loyal member of some secular organization. The aim of enlightened penology is to produce a conversion within a prisoner so that he comes to view significant slices of his existence quite differently. Certainly a "complete" change in any person is never complete; remnants of old habits persist.

Under what conditions does conversion occur? The only guides we have are the hypotheses governing gradual change, since undoubtedly the difference between the two kinds of change is quantitative rather than qualitative. Change, then, becomes rapid rather than gradual when the relevant variables are weighted in particular ways: *Hypothesis 2a: Conversion to a plan may take place when people are so basically dissatisfied that they can ignore whatever conflicts they perceive between the plan and many (but not all) of their predispositions; when they possess some (however few) predispositions and skills compatible with the plan; when they anticipate (regardless of contrary demonstrations in reality) that they thereby will achieve some significant goal; when they are convinced that they can overlook most (if not all) opposition to the plan by their peers; and/or when they have supreme (or desperate) confidence in their ability to execute the plan.* This hypothesis, I agree, is complicated, but it thus only reflects the nature of the phenomenon: conversion occurs when many personal and perhaps also cultural conditions are satisfied; there is no simple alternative way to indicate that it results from having one, many, or all of the factors weighted in the indicated manner. The phenomenon is so subtle that it can be produced by varying combinations of the factors.

Transformation refers to a more limited kind of rapid change: the individual suddenly and perhaps forever after perceives an aspect of himself or his milieu in quite a different way. In a puzzle appealing to children you do not see the human face that is hidden in the drawing of the barnyard;

suddenly you do, either because someone points it out to you or because you stumble upon it; from that moment on you transform the drawing, you can look at it and still see the barnyard but simultaneously you absolutely must also notice the face. Or that person you have admired for a long while suddenly does something dishonest, and thereafter you conceptualize him and the traits you have been admiring in a quite different light: he is a fraud, you say, hiding behind an appealing façade. The Gestalt school of German psychologists once labeled such a transformation an "ah-ha experience"; out of the blue, while attempting to solve a problem, the person grasps the point, abandons his stumbling efforts which have been leading only to futile errors, and discovers the real solution. The actions leading to the transformation, however, need not always be quick ones. As already indicated, there is a tendency for aging persons in our society to transform whatever future orientation they have had toward the present and the past, and this takes place slowly. If penology is ever successful, criminals while in prisons come slowly to view themselves, their values, and challenging aspects of their society in a different light. Political leaders with a program to be realized in the future must persuade their friends and enemies to abandon the present to some degree, often bit by bit.

The generalization specifying the conditions associated with transformation is also complicated, but at least it contains no "and/or": *Hypothesis 2b: Transformation is likely to occur when people desire to perceive some feature of their environment or of themselves differently because they are dissatisfied either generally with their existence or specifically with the situation at hand; when they willingly accept tutelage from others; and when they are capable (actually or potentially) of behaving in the new manner.*

Transformation refers to internal processes, to the way in which the person perceives and evaluates the situation or himself; corresponding action may or may not occur, but the changed perception is the first step. The process may be facilitated when the label attached to an activity is altered. Thus if the designation "modern" has prestige in developing countries and if governments seek to change agricultural practices, it is important that agriculture come to be considered part of modernization and not simply an important practice associated with the traditional society. Schools, therefore, should stress the "scientific" and "modern" aspects of agriculture (Armbrester, Baughman, & Moris, 1967) so that attitudes toward this vital occupation can be quickly transformed. Persons who adopt the new labels may gradually transform their orientation toward the future and hence become less impatient over delays in the present.

Finally there is rapid change through *transplantation* on a permanent, semi-permanent, temporary, or fleeting basis: the individual moves into a new environment either because he is forced to do so (he is thrown into prison or, in a matrilocal society, he goes to live with his wife's parents after the marriage ceremony) or because he deliberately wishes to do so (he resigns from one job and takes on another). Some of the best publicized proposals concerning changes in university education in the United States —study in a foreign country for a year, getting a job in real life during part of the academic term, spending a summer doing good—stem from the belief that temporal perspective will be broadened, ethnocentrism decreased, and values altered and acquired through a change in the milieu. Similarly, reformers who would help rather than punish convicted criminals point out that, when men are forcibly transplanted from their normal environment to penitentiaries, they should be given the opportunity there to think and feel in terms of future goals. Both the enlightened student and the reformed criminal, however, eventually return to their normal milieu, where the unsolved and poignant problem is whether the changes induced during the transplantation will or can in turn be transplanted into normal life.

There are no psychological principles governing forced or involuntary transplantation since the causes must be political, economic, social, or just plain human. Concerning voluntary change the following hypothesis may be hazarded: *Hypothesis 2c: Voluntary transplantation in connection with planning is likely to occur when people view the change as a way, perhaps the only way, to achieve significant goals, regardless of the losses therefrom; when they believe the anticipated gains to be realizable and find them intelligible; and when they obtain sufficient support from their peers.*

Both children and adults know that usually the best way to remove a bandage from a minor cut is to rip off the tape with one vigorous motion rather than to tug away slowly. I suspect that the analogy is a fair one for many kinds of social change: gentleness and slowness just prolong the misery, whereas rough-and-tough methods eliminate it quickly. But objections pour forth: these methods may produce trauma which can cause lasting pain; they may lead only to surface but not to real changes; they involve too many risks. The answer to which must be: yes, but maybe no.

CONCOMITANT CHANGE

The two sets of hypotheses now set forth seek to indicate the psychological conditions facilitating the gradual or rapid adoption of a plan. Others are needed to suggest the events in medias res: what happens after the plan has been accepted and people begin to cooperate? The situation I would summarize as follows: *Hypothesis 3: The changes accompanying the execution of a plan are likely to include altered or new modes of behavior which augment (a) sensitivity to relevant aspects of the environment, (b) participation in groups supporting the plan's objectives, and/or (c) skill in adapting to novel situations.*

The hypothesis first of all makes the obvious point that people must change as they carry out plans and that these changes result from the plan and then affect its execution. To cooperate they must come to respond differently, and in responding differently they either reinforce or extinguish a proclivity thus to respond in the future as they experience interim reward or punishment. They must be oriented toward the future in some respects and, as the orientation proves successful or unsuccessful, they continue to behave in that manner. They are likely to seek reassurance and assistance from others in their milieu. Doubts concerning the desirability of a plan being followed can be softened by discovering other persons who are moving in the same direction. Well-knit and well-organized groups that seek change or are changing, therefore, probably have as members individuals with similar goals and temporal orientations.

The factor of skill in the generalization calls attention particularly to a Spiral effect noticeable in connection with change. As already indicated, people are likely to be attracted to a plan if they have confidence in their own skill to execute it. During the execution that skill, whether they have been right or wrong, is exercised and is likely to be improved. The improvement further increases their self-confidence, and so on. If a plan requires accurate timing, for example, persons can certainly learn to consult their watches more frequently, and then the habit of doing so is reinforced.

At some point any discussion of changing and planning must pay tribute to the elasticity and fallibility of human beings. That point has now been reached: *Hypothesis 4: People who accept a plan are likely to waver while executing it and to be unable to foresee all the consequences from attaining its objectives.*

This hypothesis again calls attention to what has been called "the theory of unanticipated consequences" which explicitly appears in theorists as diverse as Marx, Durkheim, and Freud (Gouldner, 1957): planning by an individual and especially by large groups involves some degree of uncertainty. For future anticipations are never inevitably achieved, everyone has had the experience of having his anticipations come to nought, moments of doubt and hesitation are bound to creep into even the most resolute person. Changes involve to some extent abandoning the security of the known. Certainly compensating for feelings of insecurity demands that doubts be repressed and that confidence to plunge forward be increased; but the point here is that these doubts exist and, consequently, have the potentiality of affecting behavior either on a conscious or unconscious level.

Much of the wavering may end when a portion of the plans succeeds, not only because success is gratifying but also because people seek, in the fashionable phrase, cognitive balance, i.e., they would bring their attitudes in line with what they have done. You do something not because you have been converted, rather you are converted while or after doing it.

Then it is difficult or impossible to foresee the eventual consequences of accepting changes. In acculturation, for example, the initial impulse may come from material changes whose effects are obvious and dramatic, such as the desire for better clothes. But to earn the money for the clothes, men may have to work in cities and there they inevitably adopt innumerable other changes in their way of living and thinking. New and different clothes by themselves produce no change in temporal orientation, but the changes they require or induce soon may lead to a noticeable shift from the present or past to the future.

Hypotheses for gradual, rapid, and concomitant change have now been presented in words and phrases directly related to planning. They purport to indicate the conditions under which deliberate planning can be successful. If you want a group of people to anticipate such-and-such changes, if you require them to renounce certain goals in the present, if you would alter their temporal orientation, then you must find the practical means to produce the changes in the indicated manner: make them see that the plan is in accord with what they want; indicate that they will be gaining, not losing important values; try to convert them to the philosophy behind the plan; etc. The advice rolls easily off the typewriter, but the route from theoretical principles to practices is tortuous and tricky. For the moment, then, this caveat: beware of easy solutions.

Any kind of planning demands that principles—or any set of variables better or worse than ours—must somehow be combined and variously weighted. This is especially difficult, we have been emphasizing, when many persons are involved. An element of intuition or artistry is likely to be helpful and should not be undervalued. People may be frustrated, they may be eager to change, they may have the capability of executing a plan, and so on through all the hypotheses nurtured here. And yet there still remains a problem of timing: when best to approach them with the innovation. Consider, for example, the question of whether in a traditional society Change A (e.g., a modern medical clinic) should be introduced before or after Change B (a source of uncontaminated water). If empirical investigation shows that the inhabitants possess greater inclination to implement A rather than B, should A come first? Not necessarily, for they may lack confidence in their ability to utilize A or its advantages may be too difficult to demonstrate. The decision may be reached to combat whatever hostility

A evokes by introducing B first, for then people's sensitivity to any kind of change may increase which could lead to greater eagerness to adopt segments of A. One change, therefore, can facilitate another. Or when subtle modifications are required in connection with conversion or transformation, one cannot always fall back on the dictum of "better later than never": tardiness may mean that the moment has gone by and that subsequently change may be so difficult that it becomes impossible.

The question arises as to which aspect of behavior will or should be changed when reforms or innovations are introduced (Doob, 1968b, p. xii). Here is another problem too broad even to be categorized. For a manufacturer may selfishly seek to have people change specific tidbits of behavior, such as the flavor of chewing gum, the style of dress, the type of furnace in their homes. The clergymen and the politicians may wish them to modify fairly basic attitudes and values. I would make only two observations. First, the objective obviously affects the method of change to be utilized: you ordinarily are not converted to a new chewing-gum flavor, but you may be converted to a different political party or religion. Then, secondly, this analysis of time has suggested again and again that temporal orientation, temporal information, and temporal standards can play a critical role in behavior. You never quite know what will happen when you jump into the mainstream of a personality but surely, if some aspect of temporal behavior is altered, repercussions are likely to be numerous. "Take it easy," relax, such bits of advice are given in a tense society, presumably to make people behave differently or perhaps to increase their joy.

14. ENVOI

This long book will end in a brief chapter without addenda. For the voluminous data, problems, and viewpoints swirl around a limited number of themes which can be succinctly expressed.

Human beings experience time: they have knowledge about objective time defined by nature and social convention; they are motivated by their own organisms and their milieu to pay attention to temporal attributes; they pass primary and secondary judgments concerning the duration of intervals; and they respond not only to the present which they experience, and often renounce, but also to the past which they recollect and the future which they anticipate. The interacting, spiraling variables have been brought together in the diagram called the Taxonomy of Time.

"The work on perception of time has been largely directed at fragmentary problems," a competent psychologist observes (Johannsen, 1967). That is the penalty which is paid when the investigator uses the precision of psychophysics. The reverse is true, the reader has by now surely observed, when a writer succumbs to the faustian urge to include everything, or almost everything, in a multivariate dissection. Precisely such an urge has given rise to the present analysis. At the very outset I warned that we dare not ignore any of the variables if we seek an overview, and therefore I have emerged with a complicated schema and a long book.

A finite number of principles, however, has sought to mitigate the difficulty: here are guides to some of the most important or crucial processes which can be considered one at a time and yet which simultaneously fall within the overall schema. Although the variables in the giant Taxonomy are surrounded by doubleheaded arrows to signify the interaction, it is possible in a pair of extended paragraphs to give a brief summary of those principles by paraphrasing them slightly and by ignoring some of the corollaries attached to many of them. For convenient reference, the numbers in parentheses refer to the principles as they have been numbered in the text.

The *behavioral potential* affects the temporal potential through (a) culture which places values upon activities (#3.4); (b) personality which means enduring traits pertaining to time or to behavior related to time (#3.6); and (c) biochemical processes resulting from changes inside or outside the organism and not easily extinguishable (#3.5). The effects of this behavioral potential upon the components of the *temporal potential* is significant, inasmuch as (a) appropriate temporal information is provided in each society (#3.3); (b) the temporal orientation, though perforce in the past, the present, and the future on different occasions (#3.1.a), tends to be focused in one direction rather than another (#3.2 and #3.2.b) as a result of the society's modal values (#3.2.a); and (c) the temporal motive is aroused in the service of goals considered significant in the society (#4.1) and in connection with intervals whose duration affects goal attainment (#4.2). The temporal potential interacts with the stimulus at hand whose relevant components are the interval, a channel, and often one or more temporal symbols, all of which are affected by changes in the external milieu (#3.1). The stimulus gives rise to *perception* (#5.1) consisting of: (a) temporal knowledge which is likely to be salient if considered reliable (#5.1.a); (b) events whose salience depends on the nature of the temporal motive and the duration of the interval (#5.1.c); and (c) drives which are salient when strong (#5.1.b). The resulting *temporal behavior* can be of various kinds: orientation in one direction rather than another; the acquiring of temporal knowledge; the passing of temporal judgment. If the behavior involves judging or estimating the duration of the interval, the *primary judgment* is affected by: (a) the scope of the interval, which means the number and kind of perceived events and the degree of intervention (#5.2); (b) the goal being attained or not attained (#6.3) which in turn involves the strength of the relevant drive (#6.1) and particularly that of a duration-dependent drive (#6.2); and (c) cues which may or may not be verbalizable (#6.4). The *secondary judgment* to which the primary one in turn may lead is affected by: (a) the standard being employed which is likely to be either an objective one if available (#7.1) or one frequently or recently reinforced (#7.2); (b) previous experience (#7.3) especially when a Guide has resulted therefrom (#7.3.a); and (c) the translation which is probably the most propitious one under the circumstances (#7.4) and whose accuracy varies with a number of factors (#7.5), including the time elapsing between the experiencing of the interval and the judgment (#7.6). Then the primary and secondary judgments influence the future temporal potential, which is a form of feedback (#7.7).

The Taxonomy may appear complicated but it is like a simple line drawing when we begin to consider some of the outside forces affecting the functioning of any of its components. Especially significant is the development of the individual: his experiences during socialization (#8.1), the fabric of his personality traits (#8.2), and numerous activities associated with aging (#8.3). At any given moment in his existence temporal behavior can be disturbed by organic states, drugs, and sleep; even more dramatic effects are evident under hypnosis, as a result of mental disorders, or in connection with socially deviant phenomena such as crime (chap. 9). From another standpoint temporal behavior can be traced to normal functioning within groups (chap. 10) or to work or any risky activity (chap. 11). But simultaneously people change or seek to prevent change in connection with their appreciation of the temporal aspects of the universe; and such control is exercised on the one hand through the arts broadly defined (chap. 12) and on the other through planning (chap. 13).

This volume has been dedicated to a rigorous, broad analysis of time, but now I would turn to more cosmic, soul-stirring issues to which only implicit, sly references have been made so far. All of us, no matter what our cultural background or philosophy, eventually and perpetually face temporal problems situated at the core of existence. Where do we turn for guidance? The sages usually pour out contradictory truths or platitudes; for example:

| To be able to communicate to his fellows the precise time of any action in which they are mutually interested is clearly a fundamental need of Man (Brandon, 1951, p. 20). | I experience inner time when I detach myself from the world . . . and when I turn over myself to me—then I experience myself as the creator of my internal time (Nebel, 1965, p. 21). |

What I feel impelled to do is to skeletonize the issues which time raises for us, to point directly if crudely to the conflicts, to indicate my private view of what I think—I guess, I believe, I hope—we should do, and to offer reasons for that view derived remotely from the philosophical values I take to be inherent in social science and psychiatry. The six issues are:

1. Should we orient ourselves toward the past, the present, or the future?
   Past: we must appreciate how we ourselves and our surroundings have come into existence; it is satisfying to know that the past is realized again and again; we must profit from experience.
   Present: only absorption in the moment counts; the past cannot really be recaptured; the future cannot be controlled.
   Future: we must anticipate; we must plan; the unanticipated surprise can also be gratifying.
   Therefore: as the occasion demands, orientation should be shifted from

one direction to another; the full life usually demands all three orientations.

Reason: recollection, experiencing, renunciation, and anticipation are necessary, desirable, and unavoidable; preoccupation exclusively with one orientation can mean that we are incapacitated with reference to the other two.

2. Should continuity be retained or sought between the past, present, and future?

Yes: without such continuity there is dissociation, and we cannot utilize past experience as the basis for anticipating and planning the future.
No: painful recollections disturb the present and future, it is more efficient and satisfying to push ahead.
Therefore: a general sense of continuity between what is recollected from the past, experienced in the present, and anticipated in the future should be retained, though there may be lapses.
Reason: too much discontinuity can lead to dissociation with disastrous consequences, too little to fatiguing self-consciousness.

3. Should we strive to acquire at all times abundant and accurate temporal information?

Yes: only in this way can we retain contact with reality; schedules are absolutely vital; a sense of certainty is necessary and comforting.
No: reality must often be avoided, timelessness is our ultimate goal; self-regulation is freedom; uncertainty is stimulating and exciting.
Therefore: generally correct information concerning objective time and temporal sequences should be acquired and, as the occasion demands, utilized.
Reason: severely deficient temporal information is likely to produce painful experience; so much social coordination depends upon such knowledge.

4. Should temporal judgments be passed?

Yes: they must be passed, otherwise social life and especially communication become impossible; how else can deadlines be met?
No: they interfere with ongoing activity; they are a necessary evil; they disturb joy and self-assessment.
Therefore: judgments should be passed when socially essential, neither more nor less frequently than our needs and those of society demand.
Reason: infrequent judgment may produce faulty coordination with others, too frequent judgments may lead to a preoccupation with time that inhibits appropriate action.

5. Should we strive to have all temporal judgments secondary rather than primary and to base them on objective time?

Yes: only in this way can consensus be achieved and subjective errors in estimation be corrected; we live with other human beings.
No: subjective judgments are private and precious, enable us to retain our individuality; we are sollipsistically isolated.
Therefore: the distinction between subjective and objective time should be clearly appreciated and, when necessary, but not always, corrections should be made on the basis of what is perceived and of Guides from the past.
Reason: discrepancies between primary and secondary judgments inevitably arise, and the pressures of social life demand frequent but not constant checking.

6. Should we be anxious about time and its passing?
Yes: time is scarce, death is close.
No: it does no good to torture ourselves with grim facts.
Therefore: sufficient anxiety to produce motivation is desirable.
Reason: too much anxiety is incapacitating, with respect to action in general and temporal activity in particular, but some anxiety can lead to constructive activity.

Both immediately and ultimately the underlying value behind the six injunctions is some kind of golden mean: if we are to remain sane and useful, temporal balance must be achieved. That balance, however, need not always be level. For some of us, those who would reach out beyond the vale in which we dwell, it must be tipped occasionally. But how frequently, and under what circumstances? We do not know.

> To everything there is a season
> And a time to every purpose under the heaven:
> A time to be born and a time to die;
> A time to plant and a time to pluck up that which is planted;
> A time to kill and a time to heal;
> A time to break down and a time to build up;
> A time to weep and a time to laugh;
> A time to cast away stones and a time to gather stones together;
> A time to embrace and a time to refrain from embracing;
> A time to get and a time to lose;
> A time to keep and a time to cast away;
> A time to rend and a time to sew;
> A time to keep silence and a time to speak;
> A time to love and a time to hate;
> A time of war and a time of peace.
> . . . that which befalleth the sons of men befalleth beasts;

> . . . as the one dieth, so dieth the other;
> . . . so that a man has no preeminence above a beast:
> For all is vanity.
> . . . there is nothing better
> Than that a man should rejoice in his own works;
> For that is his portion.
> For who shall bring him to see what shall be after him?
>
> [Ecclesiastes 3: 1–8, 19, 22]

REFERENCES

Aaronson, Bernard S. 1968. Hypnotic alterations of space and time. *International Journal of Parapsychology*, 10, 6–31.
Adamson, Robert. 1967. Anchor effect limits. *Psychonomic Science*, 9, 179–80.
Adamson, Robert and Kathleen Everett. 1969. Response modification by "irrelevant" stimulus attributes. *Psychonomic Science*, 14, 81.
Adler, Alfred. 1916. *The neurotic constitution*. New York: Moffat, Yard.
Aiken, Lewis R., Jr. 1965. Learning and retention in the estimation of short time intervals: a circuit and a study. *Perceptual and Motor Skills*, 20, 509–17.
Aisenberg, Ruth. 1964. What happens to old psychologists? *In* Robert Kastenbaum (ed.), *New thoughts on old age*. New York: Springer.
Ajuriaguerra, J. de; M. Boehme; J. Richard; H. Sinclair; and R. Tissot. 1967. Désintégration des notions de temps dans les démences dégénératives du grand age. *Encéphale*, 56, 385–438.
Albers, Robert Joseph. 1966. Anxiety and time perspectives. *Dissertation Abstracts*, 26, 4848.
Alchian, Armen and William R. Allen. 1969. *Exchange and production: theory in use*. Belmont, California: Wadsworth.
Alexander, James. 1967. Die Zeit und der metapsychologische Begriff der Anpassung. *Psyche* (Stuttgart), 21, 693–8.
Allen, F. H. 1944. Apparent time acceleration with age. *Science*, 99, 37–8.
Allport, Gordon W. 1955. *Becoming: basic considerations for a psychology of personality*. New Haven: Yale University Press.
Ames, Frances. 1958. A clinical and metabolic study of acute intoxication with *cannabis sativa* and its role in the model psychoses. *Journal of Mental Science*, 104, 972–99.
Ames, Louise Bates. 1946. The development of the sense of time in the young child. *Journal of Genetic Psychology*, 68, 97–125.
Anast, Philip. 1965. Temporal orientation: an adaptive process. *Journal of Social Psychology*, 65, 33–9.
Apter, David E. 1965. *The politics of modernization*. Chicago: University of Chicago Press.
Arieti, Silvano. 1947. The processes of expectation and anticipation. *Journal of Nervous and Mental Disease*, 106, 471–81.
Armbrester, P. Vernon; Newton M. Baughman; and Jon R. Moris. 1967. *Agricultural science for Tanzanian secondary schools*. Dar es Salaam: privately printed.
Aronson, Elliot and Eugene Gerard. 1966. Beyond Parkinson's law: the effect of excess time on subsequent performance. *Journal of Personality and Social Psychology*, 3, 336–9.
Aronson, Elliot and David Landy. 1967. Further steps beyond Parkinson's law: a repli-

cation and extension of the excess time effect. *Journal of Experimental Social Psychology,* 3, 274–85.
Aronson, H.; A. B. Silverstein; and G. D. Klee. 1959. Influence of lysergic acid diethylamide (LSD-25) on subjective time. *Archives of General Psychiatry,* 1, 469–72.
Aschoff, Jürgen. 1965. Circadian rhythms in man. *Science,* 148, 1427–32.
Aschoff, Jürgen and R. Wever. 1962. Spontanperiodik des Menschen bei Ausschluss aller Zeitgeber. *Naturwissenschaften,* 15, 337–42.
Axel, Robert. 1924. Estimation of time. *Archives of Psychology,* 12, no. 74.

Back, Kurt W. and Kenneth J. Gergen. 1963. Apocalyptic and serial time orientations and the structure of opinions. *Public Opinion Quarterly,* 27, 427–42.
Back, Kurt W.; Stephen R. Wilson; Morton D. Bogdonoff; and William G. Troyer. 1967. In-between times and experimental stress. *Journal of Personality,* 35, 456–73.
Baer, Paul E; Don Charles Wukasch; and Sanford Goldstone. 1963. Time judgment and level of aspiration. *Perceptual and Motor Skills,* 16, 648.
Bakan, Paul. 1955. Effect of set and work speed on time estimation. *Perception and Motor Skills,* 5, 147–8.
Bakan, Paul. 1962. Retrospective awareness of error in time estimation. *Perceptual and Motor Skills,* 15, 342.
Bakan, Paul and Francis Kleba. 1957. Reliability of time estimates. *Perceptual and Motor Skills,* 7, 23–4.
Bakan, Paul; Linda G. Nangle; and M. Ray Denny. 1959. Learning, transfer, and retention in the judgment of time intervals. *Papers of the Michigan Academy of Science, Arts, and Letters,* 44, 219–26.
Baldwin, Robert O.; Donald H. Thor; and Dale E. Wright. 1966. Sex differences in the sense of time: failure to replicate a 1904 study. *Perceptual and Motor Skills,* 22, 398.
Balken, Eva Ruth. 1943. A delineation of schizophrenic language and thought in a test of imagination. *Journal of Psychology,* 16, 239–71.
Bandura, L. 1936. The concept of time among children 7 to 9 years old [translation]. *Psychological Abstracts,* 1937, 11, no. 1525.
Banks, Robin and Daniel Cappon. 1962. Effect of reduced sensory output on time perception. *Perceptual and Motor Skills,* 14, 74.
Banks, Robins; Daniel Cappon; and Ross Hagen. 1966. Time estimation of psychiatric patients. *Perceptual and Motor Skills,* 23, 1294.
Banton, Michael. 1964. *The policeman in the community.* London: Tavistock.
Barabasz, Arreed Franz. 1970a. Time estimation and temporal orientation in delinquents and nondelinquents: a re-examination. *Journal of General Psychology,* 82, 265–7.
Barabasz, Arreed Franz. 1970b. Temporal orientation and academic achievement in college. *Journal of Social Psychology,* 80, 231–2.
Barach, Alvan L. and Julia Kagan. 1940. Disorders of mental functioning produced by varying the oxygen content of the atmosphere. *Psychosomatic Medicine,* 2, 53–67.
Barber, Theodore Xenophon. 1970. *LSD, marihuana, yoga, and hypnosis.* Chicago: Aldine.
Barber, Theodore Xenaphon and David Smith Calverley. 1964. Toward a theory of "hypnotic" behavior: an experimental study of "hypnotic time distortion." *Archives of General Psychiatry,* 10, 209–16.
Barndt, Robert J. and Donald M. Johnson. 1955. Time orientation and social class. *Journal of Abnormal and Social Psychology,* 51, 343–5.
Barnett, H. G., 1953. *Innovation.* New York: McGraw-Hill.
Behar, Isaac and William Bevan. 1961. The perceived duration of auditory and visual intervals: cross-modal comparison and interaction. *American Journal of Psychology,* 74, 17–26.
Beidelman, T. O. 1963. Kaguru time reckoning: an aspect of the cosmology of an East African people. *Southwestern Journal of Anthropology,* 19, 9–20.

Beilin, Harry. 1956. The pattern of postponability and its relation to social class mobility. *Journal of Social Psychology,* 44, 33–48.
Bell, C. R. 1965. Time estimation and increases in body temperature. *Journal of Experimental Psychology,* 70, 232–4.
Bell, C. R. 1966. Control of time estimation by a chemical clock. *Nature,* 210, 1189–90.
Bell, C. R. and K. A. Provins. 1963. Relations between physiological responses to environmental heat and time judgments. *Journal of Experimental Psychology,* 66, 572–9.
Bell, C. R. and Anne N. Watts. 1966. Personality and judgments of temporal intervals. *British Journal of Psychology,* 57, 155–9.
Belyaeva-Eksemplyarskaya, S. N. 1962. On the study of the process of perception and evaluation of time [translation]. *Psychological Abstracts,* 1963, 37, no. 2278.
Benda, Ph. and F. Orsini. 1959. Étude expérimentale de l'estimation du temps sous LSD-25. *Annales Médico-Psychologiques,* 117, part 1, 550–7.
Bender, Lauretta. 1950. Anxiety in disturbed children. *In* Paul H. Hoch and Joseph Zubin (eds.), *Anxiety.* New York: Grune & Stratton, pp. 119–39.
Benford, Frank. 1944. Apparent time acceleration with age. *Science,* 99, 37.
Benjamin, A. Cornelius. 1966. Ideas of time in the history of philosophy. *In* J. T. Fraser (ed.), *The voices of time.* New York: Braziller, pp. 3–30.
Berelson, Bernard and Gary A. Steiner. 1964. *Human behavior.* New York: Harcourt, Brace & World.
Bergler, Edmund and Géza Róheim. 1946. Psychology of time perception. *Psychoanalytic Quarterly,* 15, 190–206.
Berman, A. 1939. The relation of time estimation to satiation. *Journal of Experimental Psychology,* 25, 281–93.
Bernot, Lucien and René Blancard. 1953. *Nouville, un village français.* Paris: Institut d'Ethnologie.
Bettelheim, Bruno. 1943. Individual and mass behavior in extreme situations. *Journal of Abnormal and Social Psychology,* 38, 417–52.
Bezák, Jozef and Stanislav Dornic. 1968. Effect of a sensori-motor activity on the estimation of movement velocity. *Studia Psychologica,* 10, 73–4.
Bindra, Dalbir and Hélène Waksberg. 1956. Methods and terminology in studies of time estimation. *Psychological Bulletin,* 53, 155–9.
Binswanger, Ludwig. 1958. The case of Ellen West. *In* Rollo May, Ernest Angel, and Henri Ellenberger (eds.), *Existence: a new dimension in psychiatry and psychology.* New York: Basic Books, pp. 237–364.
Bliss, Eugene L.; Lincoln D. Clark; and Charles D. West. 1959. Studies of sleep deprivation: relationship to schizophrenia. *A.M.A. Archives of Neurology and Psychiatry,* 81, 348–59.
Blum, Richard H. and Associates. 1965. *Utopiates: the use and users of LSD 25.* London: Tavistock.
Boardman, William K.; Sanford Goldstone; and William T. Lhamon. 1957. Effects of lysergic acid diethylamide (LSD) on the time sense of normals. *A.M.A. Archives of Neurology and Psychiatry,* 78, 321–4.
Boas, George. 1950. The acceptance of time. *University of California Publications in Philosophy,* 16, 249–69.
Bochner, Stephen and Kenneth H. David. 1968. Delay of gratification, age, and intelligence in an aboriginal culture. *International Journal of Psychology,* 3, 167–74.
Bohannan, Paul. 1953. Concepts of time among the Tiv of Nigeria. *Southwestern Journal of Anthropology,* 9, 251–62.
Bokander, Ingvar. 1965. Time estimation as an indicator of attention-arousal when perceiving complex and meaningful stimulus material. *Perceptual and Motor Skills,* 21, 323–8.
Bonaparte, Marie. 1940. Time and the unconscious. *International Journal of Psycho-Analysis,* 21, 427–68.

Bond, N. B. 1929. The psychology of waking. *Journal of Abnormal and Social Psychology*, 24, 226–30.
Boring, Edwin G. 1933. *Physical dimensions of consciousness.* New York: Century.
Boring, Edwin G. 1936. Temporal experience and operationism. *American Journal of Psychology*, 48, 519–22.
Boring, Edwin G. 1942. *Sensation and perception in the history of experimental psychology.* New York: Appleton-Century-Crofts.
Boring, Lucy D. and Edwin G. Boring. 1917. Temporal judgments after sleep. *In Studies in psychology* (Titchener commemorative volume). Worcester, Mass., pp. 255–79.
Böszörményi, Z. and St. Szára. 1958. Dimenthyltryptamine experiments with psychotics. *Journal of Mental Science*, 104, 445–53.
Boulter, Lawrence R. and Mortimer H. Appley. 1967. Time and effort as determiners of time-production error. *Journal of Experimental Psychology*, 75, 447–52.
Bouman, L. and A. A. Grünbaum. 1929. Eine Störung der Chronognosie und ihre Bedeutung im betreffenden Symptomenbild. *Monatsschrift für Psychiatrie und Neurologie*, 73, 1–40.
Bowra, C. M. 1962. *Primitive song.* Cleveland: World Publishing.
Braddeley, A. D. 1966. Time-estimation at reduced body-temperature. *American Journal of Psychology*, 79, 475–9.
Bradley, N. C. 1947. The growth of knowledge of time in children of school-age. *British Journal of Psychology*, 38, 67–78.
Brandon, S. G. F. 1951. *Time and mankind.* London: Hutchinson.
Braud, William G. and Stephen H. Holborn. 1966. Temporal context effects with two judgmental languages. *Psychonomic Science*, 6, 151–2.
Breasted, James Henry. 1935. The beginnings of time-measurement and the origin of our calendar. *Scientific Monthly*, 41, 289–304.
Brim, Orville G., Jr. and Raymond A. Forer. 1956. A note on the relation of values and social structure to life planning. *Sociometry*, 19, 54–60.
Brock, Timothy C. and Carolyn Del Guidice. 1963. Stealing and temporal orientation. *Journal of Abnormal and Social Psychology*, 66, 91–4.
Brodsky, Stanley L. 1967. The WAYTE method for investigating self-perceptions. *Journal of Projective Techniques and Personality Assessment*, 31, no. 2, 60–4.
Bromberg, W. 1934. Marihuana intoxication. *American Journal of Psychiatry*, 91, 303–30.
Bromberg, W. 1938. The meaning of time for children. *American Journal of Orthopsychiatry*, 8, 142–7.
Bronfenbrenner, Urie. 1958. Social class through time and space. *In* Eleanor E. Maccoby et al. (eds.), *Readings in social psychology.* New York: Holt, Rinehart, & Winston, pp. 400–25.
Brower, Judith F. and Daniel Brower. 1947. The relation between temporal judgment and social competence in the feebleminded. *American Journal of Mental Deficiency*, 51, 619–23.
Brown, D. R. and Lloyd Hitchcock, Jr. 1965. Time estimation: dependence and independence of modality-specific effects. *Perceptual and Motor Skills*, 21, 727–34.
Brown, J. F. 1931. Perception in visual movement fields. *Psychologische Forschung*, 14, 233–48.
Brown, L. B. 1965. Religious belief and judgment of brief duration. *Perceptual and Motor Skills*, 20, 33–4.
Brush, Edward N. 1930. Observations on temporal judgment during sleep. *American Journal of Psychology*, 42, 408–11.
Bryan, Judith F. and Edwin A. Locke, 1967. Parkinson's law as a goal-setting phenomenon. *Organizational Behavior and Human Performance*, 2, 258–75.
Buck, John N. 1946. The time appreciation test. *Journal of Applied Psychology*, 30, 388–98.

Buckley, Jerome Hamilton. 1966. *The triumph of time*. Cambridge: Harvard University Press.
Bünning, Erwin. 1967. *The physiological clock*. New York: Academic Press.
Burns, Neal M. and E. C. Gifford. 1961. Time estimation and anxiety. *Journal of Psychological Studies*, 12, 19–27.
Burton, Arthur. 1943. A further study of the relation of time estimation to monotony. *Journal of Applied Psychology*, 27, 350–9.
Burton, Arthur. 1960. The moment in psychotherapy. *American Journal of Psychoanalysis*, 20, 41–8.
Butler, Robert N. 1963. The life review: an interpretation of reminiscence in the aged. *Psychiatry*, 26, 65–76.
Butler, Samuel. 1917. *Erewhon or over the range*. New York: Dutton, 1917.
Butters, Nelson; Evelyn Jones; Joseph Hoyle; and Caroline Zsambok. 1968. Effect of dynamic verbal and tonal stimuli on the perception of time. *Perceptual and Motor Skills*, 27, 431–7.

Cahoon, Richard L. 1967. Effect of acute exposure to altitude on time estimation. *Journal of Psychology*, 66, 321–4.
Cahoon, Richard L. 1969. Physiological arousal and time estimation. *Perceptual and Motor Skills*, 28, 259–68.
Campbell, John. 1954. Functional organization of the central nervous system with respect to organization in time. *Neurology*, 4, 295–300.
Campos, Leonard P. 1966. Relationship between time estimation and retentive personality traits. *Perceptual and Motor Skills*, 23, 59–62.
Cantril, Hadley. 1964. The human design. *Journal of Individual Psychology*, 129–36.
Cappon, Daniel and Robin Banks. 1964. Experiments in time perception. *Canadian Psychiatric Association Journal*, 9, 396–410.
Cappon, Daniel and Milo Tyndel. 1967. Time perception in alcoholism. *Quarterly Journal of Studies on Alcohol*, 28, 430–5.
Cardozo, Richard and Dana Bramel. 1969. The effect of effort and expectation on personal contrast and dissonance reduction. *Journal of Social Psychology*, 79, 55–62.
Carlson, A. J. 1943. Apparent time acceleration with age. *Science*, 98, 407.
Carlson, V. R.; D. R. Goodenough; and I. Fineberg. 1968. Time perception and sleep. *Psychophysiology*, 5, 242.
Carrell, Alexis. 1935. *Man the unknown*. London: Hamilton.
Carter, David E. and Glenn J. MacGrady. 1966. Acquisition of a temporal discrimination by human subjects. *Psychonomic Science*, 5, 309–10.
Cassirer, Ernst. 1955. *The philosophy of symbolic forms*, Vol. 2. New Haven: Yale University Press.
Cervantes, Lucius F. 1965. *The dropout: causes and cures*. Ann Arbor: University of Michigan Press.
Chadwick, N. Kershaw. 1942. *Poetry and prophecy*. Cambridge: The University Press.
Chatterjea, Ram G. 1960. Time gap in the estimations of short duration: visual presentation. *Indian Journal of Psychology*, 35, 147–58.
Chatterjea, Ram G. 1964. Temporal duration: ratio scale and category scale. *Journal of Experimental Psychology*, 67, 412–6.
Chatterjea, Ram G. and Purabi Rakshit. 1966. Estimation of temporal interval. *Perceptual and Motor Skills*, 22, 176.
Chessnick, Richard D. 1957. The sense of reality, time, and creative inspiration. *American Imago*, 14, 317–31.
Claridge, Gordon S. 1960. The excitation-inhibition balance. *In* H. J. Eysenck, *Experiments in personality*. London: Routledge and Kegan Paul, v. 2, pp. 107–56.

Clausen, Johs. 1949. Time judgment. *In* Fred A. Mettler (ed.), *Selective partial ablation of the frontal cortex*. New York: Hoeber, pp. 254–6.
Clausen, Johs. 1950. An evaluation of experimental methods of time judgment. *Journal of Experimental Psychology*, 40, 756–61.
Clauser, Günter. 1954. *Die Kopfuhr: das automatische Erwachen*. Stuttgart: F. Enke.
Cleugh, M. F. 1937. *Time and its importance in modern thought*. London: Methuen.
Cloudsley-Thompson, J. L. 1966. Time sense of animals. *In* J. T. Fraser (ed.), *The voices of time*. New York: Braziller, pp. 296–311.
Coheen, Jack J. 1950. Disturbances in time perception in organic brain disease. *Journal of Nervous and Mental Disease*, 112, 121–9.
Cohen, John. 1954. The experience of time. *Acta Psychologica*, 10, 207–19.
Cohen, John. 1964. Psychological time. *Scientific American*, 211, no. 5, 116–24.
Cohen, John. 1966. Subjective time. *In* J. T. Fraser (ed.), *The voices of time*. New York: Braziller, pp. 257–75.
Cohen, John. 1967. *Psychological time in health and disease*. Springfield: Charles C. Thomas.
Cohen, John and Peter Cooper. 1962. New phenomena in apparent duration, distance, and speech. *Nature*, 196, 1233–4.
Cohen, John; Peter Cooper; and Akio Ono. 1963. The hare and the tortoise: a study of the *tau*-effect in walking and running. *Acta Psychologica*, 21, 387–93.
Cohen, John; C. E. M. Hansel; and John D. Sylvester. 1953. A new phenomenon in time judgment. *Nature*, 172, 901.
Cohen, John; C. E. M. Hansel; and John D. Sylvester. 1954. An experimental study of comparative judgments of time. *British Journal of Psychology*, 45, 108–14.
Cohen, Louis H. and Gregory N. Rochlin. 1938. Loss of temporal localization as a manifestation of disturbed self-awareness. *American Journal of Psychiatry*, 95, 87–95.
Cohen, Samuel I. and Alexander G. Mezey. 1961. The effect of anxiety on time judgment and time experience in normal persons. *Journal of Neurology, Neurosurgery, and Psychiatry*, 24, 266–8.
Cohen, Stewart and Oscar A. Parsons. 1964. The perception of time in patients with coronary artery disease. *Journal of Psychosomatic Research*, 8, 1–7.
Cohn, Franz S. 1957. Time and the ego. *Psychoanalytic Quarterly*, 26, 168–89.
Cole, Malvin and O. L. Zangwill. 1963. *Déjà-vu* and temporal lobe epilepsy. *Journal of Neurology, Neurosurgery, and Psychiatry*, 26, 37.
Coltheart, Max and Guy von Sturmer. 1968. A serial effect in time estimation. *Psychonomic Science*, 10, 283–4.
Cooper, Linn F. 1948. Time distortion in hypnosis. *Bulletin of the Georgetown University Medical Center*, 3, 214–21.
Cooper, Linn F. and Milton H. Erickson. 1954. *Time distortion in hypnosis*. Baltimore: Williams and Wilkins.
Cooper, Linn F. and David W. Rodgin. 1952. Time distortion in hypnosis and nonmotor learning. *Science*, 115, 500–2.
Cortés, J. B. 1961. Cited by David C. McClelland, *The achieving society*. Princeton: D. Van Nostrand, p. 327.
Costa, Paul and Robert Kastenbaum. 1967. Some aspects of memories and ambitions in centenarians. *Journal of Genetic Psychology*, 110, 3–16.
Costello, C. G. 1961. The effects of meprobamate on time perception. *Journal of Mental Science*, 107, 67–73.
Cottle, Thomas J. 1967. The circles test: an investigation of temporal relatedness and dominance. *Journal of Projective Techniques and Personality Assessment*, 31, 58–71.
Cottle, Thomas J.; Peter Howard; and Joseph Pleck. 1969. Adolescent perception of time: the effect of age, sex, and social class. *Journal of Personality*, 37, 636–50.
Cottle, Thomas J. and Joseph H. Pleck. 1969. Linear estimations of temporal extension: the effect of age, sex, and social class. *Journal of Projective Techniques*, 33, 81–93.

Cowan, L. Gray; James O'Connell; and David G. Scanlon. 1965. *Education and nation-building*. New York: Praeger, 1965.
Craik, Kenneth H. and Theodore R. Sabin. 1963. Effect of covert alterations of clock rate upon time estimations and personal tempo. *Perceptual and Motor Skills*, 16, 597–610.
Crawford, M. L. J. and D. H. Thor. 1967. Time perception in children in the absence of external temporal synchronizers. *Acta Psychologica*, 26, 182–8.
Creelman, C. D. 1960. Human discrimination of auditory duration. *University of Michigan Research Institute*, Technical Report No. 114.
Cromwell, Ruel L; David Rosenthal; David Shakow; and Theodore P. Kahn. 1961. Reaction time, locus of control, choice behavior, and descriptions of parental behavior in schizophrenic and normal subjects. *Journal of Personality*, 29, 363–80.
Culbert, Sidney S. 1954. Systematic error in the estimation of short time intervals. *Journal of Criminal Law and Criminology*, 44, 684–8.
Cullmann, Oscar. 1964. *Christ and time*. Philadelphia: Westminster.
Cumming, Elaine and William E. Henry. 1961. *Growing old: the process of disengagement*. New York: Basic Books.

Dahl, Max. 1958. A singular distortion of temporal orientation. *American Journal of Psychiatry*, 115, 146–9.
Danziger, K. 1965. Effect of variable stimulus intensity on estimates of duration. *Perceptual and Motor Skills*, 20, 505–8.
Danziger, K. and P. D. du Preez. 1963. Reliability of time estimation by the method of reproduction. *Perceptual and Motor Skills*, 16, 879–84.
Davidoff, Henry. 1953. *A world treasure of proverbs from twenty-five languages*. London: Cassell.
Davids, Anthony. 1969. Ego functions in disturbed and normal children. *Journal of Consulting and Clinical Psychology*, 33, 61–70.
Davids, Anthony; Catherine Kidder; and Melvyn Reich. 1962. Time orientation in male and female delinquents. *Journal of Abnormal and Social Psychology*, 64, 239–40.
Davids, Anthony and Anita N. Parenti. 1958. Time orientation and interpersonal relations of emotionally disturbed and normal children. *Journal of Abnormal and Social Psychology*, 57, 299–305.
Davids, Anthony and Jack Sidman. 1962. A pilot study—impulsivity, time orientation, delayed gratification in future scientists and in underachieving high school students. *Exceptional Children*, 29, 170–4.
Davidson, G. M. 1941. A syndrome of time-agnosia. *Journal of Nervous and Mental Disease*, 94, 336–43.
Day, H. I. 1968. Some determinants of looking time under different instructional sets. *Perception and Psychophysics*, 4, 279–81.
de Grazia, Sebastian. 1962. *Of time, work, and leisure*. New York: Twentieth Century.
de Greeff, Etienne. 1927. Essai sur la personalité de débile mental. *Journal de Psychologie Normale et Pathologique*, 24, 400–54.
De La Garza, C. O. and Philip Worchel. 1956. Time and space orientation in schizophrenics. *Journal of Abnormal and Social Psychology*, 52, 191–4.
Dement, William and Nathaniel Kleitman. 1957. The relation of eye movements during sleep to dream activity: an objective method for the study of dreaming. *Journal of Experimental Psychology*, 53, 339–46.
Dement, William and Edward A. Wolpert. 1958. The relation of eye movements, body utility, and external stimuli to dream content. *Journal of Experimental Psychology*, 55, 543–53.
Denys, W. and M. Richelle. 1965. Regulations temporelles simples chez des malades mentaux. *Schweizerische Zeitschrift für Psychologie und ihre Anwendungen*, 24, 263–7.

De Shon, H. Jackson; Max Rinkel; and Harry C. Solomon. 1952. Mental changes experimentally produced by L.S.D. *Psychiatric Quarterly,* 26, 33–53.
De Vaux, Roland. 1964. *Studies in old testament sacrifice.* Cardiff: University of Wales Press.
De Wolfe, Ruthanne K. S. and Carl P. Duncan. 1959. Time estimation as a function of level of behavior of successive tasks. *Journal of Experimental Psychology,* 58, 153–8.
Dews, P. B. and W. H. Morse. 1958. Some observations on an operant in human subjects and its modification by dextro amphetamine. *Journal of the Experimental Analysis of Behavior,* 1, 359–64.
Dickstein, Louis S. and Sidney J. Blatt. 1966. Death concern, futurity, and anticipation. *Journal of Consulting Psychology,* 30, 11–7.
Dilling, Carole A. and Albert I. Rabin. 1967. Temporal experience in depressive states and schizophrenia. *Journal of Consulting Psychology,* 31, 604–8.
Dimond, Stuart J. 1964. The structural basis of timing. *Psychological Bulletin,* 62, 348–50.
Ditman, Keith S.; Thelma Moss; Edwin W. Forgy; Leonard M. Zunin; Robert D. Lynch; and Wayne A. Funk. 1969. Dimensions of the LSD, methylphenidate, and chlordiazepoxide experiences. *Psychopharmacologia,* 14, 1–11.
Dmitriev, A. S. and G. S. Karpov. 1967. Time perception and estimation [translation]. *Psychological Abstracts,* 1967, 41, no. 16,097.
Dmitriev, A. S. and A. M. Kochigina. 1959. The importance of time as stimulus of conditioned reflex activity. *Psychological Bulletin,* 56, 106–32.
Dobson, William R. 1954. An investigation of various factors involved in time perception as manifested by different nosological groups. *Journal of General Psychology,* 50, 277–98.
Dodge, Raymond and Eugen Kahn. 1931. *The craving for superiority.* New Haven: Yale University Press.
Doehring, D. G. 1961. Accuracy and consistency of time-estimation by four methods of reproduction. *American Journal of Psychology,* 74, 27–35.
Doehring, D. G.; J. E. Helmer; and Elizabeth A. Fuller. 1966. Physiological responses associated with time estimation in a human operant situation. *Psychological Record,* 14, 355–62.
Dondlinger, Peter T. 1943. Apparent time acceleration with age. *Science,* 98, 300–1.
Doob, Leonard W. 1951. Unpublished data.
Doob, Leonard W. 1960. *Becoming more civilized: a psychological exploration.* New Haven: Yale University Press.
Doob, Leonard W. 1961. *Communication in Africa: a search for boundaries.* New Haven: Yale University Press.
Doob, Leonard W. 1964. *Patriotism and nationalism: their psychological foundations.* New Haven: Yale University Press.
Doob, Leonard W. 1968a. Psychological aspects of planned development. *In* Art Gallaher, Jr., *Perspectives in developmental change.* Lexington: University of Kentucky Press, pp. 36–70.
Doob, Leonard W. 1968b. *The plans of men.* Hamden: Archon Books.
Dooley, Lucile. 1941. The concept of time in defence integrity. *Psychiatry,* 4, 13–23.
Driver, Tom F. 1960. *The sense of history in Greek and Shakespearean drama.* New York: Columbia University Press.
DuBois, Franklin S. 1954. The sense of time and its relation to psychiatric illness. *American Journal of Psychiatry,* 111, 46–51.
Dudycha, George J. 1936. An objective study of punctuality in relation to personality and achievement. *Archives of Psychology,* 29, no. 204.
Dudycha, George J. and Martha A. Dudycha. 1938. The estimation of performance-time in simple tasks. *Journal of Applied Psychology,* 22, 79–86.
Dunne, J. W. 1934. *An experiment with time.* London: Faber and Faber.

du Noüy, Lecomte. 1936. *Biological time*. London: Methuen.
Du Preez, Peter. 1964. Judgment of time and aspects of personality. *Journal of Abnormal and Social Psychology*, 69, 228–33.
Du Preez, Peter. 1967a. Field dependence and accuracy of comparison of time intervals. *Perceptual and Motor Skills*, 24, 467–72.
Du Preez, Peter. 1967b. Reproduction of time intervals after short periods of delay. *Journal of General Psychology*, 76, 59–71.
Durrell, Lawrence. 1960. *Clea*. London: Faber & Faber.

Earl, William Kane. 1969. The indifference interval, the kappa effect, and temporal anticipation. *Dissertation Abstracts*, 29, 4396-B.
Edmunston, William E., Jr. and John R. Erbeck. 1967. Hypnotic time distortion: a note. *American Journal of Clinical Hypnosis*, 10, 79–80.
Efron, Robert. 1967. The duration of the present. *Annals of the New York Academy of Sciences*, 138, 713–29.
Ehrensing, Rudolph and William T. Lhamon. 1966. Comparison of tactile and auditory time judgments. *Perceptual and Motor Skills*, 23, 929–30.
Ehrentheil, Otto F. 1964. Behavioral changes of aging chronic psychotics. *In* Robert Kastenbaum (ed.), *New thoughts on old age*. New York: Springer, pp. 99–115.
Ehrentheil, Otto F. and Peter B. Jenney. 1960. Does time stand still for some psychotics? *Archives of General Psychiatry*, 3, 1–3.
Ehrenwald, Hans. 1923. Versuche zur Zeitauffassung des Unbewussten. *Archiv für die gesamte Psychologie*, 45, 144–56.
Ehrenwald, Hans. 1931a. Gibt es einen Zeitsinn? *Klinische Wochenschrift*, 10, 1481–4.
Ehrenwald, Hans. 1931b. Störung der Zeitauffassung, der räumlichen Orientierung, des Zeichnens und Rechnens bei einem Hirnverletzten. *Zeitschrift für die gesamte Neurologie und Psychiatrie*, 132, 518–69.
Eiff, A. W. v.; E. M. Böckh; H. Göpfert; F. Pfleiderer; and Th. Steffen. 1953. Die Bedeutung des Zeitbewusstseins für die 24 Stunden-Rhythmen des erwachsenen Menschen. *Zeitschrift für die gesamte experimentelle Medizin*, 120, 295–307.
Einstein, Stanley. 1965. The future time perspective of the adolescent narcotic addict. *In* Ernest Harms (ed.), *Drug addiction in youth*. Oxford: Pergamon Press.
Eisenbud, Jule. 1956. Time and the Oedipus. *Psychoanalytic Quarterly*, 25, 363–84.
Eissler, K. R. 1952. Time experience and the mechanism of isolation. *Psychoanalytic Review*, 39, 1–22.
Ekman, Goesta; Marianne Frankenhaeuser; Birgitta Berglund; and Michael Waszak. 1969. Apparent duration as a function of intensity of vibrotactile stimulation. *Perceptual and Motor Skills*, 28, 151–6.
Eliade, Mircea. 1960. The yearning for paradise in primitive tradition. *In* Henry A. Murray (ed.), *Myth and mythmaking*. New York: Braziller, pp. 61–75.
Elkin, D. G. 1928. De l'orientation de l'enfant d'age scolaire dans les relations temporelles. *Journal de Psychologie*, 25, 425–9.
Elkin, D. G. 1963. Time perception and the feedback principle. *Soviet Psychology and Psychiatry*, 1, no. 4, 9–12.
Elkin, D. G. 1964. Time perception and anticipatory reflection. *Soviet Psychology and Psychiatry*, 3, no. 3, 42–8.
Elkin, D. G. 1966. Concerning time perception and the feedback principle. *In* Anon., *Psychological research in the U.S.S.R.* Moscow: Progress Publishers, pp. 450–64.
Ellis, Laura M.; Rafe Ellis; Eugene D. Mandel, Jr.; Maurice S. Schaeffer; Geraldine Sommer; and Gerhart Sommer. 1955. Time orientation and social class. *Journal of Abnormal and Social Psychology*, 51, 146–7.
Ellis, Michael John. 1969. Proprioceptive factors in operative time estimation. *Dissertation Abstracts*, 29, 2651-B.

Emley, G. S.; C. R. Schuster; and B. R. Lucchesi. 1968. Trends observed in the time estimation of three stimulus intervals within and across sessions. *Perceptual and Motor Skills*, 26, 391–8.

Engle, T. L. and Iona C. Hamlett. 1950. The use of the time appreciation test as a screening or supplementary test for mentally deficient patients. *American Journal of Mental Deficiency*, 54, 521–5.

Ennis, W. D. 1943. Apparent time acceleration with age. *Science*, 98, 301–2.

Epley, David and David R. Ricks. 1963. Foresight and hindsight in the TAT. *Journal of Projective Techniques*, 27, 51–9.

Erdös, L. 1934. Time and character [translation]. *Psychological Abstracts*, 1935, 9, no. 5765.

Erickson, Eric Homburger. 1956. The problem of ego-identity. *Journal of the American Psychoanalytic Association*, 4, 56–121.

Erickson, Milton H. and Elizabeth M. Erickson. 1958. Further considerations of time distortion: subjective time condensation as distinct from time expansion. *American Journal of Clinical Hypnosis*, 1, 83–8.

Eson, Morris E. 1951. An analysis of time perspectives at five age levels. Unpublished Ph.D. dissertation cited by Melvin Wallace and Albert I. Rabin, Temporal experience. *Psychological Bulletin*, 1960, 57, 217.

Eson, Morris E. and Norman Greenfield. 1962. Lifespace: its contents and temporal dimensions. *Journal of Genetic Psychology*, 100, 113–28.

Eson, Morris E. and John S. Kafka. 1952. Diagnostic implications of a study in time perception. *Journal of General Psychology*, 46, 169–83.

Essman, Walter B. 1958. Temporal discrimination in problem solving. *Perceptual and Motor Skills*, 8, 314.

Etienne, Fred and Wilhelm Kutschbach. 1949. Zwei Berichte von der Gefangenschaft. *Sammlung*, 4, 120–8.

Ewing, James H. 1962. Quantitative perception techniques and psychopathology. *In* John H. Nodine and John H. Moyer, *Psychosomatic medicine*. Philadelphia: Lea and Febiger, pp. 65–70.

Eysenck, H. J. 1941. An experimental study of the improvement of mental and physical functions in the hypnotic state. *British Journal of Medical Psychology*, 18, 304–16.

Eysenck, H. J. 1959. Personality and the estimation of time. *Perceptual and Motor Skills*, 9, 405–6.

Falk, John L. and Dalbir Bindra. 1954. Judgment of time as a function of serial position and stress. *Journal of Experimental Psychology*, 47, 279–82.

Farber, Maurice L. 1944. Suffering and time perspective of the prisoner. *University of Iowa Studies: Studies in Child Welfare*, 20, no. 409, 153–227.

Farber, Maurice L. 1953. Time-perspective and feeling tone: a study in the perception of the days. *Journal of Psychology*, 35, 253–9.

Farrell, Muriel. 1953. Understanding of time relations of five-, six-, and seven-year-old children of high IQ. *Journal of Educational Psychology*, 46, 587–94.

Feifel, Herman. 1957. Judgment of time in younger and older persons. *Journal of Gerontology*, 12, 71–4.

Feifel, Herman (ed.). 1959. *The meaning of death*. New York: McGraw-Hill.

Felix, Morton. 1965. Time estimates as affected by incentive class and motivational level. *Genetic Psychology Monographs*, 72, 353–99.

Fenichel, Otto. 1945. *The psychoanalytic theory of neuroses*. New York: Norton.

Ferrall, Sarah Catherine. 1935. The absolute judgment of temporal intervals. *Thesis Abstract*, Urbana, Illinois.

Festinger, Leon. 1957. *A theory of cognitive dissonance*. Stanford: Stanford University Press.

Filer, Robert J. and Donald W. Meals. 1949. The effect of motivating conditions on the estimation of time. *Journal of Experimental Psychology*, 39, 327–31.

Fink, Howard H. 1957. The relationship of time perspective to age, institutionalization, and activity. *Journal of Gerontology*, 12, 414–7.

Firth, Raymond. 1969. Social structure and peasant economy. *In* Clifton R. Wharton, Jr. (ed.), *Subsistence agriculture and economic development*. Chicago: Aldine, pp. 23–37.

Fischer, Roland. 1966. Biological time. *In* J. T. Fraser (ed.), *The voices of time*. New York: Braziller, pp. 357–82.

Fischer, Roland; Frances Griffin; and Leopold Liss. 1962. Biological aspects of time in relation to model psychoses. *Annals of the New York Academy of Sciences*, 96, 44–65.

Fisher, Seymour and Rhoda Lee Fisher. 1953. Unconscious conception of parental figures as a factor influencing perception of time. *Journal of Personality*, 21, 496–505.

Flavell, John H. 1963. *The developmental psychology of Jean Piaget*. Princeton: Van Nostrand.

Foerster, Leona Mitchell. 1969. The development of time sense and chronology of culturally disadvantaged children. *Dissertation Abstracts*, 29, 3321–2A

Folkins, Carlyle H. 1970. Temporal factors and the cognitive mediators of stress reaction. *Journal of Personality and Social Psychology*, 14, 173–84.

Forman, Henry James. 1936. *The story of prophecy*. New York: Farrar and Rinehart.

Foulkes, W. David. 1962. Dream reports from different stages of sleep. *Journal of Abnormal and Social Psychology*, 65, 14–25.

Fox, R. H.; Pamela A. Bradbury; and I. F. G. Hampton. 1967. Time judgment and body temperature. *Journal of Experimental Psychology*, 75, 88–96.

Fraenkel, Fritz and Ernst Jöel. 1927. Beiträge zu einer experimentellen Psychopathologie: der Haschischrausch. *Zeitschrift für die gesamte Neurologie und Psychiatrie*, 111, 84–106.

Fraisse, Paul. 1948. Étude comparée de la perception et de l'estimation de la durée chez les enfants et chez les adultes. *Enfance*, 1, 199–211.

Fraisse, Paul. 1952. Les conduits temporelles et leurs dissociations pathologiques. *Encéphale*, 41, 122–42.

Fraisse, Paul. 1961. Influence de la durée de la fréquence des changements sur l'estimation du temps. *Année Psychologique*, 61, 325–39.

Fraisse, Paul. 1963. *The psychology of time*. New York: Harper and Row.

Fraisse, Paul and R. Fraisse. 1937. Études sur la mémoire immédiate: I. l'appréhension des sons. *Année Psychologique*, 38, 48–85.

Fraisse, Paul and G. de Montmollin. 1952. Sur la mémoire des films. Cited by Paul Fraisse, *The psychology of time, q.v.*, p. 228.

Fraisse, Paul and Francine Orsini. 1955. Étude expérimentale des conduites temporelles: I. l'attente. *Année Psychologique*, 55, 27–39.

Francois, Marcel. 1927. Contribution à l'étude du sense du temps. *Année Psychologique*, 28, 186–204.

Frank, Lawrence. 1939. Time perspectives. *Journal of Social Philosophy*, 4, 293–312.

Frankenhaeuser, Marianne. 1959. *Estimation of time: an experimental study*. Stockholm: Almqvist and Wiksell.

Frankenhaeuser, Marianne. 1960. Subjective time as affected by gravitational stress. *Scandinavian Journal of Psychology*, 1, 1–6.

Franz, Marie-Louise von. 1966. Time and synchronicity in analytic psychology. *In* J. T. Fraser (ed.), *The voices of time*. New York: Braziller, pp. 218–32.

Fredericson, Emil. 1951. Time and aggression. *Psychological Review*, 58, 48–51.

Fress, Pol'. 1961. Adaptation of man to time [translation]. *Psychological Abstracts*, 1962, 36, no. 2BC43F.

Freud, Sigmund. 1925. The relation of poetry to day-dreaming. *In Collected papers*. London: Hogarth, v. 4, pp. 173–83.

Freud, Sigmund. 1930. *Civilization and its discontents.* London: Hogarth.
Friedman, Kopple C. 1944. Time concepts of elementary-school children. *Elementary School Journal,* 44, 337–42.
Friedman, Kopple C. and Viola A. Marti. 1945. A time comprehension test. *Journal of Educational Research,* 39, 62–8.
Friedman, Meyer and Ray H. Rosenman. 1959. Association of specific overt behavior pattern with blood and cardiovascular findings. *Journal of the American Medical Association,* 169, 1286–95.
Friel, Charles M. 1969. Cognitive style and temporal behavior. *Dissertation Abstracts,* 29, 4365-B.
Friel, Charles M. and William T. Lhamon. 1965. Gestalt study of time estimation. *Perceptual and Motor Skills,* 21, 603–6.
Frobenius, K. 1927. Ueber die zeitliche Orientierung im Schlaf und einige Aufwachphänomene. *Zeitschrift für Psychologie,* 103, 100–110.

Gardner, William A. 1935. Influence of the thyroid gland on the consciousness of time. *American Journal of Psychology,* 47, 698–701.
Garner, W. R. 1959. The development of context effects in half-loudness judgments. *Journal of Experimental Psychology,* 58, 212–9.
Gaschk, Judith A.; B. L. Kintz; and Richard W. Thompson. 1968. Stimulus complexity, free looking time, and inspective exploration. *Perception and Psychophysics,* 4, 319–20.
Gavini, Hélène. 1959. Contribution à l'étude de la perception des durées brèves: comparison des temps vides et des temps pleins. *Journal de Psychologie Normale et Pathologique,* 56, 455–68.
Gay, John and Michael Cole. 1967. *The new mathematics and an old culture.* New York: Holt, Rinehart, and Winston.
Geer, James H.; Phyllis E. Platt; and Michael Singer. 1964. A sex difference in time estimation. *Perceptual and Motor Skills,* 19, 42.
Geiwitz, P. James. 1964. Hypnotically induced boredom and time estimation. *Psychonomic Science,* 1, 277–8.
Geiwitz, P. James. 1965. Relationship between future time perspective and time estimation. *Perceptual and Motor Skills,* 20, 843–4.
Géraud, J.; P. Moron; and H. Sztulman. 1967. À propos d'une étude éxperimentale sur l'appréhension du temps et son vécu chez un groupe de schizophrènes. *Annales Médico-Psychologiques,* 125, 802–6.
Gesell, Arnold and Frances L. Ilg. 1946. *The child from five to ten.* New York: Harpers.
Gibbens, T. C. N. 1958. Sane and insane homicide. *Journal of Criminal Law and Criminology,* 49, 110–5.
Gilliland, A. R.; Jerry Hofeld; and Gordon Eckstrand. 1946. Studies in time perception. *Psychological Bulletin,* 43, 162–73.
Gilliland, A. R. and Dorothy Windes Humphreys. 1943. Age, sex, method, and interval as variables in time estimation. *Journal of Genetic Psychology,* 63, 123–30.
Gilliland, A. R. and Richard Martin. 1940. Some factors in estimating short time intervals. *Journal of Experimental Psychology,* 27, 243–55.
Globus, Gordon G. 1966. Rapid eye-movement cycle in real time. *Archives of General Psychiatry,* 15, 654–9.
Goldberger, Leo and Robert R. Holt. 1958. Experimental interferences with reality contact (perceptual isolation): method and group results. *Journal of Nervous and Mental Disease,* 127, 99–112.
Goldfarb, Joyce Levis and Sanford Goldstone. 1963. Time judgment: a comparison of filled and unfilled intervals. *Perceptual and Motor Skills,* 16, 376.
Goldfarb, Joyce Levis and Sanford Goldstone. 1964. Properties of sound and the auditory-visual difference in time judgment. *Perceptual and Motor Skills,* 19, 606.

Goldfarb, William. 1945. Psychological privation in infancy and subsequent adjustment. *American Journal of Orthopsychiatry*, 15, 247–55.
Goldman, Ronald; Melvyn Jaffa; and Stanley Schachter. 1968. Yom Kippur, Air France, dormitory food, and the eating behavior of obese and normal persons. *Journal of Personality and Social Psychology*, 10, 117–23.
Goldrich, Judith March. 1967. A study in time orientation: the relation between memory for past experience and orientation to the future. *Journal of Personality and Social Psychology*, 6, 216–21.
Goldstein, Kurt. 1948. *Language and language disturbances*. New York: Grune and Stratton.
Goldstone, Sanford. 1964. *The time sense in normal and psychopathologic states*. Mimeographed report of progress: January 1, 1960–October 31, 1964.
Goldstone, Sanford. 1967. The human clock: a framework for the study of healthy and deviant time perception. *Annals of the New York Academy of Sciences*, 138, 767–83.
Goldstone, Sanford. 1968a. Production and reproduction of duration: intersensory comparisons. *Perceptual and Motor Skills*, 26, 755–60.
Goldstone, Sanford. 1968b. Variability of temporal judgment: intersensory comparisons and sex differences. *Perceptual and Motor Skills*, 26, 211–5.
Goldstone, Sanford; William K. Boardman; and William T. Lhamon. 1958a. Effect of quinal barbitone, dextro-amphetamine, and placebo on apparent time. *British Journal of Psychology*, 49, 324–8.
Goldstone, Sanford; William K. Boardman; and William T. Lhamon. 1958b. Kinesthetic cues in the development of time concepts. *Journal of Genetic Psychology*, 93, 185–90.
Goldstone, Sanford; William K. Boardman; and William T. Lhamon. 1959. Intersensory comparisons of temporal judgments. *Journal of Experimental Psychology*, 57, 243–8.
Goldstone, Sanford; William K. Boardman; William T. Lhamon; Fred L. Fason; and Clarence Jernigan. 1963. Sociometric status and apparent duration. *Journal of Social Psychology*, 61, 303–10.
Goldstone, Sanford and Joyce Levis Goldfarb. 1962. Time estimation and psychopathology. *Perceptual and Motor Skills*, 15, 28.
Goldstone, Sanford and Joyce Levis Goldfarb. 1963. Judgment of filled and unfilled durations: intersensory factors. *Perceptual and Motor Skills*, 17, 763–74.
Goldstone, Sanford and Joyce Levis Goldfarb. 1964a. Auditory and visual time judgment. *Journal of General Psychology*, 70, 369–87.
Goldstone, Sanford and Joyce Levis Goldfarb. 1964b. Direct comparison of auditory and visual durations. *Journal of Experimental Psychology*, 67, 483–5.
Goldstone, Sanford and Joyce Levis Goldfarb. 1966. The perception of time by children. In Aline H. Kidd and Jeanne L. Rivoire (eds.), *Perceptual development in children*. New York: International Universities Press, pp. 445–82.
Goldstone, Sanford; Clarence Jernigan; William T. Lhamon; and William K. Boardman. 1959. A further note on intersensory differences in temporal judgment. *Perceptual and Motor Skills*, 9, 252.
Goldstone, Sanford and James E. Kirkham. 1968. The effects of dextroamphetamine upon time judgment: intersensory factors. *Psychopharmacologia*, 13, 65–73.
Goldstone, Sanford; William T. Lhamon; and William K. Boardman. 1957. The time sense: anchor effects and apparent duration. *Journal of Psychology*, 44, 145–53.
Goodchilds, Jacqueline; Thornton B. Roby; and Momoyo Ise. 1969. Evaluative reactions to the viewing of pseudo-dance sequences: selected temporal and spatial aspects. *Journal of Social Psychology*, 79, 121–33.
Gooddy, William. 1958. Time and the nervous system: the brain as clock. *Lancet*, 1, 1139–44.
Goodfellow, Louis D. 1934. An empirical comparison of audition, vision, and touch in

the discrimination of short intervals of time. *American Journal of Psychology*, 46, 243–55.
Goodman, Paul S. 1967. An empirical examination of Elliott Jaques' concept of time span. *Human Relations*, 20, 155–70.
Goth, Andres. 1966. *Medical Pharmacology*. St. Louis: Mosby.
Gothberg, Laura C. 1949. The mentally defective child's understanding of time. *American Journal of Mental Deficiency*, 53, 441–55.
Gottheil, Edward and Carl G. Lauterbach. 1969. Leader and squad attributes contributing to mutual esteem among squad members. *Journal of Social Psychology*, 77, 69–78.
Gouldner, Alvin W. 1957. Theoretical requirements of the applied social sciences. *American Sociological Review*, 22, 92–102.
Grassmück, Adolf. 1934. Mit welcher Sicherheit wird der Zeitwert einer Sekunde erkannt?: II. *Zeitschrift für Sinnesphysiologie*, 65, 248–73.
Green, Helen B. and Robert H. Knapp. 1959. Time judgment, aesthetic preference, and need for achievement. *Journal of Abnormal and Social Psychology*, 58, 140–2.
Greenberg, Roger P. and Ronald B. Kurz. 1968. Influence of type of stressor and sex of subject on time estimation. *Perceptual and Motor Skills*, 26, 899–903.
Greene, Joel E. and Alan H. Roberts. 1961. Time orientation and social class: a correction. *Journal of Abnormal and Social Psychology*, 62, 141.
Gregg, Lee W. 1951. Fractionation of temporal intervals. *Journal of Experimental Psychology*, 42, 307–12.
Gridley, Pearl Farwell. 1932. The discrimination of short intervals by fingertip and by ear. *American Journal of Psychology*, 44, 18–43.
Grier, Eunice S. 1967. In search of a future. *In* Daniel Schreiber (ed.), *Profile of the school dropout*. New York: Random House, pp. 140–85.
Grimm, Kurt. 1934. Der Einfluss der Zeitform auf die Wahrnehmung der Zeitdauer. *Zeitschrift für Psychologie*, 132, 104–32.
Gross, Alfred. 1949. Sense of time in dreams. *Psychoanalytic Quarterly*, 18, 466–70.
Grossman, Joel S. and Charles E. Hallenbeck. 1965. Importance of time and its subjective speed. *Perceptual and Motor Skills*, 20, 1161–6.
Guertin, Wilson H. and Albert I. Rabin. 1960. Misperception of time in schizophrenia. *Psychological Reports*, 7, 57–8.
Guilford, J. P. 1926. Spatial symbols in the apprehension of time. *American Journal of Psychology*, 37, 420–3.
Gulliksen, Harold. 1927. The influence of occupation upon the perception of time. *Journal of Experimental Psychology*, 10, 52–9.
Gunn, J. Alexander. 1929. *The problem of time*. London: Allen and Unwin.
Gurvitch, Georges. 1963. Social structure and the multiplicity of times. *In* Edward A. Tiyakian (ed.), *Sociological theory, values, and sociocultural change*. Glencoe: Free Press, pp. 171–84.
Gutheil, Emil A. 1951. *The handbook of dream analysis*. New York: Liveright.
Guyau, M. 1902. *Le genèse de l'idée de temps*. Paris: Alcan.
Guyotat, J. and R. Burgat. 1965. Approche psychologique des mécanismes d'action de deux types de psychoanaleptiques à partir des résultats d'épreuves d'estimation du temps. *Encéphale*, 54, 342–51.

Haas, William S. 1956. *The destiny of the mind*. London: Faber and Faber.
Hall, Edward T. 1959. *The silent language*. Garden City: Doubleday.
Hall, W. Winslow. 1927. The time-sense. *Journal of Mental Science*, 73, 421–8.
Halstead, Ward C. 1947. *Brain and intelligence*. Chicago: University of Chicago Press.
Hamner, Karl C. 1966. Experimental evidence for the biological clock. *In* J. T. Fraser (ed.), *The voices of time*. New York: Braziller, pp. 281–95.
Hardy, Charles O. 1923. *Risk and risk-bearing*. Chicago: University of Chicago Press.

Hare, Robert D. 1963a. Anxiety, temporal estimation, and rate of counting. *Perceptual and Motor Skills*, 16, 441–4.
Hare, Robert D. 1963b. The estimation of short temporal intervals terminated by shock. *Journal of Clinical Psychology*, 19, 378–80.
Hárnik, J. 1925. Die triebhaft-affektiven Momente im Zeitgefühl. *Imago*, 11, 32–57.
Harrell, Thomas Willard. 1937. Factors influencing preference and memory for auditory rhythms. *Journal of General Psychology*, 17, 63–104.
Harrison, M. Lucile. 1934. The nature and development of concepts of time among young children. *Elementary School Journal*, 34, 507–14.
Hart, Albert Gailord. 1940. *Anticipations, uncertainty, and dynamic planning*. Chicago: University of Chicago Press.
Harton, John J. 1938. The influence of the difficulty of activity on the estimation of time. *Journal of Experimental Psychology*, 23, 270–87, 428–33.
Harton, John J. 1939a. The influence of the degree of unity of organization on the estimation of time. *Journal of General Psychology*, 21, 25–49.
Harton, John J. 1939b. An investigation of the influence of success and failure on the estimation of time. *Journal of General Psychology*, 21, 51–62.
Harton, John J. 1939c. The relation of time estimates to the actual time. *Journal of General Psychology*, 21, 219–24.
Harton, John J. 1942. Time estimation in relation to goal organization and difficulty of tasks. *Journal of General Psychology*, 27, 63–9.
Hauty, George T. and Thomas Adams. 1965. Pilot fatigue: intercontinental jet flight. *Office of Aviation Medicine Report*, No. 65-16.
Hawickhorst, Liselotte. 1934. Mit welcher Sicherheit wird der Zeitwert einer Sekunde erkannt?: I. *Zeitschrift für Sinnesphysiologie*, 65, 58–86.
Hawkes, Glenn R.; Robert W. Bailey; and Joel S. Warm. 1960. Method and modality in judgments of brief stimulus intervals. *Journal of Auditory Research*, 1, 133–44.
Hawkes, Glenn R.; Robert J. T. Joy; and Wayne O. Evans. 1962. Autonomic effects on estimates of time: evidence for a physiological correlate of temporal experience. *Journal of Psychology*, 53, 183–91.
Hawkins, Nancy E. and Merle Meyer. 1965. Time perception of short intervals during finished, unfinished, and empty task situations. *Psychonomic Science*, 3, 473.
Hearnshaw, L. S. 1956. Temporal integration and behaviour. *Bulletin of the British Psychological Society*, no. 30, 1–20.
Heath, Louise Robinson. 1936. *The concept of time*. Chicago: University of Chicago Press.
Heckhausen, H. 1959. Einige Zusammenhänge zwischen Zeitperspektive und verschiedenen Motivationsvariablen. Cited in David C. McClelland, *The achieving society*. Princeton: D. Van Nostrand, 1961, pp. 337, 442.
Heimann, Hans. 1952. *Die Scopolaminwirkung*. Basel: Karger.
Heirich, Max. 1964. The use of time in the study of social change. *Sociological Review*, 29, 386–97.
Helson, Harry and Samuel M. King. 1931. The tau effect: an example of psychological relativity. *Journal of Experimental Psychology*, 14, 202–17.
Henrickson, Ernest H. 1948. A study of stage fright and the judgment of speaking time. *Journal of Applied Psychology*, 32, 532–6.
Henry, Franklin M. 1948. Discrimination of the duration of a sound. *Journal of Experimental Psychology*, 38, 734–43.
Heron, W. T. 1949. Time discrimination in rats. *Journal of Comparative and Physiological Psychology*, 42, 27–31.
Herzog, George. 1936. *Jabo proverbs from Liberia*. London: Oxford University Press.
Hicks, Robert A.; William Bramble; and Sandra Ulseth. 1967. Socialization and time perception in aged Ss. *Perceptual and Motor Skills*, 24, 1170.

Hilgard, Ernest R. 1965. *Hypnotic susceptibility*. New York: Harcourt, Brace.
Hindle, Helen Morris. 1951. Time estimates as a function of distance traveled and relative clarity of a goal. *Journal of Personality*, 19, 483–501.
Hirsh, Ira J.; R. C. Bilger; and B. H. Deathrage. 1956. The effect of auditory and visual background on apparent duration. *American Journal of Psychology*, 69, 561–74.
Hirsh, Ira J. and Carl E. Sherrick, Jr. 1961. Perceived order in different sense modalities. *Journal of Experimental Psychology*, 62, 423–32.
Hoagland, Hudson. 1933. The physiological control of judgments of duration: evidence for a chemical clock. *Journal of General Psychology*, 9, 267–87.
Hoagland, Hudson. 1935. *Pacemakers in relation to aspects of behavior*. New York: Macmillan.
Hoagland, Hudson. 1943. The chemistry of time. *Scientific Monthly*, 56, 56–61.
Hoagland, Hudson. 1966. Some biochemical considerations of time. In J. T. Fraser (ed.), *The voices of time*. New York: Braziller, pp. 312–29.
Hoch, Paul H.; James P. Cattell; and Harry H. Pennes. 1952. Effects of mescaline and lysergic acid (d-LSD-25). *American Journal of Psychiatry*, 108, 579–89.
Hoche, A. 1923. Langweile. *Psychologische Forschung*, 3, 258–71.
Hoffer, A. and H. Osmond. 1967. *The hallucinogens*. New York: Academic Press.
Holt, Robert R. and Leo Goldberger. 1960. Research on the effects of isolation on cognitive functioning. *U.S. Air Force, Wright Air Development Division Technical Report* 60-260.
Homack, Walter. 1935. *Ueber das subjektive Abgrenzen von Intervallen*. Zeulenroda: Sporn.
Honigfeld, Gilbert. 1965. Temporal effects of LSD-25 and Epinephrine on verbal behavior. *Journal of Abnormal Psychology*, 70, 303–6.
Horányi-Hechst, B. 1943. Zeitbewusstsein und Schizophrenie. *Archives de Psychologie*, 116, 287.
Horst, L. van der. 1932. Ueber die Psychologie des Korsakowsyndroms. *Monatschrift für Psychiatrie und Neurologie*, 83, 65–84.
Horstein, Alan D. and George S. Rotter. 1969. Research methodology in temporal perception. *Journal of Experimental Psychology*, 79, 561–4.
Hovey, Alvah. 1859. *The state of the impenitent dead*. Boston: Gould and Lincoln.
Hovland, Carl I.; Irving L. Janis; and Harold H. Kelley. 1953. *Communication and persuasion*. New Haven: Yale University Press.
Hovland, Carl I. and Irving L. Janis. 1959. *Personality and persuasibility*. New Haven: Yale University Press.
Hubert, Henri and Marcel Mauss. 1964. *Sacrifice: its nature and function*. Chicago: University of Chicago Press.
Hyatt, J. Philip. 1947. *Prophetic religion*. New York: Abingdon-Cokesbury Press.
Hyde, Robert W. and A. C. Wood. 1949. Occupational therapy for lobotomy patients. *Occupational Therapy and Rehabilitation*, 28, 109–34.

Iacono, Gustavo. 1956. La perception de la durée. *Journal de Psychologie*, 53, 307–14.
Ikeda, Sadami. 1957. A study of the development of children's concept of past [translation]. *Japanese Journal of Educational Psychology*, 4, 203–10 (from *Psychological Abstracts*, 1959, 33, no. 4573).
Inkeles, Alex. 1969. Making men modern. *American Journal of Sociology*, 75, 208–25.
Inman, W. S. 1967. Emotion, cancer, and time. *British Journal of Medical Psychology*, 40, 225–31.
Irvine, S. H. 1970. Affect and construct: cross-cultural check on theories of intelligence. *Journal of Social Psychology*, 80, 23–30.
Irwin, Francis W.; Fannie Armitt; and Charles W. Simon. 1943. Studies on object-prefer-

ences, I: the effect of temporal proximity. *Journal of Experimental Psychology*, 33, 64–72.

Irwin, Francis W.; Carlton W. Orchinik; and Johanna Weiss. 1946. Studies in object-preferences: the effect of temporal proximity upon adults' preferences. *American Journal of Psychology*, 59, 458–62.

Israeli, Nathan. 1930. Illusions in the perception of short time intervals. *Archives of Psychology*, 18, no. 113.

Israeli, Nathan. 1932. The social psychology of time. *Journal of Abnormal and Social Psychology*, 27, 209–13.

Israeli, Nathan. 1935. Distress in the outlook of Lancashire and Scottish unemployed. *Journal of Applied Psychology*, 9, 67–9.

Israeli, Nathan. 1936. *Abnormal personality and time.* New York: Science Press.

Jacoby, Jacob. 1969. Time perspective and dogmatism: a replication. *Journal of Social Psychology*, 79, 281–2.

Jacques, Elliott. 1965. Speculations concerning level of capacity. *In* Wilfred Brown and Elliott Jacques (eds.), *Glacier project papers.* London: Heinemann.

Jaensch, E. R. and Adalbert Kretz. 1932. Experimentell-strukturpsychologische Untersuchungen über die Auffsassung der Zeit unter Berücksichtigung der Personaltypen. *Zeitschrift für Psychologie*, 126, 312–75.

Jahoda, Marie. 1941. Some socio-psychological problems of factory life. *British Journal of Psychology*, 31, 191–206.

James, E. O. 1962. *Sacrifice and sacrament.* New York: Barnes and Noble.

James, William. 1890. *The principles of psychology*, Vol. 1. New York: Holt.

Janet, Paul. 1877. Une illusion d'optique interne. *Revue Philosophie de la France et l'Étranger*, 3, 497–502.

Janet, Pierre. 1928. *L'evolution de la mémoire et de la notion de temps.* Paris: Chahine.

Jasper, Herbert and Charles Shagass. 1941. Conscious time judgments related to conditional time intervals and voluntary control of the alpha rhythms. *Journal of Experimental Psychology*, 28, 503–8.

Jaspers, Karl. 1962. *General psychopathology.* Manchester: Manchester University Press.

Jenner, F. A.; J. C. Goodwin; M. Sheridan; Ilse J. Tauber; and Mary C. Lobban. 1968. The effect of an altered time regime on biological rhythms in a 48-hour periodic psychosis. *British Journal of Psychiatry*, 114, 215–24.

Jerison, Harry J. and Arden K. Smith. 1955. *U.S. Air Force, Wright Air Development Division Technical Report* 55–358.

Jirka, Zkehĕk and Chrudŏs Valoušek. 1967. Time estimation during prolonged stay underground. *Studia Psychologica*, 9, 176–93 (*Psychological Abstracts*, 1968, no. 18,149).

Joerger, Konrad. 1960. Das Erleben der Zeit und seine Veränderung durch Alkoholeinfluss. *Zeitschrift für experimentelle und angewandte Psychologie*, 7, 126–61.

Johannsen, Dorothea. 1967. Perception. *Annual Review of Psychology*, 18, 1–40.

Johnson, Edward E. 1964. Time concepts as related to sex, intelligence, and academic performance. *Journal of Educational Research*, 57, 377–9.

Johnson, Glen L.; Albert N. Halter; Harald R. Jensen; and Thomas D. Woods. 1961. *Managerial processes of midwestern farmers.* Ames: Iowa State University Press.

Johnston, Harriet M. 1939. A comparison of the time estimation of schizophrenic patients with those of normal individuals. Unpublished M.A. thesis cited by Johs Clausen, 1950, An evaluation of experimental methods of time judgment. *Journal of Experimental Psychology*, 40, 756–61.

Jones, Austin and Marilyn MacLean. 1966. Perceived duration as a function of auditory stimulus frequency. *Journal of Experimental Psychology*, 71, 358–64.

Jones, Ernest. 1951. *Essays in applied psycho-analysis.* London: Hogarth Press.

Jones, Robert E. 1949. Personality changes in psychotics following prefrontal lobotomy. *Journal of Abnormal and Social Psychology*, 44, 315–28.
Judson, Abe J. and Cynthia E. Tuttle. 1966. Time perspective and social class. *Perceptual and Motor Skills*, 23, 1074.
Juster, F. Thomas. 1964. *Anticipations and purchases: an analysis of consumer behavior*. Princeton: Princeton University Press.

Kafka, John S. 1957. A method for studying the organization of time experience. *American Journal of Psychiatry*, 114, 546–53.
Kahler, Erich. 1956. *Man the measure*. New York: Braziller.
Kahn, Paul. 1965. Time orientation and reading achievement. *Perceptual and Motor Skills*, 21, 157–8.
Kahn, Paul. 1966. Time orientation and perceptual and cognitive organization. *Perceptual and Motor Skills*, 23, 1059–66.
Kahn, Paul. 1967. Time span and the Rorschach human movement responses. *Journal of Consulting Psychology*, 31, 92–3.
Kahnt, Otto. 1914. *Ueber den Gang des Schätzungsfehlers bei der Vergleichung von Zeitstrecken*. Leipzig: Wilhelm Engelmann.
Kaiser, I. H. and Franz Halberg. 1962. Circadian periodic aspects of birth. *Annals of the New York Academy of Sciences*, 9, 1056–68.
Kar, S. B. 1971. Individual aspirations as related to acceptance of family planning. *Journal of Social Psychology* (in press).
Kastenbaum, Robert. 1959. Time and death in adolescence. *In* Herman Feifel (ed.), *The meaning of death*. New York: McGraw-Hill, pp. 99–113.
Kastenbaum, Robert. 1960. The dimensions of future time perspective: an experimental analysis. *Journal of General Psychology*, 65, 203–18.
Kastenbaum, Robert. 1964. The structure and function of time perspectives. *Journal of Psychological Researches*, 8, 97–105.
Kastenbaum, Robert. 1965. The direction of time perspective: I. the influence of affective set. *Journal of General Psychology*, 73, 189–201.
Kastenbaum, Robert. 1966a. As the clock runs out. *Mental Hygiene*, 50, 332–6.
Kastenbaum, Robert. 1966b. On the meaning of time in later life. *Journal of Genetic Psychology*, 109, 9–25.
Kastenbaum, Robert. 1967. The impact of experience with the aged upon the time perspective of young adults. *Journal of Genetic Psychology*, 10, 153–67.
Kastenbaum, Robert and Nancy Durkee. 1964a. Elderly people view old age. *In* Robert Kastenbaum (ed.), *New thoughts on old age*. New York: Springer, pp. 250–64.
Kastenbaum, Robert and Nancy Durkee. 1964b. Young people view old age. *In* Robert Kastenbaum (ed.), op. cit., pp. 237–49.
Katona, George. 1951. *Psychological analysis of economic behavior*. New York: McGraw-Hill.
Kaunda, Kenneth D. 1966. *A humanist in Africa*. London: Longmans.
Kelm, Harold. 1962. Consistency of successive time estimates during positive feed-back. *Perceptual and Motor Skills*, 15, 216.
Kendall, Martha and Ralph F. Sibley. 1970. Social class differences in time orientation: artifact. *Journal of Social Psychology*, 82, 187–91.
Kenna, J. C. and G. Sedman. 1964. The subjective experience of time during lysergic acid diethylamide (LSD-25) intoxication. *Psychopharmacologia*, 5, 280–8.
Kerr, Willard A. and Rudolph C. Keil. 1963. A theory and factory experiment on the time-drag concept of boredom. *Journal of Applied Psychology*, 47, 7–9.
Kety, Seymour S. 1956. Human cerebral blood flow and oxygen consumption as related to aging. *Research Publications, Association for Research in Nervous and Mental Disease*, 35, 3–45.

King, H. E. 1963. The retention of sensory experience: III. duration. *Journal of Psychology*, 56, 299–306.
Kipnis, David. 1968. Studies in character structure. *Journal of Personality and Social Psychology*, 8, 217–27.
Kirkham, James; Sanford Goldstone; William T. Lhamon; William K. Boardman; and Joyce L. Goldfarb. 1962. Effects of alcohol on apparent duration. *Perceptual and Motor Skills*, 14, 318.
Kleber, Ronald J.; William T. Lhamon; and Sanford Goldstone. 1963. Hyperthemia, hyperthyroidism, and time judgment. *Journal of Comparative and Physiological Psychology*, 56, 362–5.
Kleiser, John Raymond. 1953. The effect of habit formation, mental activity, and gross muscular movements on the reproduction of time. *Dissertation Abstracts*, 13, 886.
Kleitman, Nathaniel. 1963. *Sleep and wakefulness*. Chicago: University of Chicago Press.
Klien, H. 1917. Beitrag zur Psychopathologie und Psychologie des Zeitsinns. *Zeitschrift für Pathopsychologie*, 3, 307–62.
Klineberg, Stephen L. 1967. Changes in outlook on the future between childhood and adolescence. *Journal of Personality and Social Psychology*, 7, 185–93.
Klineberg, Stephen L. 1968. Future time perspective and the preference for delayed reward. *Journal of Personality and Social Psychology*, 8, 253–7.
Kloos, Gerhard. 1938. Störungen des Zeiterlebens in der endogenen Depression. *Nervenarzt*, 11, 225–44.
Kluckhohn, Clyde. 1960. Recurrent themes in myths and mythmaking. *In* Henry A. Murray (ed.), *Myth and mythmaking*. New York: Braziller, pp. 46–60.
Kluckhohn, Florence Rockwood and Fred L. Strodtbeck. 1961. *Variations in value orientation*. Evanston: Row, Peterson.
Knapp, Robert H. 1958. n Achievement and aesthetic preference. *In* John W. Atkinson (ed.), *Motives in fantasy, action, and society*. New York: D. Van Nostrand, pp. 367–72.
Knapp, Robert H. 1962. Attitudes toward time and aesthetic choice. *Journal of Social Psychology*, 56, 79–87.
Knapp, Robert H. and John T. Garbutt. 1958. Time imagery and the achievement motive. *Journal of Personality*, 26, 426–34.
Knapp, Robert H. and John T. Garbutt. 1965. Variation in time descriptions and need achievement. *Journal of Social Psychology*, 67, 269–72.
Knapp, Robert H. and Helen B. Green. 1961. The judgment of music-filled intervals and n achievement. *Journal of Social Psychology*, 54, 263–7.
Knapp, Robert H. and Paul S. Lapuc. 1965. Time imagery, introversion, and fantasies preoccupation in simulated isolation. *Perceptual and Motor Skills*, 20, 327–30.
Knight, Frank H. 1921. *Risk, uncertainty, and profit*. Boston: Houghton Mifflin.
Koehnlein, Helmut. 1934. Ueber das absolute Zeitgedächtnis. *Zeitschrift für Sinnesphysiologie*, 65, 35–57.
Kogan, Nathan and Michael A. Wallach. 1964. *Risk taking*. New York: Holt, Rinehart, and Winston.
Kohlberg, L. 1966. Cognitive stages and preschool development. *Human Development*, 1966, 5–17.
Kohlmann, T. 1950. Das psychologische Problem der Zeitschätzung und der experimentelle Nachweis seiner diagnostischen Anwendbarket. *Wiener Zeitschrift für Nervenheilkunde und deren Grenzgebiete*, 3, 241–60.
Kolaja, Jiri. 1968. Two processes: a new framework for the theory of participation in decision making. *Behavioral Science*, 13, 66–70.
Korngold, S. 1937. Influence du genre de travail sur l'appréciation des grandeurs temporelles. *Travail Humain*, 5, 18–34.
Korzybski, Alfred. 1941. *Science and sanity*. Lancaster: Non-Aristotelian Library Publishing Company.

Kowalski, Walter J. 1943. The effect of delay upon the duplication of short temporal intervals. *Journal of Experimental Psychology*, 33, 239-46.
Krauss, Herbert H. 1967. Anxiety: the dread of a future event. *Journal of Individual Psychology*, 23, 88-93.
Krauss, Herbert H. and Rene A. Ruiz. 1967. Anxiety and temporal perspective. *Journal of Clinical Psychology*, 23, 340-2.
Krauss, Herbert H.; Rene A. Ruiz; Gerald J. Mozdzierz; and Jesse Button. 1967. Anxiety and temporal perspective among normals in a stressful situation. *Psychological Reports*, 21, 721-4.
Kruup, Kalev. 1968. The effect of corrective practice on the reliability of counting seconds. *Psychological Record*, 18, 59-62.
Kubler, George. 1962. *The shape of time*. New Haven: Yale University Press.
Kurz, Ronald B. 1963. Relationship between time imagery and Rorschach human movement responses. *Journal of Consulting Psychology*, 27, 273-6.
Kurz, Ronald B.; Robert Cohen; and Susan Starzynski. 1965. Rorschach correlates of time estimation. *Journal of Consulting Psychology*, 29, 379-82.
Kyle, Lyle C. 1955. *Planning your community*. Lawrence: Governmental Research Center, University of Kansas.

Langen, Hermann. 1935. *Experimentelle Untersuchungen bei Zeitvergleichungen*. Berlin: Triltsch and Huther.
Langer, Jonas; Seymour Wapner; and Heinz Werner. 1961. The effect of danger upon the experience of time. *American Journal of Psychology*, 74, 94-7.
Lanzkron, John and W. Wolfson. 1958. Prognostic value of perceptual distortion of temporal orientation in chronic schizophrenics. *American Journal of Psychiatry*, 114, 744-6.
Laties, Victor G. and Bernard Weiss. 1962. Effects of alcohol on timing behavior. *Journal of Comparative and Physiological Psychology*, 55, 85-91.
Lawrence, Douglas H. and Leon Festinger. 1962. *Deterrents and reinforcement*. London: Tavistock.
Lazarsfeld-Jahoda, Marie and Hans Zeisl. 1933. *Die Arbeitslosen von Marienthal*. Leipzig: Hirzel.
Leach, E. R. 1961. *Rethinking anthropology*. London: Athlone Press.
LeBlanc, Arthur F. 1969. Time orientation and time estimation: a function of age. *Journal of Genetic Psychology*, 115, 187-94.
Lefcourt, Herbert M. 1966. Interval versus external control of reinforcement. *Psychological Bulletin*, 65, 206-20.
Legg, C. F. 1968. Alpha rhythm and time judgments. *Journal of Experimental Psychology*, 78, 46-9.
Le Guen, Cl. 1958. Le temps figé du schizophrène. *Evolution Psychiatrique*, 699-735.
Lehmann, Heinz. 1967. Time and psychopathology. *Annals of the New York Academy of Sciences*, 138, 789-821.
Lehr, Ursula. 1967. Attitudes toward the future in old age. *Human Development*, 10, 230-8.
Leister, Georg. 1933. Zeitschätzungen an disparaten Reizen. *Archiv für die gesamte Psychologie*, 88, 257-300.
Le Shan, Lawrence L. 1952. Time orientation and social class. *Journal of Abnormal and Social Psychology*, 47, 589-92.
Lessing, Elise E. 1968. Demographic, developmental, and personality correlates of length of future time perspective (FTP). *Journal of Personality*, 36, 183-201.
Lessing, Gotthold Ephraim. 1766. *Laocoön*. New York: Dutton, 1930.
Levine, D. 1960. Rorschach genetic level and psychotic symptomatology. Unpublished manuscript cited by Judith March Goldrich, A study in time orientation, *Journal of Personality and Social Psychology*, 1967, 6, 216-21.

Levine, Murray and George Spivack. 1959. Incentive, time conception, and self control in a group of emotionally disturbed boys. *Journal of Clinical Psychology*, 15, 110–5.

Levine, Murray; George Spivak; Jean Fuschillo; and Ann Tavernier. 1959. Intelligence, and measures of inhibition and time sense. *Journal of Clinical Psychology*, 15, 224–6.

Lewin, Kurt. 1933. Environmental forces. *In* Carl Murchison (ed.), *A handbook of child psychology*. Worcester: Clark University Press, pp. 94–128.

Lewin, Kurt. 1942. Time perspective and morale. *In* Goodwin Watson (ed.), *Civilian morale*. New York: Houghton Mifflin, pp. 48–70.

Lewin, Kurt. 1952. *Field theory in social science*. London: Tavistock.

Lewis, Aubrey. 1932. The experience of time in mental disorder. *Proceedings of the Royal Society of Medicine*, 25, 611–20.

Lewis, M. M. 1937. The beginning of reference to past and future in a child's speech. *British Journal of Educational Psychology*, 7, 39–56.

Lewis, P. R. and Mary C. Lobban. 1957a. The effects of prolonged periods of life on abnormal time routines upon excretory rhythms in human subjects. *Quarterly Journal of Experimental Physiology*, 42, 356–71.

Lewis, P. R. and Mary C. Lobban. 1957b. Dissociation of diurnal rhythms in human subjects living on abnormal time routines. *Quarterly Journal of Experimental Physiology*, 42, 371–86.

Lhamon, William T.; Robert Edelberg; and Sanford Goldstone. 1962. A comparison of tactile and auditory time judgment. *Perceptual and Motor Skills*, 14, 366.

Lhamon, William T. and Sanford Goldstone. 1956. The time sense. *A.M.A. Archives of Neurology and Psychiatry*, 76, 625–9.

Lhamon, William T.; Sanford Goldstone; and Joyce L. Goldfarb. 1965. The psychopathology of time judgment. *In* Paul H. Hoch and Joseph Zubin (eds.), *The psychopathology of perception*. New York: Grune and Stratton, pp. 164–88.

Lichtenberg, Philip. 1956. Time perspective and the initiation of cooperation. *Journal of Social Psychology*, 43, 247–60.

Lieberman, Morton A. and Annie Siranne Copland. 1970. Distance from death as a variable in the study of aging. *Developmental Psychology*, 2, 71–84.

Lincoln, Jackson Steward. 1935. *The dream in primitive cultures*. London: Cresset Press.

Lindblom, Johannes. 1962. *Prophecy in ancient Israel*. Philadelphia: Muhlenberg Press.

Linder, Staffan Burenstam. 1970. *The harried leisure class*. New York: Columbia University Press.

Linn, Louis. 1967. Clinical manifestations of psychiatric disorder. *In* Alfred M. Freedman and Harold I. Kaplan (eds.), *Comprehensive textbook of psychiatry*. Baltimore: Williams & Wilkins, pp. 546–77.

Linton, Ralph. 1936. *The study of Man*. New York: Appleton-Century.

Lippitt, Ronald. 1942. The morale of youth groups. *In* Goodwin Watson (ed.), *Civilian morale*. New York: Houghton Mifflin, pp. 119–42.

Lipset, Seymour Martin and Reinhard Bendix. 1959. *Social mobility in industrial society*. Berkeley: University of California Press.

Llewellyn-Thomas, Edward. 1959. Successful time estimation during automatic positive feed-back. *Perceptual and Motor Skills*, 9, 219–24.

Lockhart, Russell A. 1966. Temporal conditioning of GSR. *Journal of Experimental Psychology*, 71, 438–46.

Loeb, M. 1957. The effects of intense stimulation on the perception of time. *USA Medical Research Laboratory Report*, no. 269.

Loehlin, John C. 1959. The influence of different activities on the apparent length of time. *Psychological Monographs*, 73, no. 474.

Longuet-Higgins, H. C. 1968. Holographic model of temporal recall. *Nature*, 217 (no. 5123), 104.

Loomis, Earl A., Jr. 1951. Space and time perception and distortion in hypnotic states. *Personality*, 1, 283–93.
Lovell, K. and A. Slater. 1960. The growth of the concept of time: a comparative study. *Journal of Child Psychology and Psychiatry*, 1, 179–90.
Lowin, Aaron; James H. Hottes; Bruce E. Sandler; and Mark Bornstein. 1971. The pace of life and sensitivity to time in urban and rural settings. *Journal of Social Psychology* (in press).
Luoto, Kenneth. 1964. Personality and placebo effects upon timing behavior. *Journal of Abnormal and Social Psychology*, 68, 54–61.
Lynen, John F. 1969. *The design of the present*. New Haven: Yale University Press.
Lynn, R. 1961. Introversion-extraversion in judgments of time. *Journal of Abnormal and Social Psychology*, 63, 457–8.

Mabbott, J. D. 1951. Our direct experience of time. *Mind*, 60, 153–67.
McCann, Willis H. 1943. Nostalgia: a descriptive and comparative study. *Journal of Genetic Psychology*, 62, 97–104.
McClelland, David C. 1961. *The achieving society*. Princeton: Van Nostrand.
MacDougall, Robert. 1904. Sex differences in the sense of time. *Science*, 19, 707–8.
McFie, John. 1960. Psychological testing in clinical neurology. *Journal of Nervous and Mental Disease*, 131, 383–93.
McGrath, J. J. and J. O'Hanlon. 1967. Temporal orientation and vigilance performance. *Acta Psychologica*, 27, 410–19.
MacIver, R. M. 1962. *The challenge of the passing of years: my encounter with time*. New York: Simon and Schuster.
MacLeod, Robert B. and Merrill F. Roff. 1935. An experiment in temporal disorientation. *Acta Psychologica*, 1, 381–423.
McNutt, Thomas H. and Kenneth B. Melvin. 1968. Time estimation in normal and retarded subjects. *American Journal of Mental Deficiency*, 72, 584–9.
McTaggart, John McTaggart Ellis. 1927. *The nature of existence*, Vol. 2. Cambridge: The University Press.
Mahrer, Alvin R. 1956. The role of expectancy in delayed reinforcement. *Journal of Experimental Psychology*, 52, 101–6.
Malcolm, Norman. 1959. *Dreaming*. London: Routledge and Kegan Paul.
Málek, Jiré; J. Gleich; and V. Malý. 1962. Characteristics of the daily rhythm of menstruation and labor. *Annals of the New York Academy of Sciences*, 98, 1042–55.
Mann, Thomas. 1928. *The magic mountain*. New York: Knopf.
Mannheim, Karl. 1941. *Man and society in an age of reconstruction*. New York: Harcourt, Brace.
Marshall, Henry Rutgers. 1907. The time quality. *Mind*, 16, 1–26.
Marum, K. D. 1968. Reproduction and ratio-production of brief duration under conditions of sensory isolation. *American Journal of Psychology*, 81, 21–6.
Matsuda, F. 1966. Development of time estimation [translation]. *Psychological Abstracts*, 1966, 40, no. 923.
Matulef, Norman J.; Roy E. Warman; and Timothy Brock. 1964. Effects of brief vocational counseling on temporal orientation. *Journal of Counseling Psychology*, 11, 352–6.
May, Rollo. 1958. Contributions of existential psychotherapy. *In* Rollo May, Ernest Angel, and Henri Ellenberger (eds.), *Existence: a new dimension in psychiatry and psychology*. New York: Basic Books, pp. 37–91.
Mayo, B. 1950. Is there a sense of duration? *Mind*, 59, 71–8.
Mbiti, John. 1968. African concept of time. *Africa Theological Journal*, 1, 8–20.
Meade, Robert D. 1959. Time estimates as affected by motivational level, goal distance, and rate of progress. *Journal of Experimental Psychology*, 58, 257–9.

Meade, Robert D. 1960a. Time estimates as affected by need tension and rate of progress. *Journal of Psychology*, 50, 173–7.
Meade, Robert D. 1960b. Time on their hands. *Personnel Journal*, 39, 130–2.
Meade, Robert D. 1960c. Time perceptions as affected by need tension. *Journal of Psychology*, 49, 249–53.
Meade, Robert D. 1963. Effect of motivation and progress on the estimation of longer time intervals. *Journal of Experimental Psychology*, 65, 564–7.
Meade, Robert D. 1966a. Achievement motivation, achievement, and psychological time. *Journal of Personality and Social Psychology*, 4, 577–80.
Meade, Robert D. 1966b. Progress direction and psychological time. *Perceptual and Motor Skills*, 23, 115–8.
Meade, Robert D. 1966c. Progress direction, avoidance motivation, and psychological time. *Perceptual and Motor Skills*, 23, 807–10.
Meade, Robert D. 1968. Psychological time in India and America. *Journal of Social Psychology*, 76, 169–74.
Meade, Robert D. and Labh Singh. 1970. Motivation and progress effects on psychological time in subcultures of India. *Journal of Social Psychology*, 80, 3–10.
Meerloo, Joost A. M. 1954. *The two faces of man*. New York: International Universities Press.
Meerloo, Joost A. M. 1962. *Suicide and mass-suicide*. New York: Grune and Stratton.
Meerloo, Joost A. M. 1966. The time sense in psychiatry. In J. T. Frazer (ed.), *The voices of time*. New York: Braziller, pp. 235–56.
Meerloo, Joost A. M. 1970. *Along the fourth dimension: man's sense of time and history*. New York: John Day.
Megargee, Edwin I.; A. Cooper Price; Richard Frohwirth; and Robert Levine. 1970. Time orientation of youthful prison inmates. *Journal of Counseling Psychology*, 17, 8–14.
Melges, Frederick Towne and Carl Edward Fougerousse. 1966. Time sense, emotions, and acute mental illness. *Journal of Psychiatric Research*, 4, 127–40.
Melikian, Levon. 1959. Preference for delayed reinforcement: an experimental study among Palestinian Arab refugee children. *Journal of Social Psychology*, 50, 81–6.
Mendilow, A. A. 1952. *Time and the novel*. London: Peter Nevill.
Meumann, Ernst. 1896. Beiträge zur Psychologie des Zeitbewusstseins. *Philosophische Studien*, 12, 127–254.
Meyer, Adolph. 1952. *The collected papers*, Vol. 4. Baltimore: Johns Hopkins University Press.
Meyer, Merle E. 1966. The internal clock hypothesis for astro-navigation in homing pigeons. *Psychonomic Science*, 5, 259–60.
Meyerhoff, Hans. 1955. *Time in literature*. Berkeley: University of California Press.
Mezey, Alexander G. and Samuel I. Cohen. 1961. The effect of depressive illness on time judgment and time experience. *Journal of Neurology, Neurosurgery, and Psychiatry*, 24, 269–70.
Mezey, Alexander G. and E. J. Knight. 1965. Time sense in hypomanic illness. *Archives of General Psychiatry*, 12, 184–6.
Michaud, E. 1949. *Essai sur l'organisation de la connaissance entre 10 et 14 ans*. Paris: Librarie Philosophique J. Vrin.
Michon, John A. 1965. Studies on subjective duration: II. subjective time measurement during tasks with different information content. *Acta Psychologica*, 24, 205–12.
Michon, John A. 1967. Magnitude scaling of short durations with closely spaced stimuli. *Psychonomic Science*, 9, 359–60.
Miller, Alan R.; Roland A. Frauchiger; and Vernin L. Kiker. 1967. Temporal experience as a function of sensory stimulation and motor activity. *Perceptual and Motor Skills*, 25, 997–1000.

Miller, Neal E. 1944. Experimental studies in conflict. In J. McV. Hunt (ed.), *Personality and the behavior disorders*. New York: Ronald, pp. 431–65.
Millikan, Robert A. 1932. *Time, matter, and values*. Chapel Hill: University of North Carolina Press.
Mills, J. N. 1964. Circadian rhythms during and after three months underground. *Journal of Physiology*, 174, 217–31.
Mills, J. N. 1966. Human circadian rhythms. *Physiological Reviews*, 46, 128–71.
Minkowski, Eugène. 1933. *Le temps vécu*. Paris: Collection de L'Evolution Psychiatrique.
Minkowski, Eugène. 1935. Le problème du temps vécu. *Recherches Philosophiques*, 5, 65–99.
Minkowski, Eugène. 1958a. Findings in a case of schizophrenic depression. *In* Rollo May, Ernest Angel, and Henri Ellenberger (eds.), *Existence: a new dimension in psychiatry and psychology*. New York: Basic Books, pp. 127–38.
Minkowski, Eugène. 1958b. Le temps en psychopathologie. *Psychologie Française*, 3, 9–20.
Mischel, Walter. 1958. Preference for delayed reinforcement: an experimental study of a cultural observation. *Journal of Abnormal and Social Psychology*, 56, 57–61.
Mischel, Walter. 1961a. Delay of gratification, need for achievement, and acquiescence in another culture. *Journal of Abnormal and Social Psychology*, 62, 543–52.
Mischel, Walter. 1961b. Father-absence and delay of gratification: cross-cultural comparisons. *Journal of Abnormal and Social Psychology*, 63, 116–24.
Mischel, Walter. 1961c. Preference for delayed reinforcement and social responsibility. *Journal of Abnormal and Social Psychology*, 62, 1–7.
Mischel, Walter and Carol Gilligan. 1964. Delay of gratification, motivation for the prohibited gratification, and responses to temptation. *Journal of Abnormal and Social Psychology*, 69, 411–7.
Mischel, Walter and Ralph Metzner. 1962. Preference for delayed reward as a function of age, intelligence, and length of delay interval. *Journal of Abnormal and Social Psychology*, 64, 425–31.
Mitchell, Mildred B. 1962. Time disorientation and estimation in isolation. *U.S. Air Force Technical Documentary Report*, ASD-TDR-62-277.
Mönks, Franz J. 1967. Zeitperspektive als psychologische Variable. *Archiv für die gesamte Psychologie*, 119, 131–61.
Mönks, Franz J. 1968. Future time perspective in adolescents. *Human Development*, 11, 107–23.
Moore, Wilbert E. 1963. *Man, time, and society*. New York: Wiley.
Morris, Larry W. and Robert M. Liebert. 1969. Effects of anxiety on timed and untimed intelligence tests: another look. *Journal of Consulting and Clinical Psychology*, 33, 240–4.
Morton, Felix. 1965. Time estimates as affected by incentive class and motivational level. *Genetic Psychology Monographs*, 72, 353–99.
Mowbray, R. M. 1954. Disorientation for age. *Journal of Mental Science*, 100, 749–52.
Mowrer, O. H. and A. D. Ullman. 1945. Time as a determinant in integrative learning. *Psychological Review*, 52, 61–90.
Murdock, George P. 1934. *Our primitive contemporaries*. New York: Macmillan.
Myers, Gary C. 1916. Incidental perception. *Journal of Experimental Psychology*, 1, 339–50.

Nakamura, Hajime. 1966. Time in Indian and Japanese thought. *In* J. T. Frazer (ed.), *The voices of time*. New York: Braziller, pp. 77–91.
Nawas, M. Mike and Jerome J. Platt. 1965. A future-oriented theory of nostalgia. *Journal of Indian Psychology*, 21, 51–7.
Nebel, Gerhard. 1965. *Zeit und Zeiten*. Stuttgart: Ernst Klett.

Nettleship, A. and Charles V. Lair. 1962. Time and disease. *Journal of Clinical and Experimental Psychopathology*, 23, 106–13.
Neumann, Walter. 1936. Die Konkurrenz zwischen den Auffassungen der Zeitdauer und deren Ausfüllung bei verschiedener Einstellung der Aufmerksamkeit. *Archiv für die gesamte Psychologie*, 95, 200–55.
Nichols, Herbert. 1891. The psychology of time. *American Journal of Psychology*, 3, 453–529.
Nilsson, Martin D. 1920. *Primitive time-reckoning*. Lund: G. W. K. Gleerup.
Nissen, H. W. 1930. A study of maternal behavior in the white rat by means of the obstruction method. *Journal of Genetic Psychology*, 37, 377–93.
Nitardy, F. W. 1943. Apparent time acceleration with age of the individual. *Science*, 98, 110.

Oakden, E. C. and Mary Sturt. 1922. The development of the knowledge of time in children. *British Journal of Psychology*, 12, 309–36.
Oberndorf, C. P. 1941. Time—its relation to reality and purpose. *Psychoanalytic Review*, 28, 139–55.
Ochberg, Frank W.; Irwin W. Pollack; and Eugene Meyer. 1964. Correlation of pulse and time judgment. *Perceptual and Motor Skills*, 19, 861–2.
Ochberg, Frank W.; Irwin W. Pollack; and Eugene Meyer. 1965. Reproduction and estimation methods of time judgment. *Perceptual and Motor Skills*, 20, 653–6.
Oléron, Geneviève. 1952. Influence de l'intensité d'un son sur l'estimation de la durée apparente. *Année Psychologique*, 52, 383–92.
Omwake, Katharine T. and Margaret Loranz. 1933. Study of ability to wake at a specified time. *Journal of Applied Psychology*, 17, 468–74.
Orme, J. E. 1962a. Time estimation and personality. *Journal of Mental Science*, 108, 213–6.
Orme, J. E. 1962b. Time studies in normal and abnormal personalities. *Acta Psychologica*, 20, 285–303.
Orme, J. E. 1964. Personality, time estimation, and time experience. *Acta Psychologica*, 22, 430–40.
Orme, J. E. 1966. Time estimation and the nosology of schizophrenia. *British Journal of Psychiatry*, 112, 37–9.
Orme, J. E. 1969. *Time, experience, and behaviour*. New York: American Elsevier Publishing Company.
Ornstein, Robert E. 1970. *On the experience of time*. Baltimore: Penguin Books.
Orsini, Francine. 1958. Étude expérimentale des conduites temporelles: IV. effet de l'apprentissage sur le réproduction d'une durée. *Année Psychologique*, 58, 339–45.
Ostheimer, John M. 1969. Measuring achievement motivation among the Chagga of Tanzania. *Journal of Social Psychology*, 78, 17–30.

Paige, D. D. (ed.). 1951. *The letters of Ezra Pound, 1907–1941*. London: Faber.
Pankauskas, Jonas. 1936. *Vergleich isolierter und rhythmisch gebundener Zeitstrecken*. Munich: Beck'ssche Verlagsbuchhandlung.
Parks, Theodore E. 1968. The spatial separation of two light-flashes and their perceived separation in time. *American Journal of Psychology*, 81, 92–8.
Partridge, Maurice. 1950. *Pre-frontal leucotomy*. Oxford: Blackwell.
Pearl, David and Paul S. D. Berg. 1963. Time perception and conflict arousal in schizophrenia. *Journal of Abnormal and Social Psychology*, 66, 332–8.
Pearson, Karl. 1892. *The grammar of science*. London: Dent, 1937.
Petrie, Asenath. 1952. *Personality and the frontal lobes*. London: Routledge and Kegan Paul.
Petrie, Asenath. 1958. Effects of chlorpromazine and of brain lesions on personality. *In* Harry H. Pennes (ed.), *Psychopharmacology*. New York: Hoeber-Harper, pp. 99–125.

Petrie, Asenath; Walter Collins; and Philip Solomon. 1960. The tolerance for pain and for sensory deprivation. *American Journal of Psychology*, 73, 80–90.
Pettit, Tupper F. 1969. Anality and time. *Journal of Consulting and Clinical Psychology*, 33, 170–4.
Pfaff, Donald. 1968. Effects of temperature and time of day on time judgments. *Journal of Experimental Psychology*, 76, 419–22.
Philip, B. R. and J. W. Lyttle. 1945. The relation between reaction time and duplication times for short intervals. *Bulletin of the Canadian Psychological Association*, 5, 40–2.
Phillips, David P. 1970. Cited by Russell Sage Foundation, *Annual report 68–69*, pp. 19–20.
Piaget, Jean. 1946. *Le développement de la notion de temps chez l'enfant*. Paris: Presses Universitaires de France.
Piaget, Jean. 1966. Time perception in children. *In* J. T. Fraser (ed.), *The voices of time*. New York: Braziller, pp. 202–16.
Pick, A. 1919. Zur Psychopathologie des Zeitsinnes. *Zeitschrift für Pathopsychologie*, 3, 430–41.
Piéron, H. 1936. Sens du temps et horloge chimique de l'abeille à l'homme. *In* H. Piéron et al., *Livres jubilaire de M. L. Bouvier*. Paris: Firmin Didot, pp. 269–72.
Pistor, Frederick. 1939. Measuring the time concepts of children. *Journal of Educational Research*, 33, 293–300.
Pivik, Terry and David Foulkes. 1968. NREM mentation: relation to personality, orientation time, and time of night. *Journal of Consulting and Clinical Psychology*, 32, 144–51.
Platt, Jerome J. and Russell Eisenman. 1968. Internal-external control of reinforcement, time perspective, adjustment, and anxiety. *Journal of General Psychology*, 79, 121–8.
Platt, Jerome J.; Russell Eisenman; and Edward DeGross. 1969. Birth order and sex differences in future time perspective. *Development Psychology*, 1, 70.
Platt, Jerome J. and Robert E. Taylor. 1967. Homesickness, future time perspective, and the self concept. *Journal of Individual Psychology*, 23, 94–7.
Polak, Fred L. 1961. *The image of the future*, Vol. 1. New York: Oceana Publications.
Pollock, Kenneth and Robert Kastenbaum. 1964. Delay of gratification in later life: an experimental analog. *In* Robert Kastenbaum (ed.), *New thoughts on old age*. New York: Springer, pp. 281–90.
Postman, Leo. 1944. Estimates of time during a series of tasks. *American Journal of Psychology*, 57, 421–4.
Postman, Leo and G. A. Miller. 1945. Anchoring of temporal judgments. *American Journal of Psychology*, 58, 43–53.
Poulet, Georges. 1956. *Studies in human time*. Baltimore: Johns Hopkins Press.
Pressey, Sidney L. and Raymond G. Kuhlen. 1957. *Psychological development through the life span*. New York: Harpers.
Priestley, J. B. 1964. *Man and time*. Garden City: Doubleday.
Prior, Arthur. 1957. *Time and modality*. Oxford: Clarendon.
Proust, Marcel. 1941. *The Guermantes way*. London: Chatto and Windus.
Pumpian-Mindlin, Eugene. 1935. Ueber die Bestimmung der bewussten Zeitschätzung bei Normalen und dementen Epileptikern. *Schweizer Archiv für Neurologie und Psychiatrie*, 36, 291–305.

Rabin, A. I. 1957. Time estimation of schizophrenics and non-psychotics. *Journal of Clinical Psychology*, 13, 88–90.
Rapoport, Anatol and Albert M. Chammah. 1965. *Prisoner's Dilemma*. Ann Arbor: University of Michigan Press.
Read, Margaret. 1966. *Culture, health, and disease*. London: Tavistock.

Rebello, S. 1937. Study on the notion of time [translation]. *Psychological Abstracts*, 11, no. 4066.
Redl, Fritz and David Wineman. 1957. *The aggressive child*. New York: Free Press.
Reed, G. F. and J. C. Kenna. 1964. Personality and time estimation in sensory deprivation. *Perceptual and Motor Skills*, 18, 182.
Reuchlin, Maurice. 1957. Le temps comme source de variation expérimentale. *Psychologie Française*, 2, 74–84.
Reutt, Natalia. 1962. Evaluation of identical time by children of school age [translation]. *Psychological Abstracts*, 36, no. 2FC41R.
Reynolds, Horace N. 1968. Temporal estimation in the perception of occluded motion. *Perceptual and Motor Skills*, 26, 407–16.
Richards, Whitman. 1964. Time estimates measured by reproduction. *Perceptual and Motor Skills*, 18, 929–43.
Riesman, David. 1950. *The lonely crowd*. New Haven: Yale University Press, 1950.
Rimoldi, Horacio J. A. 1951. Personal tempo. *Journal of Abnormal and Social Psychology*, 46, 283–303.
Rizzo, Adolfo E. 1967–8. The time moratorium. *Adolescence*, 2, 469–80.
Robbins, Michael; Philip L. Kilbride; and John M. Bukenya. 1968. Time estimation and acculturation among the Baganda. *Perceptual and Motor Skills*, 26, 1010.
Roelofs, Otto and W. P. C. Zeeman. 1949. The subjective duration of time-intervals. *Acta Psychologica*, 6, 126–77, 289–336.
Roelofs, Otto and W. P. C. Zeeman. 1951. Influential sequences of optical stimuli on the estimation of duration of a given interval of time. *Acta Psychologica*, 8, 89–128.
Rogers, Everett M. 1962. *Diffusion of innovations*. New York: Free Press.
Rogers, K. W. 1967. Concepts of time in secondary school children of above IQ. *British Journal of Educational Psychology*, 37, 99–109.
Rokeach, Milton and Richard Bonier. 1960. Time perspective, dogmatism, and anxiety. In Milton Rokeach (ed.), *The open and closed mind*. New York: Basic Books, pp. 366–75.
Roos, Philip and Robert Albers. 1965a. Performance of alcoholics and normals on a measure of temporal orientation. *Journal of Clinical Psychology*, 21, 34–6.
Roos, Philip and Robert Albers. 1965b. Performance of retardates and normals on a measure of temporal orientation. *American Journal of Mental Deficiency*, 69, 835–8.
Rose, Annelies Argelander. 1947. A study of homesickness in college freshmen. *Journal of Social Psychology*, 26, 185–202.
Rosenfelt, Rosalie H.; Robert Kastenbaum; and Philip E. Slater. 1964. Patterns of short-range time orientation in geriatric patients. In Robert Kastenbaum (ed.), *New thoughts on old age*. New York: Springer, pp. 291–9.
Rosenzweig, Saul and Aase Grude Koht. 1933. The experience of duration as affected by need-tension. *Journal of Experimental Psychology*, 16, 745–74.
Ross, Sherman and Leon Katchmar. 1951. The construction of a magnitude function for short time-intervals. *American Journal of Psychology*, 64, 397–401.
Rossomando, Nina A. and Walter Weiss. 1970. Attitude change effects of timing and amount of payment for counterattitudinal behavior. *Journal of Personality and Social Psychology*, 14, 32–8.
Roth, Julius A. 1963. *Timetables*. Indianapolis: Bobbs-Merrill.
Rotter, Julian B. 1966. Generalized expectancies for internal versus external control of reinforcement. *Psychological Monographs*, 80, no. 609.
Rouart, J. La temporisation comme maîtrise et comme défense. *Revue Française Psychanalyse*, 26, 383–422.
Rubin, Edgar. 1934. Some elementary time experiences. *British Journal of Psychology*, 24, 445–9.

Ruiz, Rene A. and Herbert H. Krauss. 1968. Anxiety, temporal perspective, and item content of the incomplete thoughts test (ITT). *Journal of Clinical Psychology*, 24, 70–2.

Ruiz, Rene A.; Ronald S. Reivich; and Herbert H. Krauss. 1967. Tests of temporal perspective: do they measure the same construct? *Psychological Reports*, 21, 849–52.

Rutschmann, Jacques and Leo Rubinstein. 1966. Time estimation, knowledge of results, and drug effects. *Journal of Psychiatric Research*, 4, 107–14.

Sattler, Jerome M. 1964. Counselor competence, interest, and time perspective. *Journal of Counseling Psychology*, 4, 357–60.

Sattler, Jerome M. 1965. Time estimation as a function of knowledge and achievement. *Psychological Record*, 15, 529–34.

Sattler, Jerome M. 1967. Counselor competence, interest and time perspective: a follow-up note. *Counselor Education and Supervision*, 6, 185–6.

Sauvageot, A. 1936. La notion de temps et son expression dans le langue. *Journal de Psychologie Normale et Pathologique*, 33, 19–27.

Schachter, Stanley and Larry P. Gross. 1968. Manipulated time and eating behavior. *Journal of Personality and Social Psychology*, 10, 98–106.

Schaefer, Vernon and A. R. Gilliland. 1938. The relation of time estimation to certain physiological changes. *Journal of Experimental Psychology*, 23, 545–52.

Schaltenbrand, Georges. 1967. Consciousness and time. *Annals of the New York Academy of Sciences*, 138, 632–45.

Schechter, David E.; Martin Symonds; and Isidor Bernstein. 1955. Development of the concept of time in children. *Journal of Nervous and Mental Disease*, 121, 301–10.

Schiff, William and Stephen Thayer. 1968. Cognitive and affective factors in temporal experience: anticipated and unpleasant sensory events. *Perceptual and Motor Skills*, 26, 799–808.

Schilder, Paul. 1936. Psychopathology of time. *Journal of Nervous and Mental Disease*, 83, 530–46.

Schjelderup, Harald K. 1960. Time relations in dreams. *Scandinavian Journal of Psychology*, 1, 62–4.

Schlosberg, Arie. 1969. Time perspective in schizophrenics. *Psychiatric Quarterly*, 43, 22–34.

Schmid, Nelly. 1929. *Das Vergleichsurteil auf Grund der Beobatchung von Zeitstrecken.* Thayngen: Karl Augustin.

Schmidt, Else. 1935. *Mit welcher Genaugikeit werden dargebotene Zeitintervalle aus dem Gedächtnis umgestellt?* Zuelenroda: Bernhard Sporn.

Schneck, Jerome M. 1957. The hypnotic state and the psychology of time. *Psychoanalytic Review*, 44, 323–6.

Schneevoigt, W. 1934. Die Wahrnehmung der Zeit bei den verschiedenen Menschentypen. *Zeitschrift für Psychologie*, 131, 217–95.

Schneider, Daniel E. 1948. Time-space and the growth of the sense of reality: a contribution to the psychophysiology of the dream. *Psychoanalytic Review*, 33, 229–52.

Schneider, Louis and Everre Lysgaard. 1953. The deferred gratification pattern. *American Sociological Review*, 18, 142–9.

Schoeffler, Max S. and Dorothy M. Poole. 1967. Accuracy and variability in the production of short durations. *Psychonomic Science*, 7, 423–4.

Schönbach, Peter. 1959. Cognition, motivation, and time perception. *Journal of Abnormal and Social Psychology*, 58, 195–202.

Schonbar, Rosalea Ann. 1965. Differential dream recall frequency as a component of "life style." *Journal of Consulting Psychology*, 29, 468–74.

Schönpflug, Wolfgang. 1966. Arousal, adaptation level, and accentuation of judgment. *Journal of Experimental Psychology*, 72, 443–6.

Scott, W. Clifford M. 1948. Some psycho-dynamic aspects of disturbed perception of time. *British Journal of Medical Psychology*, 21, 111–20.
Seeley, John R.; R. Alexander Sim; and Elizabeth W. Loosley. 1956. *Crestwood Heights*. New York: Basic Books.
Shectman, Frederick A. 1970. Time estimation, sequence effects, and filling activities. *Perceptual and Motor Skills*, 30, 23–6.
Shelley, Percy Bysshe. 1813. Notes on Queen May, Part VIII, lines 203–7. *In* Thomas Hutchinson (ed.), *The complete poetical works of Percy Bysshe Shelley*. London: Oxford, 1935, p. 816.
Shybut, John. 1968. Time perspective, internal vs. external control, and severity of psychological disturbance. *Journal of Clinical Psychology*, 24, 312–5.
Siegman, Aron Wolfe. 1961. The relationship between future time perspective, time estimation, and impulse control in a group of young offenders and in a control group. *Journal of Consulting Psychology*, 25, 470–5.
Siegman, Aron Wolfe. 1962a. Anxiety, impulse control, intelligence, and the estimation of time. *Journal of Clinical Psychology*, 18, 103–5.
Siegman, Aron Wolfe. 1962b. Future-time perspective and the perception of duration. *Perceptual and Motor Skills*, 15, 609–10.
Siegman, Aron Wolfe. 1962c. Intercorrelation of some measures of time estimation. *Perceptual and Motor Skills*, 14, 381–2.
Siegman, Aron Wolfe. 1966. Effects of auditory stimulation and intelligence on time estimation in delinquents and non-delinquents. *Journal of Consulting Psychology*, 30, 320–8.
Siffre, Michel. 1964. *Beyond time*. New York: McGraw-Hill.
Siipola, Elsa and Vivian Taylor. 1952. Reactions to ink blots under free and pressure conditions. *Journal of Personality*, 21, 22–47.
Siipola, Elsa; W. Nannette Walker; and Dorothy Kolb. 1955. Task attitudes in word association, projective and nonprojective. *Journal of Personality*, 23, 441–59.
Simmel, Edward C. 1963. Time estimation in hospitalized patients as a function of goal distance and magnitude of reward. *Perceptual and Motor Skills*, 17, 91–7.
Simonson, Emil. 1928. Ueber das Verhältnis von Raum und Zeit zur Traumarbeit. *Imago*, 14, 469–85.
Singer, Jerome L. 1955. Delayed gratification and ego-development: implications for clinical and experimental research. *Journal of Consulting Psychology*, 19, 256–66.
Singer, Jerome L. 1961. Imagination and waiting ability in young children. *Journal of Personality*, 29, 396–413.
Singer, Jerome L.; Harold Wilensky; and Vivian G. McCraven. 1956. Delaying capacity, fantasy, and planning ability: a factorial study of some basic ego functions. *Journal of Consulting Psychology*, 20, 375–83.
Sixtl, Friedrich. 1963. Die Zeitfehler beim Schätzen der Reizzeit und als Funktion der Reizlänge, der Intervallzeit, und der Versuchswiederholung. *Zeitschrift für Experimentelle und Angewandte Psychologie*, 10, 209–25.
Skramlich, Emil von. 1934. Die Sicherheit der subjektiven Zeitbeurteilung. *Forschungen und Fortschritte*, 10, 326–7.
Small, Arnold M., Jr. and Richard A. Campbell. 1962. Temporal differential sensitivity for auditory stimuli. *American Journal of Psychology*, 75, 401–10.
Smart, Reginald G. 1968. Future time perspectives in alcoholics and social drinkers. *Journal of Abnormal Psychology*, 73, 81–3.
Smeltzer, William Edward. 1969. Time orientation and time perspective in psychotherapy. *Dissertation Abstracts*, 29, 3922-B.
Smith, Marion W. 1952. Different cultural concepts of past, present, and future. *Psychiatry*, 15, 395–400.
Smith, Robert J. 1961. Cultural differences in the life cycle and the concepts of time.

In Robert W. Kleemeier (ed.),*Aging and leisure*. New York: Oxford, pp. 83–103.
Smith, S. and W. Lewty. 1959. Perceptual isolation using a silent room. *Lancet*, 2, 342–5.
Smythe, Elizabeth J. and Sanford Goldstone. 1957. The time sense: a normative, genetic study of the development of perception. *Perceptual and Motor Skills*, 7, 49–59.
Solanes, José. 1948. Exil et troubles du temps vécu. *L'Hygiène Mentale*, 37, 62–78.
Sollberger, Arne. 1967. Biological measurements in time, with particular reference to synchronization mechanisms. *Annals of the New York Academy of Sciences*, 138, 561–99.
Sorokin, Pitrim and Robert A. Merton. 1937. Social time: a methodological and functional analysis. *American Journal of Sociology*, 42, 615–29.
Spencer, Llewellyn T. 1921. An experiment in time estimation using different interpolations. *American Journal of Psychology*, 32, 557–62.
Spengler, Oswald. 1926. *The decline of the West*, Vol. 1. New York: Knopf.
Spiegel, E. A.; H. T. Wycis; C. W. Orchinik; and H. Freed. 1955. The thalamus and temporal orientation. *Science*, 121, 771–2.
Spivack, George and Murray Levine. 1964. Consistency of individual differences in time judgments. *Perceptual and Motor Skills*, 19, 83–92.
Spivack, George; Murray Levine; and Herbert Sprigle. 1959. Intelligence test performance and the delay function of the ego. *Journal of Consulting Psychology*, 23, 428–31.
Spreen, Otfried. 1963. The position of time estimation in a factor analysis and its relation to some personality variables. *Psychological Record*, 13, 455–64.
Springer, Doris V. 1951. Development of concepts related to the clock as shown in young children's drawings. *Journal of Genetic Psychology*, 79, 47–54.
Springer, Doris V. 1952. Development in young children of an understanding of time and the clock. *Journal of Genetic Psychology*, 80, 83–96.
Stalnaker, John M. and Marion W. Richardson. 1930. Time estimation in the hypnotic trance. *Journal of General Psychology*, 4, 362–6.
Stefaniuk, B. and H. Osmond. 1952. Unpublished observations. Cited in A. Hoffer and H. Osmond, *The hallucinogens*. New York: Academic Press, p. 119.
Stein, Kenneth B. and Kenneth H. Craik. 1965. Relation between motoric and ideational activity preference and time perspective in neurotics and schizophrenics. *Journal of Consulting Psychology*, 29, 460–7.
Stein, Kenneth B.; Theodore R. Sabin; and James A. Kulik. 1968. Future time perspective: its relation to the socialization process and the delinquent role. *Journal of Consulting and Clinical Psychology*, 32, 257–64.
Steinberg, Hannah. 1955. Changes in time perception induced by an anaesthetic drug. *British Journal of Psychology*, 46, 273–9.
Stern, Muriel H. 1959. Thyroid function and activity, speed, and timing aspects of behaviour. *American Journal of Psychology*, 13, 43–8.
Stern, William. 1938. *General psychology from the personalistic standpoint*. New York: Macmillan.
Sterne, Laurence. 1759–66. *The life and opinions of Tristram Shandy*, vol. 4. Baltimore: Penguin, 1967.
Sternlicht, Manny and Louis Siegel. 1968. Time orientation and friendship patterns of institutionalized retardates. *Journal of Clinical Psychology*, 24, 26–7.
Sterzinger, Othmar. 1935. Chemipsychologische Untersuchungen über den Zeitsinn. *Zeitschrift für Psychologie*, 134, 100–31.
Sterzinger, Othmar. 1938. Neue chemipsychologisch Untersuchungen über den menschlichen Zeitsinn. *Zeitschrift für Psychologie*, 143, 391–406.
Stevens, S. S. 1962. The surprising simplicity of sensory metrics. *American Psychologist*, 17, 29–39.

Stevens, S. S. and E. H. Galanter. 1957. Ratio scales and category scales for a dozen perceptual continua. *Journal of Experimental Psychology*, 54, 377–411.
Stott, Leland H. 1935. Time-order errors in the discrimination of short tonal intervals. *Journal of Experimental Psychology*, 18, 741–66.
Strauss, Erwin W. 1947. Disorders of personal time in depressive states. *Southern Medical Journal*, 40, 254–9
Strodtbeck, Fred L. 1958. Family interaction, values, and achievement. *In* Marshall Sklare (ed.), *The Jews*. Glencoe: Free Press, 1958.
Stromberg, Gustaf. 1961. Space, time, and eternity. *Journal of the Franklin Institute*, 272, 134–44.
Strughold, Hubertus. 1962. Day-night cycling in atmospheric flight, space flight, and other celestial bodies. *Annals of the New York Academy of Sciences*, 98, 1109–15.
Strunk, O., Jr. 1960. Reliability of time estimates. *Journal of Psychological Studies*, 11, 101–3.
Sturt, Mary. 1923. Experiments on the estimate of duration. *British Journal of Psychology*, 13, 382–8.
Sturt, Mary. 1925. *The psychology of time*. New York: Harcourt, Brace.
Sweet, Alex L. 1965. Temporal discrimination by the human eye. *American Journal of Psychology*, 66, 185–98.
Swift, Edgar James and John Alexander McGeoch. 1925. An experimental study of the perception of filled and empty time. *Journal of Experimental Psychology*, 8, 240–9.

Talland, George A. 1958. Psychological studies in Korsakoff's psychosis: II. perceptual functions. *Journal of Nervous and Mental Disease*, 127, 197–219.
Talland, George A. 1961. Les troubles de l'orientation temporelle dans le syndrome amnésique. *Encéphale*, 50, 450–70.
Tanner, Trieve J.; R. Mark Patton; and Richard C. Atkinson. 1965. Intermodality judgments of signal duration. *Psychonomic Science*, 2, 271–2.
Teahan, John E. 1958. Future time perspective, optimism, and academic achievement. *Journal of Abnormal and Social Psychology*, 57, 379–80.
Tejmar, Jaroslav. 1962. After differences in cyclic motor reaction. *Nature*, 195, 813–4.
Thompson, David C.; George Spivack; and Murray Levine. 1960. The experience of time as a determinant of self-control. *Archives of General Psychiatry*, 3, 563–6.
Thomson, William A. R. 1967. *Black's medical dictionary*. London: Black.
Thor, Donald H. 1962a. Diurnal variability in time estimation. *Perceptual and Motor Skills*, 15, 451–4.
Thor, Donald H. 1962b. Time perspective and time of day. *Psychological Record*, 12, 417–22.
Thor, Donald H. and Robert O. Baldwin. 1965. Time of day estimates at six times of day under normal conditions. *Perceptual and Motor Skills*, 21, 904–6.
Thor, Donald H. and M. L. J. Crawford. 1964. Time perception during a two-week confinement: influence of age, sex, IQ, and time of day. *Acta Psychologica*, 22, 78–84.
Thornton, Harry and Agathe Thornton. 1962. *Time and style: a psycho-linguistic essay in classical literature*. London: Methuen.
Tolor, Alexander and Vincent M. Murphy. 1967. Some psychological correlates of subjective life-expectancy. *Journal of Clinical Psychology*, 23, 21–4.
Toulmin, Stephen and June Goodfield. 1965. *The discovery of time*. London: Hutchinson.
Treisman, Michel. 1963. Temporal discrimination and the indifference level: implications for a model of the "internal clock." *Psychological Monographs*, 77, no. 576.
Triplett, Dorothy. 1931. The relation between the physical pattern and reproduction of short temporal intervals. *Psychological Monographs*, 41, no. 4, 201–65.

Tscholakow, Kyrill. 1956. Ueber Dyschronosien. *Psychiatrie, Neurologie, und medizinische Psychologie,* 8, 355–63.
Turchioe, Rita M. 1948. The relation of adjacent inhibitory stimuli to the central tendency effect. *Journal of General Psychology,* 39, 3–14.
Tutuola, Amos, 1967. *Ajaiyi and his inherited property.* London: Faber and Faber.

Urban, F. M. 1907. On systematic errors in time estimation. *American Journal of Psychology,* 18, 187–93.
Usizima, Y. 1935. Character and time consciousness. *Japanese Journal of Applied Psychology,* 3, 165–70.

Varga, Erwin. 1959. Beiträge zur Pathophysiologie der Sprache. *Psychiatrie, Neurologie, und medizinische Psychologie,* 11, 307–11.
Vaschide, N. 1911. *Le sommeil et les rêves.* Paris: Flammarion.
Verdone, Paul. 1965. Temporal reference of manifest dream content. *Perceptual and Motor Skills,* 20, 1253–68.
Vernon, Jack A. and Thomas E. McGill. 1963. Time estimations during sensory deprivation. *Journal of General Psychology,* 69, 11–8.
Vischer, A. L. 1947. Psychological problems of the ageing personality. *Bulletin der schweizerischen Akademie der medizinischen Wissenschaften,* 2, 280–6.
von Sturmer, G. 1966. Stimulus variation and sequential judgments of duration. *Quarterly Journal of Experimental Psychology,* 18, 354–7.
von Sturmer, G. 1968. Time perception, vigilance, and decision theory. *Perception and Psychophysics,* 3, 197–200.

Wallace, Melvin. 1956. Future time perspectives in schizophrenia. *Journal of Abnormal and Social Psychology,* 52, 240–5.
Wallace, Melvin and Albert I. Rabin. 1960. Temporal experience. *Psychological Bulletin,* 57, 213–36.
Walker, Charles R. 1957. *Toward the automatic factory.* New Haven: Yale University Press.
Wallach, Michael A. and Leonard R. Green. 1961. On age and the subjective speed of time. *Journal of Gerontology,* 16, 71–4.
Wallis, Robert. 1967. Time—fourth dimension of the mind. *Annals of the New York Academy of Sciences,* 138, 784–97.
Wallon, Henri; Eugénie Evart-Chmielniski; and Georgette Denjean-Raban. 1957. Reproduction de durées courtes par l'enfant. *Enfance,* 10, 97–134.
Warm, Joel S.; Emerson Foulke; and Michel Loeb. 1966. The influence of stimulus-modality and duration on changes in temporal judgments over trials. *American Journal of Psychology,* 79, 628–31.
Warm, Joel S.; Lewis F. Greenberg; and C. Stuart Dube. 1964. Stimulus and motivational determinants of temporal perception. *Journal of Psychology,* 58, 243–8.
Warm, Joel S.; James R. Morris; and John K. Kew. 1963. Temporal judgment as a function of nosological classification and experimental method. *Journal of Psychology,* 55, 287–97.
Warm, Joel S.; Richard P. Smith; and Lee S. Caldwell. 1967. Effects of induced muscle tension on judgment time. *Perceptual and Motor Skills,* 25, 153–60.
Warner, W. Lloyd. 1959. *The living and the dead.* New Haven: Yale University Press.
Weber, C. O. 1927. The properties of space and time in kinaesthetic fields of force. *American Journal of Psychology,* 38, 597–606.
Weber, Dale S. 1965. A time perception task. *Perceptual and Motor Skills,* 21, 863–6.
Weber, Max. 1930. *The Protestant ethic and the spirit of capitalism.* New York: Scribner's, 1930.
Webster, Frances R.; Sanford Goldstone; and Warren W. Webb. 1962. Time judgment

and schizophrenia: psychophysical methods as a relevant contextual factor. *Journal of Psychology*, 54, 159–64.
Weil, Andrew T.; Norman E. Zinberg; and Judith M. Nelsen. 1968. Clinical and psychological effects of marihuana in man. *Science*, 162, 1234–42.
Weinstein, Alvin D.; Sanford Goldstone; and William K. Boardman. 1958. The effect of recent and remote frames of reference on temporal judgments of schizophrenic patients. *Journal of Abnormal and Social Psychology*, 57, 241–4.
Weinstein, Edwin A. and Robert L. Kahn. 1955. *Denial of illness*. Springfield: Charles C. Thomas.
Weitzenhoffer, André M. 1953. *Hypnotism: an objective study in suggestibility*. New York: Wiley, 1953.
Weitzenhoffer, André M. 1964. Exploration in hypnotic time distortions: I. Acquisition of temporal reference frames under conditions of time distortion. *Journal of Nervous and Mental Disease*, 138, 354–66.
Werboff, Jack. 1962. Time judgment as a function of electroencephalographic activity. *Experimental Neurology*, 6, 152–60.
Werner, Heinz. 1957. *Comparative psychology of mental development*. New York: International Universities Press.
Weybrew, Benjamin B. 1963. Accuracy of time estimation and muscular tension. *Perceptual and Motor Skills*, 17, 118.
White, Carroll T. and Malcolm Lichtenstein. 1963. Some aspects of temporal discrimination. *Perceptual and Motor Skills*, 17, 471–82.
Whitely, Paul L. and J. Carver Anderson. 1930. The influence of two different interpolations upon time estimation. *Journal of General Psychology*, 4, 391–401.
Whitrow, Gerald J. 1961. *The natural philosophy of time*. London: Nelson.
Whitrow, Gerald J. 1967a. Reflections on the natural philosophy of time. *Annals of the New York Academy of Sciences*, 138, 422–32.
Whitrow, Gerald J. 1967b. Some reflections on the problem of memory. *Annals of the New York Academy of Sciences*, 138, 856–61.
Whyman, Andrew D. and Rudolf H. Moos. 1967. Time perception and anxiety. *Perceptual and Motor Skills*, 24, 567–70.
Wilen, Frank. 1943. Apparent time acceleration with age. *Science*, 98, 301.
Williams, Stephen Guion. 1966. Temporal experience and schizophrenia: a study of time orientation, attitude, and perspective. *Dissertation Abstracts*, 26, 6862–3.
Wing, Leonard W. 1962. The effect of latitude on cycles. *Annals of the New York Academy of Sciences*, 98, 1202–5.
Wirth, Wilhelm. 1937. Die unmittelbare Teilung einer gegebenen Zeitstrecke. *American Journal of Psychology*, 50, 79–96.
Withey, Stephen B. 1964. Sequential accomodations to threat. *In* George H. Grosser, Henry Wechsler, and Milton Greenblatt (eds.), *The threat of impending disaster*. Cambridge: MIT Press, pp. 104–14.
Wohlford, Paul. 1968. Extension of personal time in TAT and story completion stories. *Journal of Projective Techniques and Personality Assessment*, 32, 267–80.
Wolk, Robert L.; Stanley L. Rustin; and Julius Scotti. 1963. The geriatric delinquent. *Journal of the American Geriatrics Society*, 11, 653–9.
Woodrow, Herbert. 1928. Behavior with respect to short temporal-forms. *Journal of Experimental Psychology*, 11, 167–93, 259–80.
Woodrow, Herbert. 1930. The reproduction of temporal intervals. *Journal of Experimental Psychology*, 13, 473–99.
Woodrow, Herbert. 1933. Individual differences in the reproduction of temporal intervals. *American Journal of Psychology*, 45, 271–81.
Woodrow, Herbert. 1935. The effect of practice upon time-order errors in the comparison of temporal intervals. *Psychological Review*, 42, 127–52.

Woodrow, Herbert. 1951. Time perception. *In* S. S. Stevens, *Handbook of experimental psychology*. New York: Wiley, pp. 1224–36.
Woodrow, Herbert and Leland H. Stott. 1936. The effect of practice on positive time-order errors. *Journal of Experimental Psychology*, 19, 694–705.
Woolf, Virginia. 1927. *To the light house*. London: Hogarth Press.
Woolf, Virginia. 1928. *Orlando*. London: Hogarth Press.
Wright, David J.; Sanford Goldstone; and William K. Boardman. 1962. Time judgment and schizophrenia: step interval as a relevant contextual factor. *Journal of Psychology*, 54, 33–8.
Yaryan, Ruby B. and Leon Festinger. 1961. Preparatory action and belief in the probable occurrence of future events. *Journal of Abnormal and Social Psychology*, 63, 603–6.
Yates, Sybille. 1935. Some aspects of time difficulties and their relation to music. *International Journal of Psycho-Analysis*, 16, 341–54.
Yerkes, Robert M. and F. M. Urban. 1906. Time estimation in its relation to sex, age, and physiological rhythms. *Harvard Psychological Studies*, 2, 405–30.
Young, Shirley A. and F. C. Sumner. 1954. Personal equation, frame of reference, and other observations in remote reproductions of filled and unfilled time-intervals. *Journal of General Psychology*, 51, 333–7.
Zahn, Gordon C. 1964. *In solitary witness*. New York: Holt, Rinehart, and Winston.
Zangwill, O. L. 1953. Disorientation for age. *Journal of Mental Science*, 99, 698–701.
Zatzkis, J. 1949. Unpublished thesis quoted in David C. McClelland, John W. Atkinson, Russell A. Clark, and Edgard L. Lowell, *The achievement motive*. New York: Appleton-Century-Crofts, pp. 221, 249–53.
Zborowski, Mark. 1952. Cultural components in response to pain. *Journal of Social Issues*, 8, no. 4, 16–30.
Zeff, C. 1965. Comparison of conventional and digital time displays. *Ergonomics*, 8, 339–45.
Zelkind, Irving. 1969. Time estimation as a function of stimulus intensity and experimental methodology. *Dissertation Abstracts*, 29, 4417-B.
Zelkind, Irvin and Bernard Spilka. 1965. Some time perspective-time perception relationships. *Psychological Record*, 15, 417–22.
Zern, David. 1967. The influence of certain developmental factors in fostering the ability to differentiate the passage of time. *Journal of Social Psychology*, 72, 9–17.
Zern, David. 1970. The influence of certain child-rearing factors upon the development of a structured and salient sense of time. *Genetic Psychology Monographs*, 81, 197–254.
Zuili, Nadine and Paul Fraisse. 1966. L'estimation du temps en fonction de la quantité de mouvements effectués dans une tache: étude génétique. *Année Psychologique*, 66, 383–96.
Zurcher, Louis A.; Joe E. Willis; Frederick Ikard; and John A. Dohne. 1967. Dogmatism, future orientation, and perception of time. *Journal of Social Psychology*, 73, 205–9.

INDEX OF PRINCIPLES AND HYPOTHESES

PRINCIPLES

2.1 Different methods of assessing temporal judgment produce different judgments which may be, however, psychologically consistent; judgments from Verbal Estimation and Production are likely to be negatively correlated (p. 41).

3.1 Periodic changes in the external milieu inevitably and everywhere provide the potential for acquiring knowledge concerning the duration and succession of intervals and for the arousal of temporal motives (p. 49).

3.1.a All persons everywhere are oriented periodically toward the past, the present, and the future (p. 52).

3.2 Modally within the person, within significant groups, and within the society as a whole, one temporal perspective rather than another is likely to be facilitated (p. 54).

3.2.a The modal temporal perspective of a society reflects and affects a modal philosophy of values pertaining to other behavior (p. 56).

3.2.b The stronger the temporal perspective, the weaker the orientations in other directions (p. 59).

3.3 Each society provides appropriate information for passing temporal judgment (p. 60).

3.4 The value of an activity is positively correlated with its temporal priority, the objective time devoted to it, and the frequency or precision with which its duration is judged (p. 63).

3.4.a Temporal information is likely to have an appropriate effect upon behavior (p. 67).

3.4.b Appropriate linguistic concepts and structures mediate a positive if imperfect correlation between the temporal and behavioral potentials (p. 68).

3.5 Changes in the external or internal milieu (especially periodic ones) of significance to the organism give rise to internal processes which function more or less in phase with those changes, which are not completely independent of them, which may persist or not be immediately extinguished when external conditions are altered, and which provide unverbalizable cues for arousing the temporal motive (p. 72).

3.6 The temporal potential of a person is affected by personality traits either when those traits pertain specifically to some aspect of temporal behavior or when they influence aspects of his behavior potential which in turn modifies that temporal behavior (p. 73).

4.1 A prior interim temporal motive is likely to be aroused or temporal information to be sought whenever the duration of an interval has significance for the achievement of a goal (p. 80).

4.1.a The duration of an interval is likely to have significance when goals are achievable only, or more easily, through coordinating the activities of more than one person (p. 82).

4.1.b The duration of an interval is likely to have significance when people are seeking the same or similar goals and must cooperate or coordinate their activities to achieve them (p. 85).

4.1.c The duration of an interval is likely to have significance when it involves the power relation between groups or individuals (p. 87).

4.1.d The duration of intervals involved in planning and receiving communication has significance, respectively, to the communicator and the audience (p. 90).

4.1.e The duration of an interval is likely to have (or to have had) significance when it is anticipated (or recalled) that its continuation or termination will (or did) bring the relevant goal closer or will (or did) reduce the frustration (p. 90).

4.1.f The duration of an interval is likely to have significance whenever gratification is deferred (p. 93).

4.1.f.i Gratification is likely to be deferred when the temporal orientation is toward the future, and vice versa (p. 93).

4.1.f.ii Gratification is likely to be deferred when the frustration from renunciation seems either necessary or less than what may be attained in the future and when there is appropriate reinforcement in the present (p. 96).

4.2 A temporal motive is likely to be aroused, or temporal information to be sought, when a temporal judgment or such information has instrumental value in attaining a goal (p. 97).

4.2.a After the arousal of a temporal motive, judgment is not likely to be passed when relevant temporal information is available (p. 97).

5.1 The perceptual potential is a function of the temporal knowledge, the events, and the drives perceived or associated with the interval (p. 107).

5.1.a Temporal knowledge, when communicated as part of the stimulus potential by a source considered reliable by the individual himself, is likely to be the most salient component of the perceptual potential (p. 107).

5.1.b When temporal knowledge is not salient, the stronger the drive, the more likely it is to be perceptually more salient than the events associated with the interval (p. 108).

5.1.c When temporal knowledge is not salient and when the drives associated with the interval are not relatively strong, events are likely to be more salient than drives under either one of the following conditions: (1) when the temporal motive is delayed subsequent (rather than prior, interim, or immediate subsequent) provided that the interval being judged is transitory or extended (rather than ephemeral); or (2) when the interval is ephemeral regardless of when the temporal motive is aroused (pp. 108–9).

5.2 When temporal knowledge is absent and the event potential is salient and hence affects perception, timing acceleration tends to vary at first slightly negatively, and then positively, with increases in the scope of the interval (p. 115).

6.1 The stronger the salient drive before the attainment or non-attainment of the relevant goal (other than the end of the interval itself), the greater the timing acceleration (p. 131).

6.2 The stronger the duration-dependent drive accompanying an interim temporal motive concerning the pending portion of an interval, the greater the timing acceleration when that drive involves the interval's ending, and the greater the timing deceleration when it involves the interval's continuation; the reverse is true when judgment is passed concerning the elapsed portion of an interval (p. 133).

6.2.a The strength of a duration-dependent drive varies positively with a decrease in the perceived duration of the interval or the perceived progress toward the goal (p. 133).

6.3 The stronger the anticipation, the experience, or the recollection that a non-duration-dependent interval will be, is, or has been pleasant as a result of goal-attainment, the greater the timing deceleration; the reverse is true in connection with the unpleasant (p. 137).

6.3.a The tendency for goal attainment to be associated with timing deceleration is more probable when the drive is strong rather than weak; and vice versa (p. 138).

6.3.b The tendency for goal attainment to be associated with timing deceleration is more likely to be the guiding standard than the tendency for strong drives to be associated with timing acceleration; and vice versa (p. 139).

6.4 Unverbalizable cues are more likely to affect the primary judgment than are verbalized or semi-verbalized ones under one or more of the following conditions: when the interval is ephemeral; when it is not consciously experienced; when it is not deliberately organized as an experience or activity; when it is judged speedily (p. 141).

7.1 An objective standard for passing temporal judgment provided by the Method of Chronometry is more likely to be employed than any other standard when it is available as a temporal symbol within the stimulus potential, perceived as temporal information within the perceptual potential, and evaluated as accurate (p. 153).

7.2 The standard that is employed in passing temporal judgment, if available and relevant, is likely to be the one most frequently or recently reinforced in the past (p. 155).

7.3 The accuracy of a temporal judgment in terms of objective time varies positively with the amount of reinforced experience (p. 157).

7.3.a Reinforced experience is likely to be efficacious in aiding the accuracy of temporal judgments when the individual thereby acquires a Guide which provides insight either into the significance of environmental cues or into his own tendency to commit errors (p. 162).

7.4 When motivation is adequate and a temporal motive has been aroused, the various responses associated with the perceptual, primary judgment, and other potentials interact in a form most propitious for passing and expressing the secondary judgment (p. 174).

7.4.a When the temporal motive is delayed and when a previous judgment has been translated into a verbal form, that translation is likely to be remembered and to be utilized again (p. 175).

7.4.b The translation expressing a temporal judgment is likely to be one compatible with the standard being employed and/or the one most recently or frequently reinforced (p. 176).

7.5 The accuracy of a translation expressing a temporal judgment is a function of:
a. The modality being stimulated and hence producing perception
b. The method of measurement being employed under the specified conditions to judge the interval
c. The unverbalizable cues evoked during perception and by the primary judgment

d. The duration of the interval being judged (p. 177).

7.5.b.i Accuracy varies positively with the similarity between the method of perceiving either the interval or the standard (or both) and the method of translation or expression (p. 179).

7.6 With a delayed-subsequent temporal motive, the accuracy of temporal judgment varies inversely but imperfectly with the time elapsing between that interval and the judgment (p. 181).

7.7 A reinforced temporal judgment produces an appropriate change in the temporal potential (p. 185).

8.1 The relevant processes and Guides within the behavior potential which affect the temporal potential, as well as temporal orientation and judgments concerning duration and succession, develop slowly, inevitably, and in culturally determined ways during socialization (p. 215).

8.1.a The effectiveness of the relevant processes and Guides within the behavioral potential affecting the temporal potential as well as judgments concerning duration and succession is increased when they are stored in verbal form and when they are linked to the self (p. 218).

8.1.b The ability to pass temporal judgment, to make more or less accurate temporal judgments, to follow a more or less consistent temporal orientation (which is likely to move from the present into the future or past) improves as a function of the rate at which the psychological, linguistic, and social processes affecting the temporal potential improve (p. 221).

8.2 Any nonmomentary aspect of temporal behavior is correlated with one or more enduring personality traits (p. 231).

8.3 Temporal behavior varies with age only when changes in the state of the organism and in the total potential themselves vary with age (p. 242).

HYPOTHESES

1. People are likely to accept a plan when it is in accord (or not in conflict) with the goals they seek, their predispositions, and the modal personality traits of their society (p. 397).
    1a. People are likely to accept a plan whose demonstrable advantages they can comprehend or anticipate (p. 398).
    1b. People are likely to accept a plan suggested or demonstrated in what they consider a favorable social context (p. 399).
    1c. People are likely to accept and execute a plan when they have confidence in their own ability to carry it out (p. 399).
2. Rapid acceptance of a plan results from learning induced by conversion, transformation, or transplantation (p. 401).

2a. Conversion to a plan may take place when people are so basically dissatisfied that they can ignore whatever conflicts they perceive between the plan and many (but not all) of their predispositions; when they possess some (however few) predispositions and skills compatible with the plan; when they anticipate (regardless of contrary demonstrations in reality) that they thereby will achieve some significant goal; when they are convinced that they can overlook most (if not all) opposition to the plan by their peers; and/or when they have supreme (or desperate) confidence in their ability to execute the plan (p. 401).

2b. Transformation is likely to occur when people desire to perceive some feature of their environment or of themselves differently because they are dissatisfied either generally with their existence or specifically with the situation at hand; when they willingly accept tutelage from others; and when they are capable (actually or potentially) of behaving in the new manner (p. 402).

2c. Voluntary transplantation in connection with planning is likely to occur when people view the change as a way, perhaps the only way, to achieve significant goals, regardless of the losses therefrom; when they believe the anticipated gains to be realizable and find them intelligible; and when they obtain sufficient support from their peers (p. 403).

3. The changes accompanying the execution of a plan are likely to include altered or new modes of behavior which augment (a) sensitivity to relevant aspects of the environment, (b) participation in groups supporting the plan's objectives, and/or (c) skill in adapting to novel situations (pp. 403–4).

4. People who accept a plan are likely to waver while executing it and to be unable to foresee all the consequences from attaining its objectives (p. 404).

AUTHOR INDEX

Aaronson, B. S., 219, 252, 318, 413
Adams, T., 69, 427
Adamson, R., 155, 413
Adler, A., 264, 413
Aiken, L. R., 197, 413
Aisenberg, R., 272, 413
Ajuriaguerra, J., 270, 413
Albers, R. J., 256, 257, 261, 268, 323–24, 413, 439
Alchian, A., 352, 413
Alexander, J., 320, 413
Allen, F. H., 237, 413
Allen, W. R., 352, 413
Ames, F., 313, 413
Ames, L. B., 250, 252, 413
Anast, P., 76–77, 261, 263, 413
Anderson, J. C., 117, 445
Angel, E., 415, 434, 436
Appley, M. H., 125, 416
Apter, D. E., 400, 413
Arieti, S., 50, 216, 413
Armbrester, P. V., 77, 402, 413
Armitt, F., 244, 428
Aronson, E., 65, 413
Aronson, H., 315, 414
Aschoff, J., 70, 206, 207, 414
Atkinson, J. W., 120, 431, 446
Atkinson, R. C., 202, 204, 443
Axel, R., 61, 123, 124, 209, 248, 270, 271, 443

Back, K. W., 77, 132, 414
Baer, P. E., 253, 414
Bailey, R. W., 27, 28, 180, 203, 427
Bakan, P., 124, 152, 190, 193, 195, 208, 414
Baldwin, R. O., 151, 209, 260, 443
Balken, E. R., 321, 414
Bandura, L., 251, 414

Banks, R., 27, 28, 117, 150, 326, 329, 414, 417
Banton, M., 50, 414
Barabasz, A. F., 267, 331, 414
Barach, A. L., 311, 414
Barber, T. X., 283, 290–91, 414
Barndt, R. J., 329–30, 414
Barnett, H. G., 398, 414
Baughman, N. M., 77, 402, 413
Behar, I., 155, 201, 414
Beidelman, T. O., 76–77, 414
Beilin, H., 304, 415
Bell, C. R., 150, 256, 260, 310, 311, 415
Belyaeva-Eksemplyarskaya, S. N., 107, 294, 415
Benda, P., 315, 415
Bender, L., 319, 415
Bendix, R., 246, 415
Benford, F., 236, 415
Benjamin, A. C., 5, 415
Berelson, B., 14, 415
Berg, P. S. D., 326, 437
Bergler, E., 320, 415
Berglund, B., 119, 421
Bergson, H., 383
Berman, A., 61, 148, 415
Bernot, L., 332–33, 415
Bernstein, I., 250, 251, 440
Bettelheim, B., 170, 415
Bevan, W., 155, 201, 415
Bezák, J., 126, 415
Bilger, R. C., 121, 202, 428
Bindra, D., 17, 37, 44, 45, 144, 190, 415, 422
Binswanger, L., 243, 323, 415
Blancard, R., 332–33, 415
Blatt, S. J., 269, 270, 420
Bliss, E. L., 310–11, 415
Blum, R. H., 315, 415
Boardman, W. K., 120, 161, 194, 202, 251,

453

Boardman, W. K. (continued)
  259, 271, 284, 313, 314, 315, 318, 326,
  328, 415, 425, 431, 445, 446
Boas, G., 5, 415
Bochner, S., 100, 246, 415
Böckh, E. M., 279, 421
Boehme, M., 270, 413
Bogdonoff, M. D., 132, 414
Bohannan, P., 55, 415
Bokander, I., 191, 415
Bonaparte, M., 286, 415
Bond, N. B., 72, 416
Bonier, R., 227, 261, 262, 439
Boring, E. G., 6, 12, 26, 105, 205–06, 416
Boring, L. D., 205–06, 416
Bornstein, M., 82, 434
Böszörményi, Z., 315, 416
Boulter, L. R., 125, 416
Bouman, L., 299, 324, 416
Bowra, C. M., 385, 390, 416
Bradbury, P. A., 280, 310, 423
Braddeley, A. D., 310, 416
Bradley, N. C., 220–21, 416
Bramble, W., 427
Bramel, D., 98, 417
Brandon, S. F. G., 374, 409, 416
Braud, W. G., 197, 416
Breasted, J. H., 50, 416
Brim, O. G., 245, 416
Brock, T. C., 253, 330, 364–65, 416, 434
Brodsky, S. L., 330, 416
Bromberg, W., 250, 313, 416
Bronfenbrenner, U., 246, 416
Brower, D., 275, 416
Brower, J. F., 275, 416
Brown, D. R., 28, 118, 190, 202, 204, 416
Brown, J. F., 116, 121, 416
Brown, L. B., 148, 416
Brown, W., 366, 429
Brush, E. N., 204, 416
Bryan, J. F., 65, 416
Buck, J. N., 255, 416
Buckley, J. H., 385, 417
Bukenya, J. M., 166, 232, 439
Bünning, E., 69, 70, 278, 417
Burgat, R., 314, 426
Burns, N. M., 227, 258, 417
Burton, A., 91, 143, 164, 324, 366, 417
Butler, R. N., 239, 282, 417
Butler, S., 83, 417
Butters, N., 119, 121, 417
Button, J., 261, 432

Cahoon, R. L., 258, 311, 312, 417
Caldwell, L. S., 28, 125, 203, 444
Calverley, D. S., 290–91, 318, 414
Campbell, J., 309, 417
Campbell, R. A., 203–04, 441
Campos, L. P., 259, 417
Cantril, H., 130, 417
Cappon, D., 27, 28, 117, 150, 323, 326,
  329, 414, 417
Cardozo, R., 98, 417
Carlson, A. J., 241, 417
Carlson, V. R., 316–17, 417
Carrell, A., 234, 417
Carter, D. E., 196, 417
Cassirer, E., 59, 417
Cattell, J. P., 315, 428
Cervantes, L. F., 304, 417
Chadwick, N. K., 341, 417
Chammah, A. M., 355, 438
Chatterjea, R. G., 151, 247, 417
Chessnick, R. D., 324, 417
Claridge, G. S., 226, 257, 417
Clark, L. D., 310–11, 415
Clark, R. A., 446
Clausen, J., 17, 27, 28, 38, 104, 113, 116,
  141, 178–79, 252–53, 308, 327, 418
Clauser, G., 69, 194, 205, 253, 260, 295,
  418
Cleugh, M. F., 170, 182, 418
Cloudsley-Thompson, J. L., 69, 418
Coheen, J. J., 274, 307, 398, 418
Cohen, J., 8, 10, 28, 121–22, 123–24, 150–
  01, 165, 180, 250, 267, 383, 418
Cohen, L. H., 295, 418
Cohen, R., 255, 265, 432
Cohen, S., 275, 418
Cohen, S. I., 255, 258, 329, 418, 435
Cohn, F. S., 325, 418
Cole, Malvin, 325, 418
Cole, Michael, 166, 424
Collins, W., 259, 438
Coltheart, M., 190, 418
Comte, A., 395, 418
Cooper, L. F., 291–92, 317, 318, 418
Cooper, P., 122, 418
Copland, A. S., 268
Cortes, J. B., 266, 418
Costa, P., 268, 418
Costello, C. G., 313, 418
Cottle, T. J., 201, 210, 246, 252, 269, 418
Cowan, L. G., 92, 419
Craik, K. H., 160, 262, 321, 419, 442

Crawford, M. L. J., 178, 202, 206, 207, 256, 270, 271, 419, 443
Creelman, C. D., 36, 119, 419
Crowell, R. L., 322, 419
Culbert, S. S., 189, 208, 419
Cullmann, O., 345, 419
Cumming, E., 240, 419

Dahl, M., 324, 419
Danziger, K., 119, 193, 419
David, K. H., 100, 246, 419
Davidoff, H., 146, 419
Davids, A., 194, 245, 251, 256, 257, 260, 267, 297, 322, 328, 330, 419
Davidson, G. M., 296, 325, 419
Day, H. I., 137, 419
Deathrage, B. H., 121, 202, 428
de Grazia, S., 419
de Greeff, E., 275, 419
DeGross, E., 210, 438
De La Garza, C. O., 294, 301, 419
Del Guidice, C., 253, 330, 416
Dement, W., 316, 419
de Montmollin, G., 123, 423
Denjean-Raban, G., 116, 123, 204, 270, 444
Denny, L. G., 195, 414
Denys, W., 255-56, 327, 419
De Shon, H. J., 314-15, 420
De Vaux, R., 306, 420
De Wolfe, R. K. S., 123, 124, 190, 204, 420
Dews, P. B., 195-96, 312, 420
Dickstein, L. S., 269, 270, 420
Dilling, C. A., 328, 420
Dimond, S. J., 308, 420
Ditman, K. S., 315, 420
Dmitriev, A. S., 159, 194, 226, 420
Dobson, W. R., 208, 298, 327, 420
Dodge, R., 75, 322, 420
Doehring, D. G., 124-25, 190, 198, 254, 273, 420
Dohne, J. A., 256, 262, 446
Dondlinger, P. T., 237, 420
Doob, L. W., 16, 52, 88, 100, 151, 156, 163, 175, 186, 244, 247, 337, 374, 397, 406, 420
Dooley, L., 325, 420
Dornic, S., 126, 415
Driver, T. F., 385, 420
Dube, C. S., 208-09, 444
DuBois, F. S., 296, 420
Dudycha, G. J., 163, 253, 256, 264, 420

Dudycha, M. A., 163, 420
Duncan, C. P., 123, 124, 163, 190, 204, 420
Dunne, J. W., 24, 288, 420
du Noüy, L., 235, 421
Du Preez, P., 125, 192, 193, 202, 204, 257, 260, 419, 421
Durkee, N., 268, 430
Durrell, L., 5, 421

Earl, W. K., 122, 421
Eckstrand, G., 114, 164, 424
Edelberg, R., 201, 203, 433
Edmunston, W. E., 318, 421
Efron, R., 201, 421
Ehrensing, R., 203, 421
Ehrentheil, O. F., 324, 421
Ehrenwald, H., 292, 309, 317, 421
Eiff, A. W., 279, 421
Einstein, S., 282-83, 421
Eisenbud, J., 288, 421
Eisenman, R., 210, 260, 438
Eissler, K. R., 321, 421
Ekman, G., 119, 421
Eliade, M., 75, 421
Elkin, D. G., 10, 105, 194, 209, 250, 311, 421
Ellenberger, H., 415, 434, 436
Ellis, L. M., 245, 421
Ellis, M. J., 195, 421
Ellis, R., 245, 421
Emley, G. S., 190, 191, 422
Engle, T. L., 275-76, 422
Ennis, W. D., 236, 422
Epley, D., 257, 261, 262, 267, 422
Erbeck, J. R., 318, 421
Erdös, L., 239, 422
Erickson, E. H., 319, 422
Erickson, E. M., 317, 422
Erickson, M. H., 292, 317, 418, 422
Eson, M. E., 79, 163, 253, 268, 273, 422
Essman, W. B., 192, 422
Etienne, F., 170, 422
Evans, W. O., 278, 311, 427
Evart-Chmielniski, E., 116, 123, 204, 270, 444
Everett, K., 155, 413
Ewing, J. H., 422
Eysenck, H. J., 225, 257, 317, 417, 422

Falk, J. L., 144, 190, 422
Farber, M. L., 77, 79, 184, 422
Farrell, M., 248-49, 255, 422

Fason, F. L., 259, 422
Feifel, H., 269, 271-72, 422
Felix, M., 188, 422
Fenichel, O., 320, 422
Ferrall, S. C., 105, 422
Festinger, L., 12, 78, 353, 422, 432, 446
Filer, R. J., 133, 148, 423
Fineberg, I., 316-17, 417
Fink, H. H., 269, 423
Firth, R., 358, 423
Fischer, R., 284, 292, 423
Fisher, R. L., 258-59, 423
Fisher, S., 258-59, 423
Flavell, J. H., 247, 423
Foerster, L. M., 248, 423
Folkins, C. H., 68, 423
Forer, R. A., 245, 416
Forgy, E. W., 315, 420
Forman, H. J., 341-42, 423
Fougerousse, C. E., 28, 132, 259, 327, 435
Foulke, E., 190, 444
Foulkes, W. D., 316, 423
Fox, R. H., 280, 310, 423
Fraenkel, F., 313, 423
Fraisse, P., 64, 81, 90, 111, 112, 123, 142, 151, 161-62, 165, 182, 195, 198, 202, 241, 246-47, 248, 249, 250, 252, 253, 267, 300, 423
Fraisse, R., 151, 423
François, M., 277, 423
Frank, L., 10, 54, 423
Frankenhaeuser, M., 27, 35, 37, 114, 118, 119, 124, 126, 207-08, 276-77, 313, 423
Franz, M.-L., 287, 423
Fraser, L. T., 415, 418, 423, 436
Frauchiger, R. A., 123, 126, 423
Fredericson, E., 137, 423
Freed, H., 308, 442
Freedman, A. M., 433
Fress, P., 36, 161, 423
Freud, S., 215, 282, 285-86, 290, 319-20, 404, 423
Friedman, K. C., 174, 197, 249, 255, 267, 424
Friedman, M., 275, 424
Friel, C. M., 112, 123, 170, 259, 262, 424
Frobenius, K., 194, 205, 424
Frohwirth, R., 331, 435
Fuller, E. A., 273, 420
Funk, W. A., 315, 420
Fuschillo, J., 225, 256, 433

Galanter, E. H., 104, 443

Garbutt, J. T., 229, 424
Gardner, W. A., 309, 424
Carner, W. R., 193, 424
Gaschk, J. A., 109, 424
Gavini, H., 113, 116, 253, 424
Gay, J., 166, 424
Geer, J. H., 209, 258, 424
Geiwitz, P. J., 253, 263, 292, 424
Gerard, E., 65, 413
Géraud, J., 327, 424
Gergen, J., 77, 414
Gesell, A., 250, 424
Gibbens, T. C. N., 87, 424
Gifford, E. C., 227, 258, 436
Gilligan, C., 302, 436
Gilliland, A. R., 27, 114, 127, 164, 180, 190, 191, 194, 198, 209, 253, 312, 340
Gleich, J., 71, 434
Globus, G. C., 316, 424
Goethe, J. W., 309
Goldberger, L., 314, 424, 428
Goldfarb, J. L., 26, 44, 117, 119, 120, 161, 177, 178, 201, 209, 250, 251, 314, 322, 326, 424, 425, 433
Goldman, R., 233, 425
Goldrich, J. M., 59, 425, 432
Goldstein, K., 309, 425
Goldstone, S., 5, 26, 44, 117, 119, 120, 161, 177, 178, 190, 194, 196, 198, 201, 202, 203, 209, 222, 250, 251, 253, 256, 259, 271, 284, 309-10, 313, 314, 315, 318, 325-26, 328, 414, 415, 424, 425, 426, 431, 433, 442, 444, 445, 446
Goodchilds, J., 146, 425
Gooddy, W., 278, 425
Goodenough, D. R., 316-17, 417
Goodfellow, L. D., 201, 425-26
Goodman, P. S., 366-67, 426
Goodwin, J. C., 279-80, 429
Göpfert, H., 279, 421
Goth, A., 312, 426
Gothberg, L. C., 275, 426
Gottheil, E., 340, 426
Gouldner, A. W., 404, 426
Grassmück, A., 196-97, 204
Green, H. B., 265, 266, 426
Green, L. R., 265, 444
Greenberg, L. F., 208-09, 444
Greene, J. E., 245, 426
Greenfield, N., 268, 444
Gregg, L. W., 121, 192, 426
Gridley, P. E., 190, 192, 201, 426
Grier, E. S., 304, 426

Griffin, F., 284, 423
Grimm, K., 119, 120, 124, 426
Gross, A., 233, 286, 426
Gross, L. P., 233, 440
Grossman, J. S., 228, 254, 272, 426
Grünbaum, A. A., 299, 324, 416
Guertin, W. H., 327, 426
Guilford, J. P., 39, 426
Gulliksen, H., 123, 124, 125, 426
Gunn, J. A., 5, 426
Gurvitch, G., 4, 426
Gutheil, E. A., 286, 426
Guyau, M., 188, 214, 426
Guyotat, J., 314, 426

Haas, W. S., 58, 92, 426
Hagen, R., 326, 414
Halberg, F., 71, 430
Hall, E. T., 25, 53, 63, 67, 85, 100, 176, 189, 426
Hall, W. W., 164, 204, 426
Hallenbeck, C. E., 228, 254, 272, 426
Halstead, W. C., 256, 426
Halter, A. N., 359, 429
Hamlett, I. C., 275–76, 422
Hamner, J. C., 69, 70, 426
Hampton, I. F. G., 280, 310
Hansel, C. E. M., 121–22, 267, 418
Hardy, C. O., 355, 426
Hare, R. D., 198, 227, 258, 427
Hárnik, J., 320, 427
Harrell, T. W., 198, 427
Harrison, M. L., 255, 427
Hart, A. G., 359, 427
Harton, J. J., 113, 118–19, 146–47, 209, 254, 427
Hauty, G. T., 69, 427
Hawickhorst, L., 196–97, 204, 427
Hawkes, G. R., 27, 28, 180, 203, 278, 311, 427
Hawkins, N. E., 123, 124, 209, 427
Hearnshaw, L. S., 4, 427
Heath, L. R., 4, 427
Heckhausen, H., 266, 427
Heimann, H., 36, 312, 427
Heirich, M., 64, 427
Helmer, J. E., 273, 420
Henrickson, E. H., 151, 427
Henry, F. M., 103, 427
Henry, W. E., 240, 427
Heron, W. T., 159, 427
Herzog, G., 145, 427
Hicks, R. A., 427

Hilgard, E. R., 293, 428
Hindle, H. M., 143, 428
Hirsh, I. J., 103, 121, 202, 428
Hitchcock, L., 28, 118, 190, 202, 204, 416
Hoagland, H., 277, 310, 311, 428
Hoch, P. H., 315, 415, 428, 433
Hoche, A., 300, 428
Hofeld, J., 114, 164, 418
Hoffer, A., 283, 315–16, 428
Holborn, S. H., 197, 416
Holt, R. R., 36, 314, 424, 428
Homack, W., 201, 428
Honigfeld, G., 283, 428
Horányi-Hechst, B., 300, 428
Horst, L., 276, 428
Horstein, A. D., 123, 181, 209, 428
Hottes, J. H., 82, 434
Hovey, A., 306, 428
Hovland, C. I., 89, 428
Howard, P., 252, 418
Hoyle, J., 119, 121
Hubert, H., 306, 428
Humphreys, W., 27, 180, 194, 198, 209, 424
Hunt, J. M., 436
Hutchinson, T., 441
Hyatt, J. P., 341, 428
Hyde, R. W., 428

Iacono, G., 150, 428
Ikard, F., 256, 262, 446
Ikeda, S., 250, 428
Ilg, F., 250, 424
Inkeles, A., 81, 428
Inman, W. B., 185, 428
Irvine, S. H., 145–46, 428
Irwin, F. W., 244–45, 428–29
Ise, M., 146, 429
Israeli, N., 79, 116, 321, 429

Jacoby, J., 262, 429
Jacques, E., 366, 429
Jaensch, E. R., 116, 125, 208, 429
Jaffe, M., 233, 425
Jahoda, M., 366, 429
James, E. O., 306, 429
James, W., 34, 167, 179, 189, 219, 236, 237, 280–81, 429
Janet, Paul, 33, 235, 429
Janet, Pierre, 92, 429
Janis, I. L., 89, 429
Jasper, H., 149, 429
Jaspers, K., 4, 296–97, 429
Jenner, F. A., 279–80, 429
Jenny, P. B., 324, 421

Jensen, H. R., 359, 429
Jerison, H. J., 116, 429
Jernigan, C., 120, 259, 429
Jirka, Z., 207, 429
Jöel, E., 313, 423
Joerger, K., 313, 429
Johannsen, D., 407, 429
Johnson, D. M., 329–30, 429
Johnson, E. E., 209, 255, 429
Johnson, G. L., 359, 429
Johnston, H. M., 326, 429
Jones, A., 118, 190, 429
Jones, Ernest, 319, 429
Jones, Evelyn, 119, 121, 417
Jones, R. E., 308, 430
Joy, R. J. T., 278, 311, 427
Judson, A. J., 245, 247, 430
Juster, F. T., 393, 430

Kafka, J. S., 147, 163, 164, 253, 273, 430
Kagan, J., 311, 414
Kahn, E., 75, 322, 420
Kahn, P., 259, 262, 263, 267, 298, 430
Kahn, R. L., 298, 445
Kahn, T. P., 322, 419
Kahnt, O., 198, 430
Kaiser, I. H., 71, 430
Kaplan, H. I., 433
Kar, S. B., 393, 430
Karpov, G. S., 194, 420
Kastenbaum, R., 7, 75, 77, 239, 240, 242, 243, 253, 256, 261, 262, 268, 269, 269–70, 418, 421, 430, 438, 439
Katchmar, L., 192, 439
Katona, G., 357–58, 395, 430
Kaunda, K. D., 228, 430
Keil, C., 351, 366, 430
Kelley, H. H., 89, 428
Kelm, H., 203, 430
Kendall, M., 246, 430
Kenna, J. C., 117, 118, 257, 314, 430, 439
Kerr, W. A., 351, 366, 430
Kety, S. S., 235, 430
Kew, J. K., 27, 190, 193, 325, 444
Kidd, A. H., 425
Kidder, C., 330, 419
Kiker, V. L., 123, 126, 435
Kilbride, P. L., 166, 226, 439
King, H. E., 181, 431
King, S. M., 121, 427
Kintz, B. L., 109, 424
Kipnis, D., 259, 431
Kirkham, J., 284, 313, 314, 419, 431

Kleba, F., 190, 193, 414
Kleber, R. J., 309–10, 431
Klee, C. D., 315, 414
Kleemeier, R. W., 442
Kleiser, J. R., 190, 431
Kleitman, N., 70, 316, 419, 431
Klien, H., 274, 293
Klineberg, S. L., 245, 252, 322, 431
Kloos, G., 320, 431
Kluckhohn, C., 75, 431
Kluckhohn, F. R., 56, 77, 431
Knapp, R. H., 229, 265, 266–67, 426, 431
Knight, E. J., 294, 328, 435
Knight, F. H., 353, 431
Kochigina, A. M., 71, 159, 226, 420
Koehnlein, H., 196, 431
Kogan, N., 356, 431
Kohlberg, L., 289, 431
Kohlmann, T., 27, 431
Koht, A. G., 143, 439
Kolaja, J., 392, 431
Kolb, D., 64, 441
Korngold, S., 113, 192, 253, 431
Korzybski, A., 218, 377, 431
Kowalski, W. J., 208, 432
Krauss, H. H., 11, 257, 258, 261, 262, 263, 432, 440
Kretz, A., 116, 125, 208, 429
Kruup, K., 195, 432
Kubler, G., 373, 376–77, 432
Kuhlen, G., 241, 438
Kulik, J. A., 245, 256, 330, 442
Kurz, R. B., 149, 228, 255, 264–65, 432
Kutschbach, W., 170, 422
Kyle, L. C., 394, 432

Lair, C. V., 295, 437
Landy, D., 65, 413
Langen, H., 116, 432
Langer, J., 144, 432
Lanzkron, J., 324, 432
Lapuc, P. S., 265, 431
Laties, V. G., 196, 432
Lauterbach, C. G., 340, 426
Lawrence, D. H., 12, 432
Lazarsfeld-Jahoda, M., 367, 432
Leach, E. R., 49, 85, 432
LeBlanc, A. F., 268, 271, 432
Lefcourt, H. M., 78, 432
Legg, C. F., 308, 432
Le Guen, C., 324, 432
Lehmann, H., 33, 314, 321, 325, 328, 329, 432

Lehr, U., 269, 432
Leister, G., 178, 432
Le Shan, L. L., 245, 432
Lessing, E. E., 246, 263, 432
Lessing, G. E., 376, 432
Levine, D., 296, 432
Levine, M., 225, 226, 233, 254, 256, 259, 263, 433, 442, 443
Levine, R., 331, 435
Lewin, K., 11, 24, 172
Lewis, A., 33, 36, 296, 322, 329, 433
Lewis, M. M., 252, 433
Lewis, P. R., 278, 433
Lewtz, W., 117, 433
Lhamon, W. T., 26, 112, 120, 123, 161, 170, 194, 196, 198, 201, 202, 203, 251, 259, 271, 284, 309–10, 313, 315, 318, 326, 415, 421, 424, 425, 431, 433
Lichtenberg, P., 93, 433
Lichtenstein, M., 202, 445
Lieberman, M. A., 268, 433
Liebert, R. M., 64, 436
Lincoln, J. S., 287, 433
Lindblom, J., 341, 433
Linder, S. B., 347, 349
Linn, L., 300, 433
Linton, R., 333, 433
Lippitt, R., 337, 433
Lipset, S. M., 246, 433
Liss, L., 284, 423
Llewellyn-Thomas, E., 26, 433
Lobban, M. C., 278, 433
Locke, E. A., 65, 416
Lockhart, R. A., 45, 149, 279, 310, 433
Loeb, M., 190, 433, 444
Loehlin, J. C., 187, 190, 209, 253, 433
Longuet-Higgins, H. C., 274, 433
Loomis, E. A., 317, 434
Loosley, E. W., 54, 76, 441
Loranz, M., 205, 437
Lovell, K., 222, 251, 434
Lowell, E. L., 446
Lowin, A., 82, 434
Lucchesi, B. R., 190, 191, 422
Luoto, K., 257–58, 327–28, 434
Lynch, R. D., 315, 420
Lynen, J. F., 380, 434
Lynn, R., 257, 434
Lysgaard, E., 245, 434
Lyttle, J. W., 151, 438

Mabbott, J. D., 5, 434
McCann, W. H., 264, 434

McClelland, D. C., 228–29, 266, 418, 427, 434, 446
McCraven, V. G., 259–60, 441
MacDougall, R., 209, 434
McFie, J., 307, 309, 434
McGeoch, J. A., 123, 126, 209, 271, 443
McGill, T. E., 117, 206–07, 444
MacGrady, G. J., 196, 417
McGrath, J. J., 158, 253, 434
MacIver, R. M., 5, 134, 167, 239, 244, 287, 434
MacLean, M., 118, 190, 429
MacLeod, R. B., 112, 150, 206, 434
McNutt, T. H., 181, 256, 434
McTaggart, 167, 434
Mahrer, A. R., 100, 214, 434
Malcolm, N., 289, 434
Málek, J., 71, 434
Malý, V., 71, 434
Mandel, E. D., 245, 434
Mann, T., 15, 383
Mannheim, K., 399, 434
Marshall, H. R., 25, 434
Marti, V. A., 197, 424
Martin, R., 127, 191, 198, 253, 424
Marum, K. D., 117, 198, 200, 203, 434
Marx, K., 339
Matsuda, F., 434
Matulef, N. J., 253, 364–65, 434
Mauss, M., 306, 428
May, R., 314, 323, 329, 434
Mayo, B., 35, 434
Mbiti, 74–75, 78
Meade, R. D., 139, 143, 144, 147–48, 230, 366, 434
Meals, D. W., 133, 143, 423
Meerloo, J. A. M., 54, 57–58, 96, 296, 319, 435
Megargee, E. I., 331, 435
Melges, F. T., 28, 132, 259, 327, 435
Melikian, L., 100, 246, 435
Melvin, K. B., 181, 256, 435
Mendilow, A. A., 93, 376, 377–78
Merton, R. A., 222, 442
Mettler, F. A., 418
Metzner, R., 246, 247, 436
Meumann, E., 116, 435
Meyer, A., 7, 435
Meyer, E., 27, 123, 124, 179, 311, 427, 435
Meyerhoff, H., 33, 385, 435
Mezey, A. G., 28, 151, 255, 258, 294, 328, 329, 435
Michaud, E., 249, 435

Michon, J. A., 104, 111–12, 126, 435
Miller, A. R., 123, 126, 435
Miller, G. A., 154, 438
Miller, N. E., 133, 436
Millikan, R. A., 82, 436
Mills, J. N., 207, 309, 436
Minkowski, E., 4, 6, 60, 296, 300, 321, 328, 436
Mischel, W., 100, 246, 247, 302, 436
Mitchell, M. B., 117, 207, 258, 260, 436
Mönks, F. J., 269, 329, 436
Moore, W. E., 82, 84, 87, 228, 240, 344–45, 395, 396, 436
Moos, R. H., 204, 258, 445
Moris, J. R., 77, 325, 402, 413
Moron, P., 327, 424
Morris, J. R., 27, 190, 193, 444
Morris, L. W., 64, 436
Morse, W. H., 195–96, 312, 420
Moss, T., 315, 420
Mowbray, R. M., 324–25, 436
Mowrer, O. H., 11, 436
Moyer, J. H., 422
Mozdzierz, G. J., 261, 432
Murchison, C., 433
Murdock, G. P., 55, 436
Murphy, V. M., 270, 443
Murray, H. A., 431
Myers, G. C., 115, 123, 436

Nakamura, H., 74, 436
Nangle, L. G., 195, 414
Nawas, M. M., 264, 436
Nebel, G., 409, 436
Nelsen, J. M., 313, 445
Nettleship, A., 295, 437
Neumann, W., 188, 437
Nichols, H., 4, 437
Nilsson, M. D., 49, 75, 437
Nissen, H. W., 305–06, 437
Nitardy, F. W., 235, 437
Nodine, J. H., 422

Oakden, E. C., 249, 250, 437
Oberndorf, C. P., 319, 437
Ochberg, F. W., 27, 179, 311, 437
O'Connell, J., 92, 419
O'Hanlon, J., 158, 253, 434
Oléron, G., 119, 173, 437
Omwake, K. T., 205, 437
Ono, A., 122, 418
Orchinik, C. W., 244–45, 308, 437
Orme, J. E., 29, 103, 132, 178, 209, 230, 232, 257, 281, 284, 297, 309, 328, 437

Ornstein, R. E., 127–29, 183, 437
Orsini, F., 193, 195, 246–47, 315, 415, 423, 437
Osmond, H., 283, 315, 315–16, 428, 442
Ostheimer, J. M., 229, 265, 266, 437

Paige, D. D., 232, 437
Pankaukas, J., 191, 437
Parenti, A. N., 297, 322, 330, 419
Parks, T. E., 122, 437
Parsons, O. A., 123–24, 275, 418
Partridge, M., 308, 437
Patton, R. M., 120, 202, 204, 437
Pearl, D., 326, 437
Pearson, K., 23, 437
Pennes, H. H., 315, 428, 437
Petrie, A., 259, 308, 316, 437
Pettit, T. F., 260, 438
Pfaff, D., 72, 310, 438
Pfleiderer, F., 279, 421
Philip, B. R., 151, 438
Philips, D. P., 67, 438
Piaget, P., 142, 247–49, 250, 438
Pick, A., 438
Piéron, H., 162, 438
Pistor, F., 255, 438
Pivik, T., 316, 438
Platt, J. J., 210, 260, 436, 438
Platt, P. E., 209, 258, 424
Pleck, J., 210, 246, 252, 269, 418
Polak, F. L., 6, 24, 438
Pollack, I. W., 27, 179, 311, 437
Pollock, K., 269, 438
Poole, D. M., 195, 440
Postman, L., 154, 200, 438
Poulet, G., 383, 438
Pressey, S. L., 241, 438
Price, A. C., 331, 435
Priestley, J. B., 288, 438
Prior, A., 6, 438
Proust, M., 171–72, 383, 438
Provins, K. A., 150, 311, 415
Pumpian-Mindlin, E., 209, 270–71, 327, 438

Rabin, A. I., 17, 28, 110, 217, 230, 327, 328, 420, 426, 438, 444
Rakshit, P., 417
Rapoport, A., 355, 438
Read, M., 50, 54, 96, 438
Rebello, S., 250, 439
Redl, F., 83, 303, 330, 331, 439
Reed, G. F., 117, 118, 257, 439
Reich, M., 330, 419
Reivich, R. S., 261, 440

Reutt, N., 125-26, 439
Reynolds, H. N., 123, 439
Richard, J., 270, 413
Richards, W., 103, 104, 221, 439
Richelle, M., 255-56, 327, 419
Ricks, D., 257, 261, 262, 422
Rimoldi, H. J. A., 224, 439
Rinkel, M., 314-15, 420
Rivoire, J. L., 425
Rizzo, A. E., 295-96, 439
Robbins, M., 166, 226, 439
Roberts, A. H., 245, 426
Roby, T. B., 146, 425
Rodgin, D. W., 318, 418
Roelofs, O., 115, 119, 187, 439
Roff, M. F., 112, 150, 206, 434
Rogers, E. M., 398, 439
Rogers, K. W., 250, 439
Róheim, G., 320, 415
Rokeach, M., 227, 261, 262, 439
Roos, P., 256, 257, 268, 323-24, 439
Rose, A. A., 264, 439
Rosenfelt, R. H., 269, 439
Rosenthal, D., 322, 419
Rosenzweig, S., 143, 439
Ross, S., 192, 439
Rossomando, N. A., 11, 439
Roth, J. A., 80-81, 439
Rotter, J. B., 78, 123, 181, 428, 439
Rouart, J., 319-20, 439
Rubin, E., 6, 439
Rubinstein, L., 312-13, 439
Ruiz, R. A., 257, 258, 261, 262, 263, 432, 440
Rustin, S. L., 241, 439
Rutschmann, J., 312-13, 440

Sabin, T. R., 160, 245, 256, 330, 419, 442
Sandler, B. E., 82, 434
Sattler, J. M., 148, 263, 440
Sauvageot, A., 75, 440
Scanlon, D. G., 92, 419
Schachter, S., 233, 425, 440
Schaefer, V., 190, 312, 440
Schaeffer, M. S., 245, 421
Schaltenbrand, G., 102, 440
Schechter, D. E., 250, 251, 440
Schiff, W., 148, 440
Schilder, P., 291, 440
Schjelderup, H. K., 317, 440
Schlosberg, A., 321, 440
Schmid, N., 200, 440
Schmidt, E., 192, 202, 204, 208, 440
Schneck, J. M., 293, 440

Schneevoigt, W., 257, 440
Schneider, D. E., 287, 440
Schneider, L., 245, 440
Schoeffler, M. S., 194-95, 440
Schönbach, P., 132, 150, 440
Schonbar, R. A., 263, 440
Schönpflug, 125, 440
Schreiber, D., 426
Schuster, C. R., 190, 191, 422
Scott, W. C. M., 150, 200-01, 319, 441
Scotti, J., 241, 445
Sedman, G., 314, 430
Seeley, J. R., 54, 76, 441
Shagass, C., 149, 441
Shakow, D., 322, 419
Shectman, F. A., 190, 193, 441
Shelley, P. B., 6, 281, 441
Sheridan, M., 103, 429
Sherrick, C. E., 103, 428
Shybut, J., 321, 441
Sibley, R. F., 246, 430
Sidman, J., 194, 267, 441
Siegel, L., 275, 442
Siegman, A. W., 27, 28, 113, 118, 225, 256, 258, 263, 302, 330, 331, 441
Siffre, M., 117, 207, 441
Siipola, E., 64, 441
Silverstein, A. B., 315, 414
Sim, A., 54, 76, 441
Simmel, E. C., 145, 441
Simon, C. W., 244, 428
Simonson, E., 287, 441
Sinclair, H., 270, 413
Singer, J. L., 215, 259-60, 264, 441
Singer, M., 209, 258, 424
Singh, L., 147, 435
Sixtl, F., 190, 441
Skramlich, E., 197, 441
Slater, A., 222, 251, 434
Slater, P. E., 269, 439
Small, A. M., 203-04, 441
Smart, R. G., 323, 441
Smeltzer, W. E., 323, 441
Smith, A. K., 116, 429
Smith, M. W., 79, 441
Smith, R. J., 75, 239, 441-42
Smith, R. P., 28, 125, 203, 444
Smith, S., 117, 442
Smythe, E. J., 209, 251, 256, 271, 442
Solanes, J., 188, 442
Sollberger, A., 4, 442
Solomon, H. C., 314-15, 420
Solomon, P., 259, 442

Sommer, Geraldine, 245, 421
Sommer, Gerhart, 245, 421
Sorokin, P., 222, 442
Spencer, L. T., 116, 442
Spengler, O., 8, 219, 442
Spiegel, E. A., 308, 442
Spilka, B., 28, 226, 263, 442
Spivack, G., 225, 226, 233, 254, 256, 259, 263, 433, 442, 443
Spreen, O., 190, 253, 255, 258, 442
Sprigle, H., 256, 442
Springer, D. V., 158, 249, 442
Stalnaker, J. M., 317, 442
Starzynski, 255, 265, 432
Stefaniuk, B., 315, 442
Steffen, T., 279, 421
Stein, K. B., 245, 256, 262, 321, 330, 442
Steinberg, H., 46, 150, 316, 442
Steiner, G. A., 14, 415
Stern, M. H., 309, 442
Stern, W., 4, 442
Sterne, L., 378, 442
Sternlicht, M., 275, 442
Sterzinger, O., 313–14, 316
Stevens, S. S., 104, 442, 446
Stott, L. H., 121, 191, 443, 446
Strauss, E. W., 322, 443
Strodtbeck, F. L., 56, 76, 77, 431, 443
Stromberg, G., 289, 443
Strughold, H., 69, 443
Strunk, O., 190, 443
Sturt, M., 110, 112, 142, 249, 250, 437, 443
Sumner, F. C., 204, 254, 446
Sweet, A. L., 203, 443
Swift, E. J., 123, 126, 209, 271, 443
Sylvester, J. D., 121–22, 267, 418
Symonds, M., 250, 251, 440
Szára, S., 315, 416
Sztulman, H., 327, 424

Talland, G. A., 276, 296, 443
Tanner, T. J., 120, 202, 204, 443
Tauber, I. J., 279–80, 429
Tavernier, A., 225, 256, 430
Taylor, R. E., 263–64, 438
Taylor, V., 64, 441
Teahan, J. E., 267, 443
Tejmar, J., 270, 271, 443
Thayer, S., 149, 440
Thompson, D. C., 226, 233, 443
Thompson, R. W., 109, 424
Thomson, W. A. R., 312, 443

Thor, D. H., 72, 151, 178, 202, 206, 207, 209, 256, 260, 270, 271, 414, 419, 443
Thornton, A., 385, 443
Thornton, H., 385, 443
Tissot, R., 270, 413
Tiyakian, E. A., 426
Tolor, A., 270, 443
Toulmin, S., 62, 342, 443
Treisman, M., 103, 190, 443
Triplett, D., 38, 116, 200, 443
Troyer, W. G., 132, 443
Tscholakow, K., 295, 325, 444
Turchioe, R. M., 207, 444
Tuthill, 318, 418
Tuttle, C. E., 245, 247, 430
Tutuola, A., 58, 444
Tyndel, M., 323, 417

Ullman, A. D., 11, 436
Ulseth, S., 427
Urban, F. M., 61, 209, 444, 446
Usizima, Y., 257, 444

Valousek, C., 207, 429
Varga, E., 215, 444
Vaschide, N., 205, 444
Verdone, P., 317, 444
Vernon, J. A., 117, 206–07, 444
Vischer, A. L., 239, 444
von Sturmer, G., 190, 193, 201, 418, 444

Waksberg, H., 17, 37, 44, 45, 415
Walker, C. R., 351, 444
Walker, W. N., 64, 441
Wallace, M., 17, 110, 218, 230, 321, 444
Wallach, M. A., 265, 356, 431
Wallis, R., 393, 444
Wallon, H., 116, 123, 204, 270, 444
Wapner, S., 144, 432
Warm, J. S., 27, 28, 125, 180, 190, 193, 203, 209–10, 325, 444
Warman, R. E., 253, 364–65, 434
Warner, W. L., 55, 444
Waszak, M., 119, 421
Watts, A. N., 256, 260, 415
Webb, W. W., 326, 444–45
Weber, C. O., 125, 444
Weber, D. S., 209, 444
Weber, M., 94, 361, 444
Webster, F. R., 326, 444–45
Weil, A. T., 313, 445
Weinstein, A. D., 326, 328, 445
Weinstein, E. A., 298, 445

Weiss, B., 196, 432
Weiss, J., 244–45, 428
Weiss, W., 11, 439
Werboff, J., 311, 445
Werner, H., 144, 219–20, 395, 432, 445
West, C. D., 310–11, 415
Wever, R., 207, 414
Weybrew, B. B., 125, 445
White, C. T., 202, 445
Whitely, P. L., 117, 445
Whitrow, G. J., 5, 75, 92, 445
Whyman, A. D., 204, 258, 445
Wilen, F., 239, 445
Wilensky, H., 259–60, 441
Williams, S. G., 321, 445
Willis, J. E., 256, 262, 446
Wilson, S. R., 132, 414
Wineman, D., 83, 303, 330, 331, 439
Wing, L. W., 70, 445
Wirth, W., 162–63, 445
Withey, S. B., 92, 445
Wohlford, P., 253, 445
Wolfson, W., 324, 432
Wolk, R. L., 241, 445
Wolpert, E. A., 316, 419
Woodrow, H., 35, 102, 105, 120, 126, 126–27, 150, 190, 191, 193, 198, 200, 207, 445

Woods, T. D., 359, 429
Woolf, V., 33, 383, 384, 446
Worchel, P., 294, 301, 429
Wright, D. E., 151, 209, 260, 414
Wright, D. J., 326, 446
Wukasch, D. C., 253, 414
Wycis, H. T., 308

Yaryan, R. B., 78, 446
Yates, S., 218, 446
Yerkes, R. M., 60, 209, 446
Young, S. A., 204, 254, 446

Zahn, G. C., 305, 446
Zangwill, O. L., 324, 325, 418, 446
Zatzkis, J., 266, 446
Zeeman, W. P. C., 115, 119, 187, 439
Zeff, C., 158, 446
Zeisl, H., 367, 432
Zelkind, I., 28, 119, 226, 263, 446
Zern, D., 61–62, 218, 255, 446
Zinberg, N. E., 313, 445
Zsambok, C., 119, 121, 417
Zubin, J., 415, 433
Zuili, N., 248, 446
Zunin, L. M., 315, 420
Zurcher, L. A., 256, 262, 446

SUBJECT INDEX

References to entries marked with an asterisk are so frequent that only the significant ones are listed.

Ability. *See* Skill
Abnormality, mental, 226, 258, 259, 261–62, 293–301, 314, 319–29, 409
Aborigines (Australia), 100
Absolute judgment, method of, 26, 34, 153, 154–55, 156–57, 161, 185
* Acceleration vs. deceleration: abnormality, 298; definition, 41, 42, 43; drive, 133; drugs, 284, 312–16; events, 109–10; goal, 137, 138, 139; hypnosis, 318; scope, 115; temperature, 274; thyroid, 309–10. *See also* Estimation, over- vs. under-
Acculturation, 405
* Accuracy: abnormality, 298, 325–28; biochemical clock, 204–07; cerebral lesions, 308; factors affecting, 177–81, 183, 190–99, 255–56, 270–72, 275, 317; measurement methods, 19–20, 27–28, 203–04; secondary judgment, 158–67; sense modality, 201–03
Achievement: motivation, 11, 98–100; need for, 100, 147, 228–30, 265–67
Adjustment, method of, 25
Adrenochrome, 316
Advertising, 353–55, 372
Aesthetics, 371, 374, 377, 379, 380, 382
Africa, 4, 14, 16, 61, 62, 74–75, 78, 100, 165, 228, 386, 398
Age: abnormality, 322, 324–25; biochemical processes, 32; deferred gratification, 244–45; development, 217–22; drugs, 283; grading, 333, 335, 365; young vs. old, 32, 60, 185, 194, 234–44, 267–72, 409
Aggression, 90, 326, 392
Agnosia, 296, 325

Alcohol, 313, 314
Alcoholism, 196, 266, 277, 307, 315, 323–24
Altitude, 311
Amphetamine, 128, 289
Anality, 216, 259, 260, 297, 320
Ancestors, 51, 53, 55, 93, 94, 186, 358, 364
Animals, 3, 50, 152, 305–06; conditioning, 10, 49, 159, 214; learning, 11, 133, 195, 392; rhythm, 69–71
Anthropology, 25, 48, 49, 58, 75, 85, 99, 100, 145–46, 285, 333, 339, 400
* Anticipation, 8, 33, 92, 94, 96, 175, 222, 332, 407, 409, 410; art, 373, 379; change, 398, 401, 403, 404; crime, 301, 303; decisions, 353; dreams, 285, 289; drive, 90, 98–99, 114, 130, 137, 144–45, 149, 172; drugs, 283; group, 337, 338–39; intervention, 50–52; learning, 10–11, 49–50, 62, 184, 213–14, 216, 244; motive, 72; planning, 392–95, 396; religion, 343; risks, 357, 359; role, 334; traditional societies, 53, 58; work, 347–49
Anxiety, 134, 208, 226–27, 232, 258, 261, 279, 285, 288, 298, 312, 323, 325, 411
Aphasia, 309
Arabs, 67, 100
Archaeology, 50, 75, 186
Architecture, 376–77, 378
Argentina, 81
Aristotle, 82, 377
Arts, 371–391; arresting of time, 374–77; suspension, 377–83
Association, free, 269–70
Astrology, 342
Asynchrony, 69
Attention, 104, 188

* Attitude, 13, 32; activity-passivity, 56–57; death, 269–70; effects, 28, 65, 104, 283, 296, 395; time, 73, 222, 223, 227–28, 240–41, 264–65, 366–67
Audition, 102, 120, 121, 155, 178, 201–03, 310, 313, 318
Austria, 246, 367
Autism, 294
Autokinetic effect, 147
Average error, method of, 25
Awakening: from sleep, 69, 72, 179–80, 194, 204–05; temporal judgment after, 140, 141–42, 205–06
Awareness, temporal, 24, 167–72, 184, 315
Aztecs, 306

Balance: cognitive, 405; temporal, 411
Behavior. See Potential, behavioral
* Belief, 13, 32, 148, 341
Bergson, H., 92–93, 383
Biochemical: clock, 32, 45, 68–73, 82, 85, 153, 179–80, 204–07, 234–35, 253, 260, 277–80, 284, 292, 308, 309, 311–12; processes, 31, 32, 234–35, 243, 273–82, 280–82, 284–85, 408
Biology, 4
Birth, order of, 210
Bondei (Tanzania), 265, 266
Boredom, 65, 110, 128, 166, 183, 184, 187, 192–93, 292; impatience, 167–72; repetition, 366; work, 350–51
Borrowing and lending, 354, 360–61
Brain, 280, 295; localization, 273–74, 307–09
Budgeting, 394

Caffeine, 313
Calendar, 4, 49, 61, 64, 67, 84, 153, 154, 156, 157–58, 221, 249, 344, 347
Canada, 54, 76
Capital, 360–62
Carbon-dating, 62
Carving, 371, 373, 375, 379, 386, 389
Catatonia, 295, 319
Catholicism, 64
Chagga (Tanzania), 265, 266, 267
Change, 392, 409; concomitant, 403–05; gradual, 397–400, 405; rapid, 400–03, 405; traditions, 336–37
Channel, communication, 31, 33, 101, 337, 408
Chemistry, 391
Chlordiazepoxide, 315

Chlorpromazine, 316
Christianity, 57, 74–75, 345–46, 375, 398
Chronaxie, 4
Chronometry, method of, 17, 153–54, 157–59, 204, 298
Chronos, 382
Class, social, 244, 245–46, 269, 302, 304
* Clock. See Biochemical, clock; Timepieces
Comic strips, 389
Communication: age, 244; art, 374, 375, 391; leaders, 340; mass, 263, 354; reactions, 399, 410; temporal factors, 87–90, 409; time as medium, 63–66; traditional societies, 51
Communism, 338, 401
* Comparison, method of, 18–19, 25, 27–29, 38, 42, 43–45, 116, 120, 127, 155
Compensation, 164, 198
Conditioning, 10, 36, 49, 70–71, 141, 149, 152, 159, 161, 227
Confabulation, 298
Confidence, 399–400, 401, 404
Conformity, 81–85, 96, 335–36, 337, 344–45
Consequences, unanticipated, theory of, 404
Conservation, 362–64
Conservatism, 393
Consistency, 68, 204, 223–24, 252–55, 261, 274
Contemplation, 371–72, 373
Context, stimulus, 189
Control: experimental, 21, 102; external vs. internal, 260
Conversion, 401, 406
Coordination, 81–85, 347, 396, 410
Counting, 39, 46, 114, 124, 164, 192, 196, 198–99, 251, 256, 292, 310, 313
Craniotomy, 297
Crime, 87, 184, 301–03, 329–31, 402, 403, 409. See also Delinquency
Cues, 31; unverbalizable, 71–72, 108, 140–42, 149–51, 166, 177, 179–80, 196, 205–06, 235, 277, 292, 408
Culture, 31, 59, 243–44, 332, 401, 407–08; definition, 32; effects, 99, 147, 169, 178, 189, 215, 246–47; universal traits, 48–52
Cycles, 56–57, 93

Dance, 371, 374, 386, 387, 389, 390, 391
Death, 67, 68, 80, 281, 289, 332, 392, 411;

Death (*continued*)
attitude, 242–43, 269–70, 344; knowledge, 239–40, 296, 343
Deceleration. *See* Acceleration vs. deceleration
Decision-making, 352–55
Defecation, 278, 315, 320
Déjà vu, 297, 325
Delinquency, 113, 172, 256, 263, 283, 302, 329–331. *See also* Crime
Depression, 320, 322, 323, 327, 328, 329
Deprivation: relative, 139–40; sensory, *see* Isolation
Determinism, 47, 140, 356
Dextroamphetamine, 202, 284, 312, 313, 314
Differences, individual. *See* Personality, traits, effects of
Disease, 80–81, 185–86, 275, 277, 282
Disorientation, 270, 293–301, 320–25
Disparity, temporal, 392
Dissatisfaction, 397, 398, 401, 402. *See also* Frustration
Dissociation, 259, 410
Dissonance, 98–99, 185–86, 353
Distraction, 207
Disturbance, emotional. *See* Abnormality, mental
Diurnal fluctuations, 151
DMT, 315
Dogmatism, 262, 263
Drama, 374, 376, 379, 380, 385, 389
Dreams, 72, 217, 219, 263, 285–90, 307, 316–17, 320; day, 264, 286, 290, 351
* Drives, 31, 114, 127, 130–40, 224, 300, 350, 371; duration-dependent, 132–36, 143–45, 169, 199, 238–40, 351, 408; events, 107–09, 113–14; strength, 108; 131–32, 138–39, 142–43, 181, 187–88, 381–82, 408
Dropouts, school, 303–04
Drugs, 45–46, 128, 280, 282–85, 286, 307, 311, 312–16, 380, 409; depressants, 128, 284, 312; pyretogenic, 284; sedatives, 312; stimulants, 128, 284, 312; tranquilizers, 284, 312
* Duration, 7, 8, 25, 63, 90–91, 218, 378–79; dependence, 133–36, 143–45, 169; effects, 110, 120, 148–49, 177, 180–81, 203–04, 224, 254, 271, 327, 329; optimal, 103, 161; perception, 102, 202; significant, 81–95

Economics, 352, 353
Education, 50, 64, 89, 91–92, 197, 283, 333, 396, 400, 403
Efficiency, 349
Egocentricity, 247
Einstellung. *See* Set
Electroencephalogram, 149, 278, 308, 311, 312, 316–17
Energy, 372; direction, 58–60
England, 367
Epilepsy, 270–71, 322, 327, 329
Epinephrine, 283
Epistemology, 7
Error, time-order, 104, 120, 128
Eschatology, 74–75
* Estimation, over- vs. under-: abnormality, 226–27, 325–28; boredom, 169; brain localization, 307; consistency, 223–24; crime, 331; definition, 37–44; delay, 207–08; drives, 113, 132–33, 134, 143, 148, 150; drugs, 312–16; duration, 180–81; events, 109–11, 117, 120–21; experience, 163, 165–66, 190–91, 193, 194–95; extroversion-introversion, 257; personality traits, 258–59, 263; temperature, 277; thyroid, 309–10
Ethics, 186–87
Ethnocentrism, 403
Eurasia, 61
* Events, 31, 131, 146, 147, 188, 199; age, 236–38, 248, 408; drive strength, 131–32, 142–43, 182–83; effects, 107–27, 375
Excretion. *See* Defecation
Exile, 188
Existentialism, 297, 322–23
* Experience, 7, 10, 137, 157–67, 400, 407, 408, 409, 410. *See also* Learning; Reinforcement
Experiments: classical, 21, 111, 113, 131, 162, 173, 178, 188, 189, 219; laboratory, *see* Laboratory, psychological
Extrapolation, 395
Extroversion-introversion, 118, 225–26, 257–58, 263, 264, 265, 328

Fatalism, 56, 269, 397
Fatigue, 72, 104, 182, 205, 206, 242, 348, 366
Fear, 132–36, 144–45, 184, 258, 262, 269, 270, 325
Feeblemindedness, 181, 251, 256, 257, 275, 327, 395

Feedback, 37, 128, 213, 311, 313, 328, 408; communication, 88, 89; experience, 159, 183–86, 188, 194, 197
Fencamfamin, 314
Fiction. *See* Literature
Field-dependence, 259, 262
Flying, effects of, 69
France, 216, 246–67, 332–33
Free association, 262
Freedom, 410
Frequency, 63, 64–65, 176, 194
* Frustration, 36, 59, 90–93, 131, 138, 169, 199, 214, 238, 260, 319–20, 351, 405; relative deprivation, 139–40

Galileo, 82
Galvanic skin response, 149
Gambling, 78, 355, 357, 361
Ganda (Uganda), 166, 226, 247
Generalization, 65, 176
Geriatrics. *See* Age, young vs. old
German, 68, 76, 110
Ghana, 94
Glands, endocrine, 274. *See also* Thyroid
* Goal: attainment, 31, 84, 85, 90, 97, 137–40, 146–49, 337–38, 408; distance, 133, 143–45, 147–48; drive, 131–32, 187–88; gradient, 133; planning and change, 396, 397, 401, 403
Government, 3, 86–87
* Gratification, 16, 51, 53, 182; deferred, 12, 59, 93–96, 100, 213–16, 228, 244–47, 263, 264–65, 302–03, 304, 347–49, 360–61, 363, 394; nature of, 130–31
Gravity, 276–77
Greece, 345, 385
Grenada, 100, 247
Groups, 332–33, 409; associations, 359; ideals, 337–40; leaders, 340–43; prophets, 341–43; reference, 333–35, 339–40; support, 404; tradition, 335–37, 339
Guide, 152, 165, 196, 215, 218, 248, 249, 289, 381, 408, 411; attainment, 137; compensation, 164; definition, 109; drives, 131–32; events, 109; experience, 157, 159, 162, 163; insight, 162–63

Hallucinogen, 314
Hashish, 34
Health, 358
Heart, 273, 278, 311
Hebephrenia, 324

Hemiplegia, 298
Herodotus, 24
Hindus, 79
History, 6, 49, 62, 64, 69, 110, 113–14, 186, 255, 267, 333, 341, 373; sense of, 55
Homesickness, 263–64
Hope, 42, 130, 132–36
Hopi Indians, 25
Humanities, 14–15
Hypnosis, 39, 87, 290–93, 317–18, 335, 380, 381, 401
Hypothalamus, 309
Hypotheses, for change, 397, 398, 399, 401, 402, 403, 404
Hysteria, 226, 231–32, 322, 329

Ideals, group, 337–40
Identity, 217–18, 260
Illness, mental. *See* Abnormality, mental
Imagery, 200–01; eidetic, 174
Impatience, 168–72, 228, 242, 319
Impulse, control of, 259, 263, 269, 331
India, 59, 147, 246
Indifference point (interval, zone), 103, 105, 106, 122, 191
"Informal time," 67
Information, temporal, 9, 31, 32, 101, 127, 130, 406; abnormality, 297, 324–25, 327; effects, 63, 67–68, 255, 301, 408; need for, 80–81, 85, 97, 249, 410. *See also* Knowledge, temporal
Inheritance, 394
Inhibition, 260; retroactive, 183; theory of, 226
Innovation, 358, 392, 393, 405
Input. *See* Interval
Insight, 162
Installment buying, 353–54
Insurance, 356–58, 394
Intelligence, 64, 78, 143, 147, 224–25, 245, 248, 252, 272, 275, 281, 350, 395; judgment, 255–56; knowledge, 255; orientation, 256–57, 263
Intensity, stimulus, 104, 111, 112, 119
* Interval, 8, 31, 32, 101, 408; content, 118–21, 124–27, 167, 187, 208–09, 254, 279; duration types, 9; empty vs. filled, 110–18; intervention vs. passivity, 123–24
* Intervention, 51–52, 53, 83–84, 136, 214, 216, 219, 347, 392; attitude, 56, 394–

Intervention (continued)
95; effects, 123 27; experience, 99; perspective, 77–78; scope, 111–15
Introspection, 106, 173, 200–01, 205–06
Introversion. *See* Extroversion-introversion
Irreversibility, 92, 134–35, 385–86, 391
Islam, 53–54, 57
Isocarboxazid, 314
Isolation, 36, 117, 150, 206–07, 256, 259, 270, 271, 278–80
Italians, 76, 95, 266

Jabo (Liberia), 145
Jamaica, 163, 247
Jamais vu, 297, 325
Jews, 57, 67, 76, 95, 268
* Judgment, temporal, 8, 20, 31, 73; external, 39, 42, 43–44; primary vs. secondary, 31, 34–37, 40–41, 42, 43–44
Just noticeable differences, method of, 25, 155

Kachin (Burma), 39
Kaguru (Tanzania), 76
Kamba (Kenya), 74
Kant, I., 383
Kappa-movement, 121–22, 248
Kikuyu (Kenya), 74
Knowledge, temporal, 20, 31, 230, 249, 255, 275, 295, 296, 308, 408; effects, 107–09; society, 60–62
Korsakoff's psychosis, 277, 296, 324–25
Kpelle (Liberia), 166

Laboratory, psychological, 4, 5, 19, 21, 48, 102, 156, 161–62, 298, 355
Language, 3, 218, 283; development among children, 216–17, 221; "silent," 63–64, 66; tenses, 52, 58, 68, 74–75, 257, 266, 321, 323; vocabulary, 148, 176, 209, 250
Leaders, 83, 186, 340–43
* Learning, 11, 146, 207–08, 318, 397; accuracy, 160–67; children, 213–22; deferred gratification, 244–47; temporal potential, 247–52. *See also* Experience; Reinforcement
Leisure, 54, 87
* Length of time, definition, 39–46
Leucocytes, 71
* Levels, primary vs. secondary, 33–37, 40, 43, 168–69
Limitation, arbitrary, 10, 13, 24, 31, 48, 82, 98, 107, 134, 169, 216, 243, 343, 361

Limits, method of, 25
Limits-verbal, method of, 26, 44–45, 251
Literature, 4, 15, 93, 223, 375, 376, 377–78, 379, 380, 383–86
Lobe: frontal, 309; temporal, 273–74, 308–09
Lobotomy, 274, 308
Logic, 6
LSD, 128, 283, 284, 314–15
Luo (Kenya), 247

Magic, 286
Manic-depressive psychosis, 279–80
Marijuana, 313
Martyrs, 304–06
Marxism, 338, 360
Mashona (Rhodesia), 145–46
Melancholia, 29, 321
Menarche, 281
Menigo-encephalitis, 307
Menstruation, 49, 71, 281
Meprobamate, 313
Mescaline, 284
Messianic cults, 338
Metamphetamine, 313
Metaphysics, 3, 5, 99, 295
* Methods, measurement, 23–25, 42, 284, 294–95; accuracy, 20, 27–28, 203–04; effect, 110, 115, 119, 125, 150, 177, 178–79, 180–81, 198, 202, 224, 230, 254, 284, 322, 325, 329; personality, 223; projective, 20; reliability, 28–29, 224; types, 17–21, 25–27
Methylphenidate, 315
Middle Ages, 342
Minimal causes, method of, 25
Modality, sense, 102–03, 104, 110, 118, 177–78, 219; accuracy, 201–03
Modernization, 81, 337, 402
Monism, 273
Monotony, 113, 146, 193, 259, 349–52, 366–67
Mood, 88, 280, 375
Moon, 4, 8, 50, 61
Mormons, 77
* Motive, temporal, 31, 32, 40, 106, 113, 127, 407; arousal, 23, 80–100, 108–09, 118, 169, 174, 181–83, 185, 208, 224, 254, 277, 325, 408; orientation, 22
Multivariance, 105, 254, 257, 348, 356, 381
Music, 24, 107, 112, 153, 157, 178; effects, 125, 174, 266, 350; experience, 160;

reactions, 371, 372–73, 375–76, 377, 379–80, 382, 387, 388, 391
Mystics, 306–07
Myth, 75, 286, 383

Narcissism, 262, 319
Nation, 337, 339
Nationalism, 186, 337
Navaho, 77, 100
Nazis, 305
Near East, 232
Neuroticism, 193, 208, 216, 251, 257, 260, 264, 296, 298, 319, 321, 325, 326–27, 327–28, 395
Nigeria, 81
Nitrous oxide, 150, 313
Nostalgia, 264
Nutmeg, 315–16

Observation vs. participation, 44, 136
Obsession, 319
Occupation, 283, 333, 348–49, 364–66
Oceania, 61
Ojibwa Indians, 287
Ololiuqui, 316
Ophthalmology, 185
Opium, 314
Optimism, 263, 393
Organization, cognitive, 262
* Orientation, temporal, 8, 21–22, 31, 32, 77, 308, 392, 408, 409–10; abnormality, 295–96, 297, 320–24; achievement, 266; age, 238–41, 268; art, 372–74; capital, 360; children, 215, 221, 251–52; consistency, 253; crime and delinquency, 302–03, 330–31; decision-making, 352–53; dreams, 287; functions, 7, 21, 93–94, 186, 347; groups, 336, 341, 343–44; intelligence, 256–57; measurement, 261; occupation, 366; periodic, 52; personality, 223, 226, 227, 229–30, 261–64; planning and change, 392–96, 398, 399, 404, 405, 406; social class, 245–46; strength, 59
Output. See Intervention
Overestimation. See Estimation, over- vs. under-
Ovid, 188
Oxygen, 235, 279, 289, 311

Pacemaker, chemical. See Biochemical, clock
Pain, 306
Painting, 372, 373, 374, 375, 376–77, 378, 379, 380, 381, 388, 389
Palestine, 246
Panoramic death vision, 288–89
Paranoia, 321, 327
Parkinson's Law, 65
Penology, 401, 402
Pentobarbital, 313
* Perception, 6, 31, 34, 35, 91, 402, 407, 408; event potential, 109–15; psychophysics, 101–06; salience, 106–09
* Personality, 31, 32, 64, 293, 301, 323, 408; modal, 56, 397; traits, effects of, 73–74, 118, 191, 222–33, 255, 257–67, 269, 274, 395, 408, 409; variability, 122, 128–29, 205, 232, 239, 294, 315, 333, 351, 356, 393
* Perspective, temporal, 8, 24, 31, 32; age, 240, 252; intervention, 77–78, 403; modal, 54, 56, 58–60, 63; personality, 73, 166, 223, 227; strength, 59; traditional societies, 52–55, 78
Phenomenology, 43
Philosophy, 4, 5–6, 24, 79, 92–93, 99, 167, 373
Phobia, 320
Physics, 4, 102, 121
Physiology. See Biochemical, processes
Placebo, 283, 311, 312, 313, 314
Planning, 76, 92, 131, 392–96, 409
Plants, 70
Poetry, 74, 344, 371, 372, 376, 377, 389, 390, 391
Poland, 266
Police, 50
Political science, 6
Portrayal, time, 383–86
Postencephalitis, 307
Postponement. See Gratification, deferred; Renunciation
Pot, watched, 13, 128, 171, 199–200, 351
* Potential: behavioral, 31, 32, 47–79, 105, 110, 215, 218, 222, 231, 407, 408; definition, 30; event, 109–15; primary judgment, 34–37, 101–51, 152, 171, 174, 299, 407, 410; secondary judgment, 34–37, 106, 152–210, 299, 407, 410; stimulus, 32, 101, 104, 126, 153–54, 187; temporal, 7–9, 31, 32, 48, 50, 56, 68, 73, 107, 183, 184–85, 213, 215, 218, 231, 251, 408; total, 31, 32, 36, 183–84, 187, 242
Power, 86–87
PPT method, 27
Predestination, 51, 92

Predisposition, 397–98, 401
Premonition, 288
Present: psychological, 5–6, 8, 10, 13, 24, 174, 280–81; specious, 5
Pressure, blood, 72, 311
Prestige, 259, 291, 398
Primacy, 88
Primogeniture, 394
* Principles: nature of, 13–17; need for, 405–06; summary, 407–08; suspension of, 381–82
Priority, temporal, 63–65
* Production, method of, 18–20, 38–41, 42, 43–45, 46, 104, 117, 123, 159, 178, 181, 192–93, 204, 254, 310–11
Productivity, 348–49, 366
Progress, 56–57, 66
Prophecy, self-fulfilling, 51–52, 338–39, 341
Prophets, 341–43, 385
Protestantism, 361
Proverbs, 4, 94, 145–46, 386, 387–88
Psilyocybin, 314, 315
Psychiatry, 7, 9, 24, 33, 35–36, 50, 57–58, 59, 60, 83, 91, 223, 243, 277, 292, 293–301, 303, 335, 400
Psychoanalysis, 74, 83, 216, 218, 286–87, 294, 297, 319–20, 400
Psycholinguistics, 68, 74–75
Psychology, 7, 71, 87, 121, 223, 272, 322, 332, 333, 366, 407; Gestalt, 402
Psychopaths, 231–32, 322
Psychophysics, 101–06, 120, 173, 202, 203–04, 221, 267, 407
Psychoses, 275–76, 295, 297, 299, 300, 307, 321–22, 324, 325, 327, 329, 395
Psychotherapy, 323, 400
Public relations, 354, 359
Pulse, 311
Punctuality, 158, 166, 253, 256, 264, 320

Quinine, 313, 314

Rapid eye movements (REM), 288, 289, 316
Rating-verbal method, 26, 251
Ratio, interval-age, 235–36, 238, 267–68
Recency, 88, 176
* Recollection, 8, 10, 11, 22, 33, 90, 137, 172, 213, 237, 379, 381, 407, 410
Redundancy, 390
Regression, 294, 324
* Reinforcement, 11–12, 98, 155–56, 158–67, 184, 194–97, 271, 317, 327, 343,
363–64, 404, 408; absence, 159, 190–93, 198; differential, 195 96, 312, 327 28. See also Experience; Learning
Reliability, 28–29, 190, 193, 224
Religion, 3, 4, 51, 59, 94, 286, 304–07, 334, 341, 343–46, 388, 400, 401
* Renunciation, 8, 12–13, 290, 332, 410; age, 238–39, 241, 269; art, 380, 382, 383, 386; capital, 361–62; change, 397, 400, 407; children, 214–16, 222, 252; conservation, 363–64; decision-making, 352–53, 354; effect, 93–96; function, 9, 51, 59, 185; group, 335, 337–38, 339; personality, 231, 259; planning, 392, 393, 396; prophets, 341–42; religion, 343; risks, 356–57, 359; school dropouts, 303–04; universality, 53–54; work, 347, 365
Repetition, 88, 195, 349–50, 366–67, 389, 390–91, 397
Repression, 59, 90, 97, 131, 286
* Reproduction, method of, 18–19, 25–26, 27–29, 38, 42, 43–45, 104, 110, 120, 123, 149, 150, 159, 166, 178–79, 181, 192–93, 225–26, 294, 313
Research, market, 359
Respiration, 258, 273, 278, 308, 311, 312
Rhythm, 102, 146, 160, 218, 270, 311, 313; circadian, 69–70, 206
Right and wrong cases, method of, 25
Risk: diminishing, 355–60; taking, 395–96
Risky shift effect, 356
Role, 269, 334, 392, 399
Rome, 385
Rounding, temporal judgments, 60–61

St. Augustine, 5
Saints, 213, 400
Salience: groups, 334–35; perception, 106–09
Sampling, 19, 20–21, 100, 220, 229, 293–94
Saving, 95–96, 360–62, 394
Scarcity, temporal, 64, 84, 87, 135
Schizophrenia, 294, 380; attitude, 296; consistency, 252; knowledge, 308; orientation, 263, 321; temporal judgment, 27, 104, 116, 145, 196, 202, 208, 226, 314, 325, 326–27, 328, 381; time sense, 298–301, 315, 319, 324, 329
Science, 14, 31, 82, 273, 377, 381, 387, 391; social, 15, 24, 339

Scope, 31, 115, 129, 169; definition, 114
Scopolamine, 36, 312
Scotland, 367
Secobarbital, 202, 284, 312, 313
Semantic differential, 266
Semantics, 218, 377
Senile dementia, 27
Sensation, 4, 102
Sense of time, 165, 314–15, 318, 319, 322; development, 158, 219–22
Sensitivity, 404, 406; training, 74
Sentence, completion of, 263
Serial exploration, method of, 25
Set, 126–27
Sex differences, 149, 189–90, 209–10, 218, 239, 250, 263, 264, 269, 270
Shakespeare, 285, 376
Shandy, Tristram, 34, 378
Single stimuli, method of, 26
Skill, 396, 399–400, 401, 402, 404
Sleep, 34, 64, 66, 70, 72, 278, 279, 285–90, 316–17, 409; deprivation, 310–11
Sleeper effect, 89
Socialization, 32, 58, 61–62, 96, 213–22, 238, 252, 365, 392, 409
* Society: perspective, 53–55, 58; restrictive security, 53–54; sense of history, 55; temporal knowledge, 60–62; traditions, 79, 81, 157, 219–20, 371, 374, 385, 386, 400
  Sociology, 5, 82, 87, 99, 167, 285, 333, 384–85
Sodium amytal, 314
Somatotype, 283
Soviet Union, 71, 77, 338
Space, 23, 25, 91, 111, 121–23, 219, 309
Span, time, 396
Spanish-Americans, 77
* Speed of time, definition of, 39–46
Speleology, 117
Spinoza, B., 273
Spiral, 169, 216, 231–32, 274, 279, 385, 404, 407; definition, 66–68
Standard, 9, 18–19, 25–26, 31, 63, 109, 172, 176, 179, 199, 406, 408; source, 152–57, 238
Storage size, 113, 127–29
Stress, 68, 139, 145, 148–49, 258
Stuttering, 261
Style, 388–91
Succession, 8, 19, 62, 102, 175, 218, 250
Successive time estimation, method of, 26

Suicide, 54, 281, 296, 309
Surgency, emotional, 259–60
* Symbols, 18, 31, 33, 57–58, 59, 107, 153, 177, 196, 285, 338, 352, 363–64, 399, 408

Tanzania, 77, 229, 265
Tau-movement, 121–22
Taxonomy of time, 31, 47, 101, 187, 223, 234, 407, 408; chart, 31
Tea, 313–14
Temperature, 32, 206, 310; body, 150, 277–79, 282, 284, 308, 310–13
Testimony, 189
* Tests: MMPI, 263; Rorschach, 256, 259, 260, 262–63, 265, 269; Scottish Tartan, 229, 266–67; Thematic Apperception (TAT), 62, 100, 218, 229, 230, 253, 256–57; Time Metaphor, 228, 260, 263, 264–65, 266; Time Scale, 260
Texans, 77
Thalamotomy, 308
Threshold, 102, 106
Thyroid, 274, 280, 309–10
* Time, objective vs. subjective, 8–9, 42, 407, 409, 410–11; abnormality, 329; age, 235; art, 379, 382; daydreaming, 290; delinquency, 331; development of, 250; discrepancy, 42, 48, 97, 103, 153, 168, 206–07, 233; dreams, 288; experience, 157–59; hypnosis, 291; significance, 63, 84, 135, 161
Timelessness, 410
* Timepieces, 8, 17; confidence in, 101, 106, 107, 153, 160, 164, 165, 176, 179–80, 293, 360; effects, 67, 156, 169, 172, 185, 205, 249, 278, 279; experience, 157–58, 162, 221; functions of, 47–48, 49, 60, 81–82, 83, 344–45, 347
Time reckoning, 61–62, 75–76
Timing, 83–84, 399, 404, 405
Tiv (Nigeria), 54–55
Touch, 192, 203, 311
Traditions, 335–37, 339–40, 397
Traits, material vs. nonmaterial, 398. See also Personality, traits, effects of
Transformation, 388, 401–02
Translation, 31, 172–81, 199, 408
Transplantation, 401, 402–03
Trinidad, 100, 246, 302
Truth, in art, 386–88
Turkey, 266
Tyrol, South, 95, 345, 362–63

* Unconscious, 285–86, 295, 319, 320. *See also* Cues, unverbalizable
Underestimation. *See* Estimation, over- and under-
Unemployment, 350, 351, 367
Universals, human, 48–52
Urination, 206, 215, 278–79, 280
Utopia, 338–39

Value: activity, 63–65; instrumental, 98; modal, 56–58, 66–67, 78–79, 498
* Verbal estimation, method of, 18–20, 26, 27–29, 38–41, 42, 43–45, 104, 110, 115, 116, 123, 141, 149–50, 161, 166, 172, 178, 180, 193, 209, 310
Vierordt's law, 164, 198
Vision, 102–03, 120, 121, 155, 178, 201–03, 311, 313, 318

War, 86–87, 237, 339
Weber's law, 103, 181
Work, 347–52, 409

Yoruba (Nigeria), 58

Zulu (South Africa), 247
Zuni Indians, 77